T0396505

Statistical Mechanics
—and—
Scientific Explanation
Determinism, Indeterminism
and Laws of Nature

Statistical Mechanics
—and—
Scientific Explanation

Determinism, Indeterminism and Laws of Nature

Editor

Valia Allori

Northern Illinois University, USA

 **World Scientific**

NEW JERSEY · LONDON · SINGAPORE · BEIJING · SHANGHAI · HONG KONG · TAIPEI · CHENNAI · TOKYO

Published by

World Scientific Publishing Co. Pte. Ltd.
5 Toh Tuck Link, Singapore 596224
USA office: 27 Warren Street, Suite 401-402, Hackensack, NJ 07601
UK office: 57 Shelton Street, Covent Garden, London WC2H 9HE

British Library Cataloguing-in-Publication Data
A catalogue record for this book is available from the British Library.

STATISTICAL MECHANICS AND SCIENTIFIC EXPLANATION
Determinism, Indeterminism and Laws of Nature

ISBN 978-981-121-171-3 (hardcover)
ISBN 978-981-121-172-0 (ebook for institutions)
ISBN 978-981-121-173-7 (ebook for individuals)

For any available supplementary material, please visit
https://www.worldscientific.com/worldscibooks/10.1142/11591#t=suppl

Typeset by Stallion Press
Email: enquiries@stallionpress.com

Contents

Introduction

Valia Allori

Department of Philosophy, Northern Illinois University
vallori@niu.edu

While I am writing this chapter, I am sitting in my room at a table by a window from which I see a river. While I take a sip of my coffee, I can also see the clouds in the sky coming towards my house, which are about to ruin my weekend. Now that it is getting dark, I can also spot the Earth's silver satellite by the tree, which is gently moving in the wind. The objects of my experience (the river, the clouds, the wind, the coffee, the Moon, and the tree) are thought of as being composed by a multitude of much smaller entities. They are called macroscopic objects because they are big with respect to the dimension of their components, which are thus labelled as microscopic entities. My coffee, for instance, fits into a container which has an 8 cm (3.25″) diameter and a height of about 9.5 cm (3.75″). However, my mug contains a large number of molecules, about 10^{25}, whose size can be estimated to be about $3 \cdot 10^{-8}$ cm. All the objects that I see may change in complicated ways; that is, their temporal evolution may be governed by a very complex dynamics. Despite this, however, their behaviour is governed by general physical laws, such as the laws of thermodynamics, which govern many different physical phenomena, from the transformation of dying stars into black holes to the process of combustion in the water boiler in our houses. If one is convinced that the motion of the microscopic entities is governed by a complete theory of the world (in the sense that they apply to everything), then one is led to think that the macroscopic objects, in virtue of having such microscopic composition, have a behaviour which can in principle be explained in terms of the microscopic dynamics. This is to say that one should be able to in principle recover, under suitable approximations, the laws of thermodynamics from the more fundamental laws governing the microscopic entities. This is indeed what Ludwig Boltzmann wanted to do. Assuming that classical mechanics is the fundamental and complete theory of reality, then everything is made of microscopic Newtonian particles. In principle, it is possible to account for the behavior of macroscopic objects in terms of these laws. The "in principle" clause is important: there are too many particles involved, and the actual calculations would be impractical. Because of this, statistical methods (which include mathematical tools for dealing with large populations) may be used. This is, in essence, Boltzmann's statistical mechanics. It provides a framework for relating the microscopic properties and behavior of individual particles to the macroscopic properties and behavior of macroscopic objects. In particular, Boltzmann showed how to recover the laws of thermodynamics from statistical mechanics using just a few simplifying assumptions. The commonly accepted idea is that Boltzmann's derivation only needs three ingredients:

(1) The Newtonian laws of motion,
(2) The statistical postulate,
(3) The past hypothesis.

Roughly put, (1) describes the underlying microscopic dynamics; (2) specifies the probability measure, therefore grounding the meaning of the probabilities arising from the statistical derivation of thermodynamic laws; and (3) postulates that the universe initially was in a state of extremely low entropy and it has to be added to guarantee that entropy will likely increase in the future but not in the past.

The derivation, though, involves several crucial steps whose significance and legitimacy are still controversial, and it gives rise to several interesting questions. These questions are at the center of the philosophical literature on the foundations of statistical mechanics, which is immense. In this collection, the different contributors will take on different aspects of some of the debates in the philosophy of statistical mechanics, shedding new light on old debates.

First, there is a question about explanation: Assuming statistical mechanics is able to explain the macroscopic regularities in terms of the microscopic laws, what kind of explanation is it? Statistical mechanics uses statistical methods, which provide results that are statistical in nature. As such, they do not specify that a given physical system will certainly do such and such, but merely what the probability that it will do such and such is. Therefore, the explanation is probabilistic. However, it is far from clear how to interpret this, and many questions present themselves. How should we understand probabilistic explanations? How should we understand probabilities in a deterministic theory like Newtonian Mechanics, in which no probability appears at the fundamental level? Should we conclude that these probabilities are epistemic (a result of our ignorance of the exact details) and thus 'subjective'? These questions are among the ones that are taken up by Jean Bricmont in his paper "Probabilistic Explanations and the Derivation of Macroscopic Laws," in which he argues in favor of a Bayesian account of probabilities. Barry Loewer in "The Mentaculus Vision" argues that the best way to understand probabilities in statistical mechanics is a Humean account along the lines of the best systems account of laws and chances developed by David Lewis.

Alternatively, in the physics literature some have proposed that when deriving the macroscopic laws from the microscopic ones, rather than probability, one should focus of the weaker notion of typicality. Roughly, the idea is that one has explained a behavior when one has shown that such behavior is typical. The notion of typicality, however, is notoriously difficult to spell out. Saying that something is typical is related to saying something is highly probable, but not the same. However, what is the difference? The notion of typicality in itself is first discussed in Tim Maudlin's paper "The Grammar of Typicality," in which he discusses the relationship between typicality and probability and shows in what sense typicality can be explanatory. Valia Allori, in her "Some Reflections on the Statistical Postulate," present an argument to a similar conclusion; however she uses it to argue against the claim that certain quantum theories should be preferred based on statistical mechanics (see later). Moreover, despite the prominence of typicality in a variety of discourses, no formal logic for typicality has been proposed. In "The Logic of Typicality", Harry Crane and Isaac Wilhelm do exactly that: they propose two formal systems for reasoning about typicality; one based on propositional logic (formalizing objective facts about what is and is not typical) and the other based on intuitionistic logic (formalizing subjective judgments about typicality). Finally, in "Reassessing Typicality Explanations in Statistical Mechanics", Massimiliano Badino surveys, compares and contrasts several typicality accounts, focusing on how they are supposed to be explanatory. He concludes that typicality explanation is not an all-or-nothing affair, but rather a business of balancing multiple epistemological constraints and metaphysical requirements.

A further question which has been asked in connection to probability is about the justification of the probability measures used in statistical mechanics. To understand, let's take a step back. Statistical mechanics deals with macrostates and microstates. Microstates are the states of the microscopic particles composing a macroscopic object such as a gas, in the case of thermodynamics. The macrostates describe instead the physical state of the macroscopic object itself using thermodynamic variables like volume, pressure and temperature. Each macrostate contains a very large number of microstates: two microstates A and B can differ from one another by the position of a particle, but that difference macroscopically would not count at all. In fact, the thermodynamic variables of a gas computed starting from microstate A and from microstate B would be macroscopically the same if the only difference between A and B is having a particle "here" in A and "there" in B. Roughly, statistical mechanics will tell us how the macrostate behaves in terms of what the majority of its microstates will do. Because of this, one would need a way to count such microstates, and this is where the so-called probability measures are involved: they allow us to determine what notions, like "the great majority of microstates" mean. A problem is that there are infinitely many possible probability measures, and not all of them provide the same results! Boltzmann took a particular one, the uniform distribution, as a statistical postulate. Now the question is, what is his justification of this choice? Many have argued for or against this choice, and the issue is still unsettled. In addition, some scholars have contended that within the framework of an indeterministic version of quantum mechanics, Boltzmann's scheme of explanation would go through in a similar fashion but we would not need this additional postulation. Their claims are interesting but controversial, and therefore are worth further investigation. Loewer argues that this measure can be naturally understood in Humean terms, while Allori argues that it is derivable from dynamical considerations.

Moreover, there are debates on the nature of the past hypothesis, and its role in the explanatory machine of statistical mechanics. There seems to be a tension in Boltzmann's explanation of thermodynamics in terms of statistical mechanics: Newtonian laws are reversible in time, while macroscopic processes typically are not. In other words, when two particles (roughly thought of as two rigid billiard balls) collide elastically according to the classical laws, both the forward-in-time and the backward-in-time collisions are possible states of affairs that can happen in nature. This is what is meant when it is claimed that Newtonian mechanics is time reversal invariant. This is clearly not the case for typical macroscopic phenomena: a cup of coffee gets cold but never heats up; an ice cube melts though the water never re-solidifies; cooked sugar forms caramel but caramel cannot be transformed back into sugar, and so on. Clearly, macroscopic phenomena show a privileged temporal direction which the microscopic phenomena lack: they are irreversible. The irreversibility of macroscopic phenomena is usually described by the second law of thermodynamics in terms of entropy: the entropy of a macroscopic system never decreases. This specifies a temporal directionality in every macroscopic phenomenon which is absent in the microscopic ones. So the tension between these two behaviors is evident: how is it possible to account for irreversible processes starting from reversible ones? To cut a very long story short, Boltzmann could prove that in order to solve this tension one would have to impose some restrictions on the possible initial conditions of the universe. The postulate that selects which kind of initial condition the universe had to begin with is known as "the past hypothesis," which establishes that the universe began with extremely low entropy. Since in Boltzmann's analysis the amount of entropy of a system is connected with the size of its macrostate-the bigger the macrostate, the higher the entropy-it follows that the universe began in a very small macrostate. But how "unlikely" is such a state? If God were to choose an initial condition for the universe amongst all the possible macrostates, it would have been more likely that he picked a larger one. Does that mean that the initial low entropy macrostate requires a further explanation? Loewer argues that the past hypothesis does not need an explanation and should be seen as a law of nature. In virtue of that, it is able to explain the second law of thermodynamics that entropy always increase and also other arrows, including the physiological arrow of time. In contrast, Meir Hemmo and Orly Shenker, in "Can the Past Hypothesis Explain the Psychological Arrow of Time?", propose that the psychological arrow of time is fundamental and the entropic one is derived from it, rather than the other way around. Among those who believe that the past hypothesis requires an explanation are Dustin Lazarovici and Paula Reichert, who in their "Arrow(s) of Time without a Past Hypothesis" discuss and assess a recent proposal to explain the time arrow without a past hypothesis, and propose a generalization of the definition of Boltzmann's entropy for a gravitational system. On a similar note, Michael Kiessling, in his "The Influence of Gravity on the Boltzmann Entropy of a Closed Universe", is concerned with the extension of Boltzmann's work to a universe with gravity and provides several insightful examples. Finally, Ryan Olsen and Christopher Meacham, in "Eternal Worlds and the Best System Account of Laws", apply the Humean account of laws to the models of the universe posited by contemporary cosmology. They show how these models, which do not have a past hypothesis, are compatible with the Best system account, contrarily to what it had been previously argued. Be that as it may, most philosophers have tried to explain temporally asymmetric process, like the convergence of equilibrium, in terms of time symmetric concepts, presumably on the basis of the conviction that a satisfactory explanation is obtained only when temporal asymmetries are explained away by showing they are merely apparent. In contrast to this view, Wayne Myrvold in "Explaining Thermodynamics: What Remains to be Done?" proposes an account of how to explain temporal asymmetries invoking only time symmetric dynamical laws, because of a temporal asymmetry in the very notion of explanation.

Another set of questions arises from the generalization of statistical mechanics to the quantum domain. If the fundamental theory of the world is quantum mechanics, how does the situation change? The situation is particularly tricky because quantum mechanics itself is riddled with problems and paradoxes. Luckily, many 'realist' quantum theories, that is theories which do not suffer from the conceptual problems of the so-called orthodox quantum mechanics, have been proposed. Therefore, it has been argued that if classical mechanics is replaced by any of these realist quantum theories, the situation would not change dramatically. Indeed, it has been maintained that spontaneous localization theory, which is a realist indeterministic quantum theory, should be preferred to its deterministic alternatives in virtue of the fact that in theory, probabilities occur only once and not twice in the dynamics and the statistical postulate as well. Allori argues instead that this is not the case, showing how the statistical postulate, namely the probability measure, is actually a typicality measure which can be derived from the dynamics. Eddy Chen, in "Time's Arrow in a Quantum Universe: On the Status of Statistical Mechanical Probabilities", also proposes to dispense with the statistical postulate but in a different way. He focuses on the density matrix, rather than the wave function, and shows that there is a unique choice of the initial density matrix, which therefore allows us to eliminate statistical mechanical probabilities. Antony Valentini, in "Foundations of Statistical Mechanics and the Status of the Born Rule in de Broglie-Bohm Pilot-wave Theory", also discusses the status of relevant probability measure, this time in the framework of the pilot-wave theory, which is a deterministic quantum theory of particles. He compares and contrasts two approaches to understanding the Born rule, namely the rule which provides with probabilities in quantum theories. One based on dynamical relaxation over time and the other based on typicality of initial conditions, and argues that the latter is fundamentally misleading, since it restrict the attention only to special states.

A final topic discussed in this book is the relation between the work of Boltzmann and the one of Gibbs. Josiah Willard Gibbs provided a general account of statistical mechanics, which is alternative to the one of Boltzmann. Therefore, some natural questions arise about their relationship: Are they equivalent? If so, which account is fundamental? If not, how are they compatible? Sheldon Goldstein, Joel Lebowitz, Roderich Tumulka and Nino Zanghì in their "Gibbs and Boltzmann Entropy in Classical and Quantum Mechanics" argue, following perhaps the majority of the philosophy of physics community, that the Gibbsian approach may be calculationally preferable but only the Boltzmannian approach is conceptually satisfactory. On a similar note, Roman Frigg and Charlotte Werndl in "Taming Abundance: On the Relation between Boltzmannian and Gibbsian Statistical Mechanics" also investigate the relationship between the two accounts and argue that the Boltzmann approach is foundationally superior. In contrast, the Gibbsian approach should be best regarded as an effective theory, which works well for practical purposes but should not be used when asking interpretative questions. On the other hand, David Wallace, in "The Necessity of Gibbsian Statistical Mechanics," argues for the opposite. First, he maintains that Gibbsian statistical mechanics is more general than Boltzmannian statistical mechanics. Moreover, Wallace aims to show that the supposed conceptual problems with the Gibbsian approach are either misconceived, or apply only to certain versions of the Gibbsian approach, or apply with equal force to both approaches. Wallace, therefore, argues that Boltzmannian statistical mechanics is best seen as a special case of, and not an alternative to, Gibbsian statistical mechanics.

I have divided the volumes in five parts: "The Different Faces of Explanation", "The Language of Typicality", "The Past Hypothesis, the Arrows of Time, and Cosmology", "Some Considerations from Quantum Mechanics", and "Boltzmann and Gibbs". As the narrative above, has hopefully made clear, this division is artificial as the various papers overlap in topics.

This collection has covered only some of the questions that one can ask in the foundations of statistical mechanics, and only from some of the possible perspectives. Obviously, more collections will be necessary. However, the contributors have engaged with the questions they have chosen to write about with a deepness and a carefulness that will certainly generate many discussions to come (in addition to having made me a proud editor!).

Part I

The Different Faces of Explanation

Chapter 1

The Mentaculus Vision*

Barry Loewer

Department of Philosophy, Rutgers University
loewer@philosophy.rutgers.edu

Building on the Boltzmann's approach to statistical mechanics, David Albert proposed a framework for a complete physical theory that entails a probability distribution over all physical possible worlds. Albert and I call this framework "the Mentaculus." In this paper, I provide reasons to think that the Mentaculus entails probabilistic versions of the laws of thermodynamics and other special science laws. In addition, it is the basis for a scientific account of the arrows of time and an account of counterfactuals that express causal relations. I then argue that the best way to understand the laws and probabilities that occur in the Mentaculus are along the lines of David Lewis' best system account.

Thermodynamic laws apply to the melting of ice, the dispersal of smoke, the mixing of milk in coffee, the burning of coal and a myriad of other familiar processes. All of these manifest an arrow of time pointing from past to future. We never see ice spontaneously forming out of warm water, milky coffee separate and so on. The second law of thermodynamics, which covers these processes, says that the entropy of an energetically isolated system never decreases and typically increases until the system reaches a state of maximum

---

*Thanks to Eddy Chen, Isaac Wilhelm, Denise Dykstra, Dustin Lazarovici and especially David Albert for comments and discussion of these issues.

entropy, its equilibrium state. If all one knew about the physics of
the world was that all systems are composed of material particles
(atoms) whose motions conform to classical mechanics, then the
temporal asymmetry of the second law would or should be deeply
puzzling, since the dynamical laws of classical mechanics that govern
the motions of particles are deterministic and temporally symmetric.
Their temporal symmetry consists in the fact that if a trajectory of a
configuration of particles is compatible with the classical mechanical
laws, then the temporally reversed trajectory of the configuration
is also compatible with the laws.[1] Therefore, since the trajectories
of the particles of ice and water that comprise the melting of ice
are described by the laws, so are the trajectories that comprise the
spontaneous formation of ice out of water. The puzzle is how temporally asymmetric laws emerge from fundamental temporally symmetric laws. The development of statistical mechanics (SM) by Maxwell
and Boltzmann partly answered this question. They claimed that
a thermodynamic system, such as a gas in a box, is composed of
an enormous number of particles moving randomly under the constraints imposed by the macroscopic properties of the system. To
capture the random motion, they posited probability distributions
over the positions and velocities of the particles compatible with a
system's macroscopic state. They and their followers were able to
show that for some systems, if the system is not in equilibrium, it
is enormously likely that it will evolve in conformity with the classical mechanical laws to states of higher entropy until equilibrium is
reached. In this way, they hoped to demonstrate that a probabilistic
version of the second law is compatible with the fundamental time
symmetric laws.

Boltzmann's attempt to account for the second law encountered
a famous obstacle. If the universe is in a certain macroscopic state
now, then it follows from Boltzmann's probabilistic posit and the
temporal symmetry of the laws that while it is enormously likely
that the entropy of the universe will increase toward the future, it is

---

[1]The same symmetry is also exhibited by the laws contained in quantum mechanics, general relativity, and plausibly the dynamic laws of any future physics.

also enormously likely that its entropy was greater in the past. The same holds for approximately isolated subsystems like the ice cube. While the statistical mechanical probability assumption implied that it is likely that the ice cube will be more melted in the future, it also implied that it was more melted in the past. But we know that this isn't so. How to remove this bad consequence from Boltzmann's account while preserving its explanation of the second law is called "the reversibility problem".

Recently, David Albert in *Time and Chance* takes Boltzmann's approach to SM and building on ideas from Eddington, Feynman and many others describes a framework that solves the reversibility problem.

Albert's framework contains three ingredients:

1. The fundamental dynamical laws that describe the evolution of the fundamental microstates of the universe (and the fundamental microstates of its isolated sub-systems) are assumed to be deterministic and temporally symmetric.
2. The Past Hypothesis (PH): A specification of a boundary condition characterization of the macrostate M(0) of the universe at a time shortly after the Big Bang. In agreement with contemporary cosmology, M(0) is a state whose entropy is very small.
3. The Statistical Postulate (SP): There is a uniform probability distribution specified by the standard Lebesgue measure over the physically possible microstates that realize M(0).

Albert and I have dubbed this framework "the Mentaculus".[2] The Mentaculus is a probability map of the universe in that it determines a probability density over the set of physically possible trajectories of microstates emanating from M(0), and thereby probabilities over all physically specifiable macro histories. Once the dynamical laws are specified, the Mentaculus contains an answer to every question of the form "What is the objective probability of B given A?" for all

---

[2]Discovering these conditional probabilities faces two obstacles. One is specifying the sets of microstates corresponding to the macro conditions mentioned in the conditional probability and the second is doing the calculation.

physically specifiable propositions A and B. For example, it specifies
the probability that an ice cube will completely melt in the next five
minutes given a description of the macrostate of its environment,
and the probability that Trump will be impeached given the current
macrostate of the universe. Though, to be sure, we cannot extract
these conditional probabilities from the Mentaculus.[3] Further, Albert
has argued that the Mentaculus not only grounds the thermodynamic
arrow of time but also explains other arrows of time. While many of
the details of arguments supporting these claims are yet to be worked
out, the Mentaculus is sufficiently bold and promising to warrant
taking it seriously.[4]

    This paper has two aims. One is to make a case that the Mentac-
ulus should be taken seriously by reviewing reasons to believe that it
explains the second law of thermodynamics, various arrows of time,
and why it is arguably a complete scientific theory of the universe.
Aim number two is to argue that the best way to understand the
metaphysics of Past Hypothesis and probabilities in the Mentaculus
is a Humean account along the lines of the best systems account of
laws and chances developed by David Lewis.

    Before explaining how the Mentaculus accounts for the second
law and other temporal arrows, I want to address two worries. The
first is that my discussion assumes the ontology and dynamics of
classical physics. But classical mechanics has been superseded by
quantum field theory and general relativity, and these are likely to
be replaced by a quantum theory of gravity. The second worry is that
$M(0)$ is characterized along the lines of current cosmology: shortly

---

[3] "Mentaculus" comes from the Coen Brothers' film "A Serious Man", where it is
used by a mentally disturbed characters as a name of what he calls "a probability
map of the world." Hence, with the permission of Ethan Coen and in the spirit
of self-mockery, we appropriated the name.
[4] Rigorous proofs that systems that satisfy the statistical mechanical probability
distribution evolve in conformity to probabilistic modifications of thermodynamic
laws (e.g. the entropy of an isolated macroscopic system is very likely to increase)
have been produced only for systems that satisfy very special constraints, ergodicity (Sklar 1993) is one such example. Arguments for such claims make them
plausible.

after the moment of the big bang, the universe consisted of a rapidly expanding space-time that contained a soup of elementary particles and fields that was very dense, very hot with almost uniform density and temperature, and the entropy associated with M(0) is very low.[5] One problem is that it is vague exactly what counts as the macrostate of the early universe. More worrisome is that it is not obvious how entropy of the early universe should be characterized especially if the ontology of early universe is specified by a yet to be formulated theory of quantum gravity.

The reply to the first worry is that any future plausible candidate for fundamental ontology and laws will posit states that are composed of particles and/or fields which possess an enormous number of degrees of freedom and laws that are capable of grounding the motions of material bodies. As Eddington implied the second law will survive no matter the developments of future physics. As for the second worry, although it is not known how to characterize entropy for the very early universe, all that the Mentaculus requires is that at some point in the distant past the entropy of the universe is very low. Whatever the ontology of future fundamental theories may be, it is plausible that it will sustain a characterization of entropy analogous to the usual Boltzmannian account.

The claim that the Mentaculus is a probability map of the world is outrageously ambitious. Why should the Mentaculus be taken seriously? The line of reasoning supporting it originates in Boltzmann's account of how the temporally directed second law of thermodynamics emerges from the temporally symmetric

---

[5]The claim that the entropy of the early universe was very low may seem surprising, since an ordinary gas at a high temperature uniformly spread out in a container is a system whose entropy is high. But while the effects of gravity are miniscule for the gas in the container in the very dense early universe, the contribution of gravity to entropy is significant. It is thought that taking gravity into account, the state of the early universe has very small entropy. The reason for this is that in the presence of gravity, a very dense uniform distribution of matter/energy is very special (i.e. low entropy) and will likely evolve to higher entropy states as matter/energy clump to form larger matters such as stars. See Penrose (2005) Ch. 27 and Carroll (2010) for discussion of this point. For complications and some dissent, see Callender (2009), Wallace (2010), and Earman (2006).

fundamental dynamics. The story is familiar but a quick retelling will help to explain the Mentaculus.

Clausius introduced the concept of entropy to characterize the fact that in thermodynamic processes energy involved in the performance of work tended to be transformed into energy that is no longer available for work. For example, the operation of an engine produces heat that cannot be used to do work. Thermodynamic entropy is, roughly, a measure of the energy in a system that is unavailable for work. The original formulation of the second law of thermodynamics says that the entropy of an energetically isolated system never decreases and typically increases until the system reaches equilibrium.

Boltzmann's explanation of the second law in terms of classical mechanics begins with his characterization of entropy as a property of the macrostate of a system. The macrostate M of a system specifies its total energy, temperature, pressure, density and so on. For each macrostate M there is a set of microstates (positions and momenta of the particles that compose the system) that realize M. Boltzmann identified the entropy of M with the logarithm of the volume on the standard Lebesgue measure of the set of microstates that realize M multiplied by a constant (Boltzmann's constant). We can think of the inverse of the hyper-volume of a system's macrostate M as specifying how much information M contains about the microstates that realize it. So greater entropy corresponds to less information. The second law says that an isolated system's macrostate typically evolves from specifying more to specifying less information about its microstate. Boltzmann's next step was to interpret the Lebesgue measure as specifying a probability measure over the space of the microstates that realize M. He then argued that a probabilistic version of the second law followed.[6] His argument, in a nutshell, is that since overwhelmingly most (on the measure) of the microstates that realize the macrostate M of a system not at equilibrium are sitting on trajectories that evolve according to the dynamical laws to realize

---

[6]The question arises of what "probability" means in this context especially since the fundamental dynamical laws are assumed to be deterministic. I will discuss this issue later in the paper.

macrostates of greater entropy, it follows that it is overwhelmingly likely that entropy will increase as the system evolves. The qualification "overwhelmingly likely" is required since there will be some microstates that realize M which don't evolve the system to states of higher entropy. However, these states are rare (on the measure) and scattered among the set of states that realize M; that is, in every very small (but of a certain finite size) convex region of phase space, almost all states evolve to higher entropy. The neighborhood of each "bad" microstate that doesn't evolve to higher entropy contains mostly "good" microstates that do. A slight disturbance of the particles comprising a "bad" microstate will change it into a "good" one. Boltzmann and those following in his footsteps have been able to show, it is plausible that if it is a thermodynamic law that states system S at time t in macro state M will evolve to be in macro state M* at t*, then Boltzmann's account will recover a probabilistic version of this regularity. Boltzmann's probability hypothesis is vindicated by its success in accounting for the second law and other thermodynamic laws.

As mentioned earlier, Boltzmann's account ran smack into the reversibility problem. While the account entailed that it is enormously likely that the entropy of the universe (or an isolated subsystem) will increase toward the future, because of the temporal symmetry of the dynamical laws, it also entailed that it is enormously likely that entropy increases toward the past.[7] One reaction to the reversibility problem is to construe the Boltzmann probability distribution solely as an instrument for making *predictions* for the future behavior of a thermodynamic system, but refrain from using it for retrodictions. This approach is usually combined with the views that

---

[7]It is not only absurd but, as Albert points out, leads to cognitive instability. If the Boltzmann probability posit is applied to the macro-condition of the universe at the present time, since it implies that it is likely that this macro-condition arose out of higher entropy states, in particular this means that the "records" in books and other documentation more likely arose as fluctuations out of chaos rather than as accurate records of previous events. This undermines the claim that there is evidence reported in those books that supports the truth of the dynamical laws, and so results in an unstable epistemological situation.

statistical mechanical reasoning should only be applied piecemeal to subsystems of the universe, and the view that probability should be understood epistemically as a measure of an experimenter's knowledge (or lack of knowledge) of the system's microstate. The package of these views amounts to an instrumentalist understanding of statistical mechanics. It avoids the problem but it leaves us completely in the dark as to why Boltzmann's probability posit works for predictions and in what sense it provides *explanations* of physical phenomena.[8] And unlike the Mentaculus, it does not apply to the universe as a whole. In contrast, the Mentaculus is a realist account on which statistical mechanics applies to the entire universe and, as I will later argue, it is compatible with and objective non-epistemic account of statistical mechanical probabilities.

The Mentaculus solves the reversibility problem by conditionalizing the Lebesgue probability distribution on the low entropy initial condition M(0); i.e., the PH.[9] This results in a probability distribution that gives the same *predictions* (inferences from M(t) to times further away from M(0)) as Boltzmann's prescription when applied to the universe as a whole and to its energetically isolated subsystems while avoiding the disastrous retrodictions we found without the PH. The Past Hypothesis and the Statistical Postulate work together to eliminate or make very unlikely trajectories of configurations of the particles in the universe that are compatible with the dynamical laws although they violate thermodynamical laws. The Past Hypothesis eliminates all trajectories that fail to start in the very low entropy macrostate M(0), and the Statistical Postulate implies that as the universe evolves, it is very likely that the entropy of the macrostate that the configuration realizes will increase.

The probabilistic version of the second law not only says that the entropy of the entire universe likely increases as long as the universe is

---

[8]Boltzmann's prescription leads to instrumentalism since it couldn't be literally true as it prescribes incompatible probabilities at different times, since the uniform distribution over the macrostate at t will differ from the uniform distribution over the macrostate at other times.

[9]See Sklar (1993) for a discussion of some other proposals for responding to the reversibility paradox.

not at equilibrium, but also that this holds for typical energetically isolated or almost isolated sub-systems. For example, the entropy of an ice cube in a glass of warm water will likely increase toward the future but decrease toward the past. It may seem astonishing that conditionalizing on the macrostate of the universe 13.8 billion years in the past results in the second law holding for subsystems at much later times and that, for example, the explanation why ice cubes melt but never unmelt, ultimately involves cosmology. Here is a "seat of the pants" argument that makes it plausible that the Mentaculus has this consequence. Suppose that S is a subsystem of the universe that at time t "branches off" from the rest of the universe so as to become approximately energetically isolated. As an example, consider an experimenter that places an ice cube in a tub of warm water. As mentioned earlier, "bad" microstates (i.e. those on a trajectory that is not entropy increasing) are scattered throughout the phase space and surrounded by "good" microstates. Because of this, preparing the microstate of the ice cube in warm water system in a bad state would require incredible precision beyond the ability of the experimenter. It is plausible then that the microstate of the ice cube in warm water system at the time it branches off is not bad, and as long as it is (approximately) energetically isolated, it is very likely that its entropy will increase.[10]

Due to the sparseness of bad microstates in the phase space, it is also plausible that almost all (on the Lebesgue measure) the microstates of subsystems that are not prepared by an experimenter, but branch off and become approximately energetically isolated as the universe evolves are not bad. Of course this doesn't mean that the entropy of *every* subsystem of the universe is likely to increase. Some subsystems are not isolated, but are interacting with other parts of the universe so as their entropy likely decreases (e.g. a glass of water in a freezer) while the entropy of the larger system increases (the environment of the freezer). Also, there may be systems that are specially prepared so that even when they become isolated their entropy will

---

[10]The entropy of a system that is not energetically isolated from its environment may decrease while the entropy of the system together with its environment increases. This is the case, for example with typical living systems.

very likely decrease.[11] In these cases, the second law doesn't hold. But that is as it should be. The job is to get the second law (and other thermodynamic laws) from the Mentaculus *in so far as* the second law is correct and, arguably, the Mentaculus does exactly that.

Albert pointed out that entropy decreasing microstates are not the only "bad" microstates. There are also microstates compatible with macrostates of macro-objects that, while they need not be entropy decreasing, are on trajectories that manifest bizarre behavior. Consider, for example, a comet in orbit around the sun. Almost all the microstates of the particles that compose the comet and are compatible with its macrostate M are ones that maintain its integrity as it travels on its orbit. But there will be some microstates compatible with M whose particles' positions and momenta are so arranged that the comet disintegrates, or in Albert's example takes the shape of a statue of the royal family. Of course, nothing like this ever occurs. The reason is that the set of such arrangements has measure 0 (or indiscernible from 0) on the Lebesgue measure and so, according to the Mentaculus, has probability 0 (or indiscernible from 0). The application of classical mechanics to macroscopic objects like comets or baseballs implicitly assumes that the microstate of the enormous number of particles that constitute them is not one of these bad states. Unbeknowst to Newton and Haley, the Mentaculus justifies this assumption.

The Mentaculus provides an account not only of the laws of thermodynamics, but also an important ingredient in an account of other special science laws. Consider a true regularity of the following form: a system which is F at time t evolves to be G at time t'. Since the Mentaculus entails probabilities over all physically possible macro-histories and since F and G are, to the extent they are precise, correspond to sets of micro-histories, it entails the value of $P(G(t')/F(t))$.[12] For the regularity to be a law, the value must be

---

[11]See *Time and Chance* Ch. 5 for a discussion of how a system may be prepared so that its entropy reliably decreases.

[12]The reason I have for saying that $P(G/F)$ is near 1 is a necessary but not a sufficient condition for Fs evolve to Gs to be a law is that further conditions, for

near 1. If it were much less than 1, we should conclude that the regularity was accidental or that it only held under certain conditions which when added to the antecedent results in a conditional probability with a value near 1. The reason for believing that this is a necessary condition for a special science law is as strong as our reasons for believing that the probabilities entailed by the Mentaculus are the objective probabilities. The primary reason for believing that is the Mentaculus success in accounting for thermodynamics.

Albert also makes a case that the Mentaculus can account for the "arrows of time" in addition to the thermodynamic arrow. I want to briefly review parts of that case and add a few wrinkles to it.

Time's arrows are pervasive features of our world. The epistemic arrow is that we can know much more about the past than about the future, since there are records (including memories) of past events but nothing like that for future events. The influence arrow is that we can have some influence or control over future events but no control over the past. Both these temporal asymmetries are closely related to the fact that causes precede their effects, which itself is closely connected to the temporal asymmetry of counterfactuals. The question is where do these arrows come from? In view of the temporal symmetry of the dynamical laws, the issue is as puzzling for them as for the thermodynamic arrow.

There are two main views that are held about the status of the temporal arrows. Primitivism is the view that time itself (unlike space) is equipped with a primitive direction.[13] Somehow, the "flow of time" is behind the temporal arrows. Reductionism is the view that time (like space) has no fundamental intrinsic direction in itself, but that the temporal asymmetries arise from the fundamental laws. I won't say anything more about primitivism except that it faces the problem of connecting the primitive arrow to the familiar arrows.[14]

---

example being a component of an optimal systematization of the truths expressible in the vocabulary of a science, is plausibly an additional requirement for special science law hood. See Callender and Cohen (2009).

[13]Tim Maudlin (2007) is the most effective proponent of this view.

[14]See Loewer (2012).

Reduction of the temporal asymmetries to the laws encounters the problem posed by the temporal symmetry of the fundamental dynamical laws. The Mentaculus aims to remedy this by adding to the dynamical laws two non-dynamical laws — the PH and the SP — which introduce a temporal asymmetry. The reductionist program is to make it plausible that with these additional laws, the Mentaculus can explain the other arrows. One might worry that by including the PH, the Mentaculus is assuming a past/future distinction, but this is not correct. The Past Hypothesis is a boundary condition of the universe obtained at a time close to the time of the Big Bang. If the Mentaculus program to explain time's arrows is successful, then that time be shown to in the past and the boundary condition will earn its name.

Suppose that at time $t(1)$, you come upon a closed box in which there is some blue smoke partly spread in the box with most of the smoke near the bottom right hand corner of the box. The usual statistical mechanical probability distribution (SP) is the best you can practically do to predict the future of the smoke. The SP will predict that the smoke is likely to become more defuse over time and eventually more or less uniformly fill the entire box, but it won't enable you to predict exactly what shapes it will take as it spreads. However, the SP would be a terrible way of inferring the state of the smoke at a time $t(0)$ earlier than $t(1)$. The SP implies that at an earlier time the smoke was spread uniformly throughout the box and followed a course that led to the state in which you found it at $t(1)$, i.e., that it would evolve in the time reverse of the way you predict it will evolve from its state at $t(1)$. But if you were informed that the smoke was released into the box from a small hole at the right hand corner at $t(0)$, then conditionalizing on this would lead you to infer that the smoke spread out from the hole to its present distribution. Unlike predictions of the future evolution of the smoke, we would be justified in inferring a great deal of detail about how the smoke evolved from the hole to its condition at $t(1)$, since the way it could evolve would be restricted by its having to get from the hole to its state at $t(1)$ in a certain amount of time. The idea is that the PH plays the role for the universe analogous to the fact that the smoke was introduced from the hole at a certain time plays for inferences from $t(1)$ to times between $t(1)$ and $t(0)$.

We can think of the condition of the smoke at $t(1)$ as a record of its condition at earlier times and the state of the smoke being released from the hole as a "ready condition" for the record. Albert observes that this reasoning is common to all inferences from what we take to be a record to the state it records. The inference from a record at time t to a condition at an earlier time t' involves the assumption of a ready condition at a yet earlier time t''. But how can we know at t information on the ready condition obtained at t''? We would know it if we had a record at $t(1)$ in the ready condition obtained on but that would require knowing the state at a time before t''. This begins a regress that Albert says can be stopped only by an assumption concerning the state at the earliest time. The PH is such an assumption. It constrains the way the universe has evolved from its very low entropy condition at $t(0)$, so as to allow records to be produced of conditions at times between the time of the record and the time of the PH. It is important to note that this proposal doesn't require justification in inferring from a record (e.g. tracks in the snow) to what it is a record of (a bear having recently passed by) what one knows of the ready condition (that the snow was fresh prior to the bear walking on it) obtained or that one knows the truth of the PH. It is sufficient that the ready condition did obtain.[15]

By assuming that the PH is true, the Mentaculus provides an explanation of why there can be records of events during the interval between the time of the record and the beginning of the universe. But because there is nothing like the PH assumed about the state of the universe during the time we call "the future", there are no "records" of events occurring during that interval. For predicting the future, the best we can do is to infer on the basis of the SP and that will

---

[15] In fact, cosmologists infer from observations of the state of the universe now that at a time shortly after the big bang, the entropy of the universe was very small in agreement with the PH. The cosmological inference are themselves justified by the PH. If it had turned out that cosmologists inferred from the present state that the state of the universe was incompatible with the PH, then Albert's explanation of the existence of records would be sunk.

leave a great deal of uncertainty. The role of the PH has in underlying the account of why we can have records of events during the interval from the big bang until now justifies calling that interval "the past" in and the fact that the SP leaves the time from now epistemically much more open justifies calling that time "the future."

The other arrow of time I want to discuss is the counterfactual arrow, since it is closely connected to the arrows of influence and causation.

The Mentaculus imposes a probabilistic structure on the possible macro-histories of the universe compatible with the PH. While the possible micro-histories evolve deterministically, macro-histories evolve probabilistically. Micro-histories that are macroscopically identical from the time of the PH until t can diverge. Macro-histories can also converge (eventually if the universe has an equilibrium state, all will converge to it), but during this epoch of the universe, since it is near the big bang, divergence overwhelmingly predominates. So from the macroscopic perspective, the evolution of the universe (and typical isolated subsystems) appears to be indeterministic. The resulting branching structure is pictured in Fig. 1.1.

Counterfactuals are temporally asymmetric. Counterfactual suppositions typically lead to large departures from the future of the actual world but not from the past. For example, if Nixon had pushed the button in 1974, the subsequent course of history since then would have been very different, but the prior history would have been pretty much the same. David Lewis proposed an account of counterfactuals that he thought captured this temporal asymmetry.[16] But as Adam Elga pointed out, he was wrong.[17] The Mentaculus suggests a better account that captures this asymmetry. Unlike Lewis' account, it doesn't involve "miracles" but it is restricted to counterfactuals whose antecedents are compatible with the Mentaculus.[18] As a first

---

[16]Lewis "Times Arrow and Counterfactual Dependence" in Lewis (1986).

[17]See Elga (2001) and Loewer (2007).

[18]This account is approximately coextensive with an account that Jonathan Bennett proposes (2003).

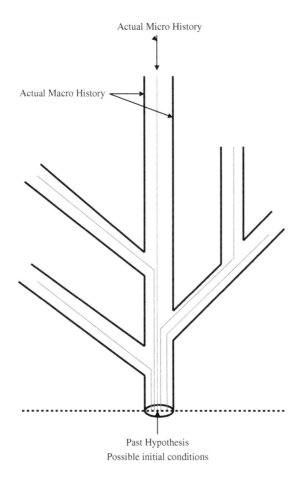

Fig. 1.1.   Evolution of micro and macro states.

stab, consider this:

> (COUNT): A(t) > B is true if conditional probability
> P(B/Disjunc(A(t)) is close to 1. Where t is the time of
> A and Disjunc(A(t)) is the disjunction of all the worlds at
> which A(t) is true. P(B/Disjunc(A(t)) is the conditional
> probability of B according to the Mentaculus given the set
> of all the worlds most similar to the actual world at which
> A(t) and the Mentaculus are true.

Suppose that at the actual world at the time referred to in
the antecedent, call it t, Nixon is distraught by the prospect of

impeachment and is in the Lincoln bedroom down the hall from the location of the button. The most similar worlds to the actual world are those whose macroscopic states are most similar to the actual macroscopic state at t except that Nixon is by the button at t pushing it. According to the Mentaculus, the probability that these evolve so that there is a nuclear war is close to 1 (assuming that the presidential launch system is working properly)[19] and so the macroscopic future from t is very different from the actual future. But the probability that the macro-history of these worlds, just like the history from a time shortly before t, will also be high. There is some back tracking since in the counterfactual worlds, Nixon will have to have gotten from the Lincoln bedroom to the location of the button. This backtracking seems right since we think that if Nixon had pushed the button at t, he would have had to have left the Lincoln bedroom a few minutes before t. In general, the more dissimilar the most similar counterfactual states in which the antecedent is true is from the actual state, the more back tracking is required. But the back tracking is limited since there are records of the actual macroscopic history that remain in the counterfactual worlds. So the macroscopic histories of the counterfactual world will be much like the actual history as is possible. If this account is on the right track, we see how the PH plays a crucial role in accounting for the temporal asymmetry of counterfactuals. It is due to the fact that the PH underwrites the existence of records of the past, the counterfactual past is similar to the actual past. But since there is no "future hypothesis" constraining the future, the counterfactual future may be very different from the actual future.

If the relation between counterfactuals and causation is anything along the lines of Lewis' counterfactual account, then it is reasonable to expect that the Mentaculus also accounts for the temporal asymmetry of causation, but it is beyond the scope of this paper to fully

---

[19] Not 1 but close to 1, since as was discussed earlier, there is a small statistical mechanical probability of a fluctuation even if the launch mechanism is working properly.

develop this suggestion. However, we can see how the Mentaculus can account for the causal influence of decisions.

Here is how COUNT accounts for the fact that we can have influence and control over the future but not the past. Suppose at t, Nixon has his finger on the button, deciding whether or not to push it. At t, COUNT will plausibly whether judge each of these counterfactuals true

1) If Nixon were to decide to push the button, there would be a nuclear war after soon after t
2) If Nixon were to decide not to push the button, there would not be a nuclear war soon after t

So at t Nixon can influence whether or not there is a nuclear war soon after t. Furthermore, since Nixon knows that these subjunctives are true, he has control over whether there will be a nuclear war. But he has no similar control over any past events. The reason is that decisions correspond to brain events involve relatively small number of degrees of freedom. For this reason, it is plausible that either decision can be added to the macroscopic state at t while keeping the remainder of the macroscopic state with all its records of the past almost exactly the same. This means that decisions made at t have no almost no effect on the macro-history prior to t.[20] Of course, the PH plays a crucial role here as well, since it is involved in the existence of records.

It has often been suggested that the thermodynamic arrow is at the bottom of the other arrows of time. That is not quite right.

---

[20]There are some complications. Alison Fernandes (2019) and Matthias Frisch (2010) pointed out that if an actual decision is itself a record of prior events, the most similar worlds that contain the counterfactual decision will differ from the actual world in what records they contain. Adam Elga pointed out that there is a complication caused by decisions to influence a past event when there is no record of that event in the actual macrostate. Say that Atlantis once existed but there is no record in the actual macrostate. Then, if D is a counterfactual decision it will be true that if D had been made, then Atlantis would not have existed. This is influence but not control, since the decider doesn't know what his decision might influence.

Rather it is the Mentaculus that accounts for both the entropic arrow and the other arrows, and thus makes for a unified account of temporal direction. While more needs to be done to develop and defend this claim, I think it is sufficiently promising to provide good reasons to take the Mentaculus seriously.

Taking the Mentaculus seriously means addressing the questions of interpreting the probabilities posited by the SP and how PH can be understood to be a law. The question concerning probability is vexing since it is widely believed that if determinism holds, then probabilities cannot be objective features of the physical world but must be construed epistemically or subjectively. Here is a famous expression of this view by Karl Popper:

> "Today I can see why so many determinists, and even ex-
> determinists who believe in the deterministic character of
> classical physics, seriously believe in a subjectivist inter-
> pretation of probability: it is in a way the only reason-
> able possibility which they can accept: for objective phys-
> ical probabilities are incompatible with determinism; and
> if classical physics is deterministic, it must be incompati-
> ble with an objective interpretation of classical statistical
> mechanics"[21]

An epistemic understanding of statistical mechanical probabilities raises the question of what is it that makes the credences specified by the SP the right ones. Sometimes appeal to an *a priori* principle, like the principle of indifference, is deployed to answer this question but aside from all the problems with such principles, it is easy to imagine a world in which these credences are not the right ones; that they are right seems to be a objective feature of our world. Also the Mentaculus probabilities must be understood as objective if they are to be part of the explanation of the laws of thermodynamics and the other temporal arrows. An ice cube doesn't melt because we believe that its microstate is a good one.

Popper thought of the objective physical probability to evolve into another state as the degree of the propensity for it to evolve into that

---

[21] Karl Popper (1959).

state, and he thought of propensities as ontologically fundamental. It is not surprising then that he concluded that statistical mechanical probabilities must be subjective if determinism is true since a state has a unique evolution. But there is another way of understanding objective probabilities so that they are non-fundamental physical features of the world that can be understood in explanations but are compatible with determinism. This other way is a development of an account originally due to David Lewis. Lewis' account of probabilities is part of his best system account (BSA) of laws. The BSA says that fundamental laws are expressed by certain generalizations or equations that are implied by the best true systematization of the "the Humean mosaic' (HM), i.e., the totality of fundamental events throughout all of space-time.[22] The best systematization is the one that best balances certain virtues that physicists prize in a theory. Lewis mentions informativeness and simplicity as the virtues that should be balanced, but there may be others.[23] Objective probability enters the account as a device to abet systematization. The idea is this: Consider a long sequence of outcomes of coin flips hthhthhhtththththththththththt..... There may be no simple sentence that is very informative about the sequence. But if a probability function is added to the system, then "the outcomes are independent with a probability .5" is both simple and informative. Lewis proposed that a probabilistic law should be evaluated in terms of "fit", by which he means the probability assigned by the laws of candidate systems to the HM.[24] Fit is best understood as a measure of

---

[22]Or at least throughout the observable universe and for 100 billion years after the time of the big bang.

[23]Lewis does not say much about how to measure simplicity and his suggestion for how to measure informativeness is defective. Other virtues may be that the system avoids fine tuning, that the best systematizations of typical regions of space time are the same as the best systematization of the entire universe, and that the system is informative about certain macroscopic phenomena, e.g., thermodynamic phenomena.

[24]Fit may not be the best way to evaluate a system with probabilities. It is not clear that it is the best way to capture the informativeness of a system that involves probabilities and it runs into trouble if the HM is infinite.

informativeness provided by probability. The system that best balances simplicity and informativeness and fit is the law and objective probability giving system.

The BSA propositions about probabilities and laws are objective features of the HM since they supervene on the HM. They are also contingent and not knowable *a priori*, since whether they are true or not depends on features of the actual HM. In particular, actual frequencies are especially relevant to probabilities although systematization allows for probabilities and associated frequencies to diverge. It has been suggested that just because BSA laws and probabilities supervene on the HM, they can't play the role that they are usually assigned in explanations. The objection is that since, the BSA laws and probabilities depend on the HM calling on them to explain parts of the HM, it would make the explanation defectively circular in the way that circular causation would be. I think this is to misunderstand how BSA laws and probabilities explain. They explain not by causing parts of the HM, as Popper seems to think propensities do, but by their role in unifying the mosaic. More needs to be said about this issue but I leave that for elsewhere.[25]

Although their views about the metaphysics of probability are miles apart, Lewis agreed with Popper that non-trivial probabilities are incompatible with determinism. But as we will see his reasons differed. Lewis says

> "To the question of how chance can be reconciled with determinism ... my answer is it can't be done.... There is no chance without chance. If our world is deterministic there is no chance in save chances of zero and one. Likewise if our world somehow contains deterministic enclaves, there are no chances in those enclaves."[26]

The reason Lewis held that objective probability and determinism are incompatible is connected with his claim that credences (and the actions they rationalize) should aim at or be guided by objective probabilities. He initially formulated this as the Principal Principle

---

[25]See Loewer (2012), Lange (2018) Emery (2019), Hicks and Van Elswyk (2015).
[26]Lewis (1982).

(PP) which says

$$(PP) : C(A/\mathrm{P}(A) = x \& E) = x^{27}$$

Where C is a credence function, $\mathrm{P}(A)$ is the objective probability of A and E is information that is "admissible". Lewis doesn't define "admissible information", but he includes laws and information temporally prior to $\mathrm{P}(A)$ as admissible. It immediately follows that if determinism is true, the laws and the past imply either A or $-A$, and so $\mathrm{P}(A)$ equals 1 or 0. However, as we will see, it turns out that the BSA can be naturally extended, and a rule that does the work of the PP devised so that non trivial objective probabilities are compatible with determinism.

To extend Lewis' BSA of laws and probabilities so that non-trivial probabilities are compatible with determinism, all that needs to be done is to allow probabilities to be assigned to initial conditions (or equivalently to physically possible histories). This fits in quite well with Lewis idea that informativeness is one of the criteria for evaluating systems. By adding a probability distribution, we might be able to greatly increase the informativeness of a system with deterministic dynamical laws with little cost in simplicity. This is exactly the case for the Mentaculus. To apply this to the Mentaculus, we need to add the criteria that make for the best systematization such that the system is informative with respect to facts involving macroscopic properties. Fleshing this idea out, requires an account of informativeness that is different from Lewis'. Lewis measure information of a system in terms of the set of worlds it excludes. While that is a sense of informativeness, it is not the one that is relevant to evaluating a candidate for the best systematization. A better idea is to measure informativeness of a system in terms of the number and significance of the truths that can be deduced from it, and include among these truths ones, the vocabulary of thermodynamics.

---

[27](PP) expresses how your belief that the objective probability of A is x should guide your credence. A related principle says that your credence in A should aim to match the objective probability of A.

The BSA (as modified above) fits the Mentaculus like a glove. My proposal is that the Mentaculus (with the correct dynamical laws) is the best system for our world. It follows that the statistical mechanical probabilities it implies are to be understood in terms of the BSA.[28] The BSA also endorses the claim that the PH is a law, since adding it to the dynamical laws and the SP greatly increases the informativeness of the system. It plays the role of a law in accounting for the fact that thermodynamic regularities hold since the big bang and that they are laws. It is often claimed that since the PH is incredibly unlikely given the SP, it "cries out for an explanation" (e.g. Carroll (2010)). But if it is a law as it is for Mentaculus, then it is not unlikely and the feeling that it requires explanation is somewhat alleviated. But even if the PH is a law, that doesn't mean that it is a fundamental law. It may have an explanation in terms of fundamental dynamical laws as in the model proposed by Carroll and Chen.[29]

I will conclude by discussing a few objections to the proposal that the Mentaculus is the best system (or a framework for the best system) for our world. There are many objections that I (to say nothing of others) can think of, so I will limit myself to few that seem to me the most pressing. Most of these come from an interesting paper by Frigg and Hoefer[30] that also construes statistical mechanical probabilities in terms of the BSA but in a very different way that I pursued in this paper

Objection 1. The Mentaculus is a wild speculation. Why believe that it is true?

Reply. My aim wasn't to demonstrate that it is true, but provide reasons for taking it seriously. Nevertheless, the case that has been made that it can account for the laws of thermodynamics, for the applicability of the dynamical laws to macroscopic systems, for the arrows of time, and cosmological evidence for the PH argue in favor

---

[28]Eddington famously observed that although physicists' views about the correct fundamental ontology and dynamical laws may change, the second law will be preserved. I would make a similar claim about the Mentaculus.
[29]Carroll and Chen (2004).
[30]Frigg and Hoefer (2013).

of it being correct. Also, the BSA is the best metaphysical account of laws and probabilities has been much discussed in recent philosophy of science. I don't want to defend it more in reply except to remark that the fact that it can make sense of the probabilities that occur in deterministic theories argues in its favor.

Objection 2. Frigg and Hoefer say that in thermodynamics, "the systems under investigation are 'laboratory systems' like gases in containers, liquids in tanks, and solids on tables. This is what SM is mostly applied to in the hands of working physicists, and so there is nothing objectionable about this restriction. However, following Albert (2000), Loewer considerably extends the domain of application of the theory, and treats the entire universe as one large SM system."

Reply. This is correct. But there is nothing wrong with applying statistical mechanics to the whole universe. This follows Boltzmann and many others.

Objection 3. But even if the Mentaculus applies to the entire universe, this doesn't show that it applies to the "laboratory systems" to which thermodynamics is usually applied. Frigg and Hoefer claim that if the Mentaculus applied to "numerous small subsystems of the world, does happen to make predictions with good fit to the patterns in HM, that is a happy accident, but not something that obviously follows from the global probability rule postulated in Loewer's approach."

Reply. Earlier in this paper, I sketched an argument following Albert which makes it plausible that the Mentaculus does imply that thermodynamics applies to 'laboratory systems' that can be considered to be approximately energetically isolated. The fact that it unifies all these applications counts in its favor.

Objection 4. There are probability distributions over the initial conditions compatible with the PH that make for a better fit than the uniform distribution. Frigg and Hoefer say "one can show that the fit of a system can be improved by choosing a peaked rather than a flat (Lebesgue) distribution. A peaked distribution, nearly dirac-delta style, over the world's actual initial condition is, qua postulate, just as simple as a flat distribution, but it will assign significantly higher probabilities to the actual world's macro-state transitions, and hence give the system to which it belongs a much higher fit. So the best system is not one with the flat distribution but one which has a distribution that is peaked over the actual initial condition."

Reply. A distribution that peaks at the actual initial condition is not as simple as the flat (i.e. uniform) distribution, since specifying it in an informative way (i.e. not by "the actual initial condition") requires specifying the positions and momenta of 10 to the power of 80 particles. Smoothing the distribution out a bit may make it simpler to describe, but then it will still be more complicated than the uniform distribution, and if sufficiently smooth, will agree with the uniform distribution in its consequences for thermodynamics.

Objection 5. The SP component of the Mentaculus is not the only probability distribution over the trajectories compatible with the PH that recovers the second law and the accounts of times arrows. By claiming that the uniform distribution is the correct one, the Mentaculus is committed to probabilities for all physically specifiable propositions such as the probability that given the current macro state of the universe, the president of the U.S. will be impeached within the next year. But this is an enormous over-commitment. There is no way to derive this conditional probability from the Mentaculus, and unlike the case of repeatable events, like melting of ice cubes or the outcomes of coin flips, there is no way to have any idea on the basis of the Mentaculus as to what it is.

Reply. It is correct that many other probability distributions will agree with the uniform one on thermodynamics. But the uniform distribution is the simplest and by the lights of the BSA that qualifies it as the lawful one. As a consequence, the Mentaculus will assign a probability to every physically specifiable proposition. If one doesn't like this, one can retreat to the view that the SP should be replaced by a family of probability distributions, each one of these agreeing on thermodynamics. This revision results in an account that is closer to typicality accounts of thermodynamics.[31] But, on the other hand, one might like the idea that all physically specifiable propositions have objective probabilities. We can then think of subjective credences as aiming to match objective probabilities. For example, my credence that the president will be impeached aims to match the Mentaculus probability that he will be impeached conditionally on the information currently available to me.

Objection 6. As mentioned earlier, the Mentaculus requires a revision of Lewis' PP. How will that work?

Reply. The PP says $C(A/\mathrm{P}(A) = x\&E) = x$ where E is "admissible." Lewis says that laws and truths in the past of the time of $\mathrm{P}(A)$ are admissible. As pointed out earlier, if E is the conjunction of the microstate at some time prior to $\mathrm{P}(A)$ and the laws, then this implies that $\mathrm{P}(A)$ can only take the values 1 and 0. To revise the PP, all that is needed is to drop the notion of admissibility entirely and replace PP with the following: MP: $C(A/\mathrm{P}(A) = x) = x$ and MPP: $C(A/B \& \mathrm{P}(A/B) = x) = x$, where P is the Mentaculus distribution. These say that given only that the Mentaculus probability of A (or conditional probability of A given B) has value x, one's credence should match the Mentaculus probability (conditional probability). If the conditional probability of an ice cube completely melting in 10 minutes given that the ice cube's macrostate is x then your credence that the ice cube will be completely melted in 10 minutes should be x. There is no conflict among MPP, non trivial objective probabilities and determinism. MPP permits inference from conditions at a time t to conditions at prior times as well as inference to future times. So, for example, it licenses an inference from the current macro-condition of the smoke in the example in the paper to probabilities of its earlier macrostate and its later macrostates.

I have endeavored to show in this paper that the hypothesis that the Mentaculus is the Humean best systematization of the universe, if

---

[31]See Maudlin's paper in this volume for a discussion of typicality. However, there remain important differences between the revised Mentaculus and typicality accounts.

true, provides a unified account of thermodynamics, the other arrows of time, special science laws, and the natures of the laws and probabilities that occur throughout the sciences. Much more needs to be done to fill in the gaps in these accounts before we conclude that the hypothesis is true. But I hope enough has been said to take it seriously.

References

Albert, D. (2015) *After Physics*. Cambridge, Mass.: Harvard University Press.

Albert, D. (2000) *Time and Chance*. Cambridge Mass.: Harvard University Press.

Bennett, J. (2003) *A Philosophical Guide to Conditionals*. Oxford University Press.

Callender, C. (2004) There is not Puzzle about the Low-entropy Past. In C. Hitchcock (ed.) *Contemporary Debates in Philosophy of Science*. Blackwell, 240–255.

Callender, C., Cohen, J. (2009) A Better Best System Account of Lawhood, *Philosophical Studies*, **145**(1), 1–34.

Carroll, S. (2010) *From Eternity to Here*. Dutton.

Carroll, S., Chen, J. (2004) Spontaneous Inflation and the Origin of the Arrow of Time. arXiv:hep-th/0410270.

Elga, A (2001) Statistical Mechanics and the Asymmetry of Counterfactual Dependence, *Philosophy of Science*, **68**, S313–S324.

Emery, N. (2019) Laws and their Instances, *Philosophical Studies*: 1–27 (forthcoming).

Earman, J. (2006) The 'Past Hypothesis': Not Even False, *Studies in the History and Philosophy of Modern Physics*, **37**(6), 399–430.

Fernandes, A. (2019) Time, Flies, and Why We Can't Control the Past. In: B. Loewer, E. Winsberg and B. Weslake (eds.) *Time's Arrows and the Probability Structure of the World*. Cambridge, Mass: Harvard University Press.

Frigg, R., Hoefer, C. (2103) The Best Humean System for Statistical Mechanics, *Erkenntnis*, **80**(S3), 551–574.

Frisch, M (2010) Does a Low-Entropy Constraint Prevent us from Influencing the Past? In: A. Huttemann and G. Ernst (eds.), *Time, Chance, and Reduction*. Cambridge University Press, 13–33.

Hicks, M. (2019) Dynamic Humeanism, *British Journal for the Philosophy of Science*, **4**(1), 983–1007.

Hicks, M., van Elswyk, P. (2015) Humean Laws and Circular Explanation, *Philosophical Studies*, **172**, 433–43.

Lange, M. (2018) Transitivity, Self-explanation, and the Explanatory Circularity Argument against Humean Accounts of Natural Law, *Synthese*, **195**, 1337–53.

Lewis, D. (1986) *Philosophical Papers* (Vol. II). Oxford University Press, New York.

Loewer, B. (2007) Counterfactuals and the Second Law. In: H. Price & R. Corry (eds.) *Causation, Physics, and the Constitution of Reality: Russell's Republic Revisited*. Oxford University Press.

Loewer, B. (2012) Two Accounts of Laws and Time, *Philosophical Studies*, **160**, 115–37.

Maudlin, T. (2007) *The Metaphysics within Physics*. Oxford: Oxford University Press.

Penrose, R. (2005) *The Road to Reality*. Knopf Doubleday.

Popper, K. (1959) The Propensity Interpretation of Probability, *The British Journal for the Philosophy of Science*, **10**(37), 25–42

Sklar, L. (1993) *Physics and Chance: Philosophical Issues in the Foundations of Statistical Mechanics*. Cambridge University Press.

Chapter 2

Probabilistic Explanations and the Derivation of Macroscopic Laws

Jean Bricmont

IRMP, Université catholique de Louvain
jean.bricmont@uclouvain.be

We will discuss the link between scientific explanations and probabilities, specially in relationship with statistical mechanics and the derivation of macroscopic laws from microscopic ones.

2.1. Introduction

It is a commonplace that macroscopic laws, in particular, the second law of thermodynamics ("entropy increases"), are true only in a probabilistic sense. But if one restricts oneself to classical physics, its laws are deterministic, so one might ask: where do these probabilities come from? How to make sense of objective probabilities in a deterministic universe? And if those probabilities are in some sense "subjective", namely assigned by us to events, and not "intrinsic" to those events, how can one say that macroscopic laws are objective?

Since the flow of heat from hot to cold is a perfectly objective fact and is law-like in its universality, should one say that our explanations of this fact are unsatisfactory if they rely on non-objective probabilities?

Our goal here is to try to answer these questions and to disentangle certain confusions that those questions tend to create.

We will start with general considerations about objectivity and subjectivity in science and also in different notions of probability.

Then we will explain how probabilities enter in the explanation of both equilibrium statistical mechanics and in the approach to equilibrium. We will illustrate the latter through a simple example, the Kac ring model. For related work in the same spirit, see e.g., Ref. Baldovin *et al.* (2019); Cerino *et al.* (2016); de Bièvre and Parris (2017).

2.2. Objectivity and subjectivity

There is a constant tension in the history of philosophy (and of science) between people who think that our thoughts are produced by our minds with little connection to the world ('idealists') and those who think that they are the result of an interaction between our mind and a mind-independent "outside world" ('realists'). At one extreme, one finds solipsism, everything going on in our mind is just like a dream or an internal movie; at the other extreme, one finds naive realists, for whom reality is "objectively" what it looks like, including colors, odors etc. It is not my ambition here to resolve this issue or even to discuss it adequately, although I am on the (non-naive) realist side.

Coming back to science, one may distinguish, even from a realist point of view, different degrees of "objectivity":

1. *Facts.* It is about facts that our intuition of objectivity is strongest: I am right now writing on a computer, the moon is there even if nobody looks at it, rivers flows and the sun shines when it does. An idealist might deny all this, but it is hard to be a realist without accepting the existence of facts "out there", independent of my consciousness.
2. *Laws.* We all know that there are regularities in Nature and it is a priori reasonable to think of those regularities as being part of Nature and to call them laws of Nature. Someone who is a realist about facts might deny the reality of laws over and above the so-called Humean mosaic of empirical facts, considering them as a mere human tool to summarize the observed regularities.[1]

---

[1]See Maudlin (2007) for a discussion of realism about laws versus the "Humean" conception.

We will not analyze this issue here and simply admit, for the sake of the discussion, the objectivity of the laws of Nature. Moreover, one can introduce a hierarchy of such laws: there are fundamental laws, governing the behavior of the most microscopic constituents of matter and derived or phenomenological laws describing the behavior of aggregate sets of such particles. For example, if "heat is molecular motion", then the laws governing heat are phenomenological, but those governing the molecular motions are fundamental.[2]

3. *Explanations.* There is a common (mis)-conception according to which the role of science is to describe and to predict but not to explain. However, if one asks: "why does it rain today?", one is asking for an explanation of a given fact and the answer will involve laws of meteorology and empirical data concerning past situation of the atmosphere. So, it will be an explanation and it will be scientific.

Science is in fact full of explanations: the theory of gravitation explains the regularities in the motion of planets, moons or satellites. The atomic theory of matter explains the proportions of elements in chemical reactions. Medical science explains in principle what cures a disease, etc.

The misconception arises because people often think of "ultimate" explanations, for example: why is there something rather than nothing? Or, how did the Universe (including the Big Bang) come to exist? But, if one puts aside those metaphysical/religious questions to which nobody has an answer, science does provide explanations of observable phenomena.

Of course, the question: "what constitutes a valid explanation?", specially when probabilities are involved, is a tricky one and we will discuss it in Sections 2.5 and 2.6.

---

[2] At least to a first approximation: one may consider the molecules as being made of more fundamental entities, whose laws would be truly fundamental, while those governing the molecules would be phenomenological. But, since we do not know what are really the ultimate constituent of matter, this fundamental/ phenomenological distinction is relative to a given context.

4. *Probabilities.* If one puts oneself in the framework of classical physics (in order to avoid quantum subtleties), then laws are deterministic which means that, given some initial conditions, future events either occur or do not occur. There is no sense in which they are, by themselves, probable or improbable. Yet, scientists use probabilities all the time. Although there is a school of thought that tries to give an objective meaning to the notion of probability (we will discuss it below), there must be something "human" about probabilities in the sense that we may use them in certain ways and have good reasons to do so, but probabilities are not expressing objective facts about the world, independent of us, for example, the motion of the moon.

2.3. Two notions of probability

There are, traditionally, at least two different meanings given to the word 'probability' in the natural sciences. These two meanings are well-known but, since much confusion arises from the fact that the same word is used to denote two very different concepts, let us start by recalling them and explain how one can connect the two. First, we speak of 'the natural sciences', because we do not want to discuss the purely mathematical notion of probability as measure (the notion introduced by Kolmogorov and others). This is of course an important branch of mathematics, and the notion of probability used here, when formalized, will coincide with that mathematical notion. However, we want to focus here on the role played by probabilities in our scientific theories, which is not reducible to a purely mathematical concept.

Therefore, the first notion that comes to mind is the so called 'objective' or 'statistical' one, that is the view of probability as something like a 'theoretical frequency': if one says that the probability of the event E under condition X, Y, Z equals p, one means that, if one reproduces the conditions X, Y, Z sufficiently often, the event E will appear with frequency p. Of course, 'sufficiently often' is vague and this is the source of much criticism of that notion of probability. But, putting that objection aside for a moment and assuming that 'sufficiently often' can be given a precise meaning in concrete circumstances, probabilistic statements are, according to this view, factual statements that can be confirmed or refuted by observations or experiments.

By contrast, the 'subjective' or Bayesian use of the word 'probability' refers to a form of reasoning and not to a factual statement. Used in that sense, assigning a probability to an event expresses a rational judgment on the likelihood of that single event, based on the information available at that moment. Note that, here, one is not interested in what happens when one reproduces many times the 'same' event, as in the objective approach, but in the probability of a single event. This is of course very important in practice: when I wonder whether I need to take my umbrella because it will rain, or whether the stock market will crash next week, I am not mainly interested in the frequencies with which such events occur but with what will happen here and now. Of course, these frequencies may be part of the information that is used in arriving at a rational judgment on the probability of a single event but, in general, they are not the only information available.

How does one assign subjective probabilities to an event? In elementary textbooks, a probability is defined as the ratio between the number of favorable outcomes and the number of 'possible' ones. While the notion of favorable outcome is easy to define, the one of possible outcome is much harder. Indeed, for a Laplacian demon, nothing is uncertain and the only possible outcome is the actual one; hence, all probabilities are either zeroes or ones. But we are not Laplacian demons and it is here that ignorance enters.[3] We try to reduce ourselves to a series of cases about which we are 'equally ignorant', which is to say the information that we do have does not allow us to favour one case over the other, and that defines the number of 'possible' outcomes. The standard examples include the throwing of a dice or of a coin, where the counting is easy, but that situation is not

---

[3]This was of course Laplace's main point in Laplace (1825), although this is is often misunderstood. Laplace emphasized that human intelligence will forever remain 'infinitely distant' from the one of his demon.

typical. At the time of Laplace, this method was called the 'principle of indifference'; its modern version is the *maximum entropy principle*. Here one assigns to each probability distribution $\mathbf{p} = (p_i)_{i=1}^N$ its Shannon entropy, given by:

$$S(\mathbf{p}) = -\sum_{i=1}^{N} p_i \ln p_i.$$

One then chooses the probability distribution that has the maximum entropy, among those that satisfy certain constraints that incorporate the information that we have about the system.

The rationale, like for the indifference principle, is not to introduce bias in our judgments, namely information that we do not have (like people who believe in lucky numbers). And one can reasonably argue that maximizing the Shannon entropy is indeed the best way to formalize that notion, see Shannon (1948); Jaynes (1983) and (Section 3.11.3) Jaynes (2003).

In practice, one starts by identifying a space of states in which the system under consideration can find itself, and one assigns a prior distribution to it (maximizing the Shannon entropy, given the information available at the initial time), which is then updated when new information becomes available.[4] Note that probabilistic statements, understood subjectively, are forms of reasoning, although not deductive ones. Therefore, one cannot check them empirically, because reasonings, whether they are inductive or deductive, are either correct or not but that depends on the nature of the reasoning not on any facts.

If someones says: Socrates is an angel; all angels are immortal; therefore Socrates is immortal, it is a valid (deductive) reasoning. Likewise, if all we know about a coin is that it has two faces and that it looks symmetric, therefore the probability of 'head' is one half, it is a valid probabilistic reasoning; throwing the coin a thousand times with a result that is always tails does not disprove the reasoning, it only indicates that the initial assumption

---

[4] For an introduction to Bayesian updating, see e.g. Jaynes (1983, 2003).

(of symmetry) was probably wrong (just as watching Socrates dead leads one to reconsider the notion that he is an angel or that the latter are immortal). The main point of Bayesianism is to give rules that allow to update one's probabilistic estimates, given previous observations.

Let us now consider some frequent objections to this "subjective" notion of probability.

1. **Subjectivism.** Some people think that a Bayesian view of probabilities presupposes some form of subjectivism, meant as a doctrine in philosophy or philosophy of science that regards what we call knowledge as basically produced by "subjects" independently of any connection to the "outside world". But there is no logical link here: a subjectivist about probabilities may very well claim that there are objective facts in the world and that the laws governing it are also objective, and consider probabilities as being a tool used in situations where our knowledge of those facts and those laws is incomplete. In fact, one could argue that, if there is any connection between Bayesianism and philosophical subjectivism, it goes in the opposite direction; a Bayesian should naturally think that one and only one among the 'possible' states is actually realized, and that there is a difference between what really happens in the world and what we know about it. But the philosophical subjectivist position often starts by confusing the world and our knowledge of it (for example, much of loose talk about everything being 'information' often ignores the fact that 'information' is ultimately information about something which itself is not information).

   Besides, ignorance does enter in the computations of probabilities but, as we will see in the next section, when we discuss the connection between probabilities and physics, this does not mean that either knowledge or ignorance are assumed to play a fundamental role in physics.

2. **Determinism.** One may object that Bayesians are committed to a deterministic view of the world. Since Bayesians regard probabilities as subjective, doesn't this deny the possibility that

phenomena can be intrinsically or genuinely random? Not necessarily. A Bayesian may be agnostic concerning the issue of intrinsic randomness and point out that it is difficult to find an argument showing the presence of intrinsic randomness in nature; indeed, it is well-known that some deterministic dynamical systems (the 'chaotic' ones) pass all the statistical tests that might indicate the presence of 'randomness'.[5] So, how can we know, when we observe some irregular and unpredictable phenomenon, that this phenomenon is 'intrinsically random' rather than simply governed by some unknown but 'chaotic' deterministic laws?

3. **(Ir)relevance to physics.** One may think that the Bayesian approach is useful in games of chance or in various practical problems of forecasting (like in insurances) but not for physics. Our answer will be based on the law of large numbers discussed in the next section.

The main point of this discussion is that there is nothing arbitrary or subjective in the assignment of "subjective" probabilities. The word "subjective" here simply refers to the fact that there are no true or real probabilities "out there". But the choice of probabilities obeys rules (maximizing Shannon's entropy and doing Bayesian updating) that do not depend of any individual's whims, although it does depend on their information.[6]

---

[5]Here is a simple example of such a system. Let $I = [0, 1]$ and let $f : I \to I$ be given by $f(x) = 2x \mod 1$. Then, writing $x \in I$ as $x = \sum_{n=1}^{\infty} \frac{a_n}{2^n}$, with $a_n = 0, 1$, we see that the map f is equivalent to the shift σ on sequences $\mathbf{a} = (a_n)_{n=1}^{\infty}$, $\sigma(\mathbf{a})_n = a_{n+1}$. Using this observation, and the fact that the Lebesgue measure on I is equivalent to the product measure on the sequences $\mathbf{a}$ giving a weight $\frac{1}{2}$ to both 0 and 1, one can check that the map f is equivalent to a sequence of "random" coin tossings with $a_n = 0$ being "head" and $a_n = 1$ being "tail". The effect is such that simple deterministic system will look as random as any apparently random system. For more fancy "chaotic" dynamical systems, see Bowen (1975); Bowen and Ruelle (1975); Eckmann and Ruelle (1985); Sinai (1972); Ruelle (1976, 1978).

[6]A further confusion arises from the fact that some probabilists, the Italian Bruno de Finetti being the best known one, do consider probabilities as expressing purely subjective degrees of beliefs that are constrained only by rules of consistency,

2.4. The law of large numbers

A way to make a connection between the two views on probability goes through the law of large numbers: the calculus of probabilities — viewed now as part of deductive reasoning — leads one to ascribe subjective probabilities close to one for certain events that are precisely those that the objective approach deals with, namely the frequencies with which some events occur when we repeat many times the 'same' experiment. So, rather than opposing the two views, one should carefully distinguish them and regard the objective one as, in a sense, derived from the subjective one (i.e. when the law of large numbers leads to subjective probabilities sufficiently close to one). Let us state the law of large numbers, using a terminology that will be useful when we turn to statistical mechanics below. Consider the simple example of coin tossing. Let 0 denote 'head' and 1, 'tail'. The 'space' of results of any single tossing, $\{0, 1\}$, will be called the 'individual phase space' while the space of all possible results of N tossings, $\{0, 1\}^N$, will be called the 'total phase space'. In statistical physics, the individual phase space will be $\mathbb{R}^3$ (if one considers only the positions or only the velocities of the particles) or $\mathbb{R}^6$ (if one considers both the positions and the velocities of the particles), and the total phase space will be $\mathbb{R}^{3N}$ or $\mathbb{R}^{6N}$ for N particles. The variables N_0, N_1 that count the number of heads (0) or tails (1) in N tossings are called *macroscopic*.

Here we introduce an essential distinction between the *macroscopic variables*, or the *macrostate*, and the *microstate*. The microstate, for N tossing, is the sequence of results for all the tossings, while the macrostate simply specifies the values of N_0 and N_1. Although this example is trivial, let us draw the following analogy with statistical mechanics: N_0 and N_1 for a given point in the total phase space (a sequence of results for all the tossings), count the

---

see e.g. de Finetti (1937, 2017). These probabilists are sometimes called "subjective Bayesians"; the view presented here is then called "objective Bayesian", see Jaynes (1983), (p. 4) and Jaynes (2003), (p. 655) for a discussion of the difference between these views.

number of 'particles' that belong to a given subset (0 or 1) of the
individual phase space.

Now, fix $\epsilon > 0$ and define a sequence of sets of microstates $\mathcal{T}_N \subset$
$\{0,1\}^N$ to be *typical*, for a given sequence of probability measures
P_N on $\{0,1\}^N$, if

$$P_N(\mathcal{T}_N) \to 1. \tag{2.1}$$

as $N \to \infty$. If the typical sets $\mathcal{T}_N$ are defined by a property, we will
also call that property typical.[7]

Let $G_N(\epsilon)$ be the set of microstates such that

$$\left| \frac{N_0}{N} - \frac{1}{2} \right| \leq \epsilon \tag{2.2}$$

Here the letter G stand for "good", because we will use the same
expression below in the context of statistical mechanics.

Then, (a weak form of) the law of large numbers states that,
$\forall \epsilon > 0$,

$$P_N(G_N(\epsilon)) \to 1 \tag{2.3}$$

as $N \to \infty$, where P_N the product measure on $\{0,1\}^N$ that assigns
independent probabilities $\frac{1}{2}$ to each outcome of each tossing. This is
the measure that one would assign on the basis of the indifference
principle, give an equal probability to all possible sequences of results.
In other words, what (2.3) expresses is that the sequence of sets
$G_N(\epsilon)$ is typical in the sense of definition (2.1), $\forall \epsilon > 0$.

A more intuitive way to say the same thing is that, if we simply
count the number of microstates that belong to $G_N(\epsilon)$, we find that
they form a fraction of the total number of microstates close to 1,
for large N.

The situation becomes more complicated but more interesting if
one tries to understand what could be a *probabilistic explanation*,
like the explanation of the second law of thermodynamics.

---

[7]This use of the word typical is not exactly the usual one, which refers to the
probability of a given set, not a sequence of sets, to be close to 1.

2.5. Explanations and probabilistic explanations

A first form of scientific explanation is given by laws. If state A produces state B, according to deterministic laws, then the occurrence of B can be explained by the occurrence of A and the existence of those laws.[8] If A is prepared in the laboratory, this kind of explanation is rather satisfactory, since the initial state A is produced by us.

But if B is some natural phenomena, like today's weather and A is some meteorological condition yesterday, then A itself has to be explained, and that leads potentially to an "infinite" regress, going back in principle to the beginning of the universe. In practice, nobody goes back that far, and A is simply taken to be "given", namely our explanations are in practice limited.

It is worth noting that there is something "anthropomorphic" even in this type of explanation. For example, if A is something very special, one will try to explain A as being caused by anterior events that are not so special; otherwise, our explanation of B in terms of A will look unsatisfactory. Both the situations A and B and the laws are perfectly objective, but the notion of explanation is "subjective" in the sense that it depends on what we, humans, regard as a valid explanation.

Consider now a situation where probabilities are involved, take the simplest example, coin tossing, and try to use that example to build up our intuition about what constitutes a valid explanation.

First observe that, if we toss a coin many times and we find approximately half heads and half tails, we do not feel that there is anything special to be explained. If, however, the result deviates strongly from that average, we'll look for an explanation (e.g. by saying that the coin is biased).

This leads to the following suggestion, suppose that we want to explain some phenomenon when our knowledge of the past is such that this phenomenon could not have been predicted with certainty (for coin tossing, the past would be the initial conditions

---

[8]This is the main idea behind the deductive nomological model, according to which scientific explanations are deductive arguments with laws as one of the premises (see Hempel (1942), Hempel and Oppenheim (1948)).

of the coins when they are tossed). We will say that our knowledge, although partial, is *sufficient* to explain that phenomenon if we could have predicted it using Bayesian probabilities and the information we had about the past. That notion of 'explanation' incorporates, of course, as a special case, the notion of explanation based on laws. Also, it fits with our intuition concerning the coin-tossing situation discussed above. Being ignorant of any properties of the coin leads us to predict a fraction of heads or tails around one-half. Hence, such a result is not surprising or, in other words, does not "need to be explained" while a deviation from it requires an explanation.

Turning to physics, consider for example, the Maxwellian distribution of velocities for a free gas of N particles of mass m. Let $\Delta(\vec{u})$, be the cubic cell of size δ^3 centered around $\vec{u} \in (\delta\mathbb{Z})^3$. Let $\mathbf{v} = (\vec{v}_1, \ldots, \vec{v}_N) \in \mathbb{R}^{3N}$, with each $\vec{v}_i \in \mathbb{R}^3$, be an element of the 'phase space' of the system (where the spatial coordinates are ignored), i.e. a configuration of velocities for all the particles, which is what we call a *microstate* of the system.

Define the *macrostate* by the set of variables $\{N_{\vec{u}}(\mathbf{v})\}_{\vec{u} \in (\delta\mathbb{Z})^3}$:

$$N_{\vec{u}}(\mathbf{v}) = |\{i | \vec{v}_i \in \Delta(\vec{u}), \ i \in \{1, \ldots, N\}\}|. \tag{2.4}$$

$N_{\vec{u}}(\mathbf{v})$ is also called *the empirical distribution* corresponding to the phase space point $\mathbf{v}$. It counts, for each $\Delta(\vec{u})$ and for a given set of velocities of all the particles, the number of particles whose velocities lie in $\Delta(\vec{u})$.

This is analogous to counting the number of heads or tails in a given sequence of coin tosses or the number of times a dice falls on a given face when it is thrown many times.

Let $G_N(\epsilon, \delta)$, for given ϵ, δ, be the set of "good" vectors $\mathbf{v}$ for which

$$\left| \frac{N_{\vec{u}}(\mathbf{v})}{N} - \frac{\exp\left(-\frac{m|\vec{u}|^2}{2kT}\right)}{(2\pi mkT)^{3/2}} \right| \leq \epsilon, \tag{2.5}$$

$\forall \vec{u} \in (\delta\mathbb{Z})^3$.

Let $S_{E,N}$ be the constant energy surface of energy E, namely the subset of $\mathbb{R}^{3N}$ defined by:

$$S_{E,N} = \left\{ \mathbf{v} = (\vec{v}_1, \ldots, \vec{v}_N) \,\middle|\, \sum_{i=1}^{N} \frac{m|\vec{v}_i|^2}{2} = E \right\} \qquad (2.6)$$

and let $\mu_{E,N}$ be the uniform measure on that surface (i.e. the restriction of the Lebesgue measure in $\mathbb{R}^{3N}$ to that surface).

Then, a variant of the law of large numbers says that for every ϵ, δ, the sequence of sets $G_N(\epsilon, \delta)$ is *typical*, in the sense of (2.1) for the sequence of probability measures $\mu_{E,N}$, if T in (2.5) is related to E in (2.6) by $kT = \frac{2E}{3N}$, which means that the set $G_N(\epsilon, \delta)$ has, for large N, a measure $\mu_{E,N}$ close to one when $kT = \frac{2E}{3N}$ holds. This is a precise way of saying that the distribution of velocities for a gas of N particles of mass m is Maxwellian.

If someone asks, how does one explain the occurrence of this Maxwellian distribution? The Bayesian answer is basically that there is nothing to explain, because this is analogous to the situation of coin tossings when the fractions of heads and tails are both close to one half. Given that we know that the energy is conserved, symmetry considerations show that the uniform measure is the most natural one and, since the Maxwellian distribution is the empirical distribution corresponding to most phase points (relative to that measure), it is exactly what we would expect if we know nothing more about the system. In fact, the only thing that would lead us *not* to predict the Maxwellian distribution would be some additional knowledge about the system (e.g. that there are some constraints or some external forces acting on it).

This answers the often heard question, "how does one justify the choice of the equilibrium measure?". It is namely here the measure $\mu_{E,N}$ of $S_{E,N}$, the natural choice on Bayesian grounds.

However, one can ask a related question, which is less trivial: how does one explain the approach to equilibrium for a closed system which starts in a nonequilibrium configuration? This is the question that Boltzmann's analysis answers and is the topic of the next section.

2.6. Time evolution and probabilistic explanations

2.6.1. *Microstates and macrostates*

We start by generalizing the notion of macrostate introduced in Section 2.5. Let $\mathbf{x}(t)$ be the *microstate* of a classical mechanical system on N particles, namely

$$\mathbf{x}(t) = (\vec{q}_1(t), \vec{q}_2(t), \ldots, \vec{q}_N(t), \vec{p}_1(t), \vec{p}_2(t), \ldots, \vec{p}_N(t)) \in \mathbb{R}^{6N},$$

where $\vec{q}_i(t) \in \mathbb{R}^3$ and $\vec{p}_i(t) \in \mathbb{R}^3$ are the position and the momentum of the ith particle at time t.

Let

$$\mathbf{x}(0) \to \mathbf{x}(t) = T^t \mathbf{x}(0)$$

denote the flow in R^{6N} induced by Hamilton's equations for that system and let Ω denote a bounded subset of R^{6N} invariant under that flow (for example, a bounded constant energy surface).

Here, a *macrostate* is simply a map $F : \Omega \to R^n$ with n, the number of macroscopic variables, being much smaller than $6N$: $n \ll 6N$.

We can give, using (2.4), a simple example of such a map, by letting $F = F(\mathbf{x}) = (\frac{N_{\vec{u}}(\mathbf{v})}{N})_{\vec{u} \in (\delta\mathbb{Z})^3}$.[9] To give another example, let, as above, $\Delta(\vec{u})$ be the cubic cell of size δ^3 centered around $\vec{u} \in (\delta\mathbb{Z})^3$ and let $\mathbf{q} = (\vec{q}_1, \ldots, \vec{q}_N) \in \mathbb{R}^{3N}$ be an element of the 'configuration space' of the system, that is a configuration of the positions for all the particles. Define

$$N_{\vec{u}}(\mathbf{q}) = |\{i | \vec{q}_i \in \Delta(\vec{u}), \ i \in \{1, \ldots, N\}\}|. \tag{2.7}$$

Assume that the particles are enclosed in a box Λ, which is a union of cubic cells of size δ^3, and let $F = F(\mathbf{x}) = (\frac{N_{\vec{u}}(\mathbf{q})}{N})_{\Delta(\vec{u}) \subset \Lambda}$. Here n is the number of cubic cells of size δ^3 in Λ, or $\frac{|\Lambda|}{\delta^3}$.

Note that the density function is simply a continuous approximation to the function F (obtained in the limit $N \to \infty$, $\delta \to 0$).

---

[9]A small caveat: here the number n of macroscopic variables seems to be infinite, since $\vec{u} \in (\delta\mathbb{Z})^3$ but the number of non-zero values of $N_{\vec{u}}(\mathbf{v})$ is finite, since, because of (2.6), $N_{\vec{u}}(\mathbf{v})$ will be 0 if $|\vec{u}| > \frac{2E}{m} + \frac{3\delta}{2}$.

One could also do that in the space $\mathbb{R}^6$, combining both positions and momenta of the particles. In that case, the continuous approximation to F is Boltzmann's f function.

Now, one can associate to the evolution $\mathbf{x}(0) \to \mathbf{x}(t) = T^t \mathbf{x}(0)$ an *induced evolution* $F_0 \to F_t$, obtained by:

$$F_0 = F(\mathbf{x}(0)) \to F_t = F(\mathbf{x}(t)). \tag{2.8}$$

A natural question is whether the evolution of F is *autonomous*, which is to say independent of the $\mathbf{x}(0)$ mapped onto F_0. If it is, then one can say that the evolution of F_t, which is called a *macroscopic law*, has been *reduced to* or *derived from* the microscopic one $\mathbf{x}(0) \to \mathbf{x}(t)$ in a straightforward way.

But such an autonomous evolution is, in general, impossible, because the evolution $\mathbf{x}(0) \to \mathbf{x}(t)$ is *reversible* meaning that, if I denotes the operation:

$$I(\mathbf{x}(t)) = (\mathbf{q_1}(t), \mathbf{q_2}(t), \dots, \mathbf{q_N}(t), -\mathbf{p_1}(t), -\mathbf{p_2}(t), \dots, -\mathbf{p_N}(t)), \tag{2.9}$$

one has:

$$T^t I T^t \mathbf{x}(0) = I \mathbf{x}(0). \tag{2.10}$$

Or, in words, if one lets the system evolve according to the dynamical laws for an amount of time t, if one then reverses the velocities (or the momenta), and if finally one lets the system evolve according those same laws for the same amount of time t, one gets the initial state with the velocities reversed.

But the evolution $F_0 \to F_t$ is often *irreversible*, for example if F is the density and if one starts with a non-uniform density, the evolution of F tends to a uniform density and will not return to a non-uniform one.

Yet the reversibility argument shows that, since changing the sign of the velocities does not change the density, for each microstate $\mathbf{x}(t) = T^t \mathbf{x}(0)$ giving rise to a given value of $F_t = F(\mathbf{x}(t))$, there may exist another microstate $I(\mathbf{x}(t))$ giving rise to the same value of $F_t = F(I(\mathbf{x}(t)))$ but such that the future time evolution of F_t will be markedly different depending on whether it is induced by $\mathbf{x}(t)$ or by $I(\mathbf{x}(t))$. Therefore, the evolution of the macrostate cannot be autonomous in the sense given here.

This seems to imply that one cannot derive a macroscopic law from a microscopic one and, in particular, that one cannot give a microscopic derivation of the second law of thermodynamics implying that the entropy monotonically increases. Yet, as we will explain now, this can be done, but not in the straightforward way suggested above.

2.6.2. Derivation of macroscopic laws from microscopic ones

The basis of the solution to the apparent difficulty mentioned in the previous subsection is that *the map F is many to one in a way that depends on value taken by F!*

To explain that, think again of the simple example of N coin tossings with $F =$ number of heads and the microstates being the sequence of results such as $(H, T, H, H, \ldots, T)$.

If $F = N$, it corresponds to a unique microstate $(H, H, H, H, \ldots, H)$

But if $F = \frac{N}{2}$ then there are approximately $\frac{2^N}{\sqrt{N}}$ microstates giving rise to that value of F.

If one considers the density function, it is easy to see that if the box Λ is divided in two equal parts, the volume in phase space where the particles are uniformly distributed in Λ will be of the order of 2^N times larger than the one where the particles are concentrated in one of those parts.

In Fig. 2.1, one has a schematic illustration of this fact[10]: each region in Ω corresponds to the set of microstates giving rise to the same value of F and we denote these regions by $\Omega_0, \Omega_1, \Omega_2, \ldots$. Let us stress that the Figs. 2.1–2.3 are highly "abstract", since the phase space Ω represented there by a two-dimensional square is in reality a subset of a space of dimension of order 10^{23}.

---

[10]Figures 2.1–2.3 are inspired by similar pictures in chapter 7 of Penrose (1989).

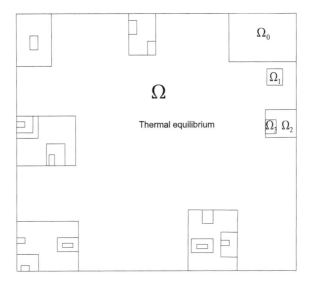

Fig. 2.1.   A partition of the phase space Ω (represented by the entire square) into regions $\Omega_0, \Omega_1, \Omega_2, \ldots$ corresponding to microstates that are macroscopically indistinguishable from one another, that is microstates that give rise to the same value of F. The region labelled "thermal equilibrium" corresponds to the value of F corresponding to the overwhelming majority of microstates.

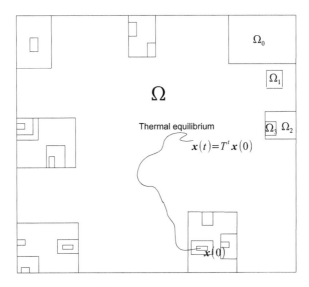

Fig. 2.2.   The curve $\mathbf{x}(t) = T^t\mathbf{x}(0)$ describes a possible evolution of a microstate, which tends to enter regions of larger volume until it enters the region of thermal equilibrium.

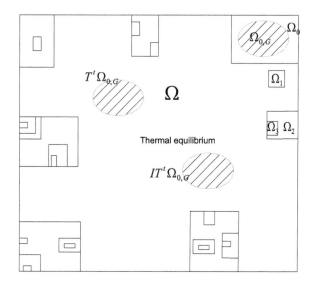

Fig. 2.3. Ω_0 are the configurations in one-half of the box at time zero; $\Omega_{0,G}$ are the good configurations in Ω_0 whose evolution lead to a uniform density at time t: $T^t(\Omega_{0,G}) \subset \Omega_t$; $I(T^t(\Omega_{0,G}))$ are the configurations of $T^t(\Omega_{0,G})$ with velocities reversed whose evolution after time t belongs to Ω_0: $T^t(I(T^t(\Omega_{0,G}))) \subset \Omega_0$.

Moreover, F usually takes a continuum of values and, in Fig. 2.1, those values were treated as discrete. But that is only in order to simplify our illustration.

In the example of coin tossing, the region labelled equilibrium in Fig. 2.1 corresponds to having approximately as many heads and tails, i.e. the set of coin tossings defined by (2.2). The region Ω_0 may correspond to having approximately one third heads and two third tails, Ω_1 may correspond to having approximately one quarter heads and three quarters tails, etc.

In the example of the gas in the box, the region labelled "thermal equilibrium" in Fig. 2.1 corresponds to an approximate uniform distribution of the particles in the box and an approximate Maxwellian distribution of their velocities.[11]

---

[11]We use the word "approximate" here, because, as for coin tossing, for a finite number of particles, the correspondence with the predicted statistical distribution is always approximate.

The region Ω_0 in Fig. 2.1 might correspond to all the particles being in one half of the box, another region might correspond to all the particles being in the other half, yet another region, say Ω_1, might correspond to all the particles being in an even smaller part of the box etc.

Of course, nothing is drawn to scale here. If the size of the region in the phase space where all the particles are concentrated in one part of the box is 2^N smaller than the one where the particles are uniformly distributed in Λ, and N is of the order of the Avogadro's number ($N \sim 10^{23}$), that region where all the particles are concentrated in one part of the box could not be seen at all if things were drawn to scale.

The thermal equilibrium region is almost equal to the entire phase space Ω and all the non-equilibrium regions put together (all the particles concentrated in one part of the box, or the distribution of the velocities being different from the Maxwellian one) occupy only a tiny fraction of Ω.

To understand how F_t can evolve irreversibly even though its evolution is induced by a reversible microscopic evolution, consider Fig. 2.2, which illustrates what one expects to happen. The microstate $\mathbf{x}(t)$ evolves towards larger and larger regions of phase space and eventually ends up in the "thermal equilibrium" region. Therefore, the induced evolution of $F_t = F(\mathbf{x}(t))$ should tend towards equilibrium. However, at the level of generality considered here, our expectation is simply based on the fact that some regions are (much) bigger than others, and so it would be natural for the microstate $\mathbf{x}(t)$ to evolves towards those bigger regions if nothing prevents it from doing so.

There are several caveats here, one is that this scenario is what one expects or hopes for. We will give in the next section an example, a rather artificial one, where this scenario can be demonstrated in detail, which shows that this scenario is certainly possible and even plausible, but it is certainly not demonstrated in any degree of generality in physically natural situations.

The more important caveat is that, even if this scenario is true, the desired evolution is definitely not true for all microstates

$\mathbf{x}(t) = T^t\mathbf{x}(0)$ giving rise to a given value $F_t = F(\mathbf{x}(t))$. This follows from the reversibility argument given in subsection 2.6.1: for every microstate $\mathbf{x}(t)$ that induces the irreversible evolution of $F_t = F(\mathbf{x}(t))$, there exists another microstate $I(\mathbf{x}(t))$ so that $F_t = F(I(\mathbf{x}(t)))$ at time t, but $I(\mathbf{x}(t))$ induces a different evolution than the irreversible one for times later than t.

Let us illustrate this explanation of irreversibility in a concrete physical example. Consider the gas introduced in subsection 2.6.1 that is initially compressed by a piston in one-half of the box Λ, and that expands into the whole box. Let F be the density of the gas. Initially, it is equal to 1 (say) in one half of the box and to 0 in the other half. After some time t, it is (approximately) equal to $\frac{1}{2}$ everywhere. The explanation of the irreversible evolution of F is that the overwhelming majority of the microscopic configurations corresponding to the gas in one-half of the box will evolve deterministically so as to induce the observed evolution of F. There may of course be some exceptional configurations for which all the particles stay in the left half. All one is saying is that those configurations are extraordinarily rare, and that we do not expect to see even one of them appearing when we repeat the experiment many times. So, the microscopic configurations that lead to the expected macroscopic behavior will be *typical* in the sense of (2.1).

Let us define the *good configurations* (up to a certain time T) as being those configurations that induce the macroscopic law up to time T, and the *bad configurations* (up to a certain time T) the other configurations; we will use indices G and B for the corresponding sets of configurations.[12]

---

[12]We introduce a time upper bound T here, because, if we wait long enough, all configurations will be bad, since the Poincaré's recurrence theorem (see e.g. Arnol'd (1989)) implies that they will all come back arbitrarily close to their initial conditions. In the example of the gas in the box, this means that all the particles will come back simultaneously to the half-box in which they were initially, which is contrary to the behavior predicted by the macroscopic laws. But the time needed for a Poincaré's recurrence to occur in a large system is typically much larger than the age of the Universe, so that, from a practical point of view, we may consider $T = \infty$.

Now, take all the good microscopic configurations in one-half of the box, and let them evolve up to a time $t \ll T$, when the density is approximately uniform. Then, reverse all the velocities. We get a set of configurations that still determines a density approximately $\frac{1}{2}$ in the box, so the value of the density function F, defined by (2.7), is unchanged. However, those configurations are not good any more. Indeed, from now on, if the system remains isolated, the density just remains uniform according to the macroscopic laws. But for the configurations just described, the gas will move back to the part of the box that they started from, see (2.10), leading to a gross violation of the macroscopic law. What is the solution? Simply that those "reversed-velocities" configurations form a very tiny subset of all the microscopic configurations giving rise to a uniform density. And, of course, the original set of configurations, those coming from the initial half of the box, also form such a small subset. Most configurations corresponding to a uniform density do not go to one-half of the box, either in the future or in the past.

To express this idea in formulas, let Ω_t be the set of the configurations giving to the function F its value F_t at time t. In other words, Ω_t is the pre-image of F_t under the map F. Let $\Omega_{t,G}$ be the set of *good configurations* at time t, namely those that lead to a behavior of F following the macroscopic laws, up to some time $T \gg t$.

One expects that, in general, $\Omega_{t,G}$ is a very large subset of Ω_t (meaning that $\frac{|\Omega_{t,B}|}{|\Omega_t|} = \frac{|\Omega_t \setminus \Omega_{t,G}|}{|\Omega_t|} \ll 1$), but is not identical to Ω_t. See (2.12) below for a more precise statement.

In our example of the gas initially compressed in one-half of the box Λ, the set Ω_0 consists of all the configurations in one-half of the box at time zero, and $\Omega_{0,G}$ is the subset consisting of those configurations whose evolution lead to a uniform density at time t, which means that $T^t(\Omega_{0,G}) \subset \Omega_t$.

Microscopic reversibility says that $T^t(I(T^t(\Omega_{0,G})) = I(\Omega_{0,G}) \subset I(\Omega_0) = \Omega_0$ (this is just (2.10) applied to the set $\Omega_{0,G}$). The last equality holds because the set of configurations in one-half of the box is invariant under the change of the sign of the velocities.

A paradox would occur if $T^t(I(\Omega_t)) \subset I(\Omega_0) = \Omega_0$. Indeed, this would mean that, if one reverses the velocities of all the configurations at time t corresponding to a uniform density, and let them evolve for a time t, one would get a set of configurations in one-half of the box (with velocities reversed). Since the operation I preserves the Lebesgue measure $|I(\Omega_t)| = |\Omega_t|$, this would imply that there are as many configurations corresponding to a uniform density (configurations in Ω_t) as there are configurations that will evolve back to one-half of the box in time t (those in $I(T^t(\Omega_{0,G}))$). Or, in other words, one would have $I(\Omega_t) \subset \Omega_{t,B} = \Omega_t \setminus \Omega_{t,G}$, which combined with $|I(\Omega_t)| = |\Omega_t|$ makes the bound $\frac{|\Omega_{t,B}|}{|\Omega_t|} = \frac{|\Omega_t \setminus \Omega_{t,G}|}{|\Omega_t|} \ll 1$ impossible.

But Ω_t is *not at all equal*, in general, to $T^t(\Omega_{0,G})$, so that $T^t(I(T^t(\Omega_{0,G})) = I(\Omega_{0,G}) \subset I(\Omega_0) = \Omega_0$ *does not imply* $T^t(I(\Omega_t)) \subset I(\Omega_0) = \Omega_0$. In our example, $T^t(\Omega_{0,G})$ is a tiny subset of Ω_t, because most configurations in Ω_t were not in half of the box at time zero.

This is illustrated in Fig. 2.3: Ω_0 is the set of configurations with all the particles in one-half of the box, and $\Omega_{0,G}$ the subset of those that evolve towards a uniform distribution after some time t. Thus, $T^t(\Omega_{0,G})$ is a subset of Ω_t, which is the set of thermal equilibrium states. Reversing the velocities of every configurations in $T^t(\Omega_{0,G})$ yields the set $I(T^t(\Omega_{0,G}))$, which is also a subset of Ω_t, but a "bad" subset, namely one that does not stay in equilibrium but moves back to the half box where the particles were to start with. Therefore, F applied to those configurations will not evolve according to the usual macroscopic laws (which implies that the density stays uniform in the half box), which is what we mean by bad configurations.

Of course it should be emphasized once more that the subsets in Fig. 2.3 are not drawn to scale. The sets Ω_0, $T^t(\Omega_{0,G})$ and $I(T^t(\Omega_{0,G}))$ are minuscule compared to the set of equilibrium configurations Ω_t.

In fact, one knows from Liouville's theorem[13] that the size of $\Omega_{0,G}$ and $T^t(\Omega_{0,G})$ are equal: $|\Omega_{0,G}| = |T^t(\Omega_{0,G})|$. Since the operation I also preserves the size of a set, we have: $|\Omega_{0,G}| = |T^t(\Omega_{0,G})| = |I(T^t(\Omega_{0,G}))|$, which is illustrated in Fig. 2.3.

---

[13]Which says that the Lebesgue measure of a set is invariant under the Hamiltonian flow T^t, see e.g. Arnol'd (1989).

Since $\Omega_{0,G} \subset \Omega_0$, we have $|\Omega_{0,G}| \le |\Omega_0|$ and we already observed that the ratio $\frac{|\Omega_0|}{|\Omega_t|} \sim 2^{-N}$ (since each of the N particles can be in either half of the box in Ω_t, but only in one-half of the box in Ω_0). Thus,

$$\frac{|I(T^t(\Omega_{0,G}))|}{|\Omega_t|} = \frac{|T^t(\Omega_{0,G})|}{|\Omega_t|} = \frac{|\Omega_{0,G}|}{|\Omega_t|} \le \frac{|\Omega_0|}{|\Omega_t|} \sim 2^{-N}, \qquad (2.11)$$

which is astronomically small for $N \sim 10^{23}$.

What one would like to show is that the good configurations are *typical* in the sense of (2.1). More precisely, one want to show that, for all times T not too large, and all $t \ll T$,

$$\frac{|\Omega_{t,G}|}{|\Omega_t|} \to 1 \qquad (2.12)$$

as $N \to \infty$.

2.6.3. *Irreversibility and probabilistic explanations*

The above explanation of the irreversible behavior of F_t is again probabilistic. The vast majority of microstates $\mathbf{x}(0)$ corresponding to the macrostate F_0 induce, through their deterministic evolution $\mathbf{x}(0) \to T^t(\mathbf{x}(0)) = \mathbf{x}(t)$, the expected time evolution of $F_0 \to F_t$.

What else could one ask for? One could wish to show that the expected time evolution of $F_0 \to F_t$ is induced by *all* microstates $\mathbf{x}(0)$ corresponding to the macrostate F_0. But we showed by explicit counterexamples that, in general, this is not possible.

So, our explanation is the best one can hope for. But is it satisfactory? It is, provided that one accepts the notion of probabilistic explanation given in Section 2.5. And if one does not accept it, it is not clear what notion of explanation one has in mind and how one could justify it. Nevertheless, by speaking of a "vast majority of microstates", we did not say what we meant by the "vast majority". Since there are uncountably many such states one needs a measure on Ω in order to make sense of that notion.

The measure on subsets of Ω that we used implicitly here is the size of the set or its Lebesgue measure.[14] But one could still ask, why use that measure and not some other measure? Obviously, if one could not argue that this measure is in some sense "natural", our whole notion of explanation would collapse. Indeed, it is easy to invent measures that will give a much greater weight to, for example, the set $\Omega_0 \setminus \Omega_{0,G}$ than to $\Omega_{0,G}$. Then, if we define the probability of a set according to such a measure, the usual induced evolution $F_0 \to F_t$ becomes improbable and some other evolution becomes probable.

So, something has to be said in favor of the naturalness of the Lebesgue measure. But that is easy enough on Bayesian grounds or on the basis of the indifference principle: the Lebesgue measure is the most symmetric one, being invariant under translations and rotations, and varying naturally under scalings. So, from that point of view, there is no alternative to taking the Lebesgue measure as our natural measure and, if the set of initial states inducing the expected evolution of F_t has a high probability relative to that measure (i.e. are typical relative to that measure), then we will consider that the evolution of F_t has been explained by the microscopic laws.

Some people want to justify the naturalness of the Lebesgue measure by invoking the fact that it is time-invariant under the Hamiltonian flow (Liouville's theorem). While this is true, it is not necessarily the best argument in favor of the naturalness of the Lebesgue measure. Indeed, there exists non-Hamiltonian systems for which the Lebesgue measure is *not* invariant, and that do possess an invariant measure ν whose support is a set $\mathcal{A}$ of zero Lebesgue measure. Moreover, one can, in certain cases, prove that for almost every initial condition $\mathbf{x}_0$ with respect to the Lebesgue measure on a set of non-zero Lebesgue measure containing $\mathcal{A}$, the time evolution $\mathbf{x}_0 \to \mathbf{x}_t$ drives the trajectory towards $\mathcal{A}$ and the statistics of the

---

[14]If Ω is a set of measure zero in $\mathbb{R}^{6N}$, for example, a constant energy surface, one has to consider the restriction of the Lebesgue measure to that surface instead of the Lebesgue measure on $\mathbb{R}^{6N}$. This measure is called the Liouville measure.

time spent by the trajectory close to subsets of $\mathcal{A}$ is proportional to the ν-measure of those subsets.[15]

In the following section we will illustrate the previous ideas in a concrete situation.

2.7. The Kac ring model

2.7.1. *The model*

Let us consider a simple model, due to Mark Kac (1959), (p. 99), see also Thompson (1972), (p. 23) and Gottwald and Oliver (2009), which nicely illustrates Boltzmann's solution to the problem of irreversibility and shows how to avoid various misunderstandings and paradoxes.

We will use a slightly modified version of the model and state the relevant results, referring to Kac (1959) for the proofs.

One considers N equidistant points on a circle; M of the intervals between the points are marked and form a set called S. The complementary set (of $N - M$ intervals) will be called $\bar{S}$. We will define

$$\alpha = \frac{M}{N}. \qquad (2.13)$$

It will be convenient later to assume that

$$\alpha < \frac{1}{2}. \qquad (2.14)$$

At each of the N points there is a particle that can have either a plus sign or minus sign (in the original Kac model, one speaks of white and black balls). During an elementary time interval, each particle moves clockwise to the nearest site, obeying the following rule: if the particle crosses an interval in S, it changes sign upon completing the move but if it crosses an interval in $\bar{S}$, it performs the move without changing sign.

---

[15] We think here of certain "chaotic" dynamical systems for which ν is a Sinai-Ruelle-Bowen measure and the set $\mathcal{A}$ is a "strange attractor", see e.g. Bowen (1975); Bowen and Ruelle (1975); Eckmann and Ruelle (1985); Sinai (1972); Ruelle (1976, 1978).

Suppose that we start with all particles having a plus sign; the question is what happens after a large number of moves. After (2.18) we shall also consider other initial conditions.

To formalize the model, introduce for each $i = 1, \ldots, N$, the variable[16]

$$
\epsilon_i = \begin{cases} +1 \text{ if the interval in front of } i \in \bar{S} \\ -1 \text{ if the interval in front of } i \in S \end{cases}
$$

and we let $\eta_i(t) = \pm 1$ be the sign of the particle at site i and time t. Then, we get the "equations of motion":

$$
\eta_{i+1}(t+1) = \eta_i(t)\epsilon_i. \tag{2.15}
$$

Let us first explain the analogy with mechanical laws. The particles are described by their positions and their (discrete) "velocity", namely their sign. One of the simplifying features of the model is that the "velocity" does not affect the motion. The only reason one calls it a "velocity" is that it changes when the particle collides with a fixed "scatterer", i.e. an interval in S. Scattering with fixed objects tends to be easier to analyse than collisions between particles. The "equations of motion" (2.15) are given by the clockwise motion, plus the changing of signs. These equations are obviously deterministic and reversible. If, after a time t, we change the orientation of the motion from clockwise to counterclockwise, we return after t steps to the original state.[17] Moreover, the motion is strictly periodic. After $2N$ steps, each interval has been crossed twice by each particle; hence, they all come back to their original sign.[18]

---

[16]See, for example, Fig. 2.4 where $\epsilon_i = -1$ and $\epsilon_{i+1} = +1$.

[17]There is a small abuse here, because it seems that we change the laws of motion by changing the orientation (from clockwise to counterclockwise). But one can attach another discrete "velocity" parameter to the particles, having the same value for all of them, and indicating the orientation, clockwise or counterclockwise, of their motion. Then, the motion is truly reversible, and we have simply to assume that the analogue here of the operation I of (2.9) changes also that extra velocity parameter.

[18]This is analogous to the Poincaré cycles in mechanics except that, here, the length of the cycle is the same for all configurations (there is no reason for this feature to hold in general mechanical systems).

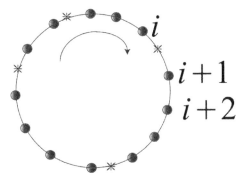

Fig. 2.4. At each site i there is a particle that has either a plus or minus sign. During an elementary time interval, each particle moves clockwise to the nearest site. If the particle crosses an interval marked with a cross (as the one between the sites i and $i+1$), it changes sign but if it crosses an interval without a cross (as the one between the sites $i+1$ and $i+2$) it does not change its sign.

Fig. 2.5. An example of distribution of crosses where every configuration is periodic of period 4.

It is also easy to find special configurations which obviously do not tend towards anything: start with all particles being "plus" and let every other interval belong to S (with $M = \frac{N}{2}$). Then, after two steps, all particles are minus, after four steps they are all plus again, etc... The motion is periodic with period 4, see Fig. 2.5 for a simple example.

Turning to the solution, one can start by analyzing the approach to equilibrium in this model à la Boltzmann.

2.7.2. *Analogue of Boltzmann's solution*

Let $N_+(t)$, $N_-(t)$ denote the total number of "plus" particles or "minus" particles at time t (i.e. after t moves, t being an integer), respectively. $N_+(t)$ and $N_-(t)$ are the macroscopic variables in this problem (since $N_+(t) + N_-(t) = N$, there is only one independent macroscopic variable).

Let $N_+(S;t)$, $N_-(S;t)$ be the number of "plus" particles or of "minus" particles which are going to cross an interval in S at time t, respectively.

We have the immediate conservation relations:

$$\begin{aligned}
N_+(t+1) &= N_+(t) - N_+(S;t) + N_-(S;t) \\
N_-(t+1) &= N_-(t) - N_-(S;t) + N_+(S;t)
\end{aligned} \qquad (2.16)$$

If we want to solve (2.16), we have to make some assumption about $N_+(S;t)$ and $N_-(S;t)$. Otherwise, one has to write down equations for $N_+(S;t)$ and $N_-(S;t)$ that will involve new variables and lead to a potentially infinite regress.

So, following Boltzmann, we introduce the assumption ("Stosszahlansatz" or "hypothesis of molecular chaos"[19]):

$$\begin{aligned}
N_+(S;t) &= \alpha N_+(t) \\
N_-(S;t) &= \alpha N_-(t),
\end{aligned} \qquad (2.17)$$

with α defined in (2.13).

The intuitive justification for this assumption is that each particle is "uncorrelated" with the event, "the interval ahead of the particle belongs to S", so we write $N_+(S;t)$ as equal to $N_+(t)$, the total number of "plus" particles, times the density α of intervals in S. This assumption looks completely reasonable. However, upon reflection, it may lead to some puzzlement: what does "uncorrelated" exactly mean? Why do we introduce a statistical assumption in a mechanical model? Fortunately here, these questions can be answered precisely

---

[19]The word "chaos" here has nothing to do with "chaos theory", in the sense of dynamical systems (see footnote 5) and, of course, Boltzmann's hypothesis is much older than that theory.

and we shall answer them later by solving the model exactly. But let us return to the Boltzmannian story.

One obtains from (2.17):

$$N_+(t+1) - N_-(t+1) = (1-2\alpha)(N_+(t) - N_-(t))$$

Thus,

$$N^{-1}[N_+t) - N_-(t)] = (1-2\alpha)^t N^{-1}[N_+(0) - N_-(0)],$$
$$= (1-2\alpha)^t. \tag{2.18}$$

since $N_+(0) = N$ and $N_-(0) = 0$.

Using (2.14) and (2.18), we obtain a *monotonic* approach to equal number of "plus" particles and "minus" particles, i.e. to equilibrium. Note that we get a monotonic approach for *all* initial conditions $(N_+(0) - N_-(0))$ of the particles.

We can see here in what sense Boltzmann's solution is an approximation. The assumption (2.17) cannot hold for all times and for all configurations, because it would contradict the reversibility and the periodicity of the motion. However, we will show now that the fact that it is an approximation does not invalidate Boltzmann's ideas about irreversibility.

2.7.3. *Microscopic analysis of the model*

Let us reexamine the model at the microscopic level, first mechanically and then statistically. The solution of the equations of motion (2.15) is:

$$\eta_i(t) = \eta_{i-t}(0)\epsilon_{i-1}\epsilon_{i-2}\cdots\epsilon_{i-t} \tag{2.19}$$

(where the subtractions in the indices are done modulo N). So we have an explicit solution of the equations of motion at the microscopic level.

We can express the macroscopic variables in terms of that solution:

$$N_+(t) - N_-(t) = \sum_{i=1}^{n}\eta_i(t) = \sum_{i=1}^{n}\eta_{i-t}(0)\epsilon_{i-1}\epsilon_{i-2}\cdots\epsilon_{i-t} \tag{2.20}$$

and we want to compute $N^{-1}(N_+(t) - N_-(t))$ for large N, for various choices of initial conditions $\{\eta_i(0)\}_{i=1}^N$ and various sets S (determining the ϵ_i's). It is here that "statistical" assumptions enter. Namely, we fix an arbitrary initial condition $\{\eta_i(0)\}_{i=1}^N$ and consider all possible sets S with $M = \alpha N$ fixed (one can of course think of the choice of S as being part of the choice of initial conditions). Then, for each set S, one computes the "curve" $N^{-1}(N_+(t) - N_-(t))$ as a function of time.

The result of the computation, done in Kac (1959), is that for any given t and for N large, the overwhelming majority of these curves will approach $(1 - 2\alpha)^t$, which is predicted by (2.18) (to fix ideas, Kac suggests to think of N as being of the order 10^{23} and t of order 10^6). The fraction of all curves that will deviate significantly from $(1 - 2\alpha)^t$, for fixed t, goes to zero as $N^{-\frac{1}{2}}$ when $N \to \infty$.

Of course, when we say "compute", one should rather say that one makes an estimate of the fraction of curves deviating from $(1 - 2\alpha)^t$ at a fixed t. This estimate is similar to the law of large numbers, since (2.20) is of the form of a sum of (almost independent) random variables.

Let us express what happens in this model in terms of the sets depicted in Fig. 2.3. The "phase space" Ω consists of all configurations of signs and scatterers (with $2M < N$), $\{\epsilon_i, \eta_i\}_{i=1}^N$.

The "thermal equilibrium" set in Fig. 2.3 corresponds to the set of configurations of particles such that $N^{-1}(N_+(t) - N_-(t))$ is approximately equal to 0, and to the set of configurations of scatterers such that $N^{-1}(N_+(t) - N_-(t))$ remains close to 0 in the future.

It is easy to compute the number of (microscopic) configurations whose number of "plus" particles is $N_+(t)$. It is given by:

$$\binom{N}{N_+(t)} = \frac{N!}{N_+(t)!(N - N_+(t))!} \tag{2.21}$$

and this number reaches its maximum value for $N_+ = \frac{N}{2} = N_-$.

We can, as in Fig. 2.1, introduce a partition of the phase space according to the different values of N_+ and N_-. What (2.21) shows is that different elements of that partition have very different number of elements, the vast majority corresponding to "equilibrium", i.e. to those near $N_+ = \frac{N}{2} = N_-$.

If one illustrates this model through Fig. 2.3, the set Ω_0 consists of all configurations of scatterers and of all particles being "plus": $\eta_i = +1, \forall i = 1, \ldots, N$. The subset $\Omega_{0,G} \subset \Omega_0$ of good configurations consists of those configurations of scatterers such that $N^{-1}(N_+(t) - N_-(t))$ tends to 0, and of all the particles having a plus sign.

Then, $T^t(\Omega_{0,G}) \subset \Omega_t$ is a set of configurations with $N^{-1}(N_+(t) - N_-(t))$ approximately equal to 0, but a set of scatterers that is special in the following sense. The configurations in $I(T^t(\Omega_{0,G}))$, where I changes the orientation of the motion from clockwise to counterclockwise, will evolve in a time t to a configuration with all particles being "plus".

Although, as far as the signs of the particles are concerned, there is nothing special about the configurations in $T^t(\Omega_{0,G})$ ($N^{-1}(N_+(t) - N_-(t))$ is close to 0), there is a subtle correlation between the configurations of the particles and the scatterers in $T^t(\Omega_{0,G})$, as shown by what happens if one applies the orientation-reversal operation I to those configurations. This is simply a "memory effect" due to the fact that the configurations in $T^t(\Omega_{0,G})$ were initially in $\Omega_0 \supset \Omega_{0,G}$ (this is similar to the memory effect of the particles of the gas in Section 2.6 that were initially in one half of the box).

But the configurations in $T^t(\Omega_{0,G})$ form a very small subset of Ω_t; indeed, $|T^t(\Omega_{0,G})| = |\Omega_{0,G}| \leq |\Omega_0|$, and $\frac{|\Omega_0|}{|\Omega|} = 2^{-N}$ (because there are two possible signs in Ω for each of the N sites but only one sign, plus, in Ω_0). By the law of large numbers, one can show that $|\Omega_t| \sim |\Omega|$ for N large, so that $\frac{|\Omega_0|}{|\Omega_t|} \sim 2^{-N}$ again for N large, and thus $\frac{|\Omega_0|}{|\Omega_t|}$ is extremely small in that limit.

Note that here, we define typical behavior by counting the number of configurations, see (2.21), which is the same as putting a probability equal to $\frac{1}{2}$ to each particle sign (note also that, in the estimates made on (2.20), we had implicitly put a probability equal to $\frac{1}{2}$ to the presence or not of a scatterer on each interval). This uniform probability is again the one following from the indifference principle.

I do not want to overemphasize the interest of the Kac model. It has many simplifying features (for example, there is no conservation of momentum; the scatterers here are "fixed"). However, it has *all* the properties that have been invoked to show that mechanical systems cannot behave irreversibly, and therefore it is a perfect counterexample that allows us to refute all those arguments (and to understand exactly what is wrong with them). It is isolated (the particles plus the scatterers), deterministic, reversible and periodic.[20]

This result, obtained in the Kac model, is exactly what one would like to show for general mechanical systems, in order to establish irreversibility. It is obvious why this is very hard. In general, one does not have an explicit solution (for an N-body system!) such as (2.15, 2.19), in terms of which the macroscopic variables can be expressed, as in (2.20).

If we prepare a Kac model many times, and if the only variables that we can control are N and M, then we expect to see the irreversible behavior obtained above, simply because this is what happens *deterministically* for the vast majority of microscopic initial conditions corresponding to the macroscopic variables that we are able to control.

2.7.4. *Conclusions*

In this paper, we recalled the more or less standard "Boltzmannian" derivation of macroscopic laws from microscopic ones. But since this derivation appeals to probabilistic notions, we tried to relate those notions with the issue of what constitutes a valid explanation in the natural sciences.

We claim that one explains a macroscopic behavior on the basis of microscopic laws if, given a macrostate F_0, the overwhelming majority of microstates corresponding to F_0 give rise, through their

---

[20]Periodicity is a stronger property than the existence of Poincaré cycles and implies that the system is not ergodic. For a discussion of why ergodicity is neither necessary to sufficient in order to justify approach to equilibrium, see (Bricmont, 1995, Section 4.2).

deterministic evolution, to an induced evolution $F_0 \to F_t$ in accordance with the macroscopic law.

We also tried to clarify the status of our "ignorance" in these explanations. It is true that we ignore the initial conditions of the microstates of our system or the details of their time evolution, but what we argued in this paper is that this ignorance does not prevent us from understanding why the system tends towards equilibrium, again because this is result of the deterministic evolution of the overwhelming majority of the microstates.

Acknowledgments

I thank Valia Allori for the interesting discussions during the preparation of this paper.

References

Arnol'd, V. I. (1989) *Mathematical Methods of Classical Mechanics*, Springer-Verlag New York Inc., 2nd ed.

Baldovin, M., Caprini, L., Vulpiani, A. (2019) Irreversibility and typicality: A simple analytical result for the Ehrenfest model, *Physica A*, **524**, 422–429.

Bowen, R. (1975) *Equilibrium States and the Ergodic Theory of Anosov Diffeomorphisms*, Lecture Notes in Mathematics, Vol. 470. Springer-Verlag, Berlin-New York.

Bowen, R., Ruelle, D. (1975) The ergodic theory of Axiom A flows, *Invent. Math.*, **29**, 181–202.

Bricmont, J. (1995) Science of chaos, or chaos in science? *Physicalia Magazine*, **17**, 159–208.

Cerino, L., Cecconi, F., Cencini, M., Vulpiani, A. (2016) The role of the number of degrees of freedom and chaos in macroscopic irreversibility, *Physica A*, **442**, 486–497.

de Bièvre, S., Parris, P. E. (2017) A rigourous demonstration of the validity of Boltzmann's scenario for the spatial homogenization of a freely expanding gas and the equilibration of the Kac ring, *J. Stat. Phys.*, **168**, 772–793.

de Finetti, B. (1937) La prévision: ses lois logiques, ses sources subjectives, *Annales de l'Institut Henri Poincaré*, **7**, 1–68.

de Finetti, B. (2017) *Theory of Probability: A Critical Introductory Treatment*, Wiley, New York.

Eckmann, J.-P., Ruelle, D. (1985) Ergodic theory of chaos and strange attractors, *Rev. Mod. Phys.*, **57**, 617–656.

Gottwald, G. A., Oliver, M. (2009) Boltzmann's dilemma: An introduction to statistical mechanics via the Kac ring, *SIAM Rev.*, **51**, 613–635.

Hempel, C. (1942) The function of general laws in history, *Journal of Philosophy*, **39**, 35–48.

Hempel, C., Oppenheim, P. (1948) Studies in the logic of explanation, *Philosophy of Science*, **15**, 135–175.

Jaynes, E. T. (1982) *Papers on Probability, Statistics and Statistical Physics*, ed. by R. D. Rosencrantz, Reidel, Dordrecht.

Jaynes, E. T. (2003) *Probability Theory: The Logic of Science*, Cambridge University Press, Cambridge.

Kac, M. (1959) *Probability and Related Topics in the Physical Sciences*, Interscience Pub., New York.

Laplace, P. S. (1825) *A Philosophical Essay on Probabilities*, Transl. by F. W. Truscott and F. L. Emory, Dover Pub., New York, 1951. Original: *Essai philosophique sur les probabilités*, C. Bourgeois, Paris 1986, text of the fifth edition.

Maudlin, T. (2007) *The Metaphysics Within Physics*, Oxford University Press, Oxford.

Penrose, R. (1989) *The Emperor's New Mind*, Oxford University Press, Oxford.

Ruelle, D. (1976) A measure associated with axiom-A attractors, *Amer. J. Math.*, **98**, 619–654.

Ruelle, D. (1978) *Thermodynamic Formalism*, Encyclopedia of Mathematics and Its Applications No. 5 Addison Wesley, New York.

Shannon, C. E. (1948) A mathematical theory of communication, *Bell System Technical Journal*, **27**, 379–423.

Sinai, J. G. (1972) Gibbs measures in ergodic theory, *Russian Mathematical Surveys*, **27**, 21–69.

Thompson, C. J. (1972) *Mathematical Statistical Mechanics*, Princeton University Press, Princeton.

Chapter 3

Some Reflections on the Statistical Postulate: Typicality, Probability and Explanation between Deterministic and Indeterministic Theories

Valia Allori

Department of Philosophy,
Northern Illinois University
vallori@niu.edu

A common way of characterizing Boltzmann's explanation of thermodynamics in term of statistical mechanics is with reference to three ingredients: the dynamics, the past hypothesis, and the statistical postulate. In this paper I focus on the statistical postulate, and I have three aims. First, I wish to argue that regarding the statistical postulate as a probability postulate may be too strong: a postulate about typicality would be enough. Second, I wish to show that there is no need to postulate anything, for the typicality postulate can be suitably derived from the dynamics. Finally, I discuss how the attempts to give preference to certain stochastic quantum theories (such as the spontaneous collapse theory) over deterministic alternatives on the basis that they do not need the statistical postulate fail.

3.1. Introduction

Boltzmann's statistical mechanics provides an explanation of the macroscopic laws of thermodynamics, such as 'entropy always increases', in terms of the microscopic Newtonian laws. In his book "Time and Chance" (2001), David Albert has made especially clear how this is done, and what problems this account faces. As a consequence, his book has been extremely influential in the discussions about the foundation and the philosophy of statistical mechanics (see also Loewer 2019). Very briefly, the view is that Boltzmann's derivation of the macroscopic laws only needs three ingredients:

1. The Newtonian law of motion,
2. The statistical postulate,
3. The past hypothesis.

Roughly put, (1) describes the dynamics of the microscopic components of the thermodynamics objects; (2) specifies the probability measure grounding the meaning of the probabilities arising from the statistical derivation of thermodynamic laws; and (3) guarantees that entropy will likely increase in the future but not in the past by postulating an initial low entropy state for the universe.

In addition, Albert argues that a corresponding quantum statistical explanation based on a universal quantum theory (such as possibly the pilot-wave theory, the spontaneous localization theory, and Everettian mechanics) would look very similar to the classical one. However, he maintains that in the case of the spontaneous localization theory, the ingredients would reduce to the following two:

1. The stochastic law of motion,
2. The past hypothesis.

In other words, contrarily to what happens in any deterministic framework, he maintains that in the spontaneous localization schema there is no need for an additional statistical postulate. The reason is that the theory, being indeterministic, has probabilities already 'built into' it. Hence, statistical mechanical probabilities in the spontaneous localization framework are just the quantum probabilities. Because of this, Albert concludes the spontaneous collapse theory should be preferred to the other alternative quantum theories.

In this paper, I focus my attention on the statistical postulate, and I wish to make three points. First, I argue that the postulate need not be understood as a postulate about probability; second, I argue that it is not a postulate after all; and third that, given the first two points, there is no argument remaining to prefer the spontaneous localization theory over the deterministic alternatives. Here is how I plan to accomplish these tasks. After having reviewed this account of Boltzmann's work in Section 2, I argue that the statistical postulate is not needed in two steps. In the first step, developed in Section 3, I assume, for the sake of the argument, that one indeed has a postulate. However, I argue that it is not necessarily a postulate about probabilities; rather, one can appeal to the weaker notion of typicality. Therefore, one is led to discuss the notion of typicality (rather than probability) measure. In the second step, discussed in Section 4, I reconstruct and use (what I take to be) Sheldon Goldstein's argument (2001, 2011) to show that the typicality measure is not postulated, but can be suitably extracted from the dynamics under certain constraints, such as stationarity and generality, which are regarded 'natural'.[1] Moreover, as I elaborate in Section 5, I reconstruct the argument developed by Nino Zanghì (2005) to show that the difference between deterministic and indeterministic theories is not as substantial as one may think. Elaborating on the concepts introduced so far, one can provide a unifying account of the statistical mechanical explanation for both kinds of theories, which therefore rely on the same ingredients. As a consequence, one cannot conclude that one type of theory or the other will be better off in explaining the macroscopic laws from a microscopic dynamics; therefore, diffusing the argument to prefer spontaneous localization theories to deterministic quantum theories.

3.2. Boltzmann's approach to statistical mechanics

The evolution of macroscopic objects is generally very complicated. Nonetheless, their behaviour is governed by simple and general physical laws such as the laws of thermodynamics. These laws are phenomenological: they merely describe regularities on the macroscopic level. A natural question, then, is whether they can be derived by a more fundamental theory. Suppose such a theory is Newtonian

---

[1]See also Dürr (2009), and Zanghì (2005).

mechanics, according to which the world is described in terms of point-like particles moving according to Newton's second law of motion. If the theory is complete, then every physical system is describable by it, and one could (at least in principle) reproduce the behaviour of macroscopic objects thought as composed of microscopic Newtonian particles. However, to do this one needs to solve Newton's equation for macroscopic bodies, and there are at least two obvious problems. First, one would need to know exactly which forces act between the particles. Moreover, since macroscopic objects are composed by an incredibly large number of particles, it is practically impossible to compute their trajectories. Nonetheless, exploiting the fact that there are so many particles, one can use statistical methods to have enough information on macroscopic systems even under these conditions. The resulting theory is called statistical mechanics, and it has been primarily developed by Ludwig Boltzmann and Josiah Willard Gibbs. In this paper, I will focus on the work of Boltzmann, which is the basis of the model I aim to discuss.

There is another challenge to recover the laws of thermodynamics from the underlying microscopic dynamics. It is connected to the time-reversibility of Newtonian mechanics, which is in stark contrast with the irreversibility of the macroscopic laws. Arguably, to say that Newtonian mechanics is time reversible is to say that if we flip the sign of the time variable in Newton's equation, the solutions of the new equation are still solutions of the original equation. This means that we cannot tell from the behaviour of an object whether it is moving forward or backward in time. Since we are assuming macroscopic bodies are a collection of microscopic Newtonian particles, this time-reversibility should be observed also at the macroscopic level. However, this is empirically false; the macroscopic phenomena of our everyday experience are all time-directed: a perfume sprayed from the corner will spread out in the whole room. The opposite, namely the perfume getting back to the initial corner, spontaneously never happens. This macroscopic irreversibility is captured by the second law of thermodynamics, according to which a quantity called 'entropy' (or 'thermodynamic entropy') always increases. Hence, macroscopic phenomena always happen in the direction of increasing entropy.

The reason why the perfume spreads in the room is that the state of spread-out perfume has a higher-entropy than the perfume-in-the-corner state. This is the explanation why this phenomenon – perfume spreading out – happens, while the opposite does not. The problem now is that if we want to derive thermodynamics from Newtonian mechanics, then we need to derive macroscopic irreversibility from microscopic reversibility. So, we need to get rid of the second solution in which the perfume comes back into the corner. But how? This is the challenge Boltzmann faced and that we will review in the next sections along the lines discussed in Albert (2001) and Loewer (2019).

3.2.1. *Microstates, macrostates and entropy*

Assuming Newtonian mechanics, the complete description at a time of any single particle is given by the pair (r, v) of its position and velocity at that time. Therefore, the complete dynamical description at a fixed time t of a body composed of N particles is given by the so-called *microstate* $X = (r_1, \ldots, r_N, v_1, \ldots, v_N)$, namely the set of the positions and velocities of all particles at time t. The set of all possible microstates constitutes *phase space*, which is in a sense the space of all possible ways the world can be. Phase space can be partitioned into sets, called *macrostates*, to describe the state of a system of particles from a macroscopic point of view. This coarse-graining is done by specifying some macroscopic thermodynamically relevant properties such as temperature, pressure and volume; each macrostate is such that all of its microstates have the same thermodynamic properties. Thus, macroscopic properties are functions on the microstate on phase space which vary very slowly on a microscopic scale, so that each macrostate has the same macroscopic property.[2]

---

[2]For instance, temperature is defined as the mean kinetic energy of the particles: $T = T(X) = \frac{1}{K_B} \sum_{i=1}^{N} \frac{1}{2} m v_i^2$, where $K_B = 1.38 \times 10^{-23} \, m^2 \, kg \, s^{-1} \, K^{-1}$ is the Boltzmann constant, which provides the macroscopic-microscopic scaling (one can think that the number of particles in a macroscopic body should be at least of the order of Avogadro number, that is $N \sim 10^{23}$).

Therefore, there are different ways in which the same macrostate (with some given macroscopic properties) can 'come from' different microstates.[3] In other words, a macrostate is the collection of all the possible ways the system can microscopically give rise to the macroscopic properties of the macrostate. In this sense, knowing the macrostate of a system does not tell us its actual microstate: all the microstates in a macrostate are macroscopically identical. And the bigger the macrostate is, the less information one has about the microscopic composition.

Different macrostates in general are composed by a different number of microstates. One can define the size of a macrostate in phase space by (suitably) 'counting' how many microstates it is composed of. For a fixed energy, there is a particular macrostate which has the following empirical feature: bodies are 'drawn to it' during their motion, and when a body finally reaches it, it will not spontaneously leave it afterwards. For these reasons, this macrostate is called the *equilibrium* state. For instance, two bodies at different temperatures put into contact will reach an equilibrium temperature, given by the mean of the initial temperatures of the two bodies, and afterwards the temperature will not change.[4] This is connected with the second law of thermodynamics, according to which entropy

---

[3]For example, a macrostate of a gas in a room with volume V is composed of all the ways in which the molecules are arranged such that one finds a gas filling the room. The gas can be in macrostate $X = (r_1, r_2, \ldots, v_1, v_2, \ldots)$. However, If we swap some positions or change a little some of their velocities such that, say, the microstate is now $X' = (r_2, r_4, \ldots, r_3, \ldots, r_1, \ldots, v'_1, v_2, \ldots)$, one still get a gas filling the room of volume V. That is, both are microstates in the same macrostate. Also, consider a generic system in microstate $X = (r_1, r_2, \ldots, v_1, v_2, \ldots)$. Now swap the velocities of particle 1 and particle 2 such that the microstate is now $X'' = (r_1, r_2, \ldots, v_2, v_1, \ldots)$. This change makes no difference to the value of macroscopic properties such as temperature; hence, both microstates belong to the same macrostate defined by temperature T. Therefore, a gas filling a room of volume V and a generic macroscopic body at temperature T, say, are each made of particles whose microstate (unknown to us) belongs to the macrostate characterized respectively by that volume and that temperature value.

[4]Likewise, consider a box divided in two regions separated by a wall, and a gas in one of the regions. When the separation wall is removed, the gas starts expanding, and in time it will occupy the whole container. This is the equilibrium state of the gas, since after having occupied the whole box the gas will not change anymore.

always increases, given that the (thermodynamic) entropy of equilibrium is maximal. From the point of view of statistical mechanics, we need to suitably define entropy and to define what the equilibrium state is, from a microscopic point of view. The equilibrium state is just a macrostate like any other. With one crucial difference: it contains incredibly many more microstates than the other macrostates. In fact, as we saw before, the number of microstates in a macrostate depends on how many ways there are to obtain the same macroscopic features. And there are so many more ways for the microstates to be distributed in order to correspond to, say, a uniform temperature macrostate than there are in order to correspond to a macrostate with non-uniform temperature.[5] This means that the size of the equilibrium macrostate is larger than any other macrostate. One can therefore understand why microstates 'tend' to equilibrium and (almost) never leave it afterwards. A microstate, in its wandering through phase space, will sooner or later fall into such a big state, and afterwards it will stay there simply because there are so many ways for a microstate to be in equilibrium.[6]

Boltzmann defined the *entropy* of a given microstate as a measure of the volume of the phase space occupied by the macrostate in which the microstate is located. It can be shown that it is equivalent to the thermodynamic entropy, so Boltzmann was left to show

---

[5]Perhaps this is best seen considering the example of the gas again. Initially, the gas is in a given macrostate, say M_1, and later it is in equilibrium state M_E. When in M_1, each microscopic particle could be anywhere in the first half of the box. That is, its microstate could be anything between (x, y, z, v_x, v_y, v_z) and $(x + L, y + L, z + L, v_x, v_y, v_z)$, where L is such that $V = L^3$ is the volume of half of the box (assuming velocities do not change). This means that the volume in phase space of each particle (namely the number of ways the particles can be) is, initially, V. At equilibrium, each particle will have double the volume to roam through, so that the volume in phase space for each particle is $2V$. This is true for all particles and, therefore, assuming that there are N particles in the gas, their volume in phase space is initially $Vol\,(M_1) = V^N$ and at equilibrium is $Vol\,(M_E) = (2V)^N$. Accordingly, $\frac{Vol(M_E)}{Vol(M_1)} = \frac{(2V)^N}{V^N} = 2^N \approx 2^{10^{23}}$, assuming that N is of the order of the Avogadro number.

[6]The view I just described has been challenged by some commentators (see Frigg and Werndl (2012), Werndl (2013), Werndl and Frigg (2015) and references therein), who have argued that the approach to equilibrium is explained only if the dynamics play a crucial role. For more on this, see Section 3.4.

that his entropy also increases in time. That is, he had to show that the volume in phase space of the macrostates a microstate will cross during its motion will increase in time. Indeed, we have just see that it does. Microstates move from regions of phase space of smaller volume to ones of bigger volume, the biggest of which is the equilibrium state; thus, entropy increases in the future. Notice, however, contrarily to the case of thermodynamic entropy, it is *possible* for Boltzmann entropy to decrease. In fact, there could be some microstates that behave 'anti-thermodynamically' in the sense that they accurately manage to avoid equilibrium and go into macrostates which are smaller than the one they are currently in. However, even if these microstates are possible, the fact that the volume of the equilibrium state is so big guarantees that such states are very few. That is, the microstates in a given macrostate that manage to avoid such a large state as equilibrium are very few. Therefore, more cautiously, one can conclude that the overwhelmingly vast majority (rather than all of them) of microstates in a given macrostate will evolve toward a bigger macrostate, hence larger entropy, in the future.

3.2.2. *The past hypothesis*

Unfortunately, as soon as this strategy is understood, one also realizes that because the underlying Newtonian dynamics is time-reversal invariant, not only the majority of microstates will go into a macrostate with higher entropy, but also most of them will *come from* a state of higher entropy as well. In fact, consider a half-melted ice cube on a kitchen table. The ice cube will likely fully melt, because the state corresponding to a fully melted ice cube has a higher entropy than the present one. However, where is the half-melted cube is most likely to come from? If we forget our memories of the past (that half melted ice cubes usually come from the freezer) and we reason merely in terms of sizes of macrostates, we should immediately see that the present state is likely to come from a past macrostate of larger size. That is, the half-melted ice cube is most likely to come from a fully-melted ice cube. This is because the laws are invariant under time reversal. Therefore, for the vast majority of microstates, entropy increases both in the past and in the future, which is not empirically adequate, assuming we do not want to say that our memories are unreliable. In fact, empirically, one needs to find that while the vast majority of microstates in a given macrostate (half-melted ice cube) will evolve into another macrostate with higher entropy (fully melted ice cube), only a very tiny minority of the microstates in the present macrostate (half-melted ice cube) comes from the initial macrostate (ice cubes in the freezer).

One needs to break this symmetry between past and future, and there are various ways to do so. One which is considered amongst the most promising is the so-called past hypothesis, which postulates that the universe begun with a microstate of extremely low entropy.[7] I will not discuss the reasons why the past hypothesis is considered to be better than the other proposed solutions, since whether the past hypothesis is true or not does not affect my arguments in this paper. However, it should be clear that if one postulates the universe begun with an extremely low entropy, one effectively 'cancels out' the possibility of it moving toward an even lower entropy, guaranteeing that it is overwhelmingly likely for the microstate to go into a macrostate of higher entropy.[8]

3.2.3. *The statistical postulate*

Let us now consider the last element in Boltzmann's account as presented by Albert and Loewer. As we have just seen, Boltzmann's account of the second law, supplemented by the past hypothesis, does not tell us how all states will evolve. Rather, it will tell us *how the vast majority* of them will move, namely toward an increasing entropy. However, if we want to have a theory which makes predictions and provides explanations, as Albert and Loewer point out, we need to

---

[7]Whether the past hypothesis is also able to account for our experience of time is discussed in Hemmo and Shenker (2019).

[8]For a discussion of the past hypothesis and the proposals on how to eliminate it, see Lazarovici and Reichert (2019) and references therein.

talk about probabilities, not about the number of microstates. In other words, we want to be able to say not only that the great majority of microstates goes toward a higher entropy state. We also want to be able to say that the *probability* of entropy increasing is high (or that entropy increase is likely). Similarly, one would not only want to say that there are very few 'abnormal' or exceptional (or anti-thermodynamical) states for which entropy decreases. One would also like to say that it is *unlikely* that entropy decreases (or that the probability of entropy decreasing is low). In this way, one can say that Boltzmann's account provides a probabilistic explanation of the second law of thermodynamics, even if the underlying theory is deterministic.

To put things differently, one could have definite results (i.e. with probability 1) if one could solve exactly all the equations for all the particles. However, since one cannot do that for practical reasons, the results obtained are only probabilistic. Therefore, the second law of *thermodynamics* is that (thermodynamic) entropy *always* increases, while the *statistical mechanics* version of the second law is that (Boltzmann) entropy *almost always* increases. This should be the same as saying that the *probability* of entropy increasing is extremely high. Here is where probabilities enter the picture. In this way one sees that a statistical mechanical explanation is characterized by providing the prediction not of what will certainly happen, but of what will probably happen. Nevertheless, as we just pointed out, in Boltzmann's theory we do not find probabilities. Rather, there are statements about what the great majority of microstates will do. Hence, we want that something like 'for the great majority of situations entropy will increase' to be equivalent to 'entropy has high probability of increasing', or to 'entropy almost always increases'. One can move from one locution to the other if all the microstates are 'equiprobable', in the sense that neither one of them is special in some way or other. This amounts to defining over phase space a 'measure' which does not privilege one microstate over another. A measure is, roughly put, a way of counting how many ways microscopic particles (i.e. microstates) can be arranged in order to give rise to the same macroscopic properties (i.e. macrostates).

The statistical postulate amounts to the assertion that the measure to be used in making inferences about the past and the future (respectively explanations and predictions) is the uniform measure (over the suitable regions of phase space), also known as the Lebesgue-Liouville measure. According to Albert and Loewer, the choice of this measure has to be introduced as a postulate because, they argue, there is no acceptable justification for the choice of such measure. In fact, there could be an infinite number of measures: the uniform measure counts all states equally, but other measures might not. So why is the uniform measure special? The usual answer found in physics books is that the measure is a reflection of our state of ignorance with respect to the microstate the system is in. In other words, the measure reflects the fact that an observer is uniformly ignorant of the true microstate the system is in; the observer will assign the same probability to each microstate because they do not know which microstate the system is in. However, this cannot be right. Albert and Loewer argue that since one should not use epistemology to guide our metaphysics, the behavior of macroscopic bodies cannot possibly be a function of how ignorant we are about their composition. For one thing, the former is objective and the latter is not;[9] thus, since no other justification is provided, the only option is to make this assumption a postulate.

In the next section, I do not dispute that the uniform measure is postulated, I instead argue that this measure need not be a probability measure. That is, I argue that one does not need to invoke the notion of probability in order to explain the macroscopic appearances. Rather, the weaker notion of typicality is sufficient. If so, then the statistical postulate should be understood as a typicality postulate. I then argue in Section 4 that one might not even need to add a postulate after all, given that the notion of typicality which explains the macroscopic appearances can be suitably derived from the microscopic dynamics.

---

[9]See also Hemmo and Shenker (2012) for additional comments on the (lack of) justification of the uniformity measure.

3.3. How to avoid probability talk: The notion of typicality

Let's discuss the statistical postulate more in detail. As we just saw in Section 2.3, the statistical postulate was introduced to bridge the gap between the talk in terms of number of microstates that we use when describing a system in terms of the microscopic dynamics, and the probability talk that we use when discussing in the predictions and explanations of macroscopic phenomena. For instance, when we see an ice cube in a glass of water we want to be able to predict that it is probable that the ice will melt in the next five minutes. Likewise, we want to be able to say that we can explain that an ice cube has melted in the last five minutes by saying that its probability of melting was high. Using statistical mechanics, we have 'merely' shown respectively that we can predict that the vast majority of microstates corresponding to an ice cube now will evolve in the next five minutes into a macrostate with higher volume corresponding to a melted ice cube, and that we can explain why the vast majority of microstates corresponding to an ice cube five minutes ago has evolved into the macrostate corresponding to a melted ice cube now. So, we need to connect the locution 'the vast majority of microstates go toward an increasing entropy', which is given by Boltzmann's account, with the locution 'the probability of entropy increasing is high', otherwise we would not have explained the regularities. The translation-rule can be provided if we count the microstates equally, using a uniform probability measure, so that we can say that 'vast majority' means 'high probability' as just discussed. Then, the question arises as to why the uniform measure is the correct probability measure. As we have seen in the previous subsection, the usual 'ignorance' justification is not tenable because the microstate will move however it will, independently of whether we know anything about it or not. Therefore, in the absence of an alternative justification, Albert and Loewer think that this measure should be postulated.

Notice that two claims are being made here:

1) The statistical postulate is needed because *probabilities* are needed to explain/predict;

2) The statistical postulate cannot be inferred from the theory but needs to be *postulated*.

In this section and the next I argue that none of them is necessarily the case: one does not need probabilities in order to get a satisfactory scientific explanation of the macroscopic phenomena (current section); and even if one did, the probabilities could be grounded in something which is not postulated but derived from the theory using symmetry considerations (Section 4).

3.3.1. *The typicality measure*

To begin with, let me reiterate that the current discussion is not an argument usually present in physics book, in which the uniform measure is justified on epistemic grounds. Moreover, in most physics books there is no mention of the statistical postulate in these terms. In the work of physicists — such as Jean Bricmont (1995, 2001, 2019), Detlef Dürr (2009), Sheldon Goldstein (2001, 2011), Joel Lebowitz (for a first example, see his 1981 paper), and Nino Zanghì (2005) — however, the postulate and the probabilities appear only indirectly as mediated by another notion, namely the notion of typicality. What is the relation between typicality and probability? In this subsection, I discuss how typicality, rather than probability, has a more fundamental role in the explanation of macroscopic laws. In the next subsection, I focus on the relations between the two concepts.

To proceed, let us assume for now that the statistical postulate holds: the uniformity measure is postulated. However, I do not assume it is a probability measure. As just mentioned, in the physics literature mentioned above, one finds that 'the vast majority of microstates approach equilibrium', say, is translated into 'the approach to equilibrium is typical' rather than into 'the approach to equilibrium is probable'. The notion of 'typicality' is notoriously controversial,[10] but I think it boils down to the following. A given behavior is typical for an object of a given type if and only if one can show that the vast majority of the systems, suitably similar to the

---

[10]See Badino (2019) and references therein for a discussion.

original one, would display that behavior.[11] For instance, consider approach to equilibrium. This behavior is typical, in the sense that the vast majority of thermodynamic bodies display this behavior. Technically, as already seen, the notion of 'vast majority' is defined in terms of a measure μ, which allows 'counting' (i.e. evaluating their number) the exceptions E to a given behavior, and fixed to a given tolerance. 'Great majority' means, therefore, that the exceptions to the given behavior, given a tolerance, are few when counted using the measure μ. This is equivalent to saying the size of the set of exceptions, as counted by μ is small; that is, the set of thermodynamically exceptional states (the ones for which entropy does not increase and thus they do not go to equilibrium) is small. One could be tempted to think that this means if the set of exception to the behavior 'go toward equilibrium' is very small, then it is highly *probable* that systems will go toward equilibrium. And thus it's only natural to interpret μ as a probability measure. However, this may be too big a step. In fact, up to this point, the measure merely does these:

1. It counts the microstates,
2. It allows us to define what it means when we say that the size of the set of exceptions E is small.

These two conditions define a *typicality* measure. In contrast, a probability measure is much richer. For one, it also has to be additive (that is, the probability of the sum of two sets is the sum of the probabilities of each set), in contrast to the typicality measure. In addition, while there is a difference between the probability being 1/2, say, and being 1/3, nothing of the sort is required by the typicality measure, whose role is to make sense of claims that describe/account for/explain phenomena as holding with some very rare exception without specifically quantify them. Similarly, the typicality measure does not have to satisfy the probability axioms.[12]

---

[11]Formally, given an object x in a set S, a property P is typical for objects in S if and only if almost all elements of S possess P. See also Wilhelm (2019) on typicality in scientific explanation.

[12]For more differences between typicality and probability, see Wilhelm (2019).

Let us now turn to the question of whether we need something more, namely whether we also need a probability measure. I think not; a typicality measure, which provides us with a rough guidance about the relative size of sets in phase space, is enough for the current purposes. In fact, the *only* reason we need a measure for the size of the macrostates is to count the number of microstates in them. The details do not count; it does not matter what precise number the measure gives us. What matters instead is that the size of the set of exceptions (i.e. abnormal or anti-thermodynamic states) is extremely small, regardless of how much it precisely is. That is, we need a typicality measure, not a probability one. Thus, postulating that the typicality measure is the uniform measure (as dictated by the statistical postulate), the second law of thermodynamics holds, *typically*. In other words, the set of the exceptional (i.e. anti-thermodynamic) microstates is very small.

Let us see in a bit more detail why typicality is able to explain everything that we seek an explanation for in recovering macroscopic laws from microscopic ones.[13]

1) We want to explain why the 'probability' that a state is in equilibrium is very high. However, we do not need a precise probability estimate about how likely it is; we do not need to distinguish between the equilibrium state being reached with 0.95 probability and 0.99 probability. All we need is a rough estimate, which is what typicality gives us. Thus, we 'merely' have to explain why the equilibrium state is typical. And to do that we need to show that it occupies the vast majority of phase space. In other words, we need to show that if A is the set of microstates that do not belong to the equilibrium macrostate, then the size of A is much smaller than the size of the equilibrium macrostate.

2) We also want to explain why systems starting from equilibrium remain there with 'high probability.' Again, details do not matter as we want to explain why the departure from equilibrium is atypical. That is, we need to show that the size of the set

---

[13] For more on the connection between typicality and explanation, see Section 3.3.

of microstates starting from equilibrium and remaining there, as measured by the typicality measure, is overwhelmingly larger that the size of the exceptions, namely the set of microstates whose temporal evolution brings them outside of equilibrium.

3) Similarly, we want to explain why systems outside of equilibrium evolve towards it with 'high probability.' Like before, this is actually the request for the reason why equilibrium-evolving behavior is typical for states outside of equilibrium. That is, we want to show that the size of the set of exceptions, namely the set of microstates starting outside of equilibrium which will not go toward equilibrium, is very small with respect to the size of the set of microstates outside of equilibrium which will go to equilibrium.

With these qualifications, one can answer to the one-million-dollar-question about entropy increase:

4) We want to explain why entropy has an extremely 'high probability' of increasing. That is, since all that is needed is an estimate of the entropy rather than a precise computation, we want to explain why typical states increase their entropy during their temporal evolution. In other words, we want to show that typical states in equilibrium will remain there (maintaining their entropy constant) and typical states outside of equilibrium will evolve towards it (increasing their entropy).

Therefore, since in Boltzmann's account using the uniform measure as a typicality measure one can show all of the above, then Boltzmann's account explains the macroscopic laws as well as the macroscopic asymmetry in terms of time-symmetric microscopic laws, even without invoking the notion of probability.

3.3.2. *Typicality and probability*

However, one could object that we still have not responded to the original question about the second law of thermodynamics. In fact, one could complain that, since typicality is not probability, one can only say that the second law is typical but not that it is probable. And this seems wrong. Thus, one should further investigate the connection between the two notions. More often than not, we express probabilistic statements and not typicality statements when describing a macroscopic phenomenon in this way.

Indeed, one may argue that the notion of typicality grounds the one of probability; that is, probability theory is a mathematical idealization developed in order to make precise the meaning of 'vast majority' through theorems like the law of large numbers. It is that notion which has to satisfy the axioms of probability, not the typicality measure from which it is derived. Be that as it may, one could also outline the connections between the various senses of probability, typicality and their explanatory role. The situation is not simple and presumably one should write another paper on this.[14] However, here are some simple considerations. First, there is a straightforward connection with the notion of subjective probabilities as degrees of belief. In fact, the measure $\mu(A)$, where A is, for example, the set of microstates for which the entropy grows, can be interpreted as the degree of belief of an observer making an inference on the probability that the entropy will increase. Thus, the measure plays a role in the justification of why we have a reasonable degree of belief about the growth of entropy. Nonetheless, since this is someone's degree of belief, it cannot help in explaining physical phenomena, given that only an objective notion has the potentiality of doing that (recall Albert's objection to epistemic accounts of the probabiliy measure).

Perhaps more interesting is the connection with probabilities as relative frequencies. When an experimenter prepares a set of repeated experiments with substantially identical initial macroscopic preparations, they will obtain empirical regularities. These empirical distributions are the relative frequencies of the various outcomes. Take a set of N gases concentrated on the corner of N similar boxes. Let them evolve freely; and, say, check what has happened after 2 hours. Record the N results: the first gas spreads out in the first box, the second spreads out in the second box, and so on. In general, the empirical distributions present statistical patterns. For instance, the vast majority of the gases spreads out in their container. Now,

---

[14]See also Maudlin (2019).

how do we explain the empirical distributions? We show that the distribution predicted by the theory describing the phenomena in question matches the one observed. In this case, we need to show that the observed distribution ρ_{emp} and the theoretical distribution predicted by statistical mechanics ρ_{theo} agree. That is, we need to show that, theoretically, the vast majority of gases expands when evolving freely. In other words, one would then have to show that, with the right measure of typicality, the theoretical distribution and the empirical one are very close. Formally, one needs to show that $\left|\rho_{emp}-\rho_{theo}\right| < \varepsilon$, with some positive constant ε small at will and with the distance measured by the typicality measure. Having said that, the connection between typicality and probability as empirical frequencies is that typicality provides the measure with which one can compare the empirical and the theoretical frequencies.[15]

3.3.3. *Explanation based on typicality*

To sum up the results of the previous subsections, in this account, to explain why a given regularity occurs is to explain why the regularity is typical. This can be done by specifying the laws of nature and the typicality measure (in addition to the past hypothesis).[16] Formally, as discussed by Isaac Wilhelm (2019), this is the explanatory schema based on typicality: if an object x has property A (or belonging to a given set), and a property B which is typical for A-type objects, then the explanation of why x has B is given by:

$$x \text{ is } A$$
$$\frac{\text{Typically, all } As \text{ are } Bs}{\therefore \ x \text{ is } B}$$

That is, the explanation of the fact that a given system has a given property is given by showing that this property is typical for objects

---

[15]For more on the relation between typicality and probability, see Volchan (2007), Goldstein (2011), Pitowsky (2012), Hemmo and Shenker (2012), Mandlin (2019), Wilhelm (2019).

[16]This is similar to what Albert and Loewer conclude, with the only difference that here we have typicality rather than probability.

of that type. In other words, if one has shown that B is typical of A-objects (the objects that belong to the set of objects having the property A), then one has explained why x has the property B too. For instance, the explanation of the fact that this gas expands is given by the fact that one can prove that free expansion is typical for gases; that is, for the vast majority of initial conditions one can show that gases expand.

A longer discussion would be appropriate, but let me observe that the explanatory schema sketched above shares some similarities with Hempel's covering law model (1965), whose main idea was that explanations are arguments with laws of nature as premises. The idea is roughly that if one finds the relevant law that made the phenomenon happen, one has found the explanation for why it happened. Here, explanations are also arguments, but the explanation is given not by nomological facts but by typicality facts. However, the difference between the two types of facts is not in kind but merely of degree. They are both nomological facts, but while laws of nature are exception-less, typicality facts are not. That is, one can think of typicality facts as nomological facts which allow for rare exceptions. Notice that this is compatible with typicality explanations being used in a macroscopic context in which it seems fine to allow for exceptions, while it is not used at the fundamental level in which exceptions are seen as problematical. If so, the parallel with the deductive-nomological (DM) model is striking:

$$x \text{ is } A$$
$$\underline{\text{All As are } Bs}$$
$$\therefore \ x \text{ is } B$$

That is, if one can show that there is an exception-less regularity such that all As are Bs, this explains why this x, which is an A, is also a B. Similarly, in the typicality schema, if one can show that B is typical of A-objects, this explain why this x, being an A, has also property B. Nonetheless, the schemas are not identical: while in the DN one can deduce that x is a B from the other premises, this is not so for the typicality schema because there are exceptions. That is, no deductive derivation can be provided, not even in principle.

In the case of Boltzmann, for instance, there will always be anti-thermodynamic states. Thus, Boltzmann proved 'merely' that most systems have increasing entropy, not that all of them did. Accordingly, the DN model could not be used to explain the laws of thermodynamics, but the typicality schema could.[17]

In addition, there is a sense in which the typicality explanation is similar to Hempel's inductive statistical (IS) model. The original idea was that if one can prove that two features occur together with high probability, then one explain why something possessing one also possesses the other. Formally, an explanation is provided by the following inductive argument:

$$\frac{\begin{array}{c} x \text{ is } A \\ Prob(A, B) = r \end{array}}{\therefore\ x \text{ is } B}[r]$$

where r is the strength of the induction. For instance, if this apple (x) is red (A), and if one can show that there is a high probability (r) that red apples are also sweet (B), then one has explained why this apple is sweet.[18] Compare this with the typicality schema [typ]. According to the IS model, a phenomenon of type A is explained

---

[17]Moreover, as in the DN model, in the typicality schema every explanation is also a prediction. For example: given gas (x) is in non-equilibrium (A); it's typical for non-equilibrium gas to expand freely (B); so this gas expands freely. This is an explanation of why the gas in this box has expanded, or a prediction that it will expand if the gas has not been observed yet. However, unlike the DN model, the typicality schema as not have the problem of asymmetry. Let's recall what this problem is. Given the height of a flagpole, the height of the sun on the horizon and the laws of optics (and trigonometry), one can deduce the length of the shadow cast by the pole. This is an adequate explanation as well as a prediction of the shadow's length of a given flagpole of that height, observed or not. However, one could also use the length of the shadow, together with the height of the sun on the horizon and the laws of optics (and trigonometry) to deduce the height of the flagpole. This is an accurate prediction, but not an explanation of the height of the pole. Critics track this down to the idea that explanation is asymmetric while deduction is not. This is not true for the typicality explanation, because the explanation is not deductive but follows the schema above.

[18]I am assuming these claims are explanatory. One can object to this, but all I am pointing out is that as long as the IS model is explanatory, so is the typicality model.

to be also a B by showing that the probability of some A being also a B is high. In typicality account, a phenomenon of type A is explained of being also a B by showing that being a B is typical of being an A. So, these schema are very similar, if not formally identical. The difference is in the logic operators (Typ vs. $Prob$), and we have seen that typicality is not probability.[19] Some probabilistic explanations are not typicality explanations: consider a nucleus of the element X having the probability of decaying being less than 0.5. Then, one can explain why this nucleus has decayed using the IS model, providing therefore a probabilistic explanation. However, it would not be typical for X to decay, since r is small.[20]

Be that as it may, I think that the reason why the apple argument in terms of probabilities appears to be explanatory (to those who think it is explanatory) is because what one has in mind is a typicality argument instead. That is, we could have said: science has shown that red apples are typically sweet, and that is why this red apple is sweet. Details on the strength of the induction (whether it is 0.98 or 0.97) do not matter much, as long as r is large. This is typical of typicality reasoning (pun intended), not of probabilistic reasoning. This is compatible with Hempel's original idea that the IS model works only for large probabilities.[21] Also, to make the point that typicality explanations are not probability ones, one could point out that probability reasoning can be counterintuitive, while typicality reasoning is not. For example, think of the probability of having a rare (one case over 10,000 people) but terrible disease when testing positive for it, assuming the testing methods has 99% of accuracy. Most people will be terrified, but they would be wrong: using Bayes

---

[19]See Crane and Wilhem (2019) for two proposals for a formal logic for typicality arguments, one based on propositional modal logic, and the other on intuitionistic logic.

[20]See Wilhelm (2019) for more examples.

[21]Later, critics of the model argued that also small probabilities can be explanatory. See Strevens (2000) for a nice summary and an argument that the strength of the induction doesn't necessarily correlate with the strength of the explanation. This does not affect my point because the argument is not that all explanations are typicality explanations.

theorem your probability of having the disease is 0.01. So, testing positive but not having the disease is, arguably, explained by the fact that Bayes theorem show that the probability of having the disease was low. However, this is extremely counterintuitive. In contrast, typicality arguments are not like that at all. They are used all the time in everyday reasoning arguably because they are very intuitive: typically, dogs are affectionate and that is why my dog is affectionate. Interestingly, this is connected to the use of stereotypes: we immediately nod in agreement when someone says that our husband's bad behavior is explained by the fact that all men are jerks. We use probability talk, but since details do not matter, we are actually using a typicality explanation: typically, men are jerks.[22] Similarly, I think, we may use probability talk to present Boltzmann's explanation of the macroscopic laws, but we actually have in mind typicality: typically, entropy increases.

On another note, as pointed out by Wilhelm (2019), notice that there seems to be the right correlation between typicality and causation. The claim that 'woodpeckers typically have a pointy beak' may be further explained, citing some genetic factor, which may play a causal role in beaks having a pointy shape. Notice, however, that this is itself a typicality explanation: those genes typically give rise to such beaks. This shows that the typicality schema is not wedded to the covering law model, even if it shares with it some important features.

Aside from the comparison with the covering laws model, let us explore whether the explanation based on typicality is adequate. Among the desiderata for a satisfactory scientific explanation, one can list the following: informativeness, predictive power, and expectability. A phenomenon is suitably explained if the account

---

[22]Notice that I am not saying that stereotypes, by themselves, can explain. In contrast, stereotypes do not really explain unless someone actually show that they hold typically. The point was merely that stereotypes are commonly used in accounting for certain phenomena happening in contrast with 'true' probabilistic explanations, in which the details mater, which are notoriously difficult to grasp (see the above mentioned use of Bayes theorem).

is able to provide an informative and concise description of the phenomenon in question and typicality is able to do that, as also pointed out by Wilhelm (2019). It is informative because it tells us about the behavior of the vast majority of systems: typical systems approach equilibrium. And it does so very succinctly, in a single sentence, merely using the uniform measure as typicality measure to count the number of microstates. Notice that it is informative only because it tells us about the typical behavior, namely the behavior that the vast majority of systems will display. Otherwise, one would not know, by merely accessing the macrostate, whether the system's microstate will evolve thermodynamically or not.

Moreover, as also emphasized by Dustin Lazarovici and Paula Reichert (2015), typicality can provide us with predictive power, which is not surprising, given what we observed above in relation with the connection with the covering law model. In fact, if one shows that a given property is typical, then one has reasons to expect to see this property in other systems similar to the original one. Given that one has proven that a typical system will approach equilibrium, this is what one should expect a system to generally do. Notice how, obviously, this predictability would not be possible if one had shown that only *some* states, not the great majority of them, approach equilibrium. Consider a gas in the corner of a box. If one could only show that some state will approach equilibrium while other will not, what can one conclude about the behavior of a given gas which has not been observed yet? Not much, since the size of equilibrium-approaching states and the size of non-equilibrium-approaching states are comparable (in our pretend-example). It is only when we can rule out the possibility of non-thermodynamic states, because the size of the corresponding states is incredibly small, that we can predict what is likely to happen in similar circumstances.

In addition, typicality provides an explanation which holds for most initial conditions. A phenomenon has been explained if it holds for typical initial conditions; that is, with rare exceptions as defined by a suitable measure of typicality. As Bricmont (2001, 2019) has suggested, if something is typical, no further explanation seems to be required. For instance, the fact that this particular gas expands is not surprising, once Boltzmann's account has been provided. What would be surprising is if it did not expand. An unacceptable explanation would be, on the contrary, one which is true only for very peculiar initial conditions. In fact, too many things could be explained by appealing to special initial conditions, for one can always find an initial condition that will account for the phenomenon. If we accept this type of explanations, allowing for fine-tuned initial conditions, then one could 'explain' everything. Let me elaborate. There is a sense in which to explain a phenomenon is to remove the surprise in seeing this phenomenon happening. This is related to what Hempel had in mind with his notion of 'expectability.' We are not surprised that a piece of salt will dissolve in water because we know the reason why it happens: the positive ions in water $(H+)$ attract the negative chloride ions, and the negative ions in water $(O-)$ attract the positive sodium ions. However, consider the case of a monkey who, by randomly hitting computer keys, ends up writing the "Divine Comedy." One can account for this fact by cherry picking an initial condition for which this actually happened. So, there is a sense in which the phenomenon is 'explained.' However, the fact that the monkey ended up writing that book is extremely surprising, and by pointing to a special, perhaps unique, initial condition that made it true does not help much to remove the surprise. So, there is a sense in which we are not completely explaining the phenomenon if we rely on special initial conditions. In other words, monkey writing books is not something that we expect. This is not something that monkeys typically do, because monkeys, typically, cannot read or write, for starters. It is not impossible that they write a wonderful book. Indeed, it could just be a very lucky set of keyboard strokes. However, if they end up writing this wonderful book, we find it surprising. That is why I used the world 'lucky' in the previous sentence: the event has happened because of a ridiculously special initial condition. A slightly different initial condition would not have brought about a similar event. Instead, if one were to point out that most initial conditions would have the same outcome, the surprise will cease. In other words, a satisfactory explanation is one which, for the majority of initial conditions, monkeys would indeed write books like the "Divine Comedy."

However, this typically does not happen: monkey randomly hitting computer keys would typically write gibberish. So, when asking for an explanation of a given phenomenon, what we are actually asking for is a reason *why we should not find the phenomenon surprising*, and the response is that the phenomenon happens for most of the initial conditions. Moreover, in this framework, if we rely on a special initial condition to account for why a phenomenon has happened, we are not truly providing an explanation for it because we are not removing the surprise element. Or, in other words, the 'explanation' lacks expectability. Something similar happens in statistical mechanics: for the great majority of initial conditions, entropy will increase, not just for a special one. Because of this, the surprise is removed and the phenomenon explained. Moreover, one can account for why anti-thermodynamic states are not observed by pointing out that they are atypical, but Boltzmann's statistical mechanics does not provide an explanation of *why* atypical states happen,[23] other than that they typically do not happen.

One could object that there is something missing, namely we need to be able to explain *all* phenomena, including the atypical ones. That is fair enough; however, let me just note that the situation remains unchanged if we use probabilities instead to explain improbable events. In any case, one could always rely on initial conditions: the only explanation of why *this* gas is not expanding, as opposed to that other one, is that this is what the initial conditions together with the dynamics bring about for it. Indeed, some have argued that the requirements for a satisfactory scientific explanation I discussed above are too strict. One would provide a satisfactory explanation of the phenomenon even if one relies on special initial

---

[23] The notion of expectability and the one of typicality go together as long as the world is typical. In fact in a typical world, entropy increases, and expectability goes with typicality. However, in an atypical world, one would expect something different than what it is typical, as in that world entropy would decrease. This is because what is expected comes from what is 'usual' in our world, and only in typical worlds, what is usual and what is typical are the same. Thank you to Katie Elliot, Barry Loewer and Tim Maudlin for making me be explicit on that.

conditions.[24] All that is required is that there is one such condition that would bring the phenomenon about. That may be so. However, as already seen, in this way, one would not be able to account for the link between expectability and explanation. It would be difficult to understand why one would find monkey writing books surprising while one would not find surprising that gases expand when evolving freely.

Moreover, let me emphasizes that the idea of a satisfactory explanation being an explanation for most initial conditions is compatible with scientific practice in physics. For example, one of the reasons why Alan Guth (1981) proposed his theory of inflationary cosmology is that the big bang model requires strict assumptions on initial condition. In contrast, inflation would explain all the phenomena without relying on these special assumptions, and for this reason is considered a better theory: "The equations that describe the period of inflation have a very attractive feature: from almost any initial conditions the universe evolves to precisely the state that had to be assumed as the initial one in the standard model" (Guth and Steinhardt 1984).

In this respect and to conclude, let me add a remark regarding the past hypothesis. The past hypothesis has been introduced in order to break the past-future symmetry of the microscopic laws, by postulating that the universe had an initial very low entropy. One may think that this is a problem that undermines the whole account based on typicality: haven't we just said that a satisfactory explanation should hold for the typical initial condition? In contrast, a low entropy state is a very atypical state. So, have we ended up explaining what is typical by postulating something atypical? Isn't that bad? One thing that can be said to mitigate the problem is that, with respect to this low entropy initial macrostate, the initial microstate of the universe is typical in regard to its future evolution, which accounts for the entropy increase.[25] However, I honestly do not think this is helping a lot, given that the initial macrostate is

---

[24]See, e.g., Valentini (2019). Also, see Myrvold (2019).
[25]See Lazarovici and Reichert (2015).

incredibly small. Indeed, many think this is a serious problem and propose mechanisms to make the initial state typical.[26] In contrast, Humeans such as Craig Callender (2004) argue that there is no need to explain the past hypothesis because, in a Lewisian fashion, it is simply one of the axioms of the best system of the world.[27] Notice, however, that from the point of view of the typicality account, the situation gets better, rather than worse. Add the typicality, rather than the probability, postulate, and everything follows without any additional need for explanation.

3.3.4. *Objections to the typicality account*

The typicality approach as applied in statistical mechanics described so far is what Massimiliano Badino (2019) dubs the "simple typicality account," or STA. In this view, the approach to equilibrium is explained entirely in terms of the size of the macrostates, by showing that the vast majority of microstates will fall into the equilibrium state. However, according to some critics, this approach dismisses the dynamics, which apparently plays no role in explaining the approach to equilibrium. Accordingly, Roman Frigg and Charlotte Werndl propose the one Badino calls "combined typicality approach," or CTA (Frigg and Werndl (2012), Werndl (2013), Werndl and Frigg (2015)). The idea is that one should show that equilibrium is approached for the typical *dynamics*, in addition to showing it is approached for the typical initial condition. Frigg and Werndl prove that typical Hamiltonians produce systems which are epsilon-ergodic, namely they are such that the time spent in a macrostate is proportional to the size of the macrostate. Because the equilibrium state is the largest of the macrostates, an epsilon-ergodic system will

---

[26]To this end, Penrose (1999), for instance, proposes his 'Weyl curvature hypothesis' as an additional law in order to explain how the low entropy initial state is not atypical. In addition, Carrol and Chen (2005) put forward a model whose purpose is to completely eliminate the past hypothesis. See Lazarovici and Reichert (2019) for a proposal built on Chen and Carrol's model.

[27]For more on this approach and its challenges, see also Olsen and Meacham (2019).

spend in it most of the time, explaining in this way the approach
to equilibrium of typical dynamics as well as for typical initial
conditions.

In this regard, let me enter into some details about the STA
and the role of the dynamics. It is not true that, strictly speaking,
in the STA the dynamics is ignored. Frigg and Werndl prove that
phenomena such as the approach to equilibrium are to be explained
for *most* (typical) initial conditions and for *most* (typical) dynamics.
Instead, Frigg and Werndl complain, the STA merely does that for
most initial conditions, forgetting about typical Hamiltonians. However, this is not so. Indeed, the STA manages to do something *more*,
rather than less, *general*. That is, in the STA, one is able to account
for phenomena such as the approach to equilibrium dynamics for
most initial conditions, *without* being specific about any feature the
dynamics needs to have. In other words, the phenomenon is explained
for *most* (typical) initial conditions and for *all* dynamics. This is why
the dynamics is never mentioned; not because it is irrelevant, but
because for most initial conditions the details of the dynamics do
not matter, and the system will reach equilibrium *regardless* of the
Hamiltonian. For more on this 'genericity' of the Hamiltonian, see
the next section.

Finally, to counter the idea that the dynamics plays no role in the
STA, in the next section I show that the statistical postulate can be
derived from the dynamics. If so, the dynamics plays a big role in the
STA, namely the role of selecting the typicality measure; henceforth,
reducing the gap between the STA and CTA.

3.4. How to dispense of the statistical postulate: The stationarity argument

Up to now we have simply assumed that the choice of the measure,
typicality or probability, had to be postulated. In this section I wish
to explore what I take to be the proposal put forward by Goldstein
(2001, 2011), Dürr (2009), and Zanghì (2005). That is, the proposal
that the typicality measure is derived from the dynamics introducing suitable symmetry constraints. I find it extremely surprising that

this argument has received very little attention in the literature,[28] because not only it provides a non-epistemic justification for the uniform measure but also shows how the dynamics plays a crucial role in the typicality account.[29]

The main idea, I take it, is that the typicality measure is the uniform measure (the Lebesgue-Liouville measure) not because it is uniform, but because of these two features: 1) it is *time-translations invariant*; and 2) it is *generic* with respect to the Hamiltonian of the system. Let's see what these features amount to, starting with the first. A measure is time-translation invariant when the volume it defines in phase space is conserved. That is, if A is any set in phase space, and A_{-t} is the set of points in phase space that evolve into A after a time t, then A and A_{-t} have the same volume (this is Liouville's theorem). The reason why time-translation invariance, also called stationarity, is a requirement for the typicality measure is connected with the idea that no temporal instant needs to be privileged. In fact, as Goldstein, Dürr and Zanghì noticed, the measure counts the space-time histories of the universe, while the phase space point is just a convenient way of representing them. So, when counting the histories, one needs to regard the initial time merely as conventional, by not privileging any particular time, and a time-translation invariance measure would guarantee that. The uniform measure is not the only time-translation invariant measure. In fact, given a conservative force, there could be other measures, whose explicit form

---

[28] Aside from the comments of Bricmont (2001, 2019), discussed in Section 4.1.

[29] Itamar Pitowsky (2012) proposes a justification of the uniform measure as the typicality measure which has been later criticized by Werndl (2013). She proposes that the choice of the typicality measure should be done using symmetry consideration, in particular the typicality measure should be invariant with respect to the dynamics. Maudlin (p.c.) maintains that there are other ways of justifying the use of the uniform measure. For instance, one could say that the uniform measure is privileged because it is the one that phase space inherits from the spatial measure. However, considerations like this will have less weight when one moves to the quantum domain, especially in the context of the pilot-wave theory. In that context in fact the choice of the typicality measure is justified by an argument which is a direct analog of the stationarity argument discussed in this section, and this is one of the reason I think it is worth exploring.

depends on the particular law of the force. However, it is argued, one can single out a unique measure by requiring an invariant measure that is also *generic* with respect to the dynamics; that is, it is *independent* of the particular law of the force. Tying the measure to the dynamics arguably makes the choice of the uniform measure *natural*: the uniform measure is the typicality measure because it is the only stationary generic measure.

This argument, which I will call from now the 'stationarity argument,' can be therefore summarized as follows:

P1: The purpose of the typicality measure is to count microstates (definition);
P2: One can do this at different times (definition/construction);
P3: One should not privilege one time over any other (the initial time is conventionally chosen)
subC : Thus, the typicality measure should be time-translation invariant or stationary;
P4: If the dynamics is Hamiltonian, the typicality measure should be independent from the specific form of the Hamiltonian (genericity);
P5: The only measure satisfying these two requirements is the uniform measure (mathematical proof);
C: Thus, the uniform measure is the typicality measure.

3.4.1. *Objections to the stationarity argument*

The stationarity argument for the uniform measure as typicality measure has been criticized most notably by Bricmont (2001), who argues that there are several problems.

First, the argument applies only to Hamiltonian systems. As we just saw, Hamiltonian systems are volume preserving. They are such that the phase-space volume, defined by the uniform measure, is preserved under time evolution. Instead, Bricmont urges us to consider certain dissipative systems like dynamical systems with a chaotic attractor. These systems usually give rise to solution flows which contract volumes in phase space. This volume contraction gives rise to a set in phase space called an *attractor*, toward which solutions ultimately evolve. Certain dissipative systems are chaotic, that is, they show sensitive dependence of initial conditions. For chaotic systems, the attractor is special because merely studying the dynamics on the attractor is sufficient to have information about the overall dynamics. As Bricmont points out, the uniform measure is not time-translation invariant for these systems (because the volume is not conserved), but it is still the typicality measure. For most initial conditions in the basin of attraction, counted using the uniform measure, the relevant empirical frequencies will be correctly reproduced. How do we justify this choice? According to Bricmont, the uniform measure is chosen as typicality measure not because it is stationary (which it is not). Rather, the justification is more generally based on what Bricmont calls "Bayesian grounds" (Bricmont 2019): *the typicality measure is whichever measure that reproduces the relevant empirical frequencies*, regardless of whether it is stationary or not. This argument, according to Bricmont, provides a more general account of the reasons why the uniform measure is the typicality measure which holds for non-Hamiltonian systems. So, why would one need another argument for Hamiltonian systems?

Another objection raised by Bricmont against the stationary account of the typicality measure is that it is too sophisticated to genuinely be part of the scientific explanation of thermodynamic phenomena. In fact, he notices that we are trying to provide an account of scientific explanation, and as such the proposed account should be not too distant from our intuitive understanding of the notion of explanation. Because of this, time-reversal invariance does not seem the right notion, since when we give our folk explanations we never use anything that remotely resembles stationarity.

To conclude, let me add an objection which challenges the premise requiring genericity for the measure. Indeed, I find this to be the weakest point of the stationarity argument as nowhere it has been argued why such a constraint is required: why would one want the measure to be independent of the Hamiltonian, hence the potential? If the whole point of caring about Hamiltonian systems was to ensure that they have the chance of describing the universe, since there is only one universe, there is also only one Hamiltonian. So why should one care about other possible dynamics?

3.4.2. *Replies to objections to the stationarity argument*

First, consider the objection based on chaotic dynamics: if for non-Hamiltonian systems the uniformity measure is chosen as the typicality measure without using stationarity considerations, why use these consideration for Hamiltonian systems? This is, in my view, a well-thought objection to the stationarity account of the typicality measures, and as such should be carefully considered. The best response to this, I think, is the following. Upon reflection, one should not find it surprising that for non-Hamiltonian systems the uniform measure is the typicality measure even if it is not stationary. Indeed, I think it would be unreasonable to expect that stationarity would pick out the correct typicality measure in these cases. In fact, while in Hamiltonian systems the volume in phase space is preserved, and the natural weighting of all points in phase space is equal, this is not so for non-Hamiltonian systems because the system is dissipative and the attractor carry more weight. And this, in turn, is not surprising because non-Hamiltonian systems are open systems. They are dissipative system, so there has to be somewhere they dissipate to. Because of this, they are *less general* than Hamiltonian systems, in the sense that while Hamiltonian systems may reproduce the behaviour of the whole universe, non-Hamiltonian systems at best can only reproduce the behaviour of open subsystems. To put it in another way, the universe is a Hamiltonian system, while non-Hamiltonian systems may be descriptions of non-isolated subsystems. If so, the stationarity constraint needs to be fulfilled to select the typicality measure for the universe, not its subsystems. Then, when describing dissipative system, which are open subsystems of the universe, the choice of the typicality measure has been *already made*. Then, one looks for other 'natural invariant' or 'physical' measures, such as the Sinai-Ruelle-Bowen (SRB) measure, defined on the attractor, which is time-translation invariant. However, when considering that these systems are merely subsystem of a more general Hamiltonian system, the apparent paradox (why the SRB measure, which is stationary, is not the typicality measure while the uniform measure is, even if it's not stationary?) is resolved. The uniform measure is selected as the typicality measure because it is the stationary measure for the universe, and whatever is stationary in a subsystem does not really matter. Bricmont's question ('why one needs to look for an additional justification for Hamiltonian systems based on stationarity if one already accepts the Bayesian justification for non-Hamiltonian systems?') seems compelling only if one considers non-Hamiltonian systems to be more general than Hamiltonian ones. However, in this context, this is not the case, and the logic is reversed. First, one looks at the universe (which is described by a Hamiltonian) to find the typicality measure using stationarity (and this is the uniform measure), and then at its subsystems (which are not Hamiltonian). When dealing with them, stationarity is no longer relevant and the typicality measure is the uniformity measure because it has been inherited by the one for the universe.

Now, let us move on to the second objection that the stationarity argument is too sophisticated to genuinely capture what we mean by explanation. In response, one could say that the ingredients of a scientific explanation need not to be familiar to us to make the theory an adequate account of scientific explanation. One judges a theory of explanation to be adequate if it is able to reproduce our intuitions regarding which truly are explanations and which are not, not necessarily using notions that we are already familiar with. For instance, as we have seen, in the DN model of explanation, an explanation is a valid deductive argument in which (at least) one of the premises is a law of nature. The adequacy of this model is judged by considering whether every explanation as given from the model is also intuitively an explanation, and *vice versa*. Indeed, one of the counterexamples of the model points to asymmetry in the account contradicts our intuition. One, in fact, can derive the length S of the shadow cast by a flagpole from the height H of the pole, the angle θ of the sun above the horizon and laws about the rectilinear propagation of light. This derivation is thus an explanation according to the deductive nomological model, and that seems right. However, the 'backward' derivation of H from S and θ, which is also an explanation according to the model, intuitively does not seem explanatory.

While it makes sense to say that the shadow of a flagpole being
a particular length is explained by the flagpole having a particular
height, we do not explain the flagpole having a certain height in
terms of the shadow being of a particular length. Rather, the flagpole has that height because it was constructed that way, for other
reasons. The model of explanation has, therefore, to pass the test of
intuition. Nonetheless, this is a test for the *outcome* of the model,
namely what counts as an explanation, rather than for the *ingredients* used to arrive to the model's outcome. In this case, one may
notice, also the ingredients used in the model to generate explanations are familiar: deductive arguments and laws of nature. However,
it does not seem to me that using other unfamiliar notions would
be problematic, unless they appear in the explanation itself. In fact,
consider Boltzmann's account of thermodynamics. We have said that
the second law, say, is explained by claiming that it can be shown that
'entropy-increasing behaviour is typical.' The sentence within quotes
is the explanation. The notions in it are 'entropy' and 'typicality,'
which have an intuitive meaning that is qualified and made precise
in the process of working out the explanation. Moreover, even if the
notions involved to derive such an explanation are far from being
intuitive, they are built bottom-up from intuitive notions into sophisticated mathematical notions. Think of the notions of macrostate,
microstate, measure, and so on. Notice that stationarity also is like
that. It is a notion which is far from intuitions and which, however,
is connected with the intuitive idea that no temporal instant should
be privileged. The explanation of the choice of the typicality measure
uses intuitive notions, through its implementation with the sophisticated notion of stationarity, to explain physical phenomena. In other
words, it is this intuition that provides the reason why stationarity enters the explanatory machinery. As Einstein (1936) said: "the
whole of science is nothing more than a refinement of our everyday
thinking". However, this refinement can lead us far from intuition
without losing its legitimacy. Be that as it may, let me conclude that,
interestingly, Bricmont agrees that 'entropy-increasing behaviour is
typical' is an explanation of the phenomena, even if it includes the
notions above (which are built from intuition into mathematical

notions without being themselves intuitive). However, he maintains that stationarity is not explanatory because it is not a notion we commonly use. Nonetheless it seems inconsistent to complain about stationarity being 'counterintuitive' and thus not explanatory, if one agrees that typicality is 'counterintuitive' but explanatory.

As far as Bricmont's own view (Bricmont (2019)), he thinks that the best way to justify the uniform measure is in a Bayesian framework. In this account, the probabilities used in statistical mechanics are seen as epistemic. They express our ignorance and are used to update one's probabilities estimate when new information becomes available. In this way, the uniform measure is taken to be a generalization of the principle of indifference, according to which one should not introduce any bias or information that one does not have. Since this account resembles the accounts that Albert criticized in his book, there may be problems as to how is it that our ignorance can *explain* the behavior of objects. In particular, in such a Bayesian account, two people may disagree about what counts as 'vast majority', and thus they may disagree about what counts as typical. Since typicality is used to explain, they may arrive to different explanation for the same physical phenomena (or may fail to explain some phenomena), and this is far from being desirable because we want explanations to be objective. Perhaps, a response to this kind of arguments would be that explanation based on typicality only requires 'coarse-grained' constraints (such as the number of non-thermodynamic states is overwhelmingly smaller than the number of thermodynamic states), and therefore no two people may actually disagree on what is typical and what is not. Otherwise, one may want to link this notion of entropy to the notion of rationality, which is what Bricmont (2019) suggests. However, this account seems to share very similar objections as the epistemic view, as well as new ones: what is rationality? More discussion on this is needed, but here I will merely recall that the debate over rationality and rational decision making in Everettian mechanics, even if the context is different, is still wide open.[30]

---

[30]See Wallace (2012) and references therein.

Finally, let me discuss the objection that it is unclear why the measure needs to be generic under the dynamics. Here is a possible answer. While it is true there is a unique true Hamiltonian H when looking at subsystems within the universe, this may be different. One can use some effective Hamiltonian H_{eff} (if the subsystem of the universe is still Hamiltonian), which is an approximation of the true one (H) for that subsystem under consideration. If we use H to find the typicality measure, we will find many stationary ones; likewise, if we use H_{eff}, we find another bunch of stationary measures. Which one should we choose? Since both H and H_{eff} should give rise to the same empirical results (otherwise we have done a bad approximation), then one should require that the stationary measure they find be the same. In addition, one could argue[1] that in order to provide a genuine explanation one would have to require genericity: it is not the specific form of the Hamiltonian that the fact is explained; rather it is explained independently of what kind of Hamiltonian we have. Moreover, and perhaps more importantly, notice that the request for genericity is what guarantees that the dynamics is 'irrelevant' in the sense discussed in Section 4.2: the phenomena are explained for typical initial conditions and for all (rather than for typical) Hamiltonians. That is, the genericity of the Hamiltonian is what guarantees that explanation is so general that the details of the dynamics do not matter at all.

3.4.3. *Boltzmann's ingredients in the typicality account*

To conclude this section, let me summarize the situation. We started from the characterization of Boltzmann's explanation of macroscopic laws in terms of the classical dynamical laws, the statistical postulate and the past hypothesis. Some reflections on the statistical postulate lead us to the conclusions that:

1) Typicality (not necessarily probability) is enough to explain;
2) The correct typicality measure (namely the one that proves empirically adequate) may be inferred from the dynamical laws using symmetries considerations, and therefore not postulated.

If the arguments presented here are sound, one could conclude that the statistical postulate is not needed because the typicality measure is suitably derivable from the dynamics. As a consequence, the ingredients of Boltzmann's explanation are now reduced to the following:

1) The laws of motion (which determine the typicality measure on phase space);
2) The past hypothesis.[31]

3.5. Quantum statistical mechanics

Now that we have discussed Boltzmann's account in the classical domain, let's discuss about its possible generalizations. If one wishes to generalize Boltzmann's explanatory schema to the quantum domain, *prima facie*, one should not expect some fundamental differences, especially in the case of deterministic theories such as the pilot-wave theory.[32] However, the spontaneous localization theory provides something new, namely intrinsic stochasticity. In fact, this is a theory in which the wave-function does not always evolve according to the Schrödinger equations. Rather, it does for some time, then at a random time, the wave-function 'collapses' into a random localization, then it continues to evolve according to the Schrödinger dynamics, and so on. Albert (2001) argues that this theory can provide a *dynamical* explanation for the statistical postulate. If so, the statistical mechanical probabilities would just be the quantum mechanical probabilities; that is, Albert argues that one can dynamically derive the statistical postulate if the spontaneous localization theory is true. In the last section I also argued that one can dynamically derive the statistical postulate, even if the arguments are very different. In this section, I plan to show that this approach can be extended to all theories, including indeterministic quantum ones.

---

[31]Interestingly, but from a very different perspective, Eddy K. Chen (2019) has recently argued that one can dispense of the statistical postulate in a quantum extension of statistical mechanics by assuming that the ontology is given by the density matrix rather than by the wave-function.

[32]See Goldstein *et al.* (2019).

Before entering into this, let us present Albert's argument for his thesis.

Since the overwhelming majority of microstates is thermodynamically normal (that is, entropy-increasing), they are stable. In fact, everything close to them is also likely to be normal. In contrast, abnormal microstates (that is, entropy-decreasing) are very unstable, since they are surrounded almost always by normal microstates. Because of this, any abnormal system is extremely close to being in a normal state. Albert's idea is that the effect of a wave-function collapse like those happening in the spontaneous localization theory, with overwhelming likelihood, will keep a normal microstate normal, and will make an abnormal microstate 'jump' into a normal one. In the theory, to technically implement the collapse, the wave-function is multiplied by a Gaussian, which effectively restricts the support of the wave-function to a random and very small region of space. In this sense, there is a set of random and small macrostates (the regions after the collapse) to which the wave-function can go to at any time, each of which with its own probability distribution, is given by the quantum rules. That is, there is *automatically* a probability distribution on each macrostate, *in contrast* with deterministic theories, where one has to add it by hand. The region after the collapse is, by construction, smaller than any region possibly representing a macrostate. However, the size of the set of abnormal states is much smaller, so one could still say that the vast majority of microstates in the collapsed region will go toward a higher entropy state, in agreement with the statistical mechanical predictions. Therefore, Albert claims that the spontaneous collapse theory can do away with the statistical postulate. It is the dynamics itself, being open toward the future and assigning a probability to each possible collapsed region, that fills in the gap between the microstate number talk and the probability talk.

3.5.1. *The indeterministic case*

Let's now explore how the approach described in this paper may translate into the quantum domain and to indeterministic theories.

Given the conclusions drawn in Section 4, namely that the statistical postulate is not needed and the relevant explanatory notion is derivable from the dynamics, one naturally wonders whether this extends to indeterministic theories as well. In Section 4, we considered Boltzmann's explanation when the fundamental theory of the world is deterministic. As already discussed, it seems that in an indeterministic theory, the dynamical laws are time-directed in the sense that while the past is determined the future is open, and each possible future has its own probabilities of happening. Albert argued that this is likely to help, getting rid of the statistical postulate. However, I argue, following Zanghì's suggestions (2005), that the differences are not as striking as one may think and do not majorly affect the structure of Boltzmann's explanatory schema (with qualifications).

Indeterministic theories are notoriously difficult to get a grip on, so one may use the simple model proposed by Ehrenfest (Baldovin *et al.* (2019), Ehrenfest (2015)). This model can be taken as the prototypical example of indeterministic dynamics, and can be used to study the differences and the similarities between deterministic and indeterministic theories (Zanghì 2005). Suppose there are N numbered balls in two boxes, one on the right and on the left. The Ehrenfest process is a discrete-time process which describes the indeterministic dynamics defined by a series of random jumps of the balls from one box to the other, one at a time. To describe this dynamics, somewhat oddly but ultimately usefully, one can think in terms of macrostates and microstates also in this context. The microstate at a given time is the list of ball locations in one or the other box at that time. That is, $X = (a_1, \ldots, a_N)$, such that the i-th entry is 0 if the i-th ball is in the right box and is 1 if it is in the left one. At another time, take a ball at random from a box and put it in the other (or, let one ball at random 'jump' from one box to the other). Consequently, the microstate changes. For instance, assuming for simplicity, there are only two balls, there are only four possible initial states: (1,1), both balls on the left; (0,0), both balls on the right; (1,0) and (0,1), respectively, the first ball on the left and the second on the right, and the other way around. If the actual initial state is, say, (1,1), then the system can evolve at time $t = 1$ into (1,0) or (0,1) with probability

$\frac{1}{2}$ each. Assuming the state at $t = 1$ is (1,0), then, it can evolve
at time $t = 2$ into either (0,0) or (1,1), again with $\frac{1}{2}$ probability
each. And so on. Now drop the simplification that there are just two
balls. If at the beginning, all balls are in the left container, that is
$X_0 = (1,1,1,1\ldots)$, it can be shown that, as the number of jumps
increases, the number of balls contained in each box will tend to be
the same. If so, an example of microstate after a sufficiently long
time T is $X_T = (1,0,\ldots,0,1\ldots)$, with an equal number of randomly
distributed zeros and ones. This is so for combinatorial reasons, given
that there are many more ways in which N balls can be distributed
half in the left and half in the right box than any other distributions.
Thereby, one can define an 'equilibrium' macrostate as the state of
all the microstates in which the balls are half in one box and half in
the other. We can call it equilibrium because after this configuration
is reached, the systems will tend to stay there, just like in statistical
mechanics.

Can we write a law for this behavior? A first possibility is to
describe the law in pure probabilistic terms. If the system is at a
given initial time $t = 0$ in the state X_0; at time t, the system will
have certain probability (namely $1/2^N$) to evolve in one of the 2^N
possible states. In this picture, the past is fixed and the future open,
and accordingly there is a fundamental difference between indeterministic and deterministic dynamics, in which both the past and the
future are fixed. Nevertheless, Zanghì points out, there is another
mathematically equivalent way of reformulating the law, avoiding
the postulation of a fundamental difference between past and future.
Let us start from the microstate X_0 at $t = 0$. At time $t = 1$, there
are N possible states; at $t = 2$, the different states are N^2; and so
on, so that at $t = n$ they are N^n. In other words, from the initial
microstate X_0, which describes how the balls are arranged in the two
boxes, the microstate, namely the balls distribution, evolves in time
as described by a branching graph (See Figure 3.1).

In a deterministic dynamics, one microstate evolves in time into
another microstate and therefore the complete history of the world is
a single curve in phase space. In contrast, in an indeterministic theory
described in this way, given that multiple states become available to

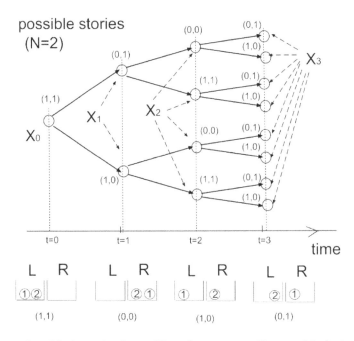

Fig. 3.1. Possible histories in an Ehrenfest process. For graphical simplicity, I have assumed that there are only two balls ($N = 2$).

the incoming microstates, the history of the world involves a continuous ramification in phase space. Accordingly, the set of all the histories of the world for all the 2^N possible initial microstates is the union Ω of the graphs corresponding to the different initial microstates, which becomes the space of all the possible state of affairs of the world in the case of an indeterministic theory. This branching scenario is a pictorial representation of what having multiple future means. However, even if there are multiple future available, the system will invariably evolve into only one of them. That is, if we assume a timeless view and look at the space of possible histories, there will *always* be single histories, namely single sequences of states at different times; one sequence for each path the system at one time will have taken at a later time.

Now, let us go back to Boltzmann's explanation. One of the ingredients was the typicality measure, which is a measure in phase space, given that phase space represents the space of all the possible histories of the world. However, in this context, the space of the possible histories of the world is Ω, namely the union of the graphs corresponding to each possible initial microstate. Accordingly then, the typicality measure will have to be taken on such space and thus applied to all the possible histories. If the stationary account of typicality is correct, then the indeterministic dynamics will determine as typicality measure the one that which is stationary on the space of possible histories. Namely, the one that does not privilege any temporal instant.

One can define the notion of time-translation invariance in this context as follows. Assuming that A is a set of possible histories and that A_{-t} is the set of these histories translated back in time, then the measure is time-translation invariance if it assigns the same size to both sets: $\mu(A) = \mu(A_{-t})$. In fact, since there is no privileged time, a possible history is a sequence of instantaneous states with both endings open, of the form: $X = [\ldots, X_{-1}, X_0, X_1, \ldots]$, which is infinite on both the right and the left hand sides. Translating this sequence back in time of, say, a single time unit means considering another way the world can be, one in which the state at time $t = 1$ is identical with the state at time $t = 0$ of the original sequence, namely X_0; the state at time $t = 2$ is identical with the state at time $t = 1$ of the original sequence, namely X_1; and so on.[33]

How does this apply to the spontaneous collapse theory? The situation is more complicated because there are not only two states to jump between, but infinitely many, since the positions in which the wave-function can be localized after a collapse form a continuum.[34] Moreover, the accessible states are constrained by the quantum probability rules rather than being accessible with the same probability. However, it seems that given the initial wave-function, the set of possible accessible states at a later time will still be a

---

[33]See also Bedingham and Maroney (2017), and Allori (2019) for a similar view to define time reversal for indeterminist theories.

[34]I am leaving aside considerations about the so-called primitive ontology approaches to GRW theory (see Allori, Goldstien Tumulka and Zanghì (2008)) but a generalization seems straightforward.

graph. Similarly, therefore, the space of possible microstates is the union of these graphs, one for each initial possible microstate. Hence, the typicality measure is the stationary, generic measure on this space as defined by the dynamics.

If this is so, there are more similarities than differences between deterministic and indeterministic theories. In deterministic theories, given an initial state, the dynamical laws determine the typicality measure on the space of possible histories of the world, namely phase space. And in indeterministic theories, given an initial state, the dynamical laws determine the typicality measure on the space of possible histories, namely Ω. Thus, in both deterministic and indeterministic cases, given an initial condition and the dynamics, 'everything follows'. In other words, the set of possible states and the typicality measure are specified by the dynamics.

3.6. Conclusions

Let us finally go back to the originally discussed Boltzmann's account of macroscopic regularities. According to Albert and Loewer's view discussed here, three ingredients are needed:

1. The laws of motion,
2. The past hypothesis,
3. The statistical postulate.

From this, one can construct an argument that the spontaneous localization theory does not need to introduce the statistical postulate, for probability appears once and, so to speak, it is 'in the right place' within the dynamics. Thus, if the spontaneous localization theory were true, one would need two, rather than three, ingredients and on this basis, such a theory should be preferred over the deterministic alternatives.

In this paper, I have done the following:

1) I have shown that, building on the work of Goldstein and collaborators, assuming there is a statistical postulate that specifies how to count states, it is not necessary for it to be about probabilities. In fact the notion of typicality is enough.

2) I have also shown that the statistical postulate, now understood as a typicality postulate, can be derived from the dynamics under symmetry constraints (stationarity and genericity).
3) Finally, I have shown, following Zanghì, that both indeterministic and deterministic theory can ground a typicality-based understanding of macroscopic phenomena. In fact, when appropriately reformulated, they both require the same two ingredients:

   a. The specification of the dynamical laws, which determine the space of possible histories as well as the typicality measure;
   b. The past hypothesis.

If that is so, then both indeterministic and deterministic theories have basically the same relevant structure, and therefore there cannot be any fundamental difference in the satisfaction of explanation of macroscopic laws.

Acknowledgements

I am particularly grateful to Jean Bricmont and Barry Loewer for their comments on an early version of this paper. Also, I thank the participants of the 11th MUST (Models of Explanation) conference in Turin, Italy (June 11–13, 2018), as well as the participants of the Rutgers Philosophy of Probability Workshop (October 24–26, 2019), in particular David Albert, Katie Elliot, Carl Hoefer, Tim Maudlin, Wayne Myrvold and Issac Wilhelm.

References

Albert, D, Z. (2001) *Time and Chance*, Harvard University Press, Cambridge.

Allori, V. (2019) Quantum Mechanics, Time and Ontology, *Studies in History and Philosophy of Modern Physics*, **66**, 145–154.

Allori, V., Goldstein, S., Tumulka, R., Zanghì, N. (2008) On the Common Structure of Bohmian Mechanics and the Ghirardi-Rimini-Weber Theory, *The British Journal for the Philosophy of Science* **59**(3), 353–389.

Baldovin, M., Caprini, L., Vulpiani. A. (2019) Irreversibility and Typicality: A Simple Analytical Result for the Ehrenfest Model, *Physica A*, **524**, 422–429.

Bedingham, D. J., Maroney, O.J.E. (2017) Time Reversal Symmetry and Collapse Models, *Foundations of Physics*, **47**(5), 670–696.

Bricmont, J. (1995) Science of Chaos or Chaos in Science? *Annals of the New York Academy of Sciences*, **775**(1), 131–175.

Bricmont, J. (2001) Bayes, Boltzmann and Bohm: Probabilities in Physics. In: J. Bricmont, G.C. Ghirardi, D. Dürr, F. Petruccione, M.C. Galavotti, and N. Zanghì (eds) *Chance in Physics: Foundations and Perspectives. Lecture Notes in Physics*, **574**, 3–21. Berlin: Springer.

Bricmont, J. (2019) Probabilistic Explanations and the Derivation of Macroscopic Laws. This volume.

Callender, C. (2004) There is No Puzzle about the Low Entropy Past. In: C. Hitchcock (ed.). *Contemporary Debates in Philosophy of Science*, 240–255. London: Blackwell.

Carroll, S., Chen, J. (2005) Spontaneous Inflation and the Origin of the Arrow of Time, *General Relativity and Gravitation*, **37**, 1671–1674.

Chen, E. K. (2019) Time's Arrow in a Quantum Universe: On the Status of the Statistical Mechanical Probabilities. This volume.

Crane, H., Wilhelm, I. (2019) The Logic of Typicality. This volume.

Dürr, D. (2009) *Bohmian Mechanics*. Berlin: Springer.

Ehrenfest, P., Ehrenfest, T. (2015) *The Conceptual Foundations of the Statistical Approach in Mechanics*. Dover Publications, New York, reprint edition.

Einstein, A. (1936) Physics and Reality, *Journal of the Franklin Institute*, **221**, 349–382.

Frigg, R. (2009) Typicality and the Approach to Equilibrium in Boltzmannian Statistical Mechanics, *Philosophy of Science*, **76**, 997–1008.

Frigg, R. (2011) Why Typicality Does Not Explain the Approach to Equilibrium. In: M. Suárez (ed*.): Probabilities, Causes and Propensities in Physics*, 77–93. Dordrecht: Springer.

Frigg, R., Werndl, C. (2011) Explaining Thermodynamic-like Behaviour in terms of Epsilon-ergodicity, *Philosophy of Science*, **78**, 628–652.

Frigg, R., Werndl, C. (2012) Demystifying Typicality, *Philosophy of Science*, **5**, 917–929.

Ghirardi, G. C., Rimini, A., Weber, T. (1986) Unified Dynamics for Microscopic and Macroscopic Systems, *Physical Review D*, **34**, 470.

Goldstein, S. (2001) Boltzmann's Approach to Statistical Mechanics. In: J. Bricmont, G.C. Ghirardi, D. Dürr, F. Petruccione, M.C. Galavotti, and N. Zanghì (eds) *Chance in Physics: Foundations and Perspectives. Lecture Notes in Physics*, **574**, 39–54. Berlin: Springer.

Goldstein, S. (2011) Typicality and Notions of Probability in Physics. In: Y. Ben-Menahem, and M. Hemmo (eds) *Probability in Physics*, 59–71. Heidelberg: Springer, The Frontiers Collection.

Goldstein, S., Lebowitz, J. L., Tumulka, R., Zanghì, N. (2019) Gibbs and Boltzmann Entropy in Classical and Quantum Mechanics. This volume.

Guth, A. H. (1981) Inflationary Universe: A Possible Solution for the Horizon and Flatness Problems, *Physical Review D*, **23**, 347–356.

Guth, A.H., Steinhardt, P.J. (1984) The Inflationary Universe, *Scientific American*, 116–128.

Hemmo, M., Shenker, O.R. (2012) *The Road to Maxwell's Demon. Conceptual Foundations of Statistical Mechanics.* Cambridge: Cambridge University Press.

Hemmo, M., Shenker, O.R. (2019) Can the Past Hypothesis Explain the Psychological Arrow of Time? This volume.

Hempel, C. G. (1965) *Aspects of Scientific Explanations and Other Essays in the Philosophy of Science.* New York: The Free Press.

Lazarovici, D., Reichert, P. (2015) Typicality, Irreversibility and the Status of Macroscopic Laws, *Erkenntnis*, **80**(4), 689–716.

Lazarovici, D., and Reichert, P. (2019) Arrow(s) of Time without a Past Hypothesis. This volume.

Lebowitz, J. L. (1981) Microscopic Dynamics and Macroscopic Laws. Annals New York Academy of Sciences, 220–233.

Loewer, B. (2019) The Mentaculus Vision. This volume.

Maudlin, T. (2019) The Grammar of Typicality. This volume.

Myrvold, W. C. (2019) Explaining Thermodynamics: What Remains to be Done? This volume.

Olsen, R.A., Meacham, C. (2019) Eternal Worlds and the Best System Account of Laws. This volume.

Penrose, R. (1999) *The Emperor's New Mind.* Oxford: Oxford University Press.

Pitowsky, I. (2012) Typicality and the Role of the Lebesgue Measure in Statistical Mechanics. In: J. Ben-Menahem and M. Hemmo (eds.) *Probability in Physics*, 51–58. Berlin and New York: Springer.

Strevens, M. (2000) Do Large Probabilities Explain Better? *Philosophy of Science*, **67**, 366–90.

Uffink, J. (2007) Compendium of the Foundations of Classical Statistical Physics. In: J. Butterfield and J. Earman (eds.) *Handbook for the Philosophy of Physics*, 923–1047. Amsterdam: Elsevier.

Uffink, J. (2008) Boltzmann's Work in Statistical Physics. *The Stanford Encyclopedia of Philosophy.*, http://plato.stanford.edu

Valentini, A. (2019) Foundations of Statistical Mechanics and the Status of the Born Rule in the de Broglie-Bohm Pilot-wave theory. This volume.

Volchan, S. B. (2007) Probability as Typicality, *Studies in History and Philosophy of Modern Physics*, **38**, 801–814.

Wallace, D. M. (2012) *The Emergent Universe.* New York: Oxford University Press.

Wallace, D. M. (2019) The Necessity of Gibbsian Statistical Mechanics. This volume.

Werndl, C. (2013) Justifying Typicality Measures of Boltzmannian Statistical Mechanics and Dynamical Systems, *Studies in History and Philosophy of Modern Physics*, **44**(4), 470–479.

Werndl, C., Frigg, R. (2019) Taming Abundance: on the Relation between Boltzmannian and Gibbsian Statistical Mechanics. This volume.

Wilhelm, I. (2019) Typical: A Theory of Typicality and Typicality Explanation. *The British Journal for the Philosophy of Science.* https://doi.org/10.1093/bjps/axz016

Zanghì, N. (2005) I Fondamenti Concettuali dell'Approccio Statistico in Fisica. In: V. Allori, M. Dorato, F. Laudisa, and N. Zanghì (eds.) *La Natura Delle Cose. Introduzione ai Fundamenti e alla Filosofia della Fisica*, 139–228. Roma: Carocci editore.

Chapter 4

Explaining Thermodynamics: What Remains to be Done?

Wayne C. Myrvold

Department of Philosophy,
The University of Western Ontario
wmyrvold@uwo.ca

In this chapter, I urge a fresh look at the problem of explaining equilibration. The process of equilibration, I argue, is best seen not as part of the subject matter of thermodynamics, but as a presupposition of thermodynamics. Further, the relevant tension between the macroscopic phenomena of equilibration and the underlying microdynamics lies not in a tension between time-reversal invariance of the microdynamics and the temporal asymmetry of equilibration, but in a tension between preservation of distinguishability of states at the level of microphysics and the continual effacing of the past at the macroscopic level. This suggests an open systems approach, where the puzzling question is not the erasure of the past, but the question of how reliable prediction, given only macroscopic data, is ever possible at all. I suggest that the answer lies in an approach that has not been afforded sufficient attention in the philosophical literature, namely, one based on the temporal asymmetry of causal explanation.

4.1. Introduction

Early in the previous century, Josiah Willard Gibbs wrote,

> A very little study of the statistical properties of conservative systems of a finite number of degrees of freedom is sufficient to make it appear, more or less distinctly, that the general laws of thermodynamics are the limit towards

which the exact laws of such systems approximate, when
their number of degrees of freedom is indefinitely increased
(Gibbs, 1902, 166).

From the vantage point of the twenty-first century, Gibbs' optimism
may seem naïve, as the relation between thermodynamics and
statistical mechanics continues to be a topic of philosophical discussion, with no consensus in sight.

In this chapter, I will argue that Gibbs is right. The laws of thermodynamics have been satisfactorily explained on the basis of statistical mechanics and, indeed, the seeds of the explanation were already
present in Gibbs' work, as (by a not entirely surprising bit of good
fortune) the relevant parts of classical statistical mechanics can be
taken over with little ado into quantum statistical mechanics. The
explanation of the laws of thermodynamics consists of finding appropriate statistical mechanical analogues of thermodynamic concepts,
and deriving relations between them that approximate the laws of
thermodynamics for systems of many degrees of freedom.[1] Along
the way, something interesting and subtle happens to the temporal
irreversibility of the second law of thermodynamics. The statistical
analogue of the second law of thermodynamics is, unlike its thermodynamic counterpart, *not* temporally asymmetric; in place of temporal asymmetry is a distinct asymmetry related, but not identical,
to it. This will be made clear in section 4.3, below. This removes
what has been seen as the chief stumbling-block for the reduction
of thermodynamics to statistical mechanics, namely, the *prima facie*
tension between a temporally asymmetric second law of thermodynamics and time-reversal invariance of the underlying dynamics.

This doesn't mean that we're out of the woods, though. There
still remains an important task, not yet fully accomplished, the task
of explaining the process of equilibration in statistical mechanical

---

[1] I hope it goes without saying that the project of investigating behaviour asymptotically approached by systems with a finite number of degrees of freedom as
the number of degrees of freedom is increased indefinitely is a distinct project
from that of investigating an idealized system with infinitely many degrees of
freedom, though, with care, the latter project might be informative about the
former.

terms. This involves both an explanation of the general tendency for systems out of equilibrium to relax to an equilibrium state unless maintained in a non-equilibrium state by an external influence, and an explanation of the paths to equilibrium and the rates at which those paths are traversed. This task has, of course, been regarded as part of thermodynamics, and a principle of equilibration has been counted by some as a law of thermodynamics (Uhlenbeck and Ford, 1963; Brown and Uffink, 2001). Compared to the other laws of thermodynamics, a law to the effect that, left to themselves, systems tend to relax to equilibrium, is an outlier. Though long recognized as an important principle, it was a late entry to the list of laws of thermodynamics: it was not referred to as a law of thermodynamics until the 1960s, more than a century after Kelvin initiated talk of laws, or fundamental principles, of thermodynamics. There is also an important conceptual distinction between the equilibration principle and the other laws of thermodynamics. It is unique among the so-called laws of thermodynamics in that its formulation does not require a distinction between energy transfer as heat and energy transfer as work. It is also the chief locus of temporal asymmetry.

Though, of course, this is ultimately a matter of choice of terminology and nothing more, it seems to me that it is helpful to highlight the differences between the equilibration principle and the more traditional laws of thermodynamics by restricting the scope of "thermodynamics" to something like what its founders intended, and to regard the equilibration principle not as belonging to thermodynamics proper, but as a *presupposition* thereof. A payoff in conceptual clarity of this choice is that it will perhaps mitigate somewhat the tendency to conflate the equilibration principle with the second law.

4.2. What is thermodynamics?

In contemporary physical parlance, to speak of the *dynamics* of a system is to speak of the laws according to which its state evolves over time. This has given rise to a folk etymology for the term "thermodynamics," according to which thermodynamics should mean the dynamical laws governing heat transfer. This folk etymology is incorrect. Understanding the actual etymology of the term is not merely a matter of historical interest. Contrary to the impression that the folk etymology would give, thermodynamics is aptly named, as the term stems from, and highlights, the two concepts that are at the core of the subject.

The term *thermodynamics* is composed from the Greek words for *heat* and *power*. It refers to the study of the ways in which heat can be used to generate mechanical action and in which heat can be generated via mechanical means. The word's first appearance is in Part VI of Kelvin's "On the Dynamical Theory of Heat" (Thomson 1857, read before the Royal Society of Edinburgh on May 1, 1854). There, he recapitulates what in Part I (read March 17, 1851; published in 1853) he had called the "Fundamental Principles in the Theory of the Motive Power of Heat", now re-labelled "Fundamental Principles of General Thermo-dynamics." The context makes clear that the new term is intended to denote the study of the relations between mechanical action and heat. It is worth noting that the term "thermodynamics" had not been used in connection with the earlier investigations, by Fourier, Kelvin, and others, of the laws of heat transport.

The term "thermodynamics" flags the distinction that is at the heart of the subject, the distinction between two modes of energy transfer: as work, and as heat. This is not a distinction that belongs to fundamental physics. Pure mechanics, whether classical or quantum, employs the concept of energy, but not this distinction between ways in which energy can be transferred from one system to another.

The two laws identified by Kelvin as laws of thermodynamics crucially invoke the heat-work distinction, and cannot be formulated without it. The first law states that total energy is conserved, whether transferred as heat or work; the net change in the internal energy of a system during any process is the net result of all exchanges of energy as work or as heat. The second law, in any of its formulations, also invokes the distinction. This is obvious in the Clausius and Kelvin formulations.

> Heat can never pass from a colder to a warmer body without some other change, connected therewith, occurring at the same time (Clausius 1856, 86, from Clausius 1854, 488).

> *It is impossible, by means of inanimate material agency,*
> *to derive mechanical effect from any portion of matter by*
> *cooling it below the temperature of the coldest of the sur-*
> *rounding objects* (Thomson 1853, 179; reprinted in Thom-
> son 1882, 265).

There is also an entropy formulation of the second law, which says that, in any process, the total entropy of all systems involved in the process does not decrease. At first glance, this might seem not to depend on the distinction between energy transfer as work and energy transfer as heat. But recall the definition of thermodynamic entropy. The entropy difference between two thermodynamic states of a system is calculated by considering some thermodynamically reversible process that links the two states and the associated heat exchanges between the system and the external world. The entropy difference between the two states of the system is the integral of $đQ/T$ over any such process (which cannot depend on which reversible process is chosen for consideration, on pain of violation of the second law in Clausius or Kelvin form).

In addition to the two laws of thermodynamics identified as such by Kelvin, there is another, more basic law, called the *zeroth law*, on which the definition of temperature depends.[2] The zeroth law has to do with the behaviour of systems when brought into thermal contact. Thermal contact (taken as a primitive notion) between two systems is a condition under which heat may flow between the systems. Under conditions of thermal contact, heat may flow from one system to the other or else there may be no heat flow, in which case the systems are said to be in thermal equilibrium with each other. This induces a relation between states of systems that can be put into thermal

---

[2]The phrase occurs in Fowler and Guggenheim (1939, 56), and became a textbook staple in the years that followed. This was not, however, its first occurrence. A few years earlier, in a note on terminology, Charles Galton Darwin (1936) mentions the "*zeroth* law of thermodynamics" in a way that suggests that he expects it to be familiar to his readers; it is brought up only to illustrate the use of the word "zeroth", in Darwin's estimation a "terrible hybrid." Sommerfeld (1956, 1) attributes the coinage to Fowler, when giving an account of Saha and Srivastava (1931, 1935).

contact with each other, which holds between the states (whether or
not the bodies are actually in thermal contact) if there would be no
heat flow if the systems were brought into thermal contact. This is
obviously a symmetric relation. If we take a system in equilibrium to
be in thermal contact with itself, it is also a reflexive relation. It is,
therefore, an equivalence relation if and only if it is transitive. The
zeroth law states that the relation is, indeed, transitive. If the zeroth
law holds, we can partition thermodynamic states of systems into
equivalence classes under this relation, which we may regard as the
relation of being of the same temperature. The zeroth law requires
the notion of thermal contact, and hence the notion of heat flow, and
thus, like the first and second laws, requires the heat-work distinction
for its formulation.

The three laws have been recognized as laws of thermodynamics,
and presented as such, in textbooks since the 1930s. There is also a
late-comer, more basic than the rest. This is what Brown and Uffink
(2001) call the *Equilibrium Principle* (though perhaps *Equilibration
Principle* would be better):

> An isolated system in an arbitrary initial state within a
> finite fixed volume will spontaneously attain a unique state
> of equilibrium (Brown and Uffink, 2001, 528).

To signal that this is more fundamental than the traditional zeroth,
first, and second laws, Brown and Uffink label this the *minus first
law*. They point out that a principle of this sort had been recognized
as a law of thermodynamics earlier, by Uhlenbeck and Ford (1963,
5).[3]

The equilibration principle has often been conflated with the second law. The two are distinct, however. It is a consequence of the
second law that, *if* an isolated system makes a transition from one
thermodynamic state to another, the entropy of the final state will
not be higher than that of the initial state. This does not entail that
its behaviour will be that dictated by the equilibration principle. As

---

[3]Uhlenbeck and Ford called it the *zeroth law* to emphasize its priority over the
first and second laws. We will continue to follow standard terminology in taking
transitivity of thermal equilibrium to be the zeroth law.

far as the second law is concerned, there might be some set of distinct thermodynamic states of the same entropy that the system cycles through. Or else it might fail to equilibrate when isolated, remaining in some quasistable nonequilibrium state until some external disturbance triggers a slide towards equilibrium.

The equilibration principle is unlike the zeroth, first, and second laws. The others have to do with exchanges of energy between systems, and rely on a distinction between energy exchange as heat and energy exchange as work. The equilibration principle, on the other hand, has to do with the spontaneous behaviour of an isolated system and, *ipso facto*, has nothing at all to do with energy exchange, in any mode.

Should we regard the equilibration principle as a law of thermodynamics? A case can be made for not doing so, as refraining from doing so provides a neat separation of two sorts of theoretical investigations.

There is a tradition, which can be traced back to Maxwell[4] and which has undergone renewed interest in recent years, of thinking thermodynamics as a *resource theory*.[5] That is, it is a theory that investigates how agents with specified powers of manipulation and specified information about physical systems can best use these to accomplish certain tasks. A theory of this sort involves physics, of course, because it is physics that tells us what the effects of specified operations will be. But not *only* physics; considerations not contained in the physics proper, having to do with specification of which operations are to be permitted to the agents, are brought to bear. On such a view, the work-heat distinction rests on a distinction between variables that the agent can keep track of and manipulate, and variables that are not amenable to such treatment. If we take this view, we would not expect to capture concepts such as heat, work, and entropy within physics proper, and would not expect there to be

---

[4]See Myrvold (2011).
[5]See del Rio *et al.* (2015) for a general framework for resource theories, Wallace (2016) and Bartolotta *et al.* (2016) for some recent work, and Goura *et al.* (2015) for a review.

statistical mechanical analogues of these concepts definable in purely physical terms.

The equilibration principle is in a different category. Though it too requires for its formulation a distinction not found in the fundamental physics — a distinction between macrovariables, used to define the thermodynamic state, and microvariables required to specify the complete physical state of a system — it doesn't require the heat-work distinction. Someone who (like Maxwell) holds that the distinction between heat and work vanishes when all limitations on knowledge and manipulation are removed would expect the zeroth, first, and second laws to make sense only in the presence of such limitations. The same cannot be said of the equilibration principle.

If we take the word *thermodynamics* in its originally intended sense, as the science of heat and work, then the equilibration principle is not a law of thermodynamics. Instead, since it delivers the equilibrium states with which thermodynamics deals, it is a *presupposition* of all thermodynamics. Though this is merely a terminological issue, regarding the scope of the term *thermodynamics*, there is something to be said for flagging the relation between the equilibration principle and the traditional laws of thermodynamics by saying that only the latter, not the former, comprise thermodynamics proper. And if we take the scope of thermodynamics to comprise only the zeroth, first, and second laws, then the task of explaining thermodynamics in statistical mechanical terms is a much less daunting one. It is of this more modest goal that Gibbs speaks in the quotation with which we started.

This, of course, leaves us with the deep, important, and interesting task of explaining equilibration. This is a matter to which much valuable and interesting work has been, and continues to be, devoted. The literature on equilibration deserves more attention from philosophers than it has heretofore received. Seeing equilibration as not a matter of thermodynamics but, rather, a process that thermodynamics presupposes, helps one to view this work more clearly, as the enterprise of studying equilibration is thereby freed of extraneous thermodynamic concepts (and in particular, of the concept of entropy).

4.3. Explaining thermodynamics in terms of statistical mechanics

Statistical mechanics deals with systems composed of a large number of interacting subsystems, and examines their aggregate behaviour, eschewing a detailed description of the microstates of the system. In the decade from 1867 to 1877, the major figures working to lay the foundations of what Gibbs would later call *statistical mechanics* — Maxwell, Kelvin, and others in Britain, Gibbs in the U.S., and Boltzmann on the continent — came to realize that what was to be recovered from statistical mechanics was not the laws of thermodynamics as originally conceived, but a modified version on which what the original version of the second law declares to be impossible should be regarded as possible but, when dealing with things on the macroscopic scale, highly improbable. Because of random fluctuations of molecules, on a given run, a heat engine operating between two reservoirs might yield more work than the Carnot limit on efficiency permits but, by the same token, it might also yield less. What we can expect from statistical mechanics is some statement to the effect that we can't *predictably and reliably* exceed the Carnot efficiency.

This suggests that we will have to traffic in probabilities and consider, in the classical context, properties of probability distributions on the phase spaces of classical systems and, in the quantum context, of density operators.

Construing a macroscopic system as composed of a large number of molecules also motivates a reconsideration of the notion of an equilibrium state. Though systems may settle into a state in which macroscopically measurable quantities are not changing perceptibly, this state cannot be a state of quiescence at the microphysical level. At the level of individual molecules, a system in thermodynamic equilibrium is seething with activity, and the observed macroscopic repose is the net result of averaging over large numbers of rapidly changing microphysical parameters. There is, of course, no sharp line between macroscopic and microscopic and, at the mesoscopic scale (illustrated by Brownian motion), the state into which a system settles is one in which some measurable parameters are continually fluctuating, with a stable pattern of fluctuations. Here, again, probabilistic considerations come into play; we want to say that large fluctuations of macroscopic parameters during a time scale typical of our observations are not impossible, but merely improbable. Considerations such as this suggest that, in statistical mechanics, what we should associate with the condition of equilibrium is not an unchanging state, but a stable probability distribution. The process of equilibration will then be one in which non-equilibrium probability distributions over initial conditions converge (in an appropriate sense) towards the equilibrium distribution.

We also need, in order to apply thermodynamic concepts to statistical mechanical systems, a way of partitioning energy exchange between systems into heat exchange and work. The standard way to do this is to treat certain parameters (think of, for example, the position of a piston) as exogenously given, not treated dynamically and not subject to probabilistic uncertainty. The Hamiltonian of a system may depend on such parameters, and so a change in an exogenous parameter can result in a change in the energy of the system. Change to the energy of a system due to changes in the exogenous parameters on which its Hamiltonian depends are to be counted as work done on or by the system, and all other changes of energy counted as heat exchanges.

In thermodynamics, *thermal states* play an important role. A system that has thermalized (relaxed to thermal equilibrium) is one from which one can extract no energy as work; the only way to extract energy from it is via heat flow. In such a state, a system has a definite temperature, uniform throughout the system. When two thermalized systems are placed in thermal contact with each other, the expected heat flow is from the hotter to the colder. These things continue to be true if we consider a number of systems at the same temperature; an assembly of thermal systems at the same temperature is also a thermal system at that temperature. There are arguments (see Maroney 2007 for a lucid exposition) that these considerations lead to the conclusion that the appropriate probability distributions to associate with thermal states are what Gibbs named *canonical distributions*. In the classical case, the canonical distribution has density function (with respect to Liouville measure),[6]

$$\rho_\beta(x) = Z^{-1} e^{-\beta H(x)}, \tag{4.1}$$

where x ranges over phase-space points, H is the Hamiltonian of the system, and Z a normalization constant chosen to make the integral of ρ_β over the accessible region of phase space equal to unity. The parameter β indicates the temperature of the thermal state, and is inversely proportional to the absolute (Kelvin) temperature.

For a quantum system, a canonical state is represented by the density operator,

$$\hat{\rho} = Z^{-1} e^{-\beta \hat{H}}. \tag{4.2}$$

We want an analogue of the second law of thermodynamics. The second law of thermodynamics is equivalent to the statement that no heat engine can operate with an efficiency exceeding the Carnot efficiency. We can state this as follows.

> If a system undergoes a cyclic process, exchanging heats Q_i with thermal systems at temperatures T_i (and exchanging no heat with any other system), and returning to its original state, then

$$\sum_i \frac{Q_i}{T_i} \leq 0. \tag{4.3}$$

> Moreover, if the process is thermodynamically reversible, it can be run in the opposite direction, with heat exchanges $-Q_i$. This entails that, for such a process,

$$\sum_i \frac{Q_i}{T_i} = 0. \tag{4.4}$$

This, in turn, entails that, if two thermodynamic states a and b of the system can be connected by a reversible process, then the heat exchanges the system makes with thermal systems must be such

---

[6]We're assuming that the system is confined to a bounded region of phase space or otherwise subjected to conditions that render this a normalizable distribution.

that the sum of Q_i/T_i over all such exchanges is the same for any reversible process connecting the states. We can use this to define a state function S, the thermodynamic entropy, such that

$$S(b) - S(a) = \sum_i \frac{Q_i}{T_i}, \qquad (4.5)$$

where the sum may be taken over any thermodynamically reversible process taking state a to state b. This uniquely determines the entropy difference of two thermodynamic states, as long as they can be connected by a thermodynamically reversible process.

It turns out that there is something analogous in the form of a theorem about probability distributions, a theorem that is provable in two versions, classical and quantum.

Since we're dealing with multiple systems that may interact, we have to consider probability distributions over the state space of the composite system. Given a probability distribution P_{AB} over the state space of a composite system consisting of disjoint subsystems A and B, we can define marginal distributions P_A and P_B as the restrictions of P_{AB} to the degrees of freedom of A and B, respectively. Now, consider a system A that interacts with a number of thermal systems $B_i, i = 1, \ldots, n$, at temperatures T_i. Suppose that at time t_0 there is no interaction between the system A and the thermal systems B_i, and that the probability distribution of the joint system consisting of A and the thermal systems B_i is such that there are no correlations between A and the thermal systems. Suppose that, between t_0 and a time t_1, A interacts with the B_is successively, possibly exchanging energy with them. During this time, the energy of A may also change via manipulation of exogenous variables, but the only heat exchanges are with the thermal systems B_i. We also assume that at time t_1 there is no interaction between A and the thermal systems. Let us take the process to be a cyclic one, in the sense that the marginal probability distribution of A at t_1 is the same as at t_0. As the system A has interacted with the thermal systems B_i in the interim, we do not assume that at t_1 A is uncorrelated with them.

As we are dealing with probability distributions, the heats Q_i that the system A exchanges with the thermal systems B_i is not determined by the setup. Instead, there will be a probability distribution over the energy exchanges. We can consider the *expectation values* of these energy exchanges. It is easy to show that, provided that A is uncorrelated with the B_is at t_0 and has the same marginal at t_0 and t_1, the expectation values $\langle Q_i \rangle$ of heat exchanges satisfy

$$\sum_i \frac{\langle Q_i \rangle}{T_i} \leq 0. \tag{4.6}$$

This means that, even if fluctuations might yield, on an individual run, a greater yield of work than permitted by the Carnot limit on efficiency of heat engines, one cannot consistently and reliably violate the Carnot limit. Suppose I operate a heat engine between two heat reservoirs, a hot one of temperature T_1 and a cooler one of temperature T_2, and I operate the engine in a cycle, restoring the marginal distribution of the engine at the end of the cycle. Prior to the beginning of the cycle my engine is uncorrelated with the heat baths. Suppose my process is such that the expectation value of heat extracted from the hot bath is $\langle Q_1 \rangle$. Then, the expectation value of work obtained must satisfy

$$\langle W \rangle \leq \left(1 - \frac{T_2}{T_1} \right) \langle Q_1 \rangle. \tag{4.7}$$

If the cycle is a reversible one, that is, if it can be run in the opposite direction with the signs of the expectation values reversed, we must have equality in (4.6) and (4.7). For a reversible cycle, the expectation value of the work obtained is proportional to the expectation value of heat extracted, with the factor of proportionality equal to the Carnot efficiency of a heat engine operating between reservoirs at temperatures T_1 and T_2.

Equation (4.6) can be thought of as a statistical mechanical analogue of the second law of thermodynamics. It should be stressed that it is a *theorem* of statistical mechanics, which follows from the stated conditions on probability distributions, that at time t_0, the distributions of the systems B_i be canonical, with temperatures T_i, and uncorrelated with A.

In thermodynamics, the fact that relation (4.4) holds for *any* reversible process linking two thermodynamic states a, b permits us to define the entropy difference $S(b) - S(a)$, provided that there is at least one reversible process linking the two states, and this uniquely defines the state function S, up to an additive constant. Similarly, in statistical mechanics, the fact that (4.6) holds for arbitrary probability distributions P_A entails that there is a functional $S[P_A]$ that takes probability distributions over the state of A as input and yields real numbers, with the following properties. If, at t_0, A is uncorrelated with thermal systems B_i, whose probability distributions are given by canonical distributions at temperatures T_i, and if A exchanges heat in the interval between t_0 and t_1 with thermal systems B_i and with nothing else, then the expectation values of these heat exchanges, $\langle Q_i \rangle$, satisfy[7]

$$S[P_A(t_1)] - S[P_A(t_0)] \geq \sum_i \frac{\langle Q_i \rangle}{T_i}. \qquad (4.8)$$

Suppose now that there is a reversible process connecting $P_A(t_1)$ and $P_A(t_0)$. In this context, this means that there is a process taking $P_A(t_0)$ to $P_A(t_1)$, and another process starting at t_1 and ending at a time t_2 such that the marginal of A at t_0 is restored at t_2 (that is, $P_A(t_2) = P_A(t_0)$), such that

$$\sum_i \frac{\langle Q_i \rangle}{T_i} = 0, \qquad (4.9)$$

with the sum taken over the entire process. For a reversible process, we have equality.

$$S[P_A(t_1)] - S[P_A(t_0)] = \sum_i \frac{\langle Q_i \rangle}{T_i}. \qquad (4.10)$$

---

[7]This is valid for both classical and quantum mechanics. The classical version is found in Gibbs (1902, 160–164), and the quantum version in Tolman (1938, 128–130).

Thus, if two probability distributions can be linked by a reversible process, this *uniquely* determines the functional S, up to an additive constant. In the classical case, this is the Gibbs entropy,

$$S_G[P] = -k \int \rho(x) \log \rho(x) \, dx, \qquad (4.11)$$

where $\rho(x)$ is a density for P with respect to Liouville measure. In the quantum case, S is the von Neumann entropy.

$$S_{vN}[\hat{\rho}] = -k \mathrm{Tr}[\hat{\rho} \log \hat{\rho}]. \qquad (4.12)$$

Thus, we find that the Gibbs entropy, in the classical context, and von Neumann entropy, in the quantum context, play the role played by thermodynamic entropy in the second law, if we consider expectation values of heat exchanges rather than actual values.[8]

The proof that (4.8) obtains relies only on fairly basic facts about evolution of states, classical or quantum, facts that are indifferent to the temporal order of t_0 and t_1. And, indeed, the result itself is indifferent. Yet, equation (4.8) itself is not symmetric under interchange of t_0 and t_1. The two times enter into the statement of the theorem asymmetrically, because it is assumed that the system A is uncorrelated with the thermal systems B_i at time t_0, and it is not assumed that this holds at time t_1. In application, we expect that this is reasonable if t_0 is the earlier time, prior to interactions between A and the thermal systems B_i. That is, it is assumed that we have available to us thermal systems, and that the process by which these come about — relaxation to equilibrium — effectively effaces any correlations there might be between the thermal system and other systems. Thus, to understand the source of temporal asymmetry in thermodynamics, we should look to the process of relaxation to equilibrium.

---

[8]It should be noted that we are not assuming that the actual values of heat exchanges are even close to the expectation values. The theorem is valid without restriction, and holds even in cases in which the variances of the random variables Q_i are appreciable.

4.4. The ubiquity of forgetfulness

Much of the discussion surrounding the relation of microscopic dynamics to the behaviour of macroscopic objects has focussed on the issue of time reversal invariance: how can we reconcile temporally irreversible behaviour at the macroscopic level with time-reversal invariance of the fundamental dynamics?

Time reversal invariance isn't really the key issue, however. A symptom of this is that, although there is literature on the proper characterization of time-reversal invariance in classical electromagnetism and in quantum mechanics (Albert, 2000; Earman, 2002; Malament, 2004; Callender, 2000), none of it (as the authors would be the first to acknowledge) gets us any closer to understanding the phenomenon of equilibration; nor does the actual breakdown of time-reversal invariance of the weak nuclear force. Though not time-reversal invariant, the weak force is thought to be invariant under a combination of charge conjugation, parity inversion, and time inversion (CPT). A universal tendency to equilibrium, encompassing both matter and antimatter, breaks CPT symmetry every bit as much as it breaks T-symmetry. Similarly, even if claims by Albert and Callender of failure of T-invariance of classical electromagnetism and quantum mechanics were correct, this would get us no closer to understanding equilibration. What is striking about equilibration is that the state of equilibrium approached by a system is independent of the precise details of its initial state. This is a temporal asymmetry of a sort different from a mere violation time-reversal invariance. It is a failure of *invertibility*: distinct initial states lead to the same final state.

This is a phenomenon that is familiar to us in the macroscopic world. The relentless processes of decay leads inexorably to erasure of evidence about the past. Put two objects at different temperatures into thermal contact with each other, allow them to equilibrate, coming to the same temperature, and you will be unable to tell, by experiments performed on the systems, which had been the hotter and which the cooler. Everything we take to be a record of the past is subject to the same decay; a book will eventually crumble into dust, and the words that its pages once held will be lost.

Moreover, not just any failure of invertibility suffices to permit this sort of forgetting. Suppose that we have some measure μ on the state space of the system that is conserved under the system's temporal evolution.[9] Then, whether or not the dynamics are invertible, any two disjoint sets will be mapped, by the dynamics, into sets whose overlap has zero measure. In this sense, they remain as distinguishable as they were before.

Equilibration requires the temporal evolution of the system to erase distinctions between initial states. If there is a measure on the state space of the system that is conserved under temporal evolution, then we have preservation of distinguishability with respect to that measure. In the quantum context, what matters is that the evolution of an isolated system cannot decrease the absolute value of the inner product of two state vectors.

The salient tension between equilibration and the underlying microdynamics, in the classical context, is not the tension between time-reversal symmetry of the microdynamics and temporal asymmetry at the macroscopic level but, rather, the tension between the existence of a conserved measure at the microphysical level and obliteration of traces of the past at the macrophysical level. In the

---

[9]Some terminology. Suppose we have a dynamics on the state space of a system; that is, for each t in some interval of the real line, there is a mapping T_t of the system's state space into itself, which takes a state at a time t_0 to the state at time $t_0 + t$. Given the dynamics, a measure μ_0 on the state of the system induces a measure μ_t on its state at $t_0 + t$. The measure μ_t assigns to any measurable set of states A the measure, on μ_0, of the set of states at time t_0 that get mapped into A.

$$\mu_t(A) = \mu_0(T_t^{-1}(A)).$$

We will say that the measure μ_0 is *invariant* under the evolution T_t if $\mu_t = \mu_0$.

Given a set of states A, we can also track the changes (if any) of the measure of its image $T_t(A)$ under evolution. We will say that the measure μ_0 is *conserved* under the evolution T_t if $\mu_0(T_t(A)) = \mu_0(A)$ for all measurable A.

Any conserved measure is an invariant measure. If T_t is an invertible map, then any invariant measure is also a conserved measure. If T_t is not invertible, then an invariant measure might not be a conserved measure, though it will be non-decreasing under the action of T_t.

quantum context, the salient tension is between forgetfulness at the macroscopic level and the existence of a conserved inner product on the Hilbert space of the system. Time-reversal invariance of the microphysics is a red herring.

4.5. The open systems approach

There are two approaches, at first glance strikingly different, towards the study of equilibration. On one approach, one considers an isolated system, but focuses attention on a limited set of dynamical variables of the system, typically thought of as its macrovariables. The other considers a nonisolated system, in interaction with its environment, and tracks the evolution of the state of the system.

The two approaches are not as different from each other as might seem at first glance. In each case, we are investigating the evolution of a limited set of degrees of freedom of a larger system and disregarding the rest. The larger system is itself treated as isolated, and hence undergoing Hamiltonian evolution. There is no forgetting in the large system; if its full set of degrees of freedom is considered, distinguishability of sets of states is preserved under evolution.

Obviously, this will not hold for subsystems. Though the present state of the whole may determine the past state of a subsystem, it is clear that the present state of a subsystem, disregarding the state of the environment with which it interacts, will not suffice to determine either the subsystem's past or future. The evolution of the subsystem can be a process of forgetting, because details relevant to its past state may have been exported to the environment. Similarly, details relevant to the past macrostate of an isolated system may become embedded in inaccessible details of its microstate.

Therefore, if we consider a nonisolated system, there is no mystery of reconciling nonconservation of distinguishability of its states with Hamiltonian dynamics of the whole of which it is a part; one would expect this to be ubiquitous. What becomes puzzling instead is the temporal unidirectionality of this phenomenon.

If the description we are working with is a partial description of a total system, it is no surprise that even though the dynamics together with a *complete* specification of the present state uniquely determine the past, a partial description contains less than complete memory of the past, and the present is compatible with more than one past. By the same token, it would not be surprising if a partial description of the present radically and drastically underdetermined the future evolution. However, in a whole host of cases, we do expect to be able to evolve the present macrostate forward and make reliable predictions about future macrostates. In some cases, researchers are able to provide reliable, autonomous equations for the forward evolution of the macrostate. What is puzzling is not how one obtains macrodynamics that does not preserve distinguishability from microdynamics that does. The puzzle is why the phenomenon is not as ubiquitous in the forward direction: why does the present macrostate of a system not underdetermine its future as radically as it does its past? To that extent that equilibration is occurring, the evolution merges distinct past states into a single present. Why, then, does this present not bifurcate into distinct possible futures? That is, the real puzzle is, how is prediction ever possible, given that we only have access to a partial description of the state of a system?

4.6. Obtaining autonomous dynamics for subsystems

To get a sense of circumstances under which one can obtain autonomous dynamics for a system in interaction with its environment, rendering prediction of the system's state possible, it is worthwhile to consider a sufficient condition of particular interest.

Consider a system, A, that is in interaction with another system, B. Suppose that, for the duration of the time interval considered, the joint system AB can be treated as isolated, at least as far as the evolution of A is concerned.[10] Given the state ρ_{AB} of the combined system at some time t_0, we can consider the reduced state ρ_A, which is the restriction of ρ_{AB} to the dynamical variables of A; the reduced

---

[10]This is a much weaker condition than the condition that the joint system be approximately isolated. It might have considerable interactions with its environment, as long as those interactions don't affect the evolution of A in an appreciable way.

state ρ_B; and also the product state $\rho_A \otimes \rho_B$, a state in which there are no correlations between the two systems. We can consider two mathematical operations:

1. Apply the system's dynamics to evolve $\rho_{AB}(t_0)$ forward to $\rho_{AB}(t)$, and then obtain from that, $\rho_A(t)$.
2. Apply the system's dynamics to the product state $\rho_A(t_0) \otimes \rho_B(t_0)$ to obtain a state that we will call $\tilde{\rho}_{AB}(t)$, and then obtain from that, $\tilde{\rho}_A(t)$.

Now it might so happen that, for some range of values of t, $\rho_A(t)$ is equal to $\tilde{\rho}_A(t)$ or, at least, sufficiently close that the difference is negligible. In such a case, we will say that the correlations between A and B are irrelevant for A's evolution. Suppose that this holds for all t in some interval $[t_0, t_1]$. Suppose also that the influence of A on its environment is such that $\rho_B(t)$ is only negligibly changed during the interval. In such circumstances, we obtain an evolution of ρ_A that can be applied to any initial state of A and which yields $\rho_A(t)$ as a function of $\rho_A(t_0)$ (holding $\rho_B(t_0)$ fixed).

Under what circumstances will this obtain? If B is a large, noisy environment that can be regarded as a heat bath, it might be the case that, though correlations build up between A and B, the traces of interactions with B become so distributed throughout the system that they play no significant role in future interactions. You may imagine B to be a gas or liquid composed of a large number of molecules undergoing chaotic motion. Molecules that interact with A wander off into the environment and interact with a great many other molecules before they interact with A again, and correlations between A and its environment become so diffused that they are irrelevant to the influence of B on A.

It is this effective lack of correlations that leads to equilibration. Suppose, for example, that the systems A and B are initially at different temperatures. Nothing in the dynamics of the system forbids a steady transfer of energy from the colder to the hotter. But, on average, molecules of the hotter system will have greater energy than those of the cooler system; thus, if the molecules that interact are anything like a random sample of molecules in the two systems, then, with high probability, on any appreciable time scale, the net effect will be a transfer of energy from the hotter to the colder system.

The presentation adopted here differs somewhat from the usual presentation in textbooks. In the usual presentations, one assumes that the state of the combined system at some time t_0 is a product state, and evolves the joint system forward.[11] Under certain conditions, one will obtain autonomous evolution for ρ_A. This raises the question of what justification one might have for taking systems A and B to be uncorrelated at time t_0.

It might seem that there's a simple answer to this. In the sorts of situations routinely studied in the laboratory, one subjects the systems A and B to independent preparations and, at some time t_0, places them into thermal contact with each other, examining the subsequent evolution of A. One could invoke some sort of principle to the effect that systems that have not yet interacted are uncorrelated.

This is too quick, however. In the situation considered, there is considerable overlap in the causal pasts of the two systems, and plenty of opportunity for correlations to have built up. What is really being assumed is that the preparation procedures efface those correlations. For example: suppose that our system A consists of a block of ice, and is placed in a bath of warm water, both sourced from the same bottle of distilled water. It is quite possible that some of the molecules that end up in the block of ice are ones that have interacted with molecules that end up in the heat bath. But any entanglement between such molecules will have been very thoroughly and effectively erased by subsequent interactions. The sense that we may have, that a product state is the default state for a pair of systems, in the absence of a process to induce and maintain entanglement, is warranted by the ubiquity of mechanisms of decoherence continually erasing such entanglement. The guiding principle should be: if A and B are two systems interacting with a large noisy environment but not directly with each other, we should expect their state to be

---

[11]This is one special case of the procedure discussed by Wallace (2011).

effectively a product state in the absence of some process counter-acting the effects of decoherence.

The rationale for employing a product state at time t_0, the time at which the systems are brought into thermal contact, is therefore, much the same as the rationale for employing a product state at later times. We are *not* committed to a full specification of the quantum state of A and its environment being a product state, at t_0 or any other time. We are only committed to the much weaker claim that any entanglement that might exist between A and B at time t_0 is irrelevant to the subsequent development of the system A. The usual textbook treatment, which gives the impression that we are assuming a product state at some time t_0, a condition that could (if there are nontrivial interactions between the systems) hold only for isolated instants, is misleading, as it suggests that the moment t_0 must be singled out as a special moment. This obscures the fact that the rationale for employing a product state (which is not the same as assuming that the state *is* a product state) at time t_0 is of the same sort as the rationale for employing a product state at other times.

4.7. Temporal asymmetry and the open systems approach

Consider again the example of a system that is in thermal contact with a large, noisy environment, that may be treated as a heat bath. Suppose that, at time t_0, the system is not in thermal equilibrium with its environment. Also, suppose that we are convinced that, for some time t_0 (not the initial moment at which the system and its environment are brought into thermal contact), employment of a product state at that time yields correct results for times after t_0, and that calculations based on that state yield an approach to equilibrium, with the hotter system cooling and the colder system warming.

Could the procedure be applied in the opposite temporal direction? Could it be the case that taking an uncorrelated state at time t_0 and applying the dynamics in the reverse temporal direction yields a correct, or approximately correct, description of the evolution of the reduced state of A? Evolving a product state backward leads to correlations prior to t_0, correlations that, moreover, are precisely balanced in such a way that the interactions between two systems leads, not merely to effective disentanglement, but to actual disentanglement at t_0. Moreover, typically this procedure would yield approach to equilibrium as one moves away from t_0 in both forward and backwards directions. If A is hotter than B, such an evolution would be, prior to t_0, one in which energy is transferred from the cooler environment to the hotter body A. Although, on average, the molecules of B have less energy than those of A, the ones that interact with A are systematically those with higher energy than average.

A natural reaction to such a scenario is that there is something uncanny and conspiratorial about the states prior to t_0. The sorts of correlations that lead to transfer of energy from B to A are not the sort that are explicable by appeal to events in their common history, and the erasure of the correlations at t_0 is not at all like the sort of erasure of correlations that we expect decoherence to produce.

Some readers will have attempted to train themselves to dismiss such judgments as misguided prejudices; the proper metaphysical attitude, it might be thought, is one of temporal democracy. Any metaphysically respectable explanation, on this mode of thought, ought to work equally well in both temporal directions.

Is this attitude warranted? The intuition that motivates it seems to be that, at some deep level, there is no real difference between past and future temporal directions, no difference that makes a metaphysical difference. We should ask what grounds, if any, we might have for believing this to be true. A conviction of this sort cannot be based on empirical evidence, as the empirical phenomena exhibit a profuse abundance of temporal asymmetries (and this may even be a precondition for the existence of empirical phenomena at all, as it may be a precondition for the existence of cognizing subjects). It might be said that a conviction of this sort is mandated by the time-reversal invariance (or CPT-invariance) of fundamental physical laws. To reach such a conclusion requires an additional premise, one that often goes unstated; the conclusion only follows if we add the stipulation that our explanation can only invoke considerations that follow from fundamental dynamical laws.

Such a stipulation would be unwarranted. To see this, let's step back a moment, and think about the nature of explanation. Consider, for vividness, an example suggested by Maxwell's remark that "The 2nd law of thermodynamics has the same degree of truth as the statement that if you throw a tumblerful of water into the sea, you cannot get the same tumblerful of water out again."[12] Imagine yourself standing by the seaside. You have a tumbler containing a half-litre of fresh water in your hand, and you toss the water into the sea. You then hold the tumbler above the surface of the sea, and wait for a half-litre of fresh water to leap from the sea into the tumbler.

Of course, we expect it to be a long wait; you could stand there until the sun burns out, and you would not expect to see the temporal reverse process of fresh water being poured into the ocean.

Why not?

Consider two possible answers to this question. One is the one that would most readily come to mind to most people. It is, I claim, the correct answer, and one that has been too hastily dismissed in the literature on the philosophy of statistical mechanics. The other is the one that is, perhaps, most prevalent in the literature on the philosophy of statistical mechanics.

For the first answer, consider what sort of process the temporal inverse of your tossing the tumbler-full of fresh water into the sea would be. It is one in which the seemingly random thermal motion of molecules becomes a coordinated one, and a very large number of water molecules (and none at all of the molecules of dissolved matter with which they are continually interacting) coalesce in one area of the sea, all possessing an upward velocity that takes them, all in the same general direction, away from the surface of the water, on trajectories that happen to land exactly in the tumbler. We are apt to find the prospect of this concatenation of events unlikely. Asked why, a natural response would be, "What would make them behave like that? The presence of the tumbler can't have that kind of effect on the water molecules."

---

[12]This appears in a letter to John William Strutt, Baron Rayleigh, dated Dec. 6, 1870. It can be found in Garber *et al.* (1995, 205) and Harman (1995, 583).

This sort of thinking has temporal asymmetry deeply embedded in it. This can be seen vividly by considering again the process of hurling the water into the sea. Consider a moment (or a brief time interval) at which the water is in the air, having left the tumbler but not yet hit the sea. In this situation, there is coordinated motion of the bits of water; the velocities of the several parts of the blob of water are such that, were all of them reversed, the water would return to the place where the tumbler was when the water left it. If one asked for an explanation of this remarkable coordination between the positions and instantaneous velocities of the bits of water, and their relation to the former position of the tumbler, an adequate explanation of the phenomenon would be: a velocity-reversal of the water would put it on a trajectory that leads back to the former location of the tumbler *because that is where it came from.*

This, I claim, is a perfectly adequate explanation and ought to be counted as such on any account of explanation.[13] The same cannot be said of its temporal inverse. If asked why a blob of fresh water that has just emerged from the sea is on a trajectory that will take it into an awaiting tumbler, an answer of *"because that's where it is going"* would not be counted as an adequate explanation.

Underlying these judgments is a concept of explanation on which an adequate explanation can be provided by citing a cause in the recent past that, in conjunction with the laws of physics, accounts for the coordination in question. The concept of cause invoked is one on which the cause-effect relation is temporally asymmetric: a cause must be in the past of its effect. That is, we are invoking a temporally asymmetric notion of what it is to explain something. A causal explanation may invoke dynamical laws that are themselves temporally symmetric; the temporal asymmetry of the explanation lies in the temporal asymmetry of the notion of cause being employed.

---

[13]One could, of course, ask further questions, such as how the water came to be in the tumbler, or why there are tumblers containing fresh water, but these would be requests for explanation of other matters. The original why-question has been answered.

Considerations of this sort are not new, of course, and an account of temporal asymmetry in physics invoking considerations of this sort has been defended by Penrose and Percival (1962), and Penrose (2001). The underlying idea is that there be no correlations between systems not attributable to common causes in their past. Acceptance of this basic idea does not necessitate acceptance of Reichenbach's formalization of it, on which the common cause screens off correlations.[14] This sort of reasoning must be applied with due caution and some finesse. It becomes empty if it can only be applied to events with no common past. The fact that there are systems that are effectively independent relies on a process by which events in their common past are rendered irrelevant to their future state.

The other sort of explanation, favoured by neo-Boltzmannians such as Lebowitz (1993, 1999a,b), Goldstein (2001), Price (1996, 2002), and Albert (2000), eschews a temporally asymmetric postulate about probabilities. One chooses a time t_0, and imposes a probability distribution over the state of the system at time t_0 that is the restriction of Liouville measure to its macrostate at that time. Little, if anything, is said about the rationale for this choice of probability distribution, other than that it seems to work, as long as one considers its implications only for events to the future of t_0.[15]

---

[14]This remark is there because, in Bell-type experiments, it does seem to make sense to say that the entanglement exhibited by distant particles can be attributed to the circumstances of the generation of the particles pairs at their common source, though there is no Reichenbachian screening-off common cause.

[15]To be sure, Liouville measure, applied to the whole of an isolated system's phase space, is distinguished from other measures in various ways. For one thing, it is a conserved (hence invariant) measure, no matter what the system's Hamiltonian is. Invariance is sometimes mentioned by neo-Boltzmannians (see Lebowitz 1999a, p. 520; 1999b, S348; Goldstein 2001, p. 53) as a reason for privileging Liouville measure. This property was invoked by Gibbs as a necessary condition for a measure to represent equilibrium. It cannot be invoked as a rationale for preference out of equilibrium. For a system that is out of equilibrium and undergoing a process of relaxation towards equilibrium, the probability to be attached to a given set of microstates is not unchanging; presumably, one wants to say, of a such system, that it is more likely to exhibit the macroscopic indications of equilibrium at later times in the process than at earlier ones.

There is a view of probability, often wrongly attributed to Laplace and often associated with the Principle of Indifference, that probability can be reduced to mere counting of possibilities. The probability of an event A is the number of ways that A can occur, divided by the total number of ways things can be. This sort of thinking has been roundly critiqued in the literature on the philosophy of probability, and has (rightly, in my view) been widely regarded as untenable. Yet, it seems to linger in the way some advocates of the neo-Boltzmannian approach talk.

The fundamental problem with a mere-possibility-counting conception of probabilities is that, as Laplace himself stressed, it requires, in order to get off the ground, a judgment about which possibilities are to be regarded as equally probable, and any such judgment will require grounds for favouring that choice over others. In the case of a continuum of possibilities, the Principle of Indifference is thought to enjoin us to adopt a probability distribution on which the parameters we are using to characterize the state of the system are uniformly distributed. This requires a choice of parameterization. Liouville measure, which plays a central role in equilibrium statistical mechanics, is uniform in canonical phase-space coordinates, but not in others.

If a probabilistic postulate of the sort invoked by neo-Boltzmannians, or any other time-symmetric postulate, is to be applied out of equilibrium, then it must be applied at some special time t_0, which is either the beginning of all things (if there is such a time), or else a turning point, with approach toward equilibrium as one moves away from this time, in both temporal directions. One could, for example, apply it to an isolated system that has equilibrated but is at the peak of a fluctuation from the equilibrium mean values of its macrovariables. This is a marked difference from the approach considered here, which involves time-asymmetric considerations that can be applied at *any* time. Moreover, the same work needs to be done on either approach, that of explaining why the sorts of correlations that have built up between systems as a result of events in their common past tend to become irrelevant for prediction of the forward evolution of the system, though they remain relevant to retrodiction of the systems' past.

Obviously, a full-scale critique of this sort of approach is beyond the scope of this chapter. I will say only that, insofar as it takes motivation from the idea that probability of an event can be thought of a matter of mere counting, it should be viewed with suspicion. In addition, making reference to a special time in the remote past to explain the ubiquitous and mundane phenomenon of equilibration strikes me as an act of desperation. For much of the period of the development of statistical mechanics, a steady-state cosmology was a live option among serious cosmologists. An advantage of the approach advocated here is that it does not make explanation of events in the laboratory or in our homes, such as the cooling of a cup of coffee, sensitive to large-scale cosmological questions, and can be applied in exactly the same way whether or not there is a cosmologically privileged instant t_0. It also fits better with nonequilibrium statistical mechanics as practiced; one will search in vain for textbooks of nonequilibrium statistical mechanics for the sorts of cosmological considerations so frequently found in the philosophical literature on statistical mechanics.

4.8. Conclusion

If one construes thermodynamics as it was construed at the time that Gibbs was writing, as the science of work and heat, then Gibbs' remark, quoted at the beginning of this chapter, does not seem so naïve. We do have satisfactory statistical mechanical analogues of the first and second laws of thermodynamics. Moreover, these analogues are not, in and of themselves, asymmetric in time. Their formulation presupposes, however, the availability of thermal systems, that is, systems that have relaxed to thermal equilibrium. Understanding circumstances in which this does and does not occur is an active and ongoing research project in physics, and it is one that philosophers would do well to pay attention to. Much of this work involves investigation of conditions under which one obtains autonomous dynamics either for the state of a subsystem of a larger system, or for a limited set of degrees of freedom of an isolated system. What is needed is an explanation of why the sorts of states that we can reliably produce in the laboratory or can reasonably expect to find in nature are states that afford the autonomous dynamics we seek.

The phenomena to be explained are temporally asymmetric. I have suggested that we may nevertheless obtain an explanation invoking only time-symmetric dynamical laws, because of a temporal asymmetry in the very notion of explanation. This is a temporal asymmetry that most philosophers have shied away from, perhaps taking it as axiomatic that we have not reached a satisfactory explanation until temporal asymmetries have not merely been explained but have been *explained away*, having been shown to be either merely apparent or else merely local facts about our limited region of the universe. As a result, proposals such as that of Percival and Penrose have received short shrift in the philosophical literature. At the very least, I hope, in this chapter, to have persuaded readers that they deserve serious consideration.

References

Albert, D. (2000) *Time and Chance*, Harvard University Press, Cambridge.

Bartolotta, A., S. M. Carroll, S. Leichenauer, J. Pollack (2016) Bayesian second law of thermodynamics, *Physical Review E*, **94**, 022102.

Bricmont, J., D. Dürr, M. Galavotti, G. Ghirardi, F. Petruccione, N. Zanghì (eds.) (2001) *Chance in Physics*, Springer, Berlin.

Brown, H. R., J. Uffink (2001) The origins of time-asymmetry in thermodynamics: The minus first law, *Studies in History and Philosophy of Modern Physics*, **32**, 525–538.

Callender, C. (2000) Is time 'handed' in a quantum world? *Proceedings of the Aristotelian Society*, **100**, 246–269.

Clausius, R. (1854) Ueber eine veränderte Form des zweiten Hauptsatzes der mechanischen Wärmetheorie, *Annalen der Physik*, **93**, 481–506. Reprinted in Clausius (1864, 127–154); English translation in Clausius (1856), and in Clausius (1867).

Clausius, R. (1856) On a modified form of the second fundamental theorem in the mechanical theory of heat. *The London, Edinburgh, and Dublin Philosophical Magazine and Journal of Science* **12**, 81–88. English translation of Clausius (1854).

Clausius, R. (1864) *Abhandlungen über die mechanische Wärmetheorie*, Volume 1. Braunschweig: Friedrich Vieweg und Sohn.

Clausius, R. (1867) *The Mechanical Theory of Heat, with its Applications to the Steam Engine, and to the Physical Properties of Bodies.* London: John van Voorst. English translation, with one additional paper, of Clausius (1864).

Darwin, C. G. (1936) Terminology in physics, *Nature*, **138**, 908–911.

del Rio, L., L. Krämer, R. Renner (2015) Resource theories of knowledge. arXiv:1511.08818 [quant-ph].

Earman, J. (2002) What time reversal invariance is and why it matters, *International Studies in the Philosophy of Science*, **16**, 245–264.

Fowler, R., E. A. Guggenheim (1939) *Statistical Thermodynamics: A Version of Statistical Mechanics for Students of Physics and Chemistry.* Cambridge: Cambridge University Press.

Garber, E., S. G. Brush, C. W. F. Everitt (eds.) (1995) *Maxwell on Heat and Statistical Mechanics: On "Avoiding All Personal Enquiries" of Molecules.* Bethlehem, Pa: Lehigh University Press.

Gibbs, J. W. (1902) *Elementary Principles in Statistical Mechanics: Developed with Especial Reference to the Rational Foundation of Thermodynamics.* New York: Charles Scribner's Sons.

Goldstein, S. (2001) Boltzmann's approach to statistical mechanics. In Bricmont *et al.* (eds.) (2001), 39–54.

Goura, G., M. P. Müller, V. Narasimhachar, R. W. Spekkens, N. Y. Halpern (2015) The resource theory of informational nonequilibrium in thermodynamics, *Physics Reports*, **583**, 1–58.

Harman, P. M. (Ed.) (1995) *The Scientific Letters and Papers of James Clerk Maxwell, Volume II: 1862-1873.* Cambridge: Cambridge University Press.

Lebowitz, J. L. (1993) Boltzmann's entropy and time's arrow, *Physics Today*, **46**(September), 33–38.

Lebowitz, J. L. (1999a) Microscopic origins of irreversible macroscopic behavior, *Physica A*, **263**, 516–527.

Lebowitz, J. L. (1999b) Statistical mechanics: A selective review of two central issues, *Reviews of Modern Physics*, **71**, S346–S357.

Malament, D. B. (2004) On the time reversal invariance of classical electromagnetic theory, *Studies in History and Philosophy of Modern Physics*, **35**, 295–315.

Maroney, O. (2007) The physical basis of the Gibbs-von Neumann entropy. arXiv: quant-ph/0701127v2.

Myrvold, W. C. (2011) Statistical mechanics and thermodynamics: A Maxwellian view, *Studies in History and Philosophy of Modern Physics*, **42**, 237–243.

Penrose, O. (2001) The direction of time. In Bricmont *et al.* (eds.) (2001), 61–82.

Penrose, O., I. C. Percival (1962) The direction of time, *Proceedings of the Physical Society*, **79**, 605–616.

Price, H. (1996) *Time's Arrow and Archimedes' Point.* Oxford: Oxford University Press.

Price, H. (2002) Boltzmann's time bomb, *The British Journal for the Philosophy of Science*, **53**, 83–119.

Saha, M. N. and B. N. Srivastava (1931). *A Text Book of Heat*. Allahabad: Indian Press.

Saha, M. N. and B. N. Srivastava (1935). *A Treatise of Heat*. Allahabad: Indian Press.

Sommerfeld, A. (1956). *Thermodynamics and Statistical Mechanics: Lectures on Theoretical Physics*, Vol. V. New York: Academic Press.

Thomson, W. (1853) On the dynamical theory of heat, *Transactions of the Royal Society of Edinburgh*, **20**, 261–288. Reprinted in Thomson (1882, 174–210).

Thomson, W. (1857) On the dynamical theory of heat. Part VI: Thermo-electric currents, *Transactions of the Royal Society of Edinburgh*, **21**, 123–171. Reprinted in Thomson (1882, 232–291).

Thomson, W. (1882) *Mathematical and Physical Papers*, Volume I. Cambridge: Cambridge University Press.

Tolman, R. C. (1938) *The Principles of Statistical Mechanics*. Oxford: Clarendon Press.

Uhlenbeck, G. E., G. W. Ford (1963) *Lectures in Statisical Mechanics*. Providence, R.I.: American Mathematical Society.

Wallace, D. (2016) Thermodynamics as control theory, *Entropy*, **16**, 699–725.

Wallace, D. (2011) The logic of the past hypothesis. Available at http://philsci -archive.pitt.edu/8894/.

Part II

The Language of Typicality

Chapter 5

Reassessing Typicality Explanations in Statistical Mechanics

Massimiliano Badino

Department of Human Sciences, University of Verona
massimiliano.badino@univr.it

Typicality exerts a steady fascination on many experts of statistical mechanics. It is simple, it sounds right, and it falls naturally into the great tradition of Boltzmann's groundbreaking work. But typicality is a tricky business. How and what exactly typicality is supposed to explain is far from clear and, for several physicists and philosophers, it is too simple to be right. In particular, typicality seems to sideline dynamical considerations, which is much too high a price to pay. In this paper, I try to make order in the confused debate on typicality. I construe a general epistemological framework in which typicality explanations can be discuss and introduce some basic distinctions. My main point is that typicality is a pluralistic concepts that calls for a pluralistic attitude. From this perspective, the confusion of the debate on typicality starts to make sense.

5.1. Introduction

Among the foundational approaches to equilibrium statistical mechanics, typicality has recently stirred the most heated and interesting debate. While fiercely supported by some prominent physicists (Dürr, 2001; Dürr, Goldstein, & Zanghí, 1992; Goldstein, 2001, 2012; Lebowitz, 1993, 1999; Zanghí, 2005), it has been repeatedly challenged by philosophers of science (Frigg, 2007, 2009; Hemmo & Shenker, 2012; Pitowsky, 2012). Roughly speaking, the typicality approach aims at explaining the qualitative features of thermal equilibrium — its being unidirectional and exceptionless — by showing that these features are 'typical' of usual statistico-mechanical systems. The notion of typicality, however, is notoriously mudded. It is related to probability but, in the intention of the 'typicalists', to say that a property and event or a behavior are typical means something different, and possibly something more than 'highly probable'.

Many writers have leveled criticisms against the physical, mathematical, and philosophical acceptability of the various steps of the typicality approach. These criticisms have revealed that, beneath its deceptive intuitiveness, typicality hosts a number of problematic assumptions. In the attempt to unravel the technical intricacies nested in the notion of typicality, philosophers have mostly — and wittingly — left aside the issue of its explanatory power. Itamar Pitowsky was possibly the first to adumbrate this question. In discussing the role of the Lebesgue measure, Pitowsky doubted that the notion of typicality added any explanatory surplus to the conclusions one can draw from measure-theoretical statements (Pitowsky, 2012, pp. 54–56). In this paper I want to reassess these questions from a somewhat higher standpoint. My primary goal is to provide a framework to productively evaluate and contextualize some of the epistemological and metaphysical issues at stake in the typicality approach in statistical mechanics. To anticipate my main conclusions, I will argue (1) that the present debate is partly marred by an incomplete understanding of what a typicality explanation is; and (2) that typicality is a pluralistic concept and, unsurprisingly, its use for explanatory purposes requires a pluralistic stance.

My argument is organized as follows. Firstly, I discuss typicality explanations in general (section 2). This discussion will be cursory because a full-fledged theory of typicality explanation would need an article in its own rights. In section 3, I move to examine typicality explanations in statistical mechanics. I consider two approaches: the "single typicality approach" or STA, which is mainstream in the debate, and the "combined typicality approach" or CTA, one version of which has been recently developed by Frigg and Werndl (2012). I discuss the relation between these approaches and claim that they only pick out one aspect of typicality. I close with some general considerations and a historical point (section 4).

5.2. Typicality Explanations

5.2.1. *Typicality-claims*

Intuitively, to say that an event or a property is typical means something less than saying that it occurs invariably and something more than saying that it is common. Typicality seems to convey two distinct albeit interrelated ideas: that something (a event, a property) is numerically overwhelmingly predominant amongst other instances of the same species, and also that this predominance is somewhat connected to being an instance of that species, i.e., a certain type. Variations or atypical instances, accordingly, are to be dismissed. I call the statement that ascribes typicality to a property a typicality-claim (TC). In other words, a TC reads roughly as follows:

(TC)   The property P is typical in the set K

In general, a typicality explanation occurs when one uses a TC for explanatory purposes. For example, I might wonder why the squirrels living near my house are vegetarian and someone might explain to me that it's typical for a squirrel to be vegetarian or, alternatively:

(1) The property of being vegetarian is typical in the set of squirrels

As an explanation, this is satisfying enough. My initial wonder about the alimentary habits of nearby squirrels has been removed by learning that these habits are, in fact, typical. Much after a Hempelian fashion, a TC acts as a surprise-remover. A previously puzzling fact becomes an example of a normal state of affair. This suggests that typicality explanations work according to the following argumentative scheme:

(TC)   The property P is typical in the set K
(FC)   x is a member of K
(EX)   x has the property P

Here, EX is the explanadum; FC is a factual claim; while the TC serves as the explanans. Once TC and FC are given, EX is no longer surprising, in fact it becomes expectable. To be sure, the explanation above is not Hempelian through and through. For one, a TC is not a nomological statement: claims of typicality are not, in themselves, laws of nature. Further, it is important to appreciate that a typicality explanation requires an *argumentative scheme*, that is, a structure which still needs to be filled in with further arguments to justify each step of the scheme. In particular, we will see that the establishment of a TC might involve a number of assumptions and sub-arguments. Hence, while typicality works explanatorily in an Hempelian way, one should not conclude that a typicality explanation is a deductive argument in strict Hempelian terms. I will come back on this point in section 3.3. One might wonder whether typicality explanations are in fact examples of probabilistic explanations, Hempel-style. It seems, in other words, that the explanandum follows with high probability from the conjunction of TC and FC. This is not the case, but to see why, we need to investigate more deeply the nature of TC, which is what I am going to do in the next section.

Before turning to that point, though, I want to mention a further inference supported by a TC, which will come in handy later on. We have seen that a TC gives us good reasons to expect that a member of set K has a property P which is typical in K. What if it has not such a property, i.e., what about atypical events? In that case, the typicality of P warrants us to conclude that x's not having P has been somewhat arranged, prepared or it is due to factors that are external to x's being a member of K. Thus, a TC grounds the following inference:

(TC)   The property P is typical in the set K

(FC)   x is a member of K and does not have P

(C)   x's not having P is due to external factors or intervention

This inference can be illustrated by a famous example due to Reichenbach. Typically, wind and sea produce shapeless and confused signs on the beach. If I spot on the sand signs that have a distinct form (for example, footprints), I tend to think that these signs, being not typical of the action of wind and sea, have been left by something or someone, i.e., they are due to external factors or intervention. I come back to these considerations in section 3.2.

5.2.2. *Unpacking typicality-claims*

I have held that a typicality explanation is the use of a TC for explanatory purposes, but what is there in a TC that grounds such a use? To rephrase the question in perhaps clearer terms: Why I can use a TC for explanation in the first place? I have claimed that typicality captures two distinct ideas: (1) that a certain property is shared by the vast majority of the members of K; and (2) that having that property is somewhat inherent to being a member of K. This suggests that a TC can be analyzed in two different ways and that these two ways must be in a connection of some sort. In this section I want to explicate further these analyses, provide examples, and investigate their relation.

The first idea encapsulated in a TC is that a typical property holds for the vast majority of the members of a set K. For example, to state that the typical sequence of 1,000 coin tosses has roughly the same number of heads and tails can be taken to mean that the ratio between the number of such sequences to the total number of possible sequences is close to 1. I call this reading a measure-theoretical (MT) analysis of a TC. More specifically, three ingredients are required for analyzing a TC in measure-theoretical terms. Firstly, we obviously need to specify the set K mentioned in the TC. As we need to count elements, the set must be completely specified, which means we must be in the position to give, at least in principle, a full list of the members of K (more on this later). Secondly, we need to distinguish various properties of interest for members of K. This can be done rigorously by defining a partition **B** of subsets $K_1, \ldots, K_n$ of K such that $K_1, \ldots, K_n$ are pairwise disjoint and $K_1 \cup \ldots \cup K_n = K$. Each subset corresponds to a property and no member of K possesses two different properties, that is, the properties are mutually exclusive. In other words, the typicality of a property is always evaluated against the background of the possible alternatives. Lastly, we need a counting procedure γ to determine how many members of K belong to each of the subsets $K_1, \ldots, K_n$.

Once we have the triple (K, **B**, γ) we can rigorously proceed to a MT analysis of the TC. In the example above, K is the set of possible sequences of 1,000 coin tosses, $K_1, \ldots, K_n$ are the possible distributions of heads and tails in a sequence, and γ is the usual operation of summing members of a set. Notice that the triple (K, **B**, γ) does not establish the truth of the TC, yet. We still need to clarify what we mean by 'vast majority'. This is not an easy concept to define. In fact, this is what philosophers call a *vague* concept. It is generally possible to point at clear examples and counterexamples of a vague concept, but it is impossible to draw a sharp line between cases in which it applies and cases in which it does not. The concept of vast majority falls in this category. Vagueness, however, does not undermine the applicability of the MT analysis. We only need to confine such use to the cases in which the majority is indisputably vast and, fortunately enough, statistical mechanics falls into this category. This leads to another important point. Intuitively, one could think that typicality is just a shortening for high probability; however, there are technical as well as philosophical reasons to keep the two notions apart. Typicality can be related to the act of *counting*. A property is typical when it is shared by the vast majority of individuals. To perform a MT analysis of a TC, one only needs to establish a counting procedure, usually a measure function. A probability function is a different mathematical object. More importantly, there are many ways in which such mathematical object can be understood: as a relative frequency, as a propensity, as a degree of belief, or just as a normalized measure function that fulfills certain axioms. In other words, the notion of probability is much richer and complex than the simple act of counting.

There is a second idea built into a TC. To claim that a certain property is typical of the members of a set might also be taken to mean that it characterizes a special natural type, 'member of the set K'. Thus, if being vegetarian is typical of squirrel, then it is a characteristic of the type 'squirrel'. In other words, typicality might refer to the metaphysical import, in terms of properties, of being a member of the set K and, consequently, typicality tells us something profoundly different from the mere distribution of the property among the members. This meaning of typicality as 'characterizing feature' requires a different sort of analysis, which I call causal-factors (CF) analysis. Briefly put: a TC can be analyzed in terms of a series of causal factors $C_1, \ldots, C_m$, where (i) is responsible to turn a generic individual x into a member of K and (ii) has a strong tendency to bring about the property P. This last statement calls for a number of clarifications.

First, although I dubbed $C_1, \ldots, C_m$ as 'causal factors', one does not need to assume any particular theory of causality. For the purpose of a CF analysis it suffices that $C_1, \ldots, C_m$ are a set of factors that must be at work for an individual x to be a member of K. However, to fix ideas, in this article I will assume James Woodward's manipulation account of causality (Woodward, 2003) which, I think, captures the key features for the typicality case. Roughly speaking, according to Woodward, causal factors are difference-makers, which is to say they are the elements that make a difference for something to happen. More precisely, C is a causal factor for the phenomenon E if and only if: (i) C and E can be represented by variables; (ii) only an intervention on the value of the variable C results in a certain change in the value of the variable E; and (iii) the relation between variables C and E is invariant. Woodward defines an intervention on some variable X with respect to some second variable Y as "a causal process that changes X in an appropriately exogenous way, so that if a change in Y occurs, it occurs only in virtue of the change in X and not as a result of some other set of causal factors" (Woodward, 2000, pp. 199–200). An intervention does not need to be carried out by a human agent. It suffices to specify, possibly in a counterfactual way, that a certain causal process can change the value of a variable in order to examine the consequence on the other variable. Furthermore, the relation must be invariant, that is, "it would continue to hold — would remain stable or unchanged — as various other conditions change" (Woodward, 2000, p. 205).

Second, the same metaphysical neutrality that holds for the notion of 'causal factors' also holds for the notion of 'bringing about'. Causal factors can be understood as producer, as mechanism, as propensities, and many other things. Once again, to fix ideas, I will speak of causal factors as generating a propensity to the manifestation of a certain property P. However, my argument does not hinge on this specific metaphysical choice.

Having said that, in order to provide a CF analysis of a TC it is necessary to determine a set of causal factors $C_1, \ldots, C_m$, where (CF1) are difference-makers for an individual x to be a member of K, and (CF2) generate a strong propensity to the property P. If both (CF1) and (CF2) are satisfied, then one can conclude that P is a 'characterizing property' of being a member of K. As in the previous case, the analysis makes use of a vague concept, i.e., 'strong propensity'. Again, the vagueness does not in itself undermines the applicability of the concept. We can have indisputable examples and counterexamples of strong propensity, although we are unable to draw a clear-cut line.

Let's try to make all this a bit clearer by means of an example. I meet a professional piano player and I can't help notice that her hands are quick and agile. This suggests to me the following TC:

(2) The property of having agile hands is typical among professional piano players.

A CF analysis of (2) would run as follows. To become a professional piano player one has to undergo a very rigorous training, which includes some specific exercises to improve the agility of the hands. These exercises make a crucial difference between an ordinary person and a professional piano player, and they also bring about a strong propensity to acquire agility. As both (CF1) and (CF2) conditions above are satisfied, we have completed a CF analysis of the TC (2).

5.2.3. *Comparing the analyses*

A systematic comparison between the MT and the CF analysis would lead our discussion too far away from the main focus of this article and I will therefore leave it for another occasion. Here, it will suffice to highlight some intuitive similarities and differences in order to grasp the key ideas behind the distinction. I will discuss three main points.

The first point concerns the relations that MT and CF analysis have with set K. From the foregoing discussion, it is apparent that the

MT analysis treats K as a collection of independent tokens, while the CF analysis treats it as a collection of instances of a certain natural type. As an example, let's consider the following TC:

(3) A typical lottery ticket is a non-winning one.

In this case, K includes all lottery tickets and the MT analysis consists in specifying a way, i.e. a counting procedure, to decide whether a ticket is a non-winning one. For example, it might be that there are 1 million tickets and only 10 of them will win a prize. Simple arithmetic considerations would thus lead us to conclude that the vast majority of tickets are non-winning. In this analysis of (3), the procedure of lottery drawing is fully irrelevant. Even more, the fact that K contains lottery tickets is in itself secondary: the only feature that matters for the MT analysis is that the elements of K are individual and distinguishable tokens with well-defined properties and can therefore be counted.

The situation is conspicuously different with the CF analysis. By definition, this analysis requires the specifications of the factors that make a difference for being a member of K. In other words, these factors characterize the 'K-type' in general. As an illustration of this point, let's consider again statement (2) of the previous section. I have suggested that a CF analysis of (2) boils down to specify a series of causal factors for being a professional piano player which, at the same time, have a strong propensity or disposition to induce the property of having agile hands. From this perspective, a CF analysis is concerned with the features of the type 'professional piano player', and the members of K, in this case, should be considered as instances of this type. The point of the CF analysis is to single out those factors that constitute the type and are also responsible for the property P of interest.

My second point concerns an analogy that might be helpful to grasp the distinction between MT and CF analyses; that is, an analogy with extension and intension of a term. To any term of our language, one can associate two notions. One notion is the collections of objects the term refers to, and this is called the extension of the term. In this case, a term is seen as a label for the collection of physical objects. The second notion is the sum total of the criteria that allows one to apply the term, that is, to include an object in the extension; this is called the intension of the term. In this case, the term is seen as a cluster of properties that characterize an object to be called by that term.

The distinction between MT and CF analysis runs along a similar line. On one hand, the MT analysis is extensional to the degree that it aims at showing that an overwhelmingly large *quantity* of members of K have the property P. In so doing, the MT analysis focuses exclusively on the individual members of the set. By contrast, the CF analysis is intensional to the degree that it aims at showing that certain *qualities* shared by all the members of K have a strong propensity to bring about the property P. In so doing, the CF analysis focuses exclusively on the features that make a difference to be a member of K.

Although I don't want to overplay this analogy, I have to note another interesting connection. Intension and extension provide different kinds of information about the same term. If I know the full intension of a term I can, in principle, determine its full extension because, for each object, I can decide whether or not it is described by the term. This is not the case if I have the extension. Even if I can produce a complete list of objects described by a certain term, that does not furnish the full set of relevant features of the objects. Something very similar happens in the case of typicality. The knowledge of the relevant causal factors that have a strong propensity to bring about P tells me also that the vast majority of members of K have P, while the latter tells me nothing about the former.

This brings me to my third point, which concerns the epistemological and metaphysical differences between MT and CF analysis. In order to perform a MT analysis it is necessary to have (i) a complete set K and (ii) a well-defined counting procedure. The first requirement means that K cannot be given in an open or incomplete form. For example, the set of lottery tickets (for a certain lottery) is complete because all lottery tickets are available for counting at the same time. In addition, one also needs to define a satisfactory counting procedure which, in many cases, is a measure function. This operation is not as obvious as it might seem at first sight, as we shall see later on (section 3.3). Thus, the MT analysis has relatively high *epistemologically constraints*; it calls for a number of rigorous conditions to be satisfied. It has, however, relatively low *metaphysical constraints*; that is to say, it does not involve special metaphysical commitments. A MT analysis merely deals with the, so to speak, phenomenological distribution of the property P over the set K; i.e., it testifies a fact of the matter (provided, of course, a counting procedure).

The situation is somewhat reversed with the CF analysis. In order to perform such analysis, we first need to settle several metaphysical commitments concerning how to interpret the causal factors and the tendency they generate. Here I have treated causal factors according to the manipulation account and I have spoken of propensity or disposition, but other metaphysical options are obviously open for consideration, which lead to other metaphysical pictures of typicality. On the other hand, the CF analysis is epistemologically lighter. It can be applied to arbitrarily complex and extended sets of individuals, and do not require a rigorous counting procedure: its metaphysical constraints are high, while the epistemological ones are low. To put it in other terms, the MT analysis can be applied only in well-defined conditions, it is rigorous within those limits, but it ultimately tells us a story about the phenomenology of the system. By contrast, the CF analysis can be applied in much looser conditions. It is, consequently, not as rigorous as the MT analysis, but it tells us a metaphysically richer story.

To conclude, I want to come back to the issue of atypical events. It is easy to see that both analyses support the inference that if an atypical event happens, then some external factors must be at work. In the case of the MT analysis, this support is provided by the fact that atypical events are very few, so they must be chosen on purpose or caused by something. This is precisely the rationale behind statistical tests, for example. The same holds true for the CF analysis: if a property is characteristic for a certain natural type, its absence can be connected with the action of external factors that have altered the natural type.

5.3. Typicality in Statistical Mechanics

5.3.1. *How Typicality is used in Statistical Mechanics*

In the previous section, I have elaborated some general points concerning the explanatory value of typicality. In particular, I have claimed that typicality explanations happen by appealing to some sort of counting procedure or set of causal factors. In this section, I show that this tension can help us make sense of the debate on typicality in statistical mechanics.

A good account of how typicality is supposed to work in statistical mechanics can be found in Lebowitz (1993). In the opening section of his article, Lebowitz distinguishes neatly between qualitative and quantitative aspects of thermal equilibrium. Firstly, he introduces the concepts of macro- and microstate. A microstate is a complete description of the state of the system in terms of the positions and momenta of each particle. If the system consists of N particles, its microstate x is a point in a $6N$-dimensional phase space. Microstates can be grouped into macrostates, i.e., regions in the phase space comprising microstates that look the same from a macroscopic point of view.

Ludwig Boltzmann showed that one can attach a quantity, called Boltzmann entropy, $S_B(M,t)$, to a macrostate M (t being time) and consequently to each of its microstates $S_B(x,t)$, where x is a microstate belonging M. Lebowitz states that "S_B typically increases in a way which explains and describes qualitatively the evolution towards equilibrium of [thermal] systems" (Lebowitz, 1993, p. 2). Thus, the explanadum of a typicality explanation is the observed thermodynamic-like behavior, i.e., the time increase of the Boltzmann entropy: $S_B(M,t) \leq S_B(M,t')$, where $t' > t$. There are also quantitative aspects of equilibrium, but they require different sorts of resources: "the quantitative description of the macroscopic evolution is given by hydrodynamical-type equations which can be derived (explicitly, in some cases) from the microscopic dynamics by utilizing the collective aspect of macrobehavior, i.e., as a law of large numbers arising from the very large macro/micro-ratio." The quantitative aspects are related to typicality, but are not directly explained by it.

How does typicality contribute to the explanation of the thermodynamic-like behavior of thermal systems? To answer, we need to introduce some technical concepts. Let Σ be the accessible phase space of a physical system S, and x_i the phase points of this space corresponding to the possible microstates of S. Furthermore, let us assume that the space can be partitioned into a series of pairwise disjoint subsets $M_1, M_2, \ldots, M_E$. These subsets contain phase points x_i and M_E is the subset of the phase point corresponding to a state of equilibrium. To each microstate x_i belonging to a certain subset M_i, one can assign a Boltzmann entropy $S_B(M_i, t)$. The Boltzmann entropy $S_B(M_E, t)$ for equilibrium is a maximum. Finally, let the dynamics of the phase points be given by an Hamiltonian flow $H(x)$, i.e., $H(x_{t-1}) = x_t$. A trajectory of the system is a sequence of microstates x_i of S generated by $H(x)$. This also means that one can gather all the phase points generated by the Hamiltonian flow, and form a new subset of the phase space T_H which represent the trajectory generated by H.

Let us now see how these ingredients combine with typicality to explain thermodynamic-like behavior. The starting point is the following fairly uncontroversial claim (with qualifications to be discussed in a minute):

(EQ) The property of being a state of equilibrium is typical in Σ.

For historical as well as technical reasons, the standard approach to (EQ) is through MT analysis. The historical reason is that a MT analysis recalls Boltzmann's combinatorial explanation of the second law of thermodynamics (Boltzmann, 1877), which many physicists see as the forerunner of the modern typicality approach. Boltzmann's basic intuition was that thermal systems are normally found in equilibrium state (or, if disturbed, tend rapidly to come back to it) simply because there are overwhelmingly more ways to be in equilibrium than otherwise. Contemporary writers have internalized this idea.

For example, Roman Frigg defines typicality in the following way:

> "Intuitively, something is typical if it happens in the 'vast
> majority' of cases: typical lottery tickets are blanks, typical
> Olympic athletes are well trained, and in a typical series
> of 1,000 coin tosses the ratio of the number of heads and
> the number of tails is approximately one." (Frigg, 2009,
> pp. 997–998)

Another notable example, even more straightforwardly related to statistical mechanics, is the following:

> "Generally speaking, a set is typical if it contains an 'over-
> whelming majority' of points in some specified sense. In
> classical statistical mechanics there is a 'natural' sense:
> namely sets of full phase-space volume." (Volchan, 2007,
> p. 803)

The influence of Boltzmann's work is clearly visible in the way typicality is understood by the practitioners of statistical mechanics. Yet, Boltzmann's real breakthrough lies in his elaboration of a way to effectively calculate how overwhelming the dominance of the equilibrium state is. Boltzmann used combinatorics to literally count the number of ways, i.e. microstates, in which thermal systems composed of very many particles can be in equilibrium and discovered that such number was enormously larger than any of the numbers corresponding to other states. Originally, Boltzmann's combinatorial approach required that the particles could assume only discrete energy values. Although he managed to extend his argument to an energy continuum, a more effective way to do so is to use a measure function. The modern approach therefore specifies a measure function and analyzes (EQ) as stating that the equilibrium state has the overwhelmingly largest measure of them all.

The choice of the measure function is not entirely innocent, though. The natural candidate is the Lebesgue measure, which is simple and has several intuitively desirable properties. However, objections have been levelled against the use of the Lebesgue measure. Perhaps, the philosophically most radical one was elaborated by Meir Hemmo and Orly Shenker (2012). They distinguish between legitimate and illegitimate forms of typicality depending on the justification of the measure function. If the measure function is defined on the basis of observed relative frequencies, then typicality expresses an incontestable fact on the phase space and should be accepted as such. But if the measure function is defined a priori, on the basis of some intrinsic preference of the Lebesgue measure, then typicality inverts the "epistemological arrow" that goes from experience to theory and becomes philosophically ungrounded (Hemmo & Shenker, 2012, pp. 182–191). Apart from these philosophical qualms, consensus is fairly widespread among physicists that the Lebesgue measure should be adopted to evaluate the extension of phase space regions. Thus, (EQ) is very naturally analyzed in MT terms: the equilibrium state is typical in the phase space, or the vast majority of phase points belong to the equilibrium region.

5.3.2. *Size Does Matter*

There is a common wisdom about how to construe a typicality explanation of equilibrium and, for reasons that will be clear in the next section, I will call this approach the "simple typicality approach" (STA). The STA rests on the contention that (EQ) alone is sufficient to construct a valid typicality explanation of the thermodynamic-like behavior. The argument goes as follows. The MT analysis of (EQ) leads to the conclusion that the equilibrium state is enormously larger than other possible macrostates (in fact, it is enormously larger than all the other states combined!). Thus, the extension of the equilibrium is such that it seems plainly impossible for a normally behaving system not to spend in it the vast majority of time. Possibly the clearest expression of this idea is due to Sheldon Goldstein:

> "[The accessible phase space] consists almost entirely of phase points in the equilibrium microstate, with ridiculously few exceptions whose totality has the volume of order of $10^{-10^{20}}$ relative to that of [the equilibrium macrostate]. For a non-equilibrium phase point $[x]$ of energy E, the Hamiltonian dynamics governing the motion $[x(t)]$ would have to be ridiculously special to avoid reasonably quickly carrying $[x(t)]$ into [the equilibrium macrostate] and keeping it there for an extremely long

time — unless, of course, [x] itself were ridiculously
special." (Goldstein, 2001, pp. 43–44)

The underlying idea of this quote is that the equilibrium macrostate
is so big that exceptions, although mathematically existent, are
physically irrelevant. Thus, there is no need to consider the Hamiltonian of the system. Nothing short of "ridiculously special" cases
would prevent the system from entering into the equilibrium state
and staying there for a very long time. In other words, atypical events
are just theoretical possibilities that do not deserve serious consideration. This thought is well expressed in this passage by Detlef Dürr:

> "What is typicality? It is a notion for defining the small-
> ness of sets of (mathematically inevitable) exceptions and
> thus permitting the formulation of law of large numbers
> type statements. Smallness is usually defined in terms of
> a measure. What determines the measure? In physics, the
> physical theory. Typicality is defined by a measure on the
> set of 'initial conditions' (eventually by the initial condi-
> tions of the universe), determined, or at least strongly sug-
> gested, by the physical law. Are typical events most likely
> to happen? No, they happen because they are typical. But
> are there also atypical events? Yes. They do not happen,
> because they are unlikely? No, because they are atypical.
> But in principle they could happen? Yes. So why don't
> they happen then? Because they are not typical." (Dürr,
> 2001, p. 130)

Note how Dürr highlights that the prime message of typicality is
the irrelevance of the exceptions. This is not a new strategy in physics
and, in the case in point, it does not lack a certain epistemological
bite. In section 2.1, we have seen that it is customary to use a TC
to infer the existence of external intervention to explain the occurrence of an atypical event. The idea is that atypical events, albeit
possible, do not take place spontaneously or naturally: they must be
caused. In their arguments, Goldstein, Dürr, and other upholders of
the STA combine the standard inferential scheme and the "interven-
tionist" inference grounded by (EQ). In brief, the typicality of the
equilibrium state allows us to conclude that equilibrium will happen
simply because it would take an external intervention (and a very

special one) to prevent it from happening. Nothing more than (EQ) is required to explain thermodynamic behavior:

> "The convergence to equilibrium of natural phenomena is neither a consequence of a new physical law, nor the effect of an attractor in the microscopic dynamics. The equilibrium macrostate does not attract anything, systems, typically, "fall in it" because, in the phase space, the equilibrium macrostate occupies a region enormously bigger than the others." (Zanghí, 2005, p. 170)

It is not difficult to see the advantages and disadvantages of this approach. On the one hand, it seems a healthy line of argument. Typicality, by its very nature, grounds an epistemological asymmetry. Typical events do not call for explanation while atypical events prompt the resort to external intervention. This is precisely the point of Dürr's quotation above. From this, it follows that the typicality of the equilibrium state is all one needs to explain why thermal systems exhibit thermodynamic-like behavior.

On the other hand, it seems odd, to say the least, that the microscopic dynamics, as Zanghí says, should play no significant role in the process. After all, as a mechanical system is completely determined by its Hamiltonian, it looks strange that the evolution toward equilibrium should fail in this respect. As Roman Frigg notes, "whether or not [the system reaches equilibrium] depends on the dynamics of the system, and whether the dynamics is of the right kind is a question that cannot be answered by appeal to measure-theoretical arguments about the system's macroscopic structure" (Frigg, 2011, p. 82). To claim, as Zanghí (2005) and Goldstein do, that dynamics is largely irrelevant to explain thermodynamic-like behavior flies in the face of the foundations of statistical mechanics.

We have reached a curious deadlock. On the one hand, it is reasonable to claim that, if something is typical, exceptions should not be taken in serious account. That is precisely what 'overwhelmingly vast majority' is supposed to mean. On the other hand, it does not look right to state that equilibrium depends on the structure of the phase space and not on the dynamics. Little wonder that critics often regard STA as a sophisticated form of handwaving. But I suggest that this tension is rooted in the nature of typicality explanation I have outlined in section 2, and that the specificities of statistical mechanics only make it more evident. On the one hand, typicality always leaves open the possibility of an alternative — and somewhat deeper — analysis in terms of causal factors; on the other, statistical mechanics has been specifically designed to minimize the impact of the microscopic dynamics about which we have very little information.

To be sure, the difficulty was intuitively felt by the supporters of STA as well and they tried occasionally to mitigate it by resorting to some sort of causal factors. For example, back in 1993, Joel Lebowitz argued that the asymmetry derived by typicality was in fact a consequence of one key feature of thermal systems, the fact that they are constituted by a very high number of degrees of freedom:

> "The central role in time asymmetric behavior is played by the very large number of degrees of freedom involved in the evolution of the macroscopic systems. It is only this which permits statistical predictions to become 'certain' ones for typical individual realizations, where, after all, we actually observe irreversible behavior." (Lebowitz, 1993, p. 3)

The equilibrium state is as dominant as the impressive number of Goldstein shows precisely because there are very many degrees of freedom. The unidirectionality of the thermodynamic-like behavior is in fact a feature that emerges only when the number of degrees of freedom increases. The dynamical reason is that when there are very many particles, trajectory are unstable and atypical behaviors tend not to show up. In order to exhibit an atypical behavior such as entropy decrease, trajectories should be perfectly aimed at the tiny regions of non-equilibrium, but such a perfect aim, although theoretically possible, is practically out of reach because the high number of particles would make the system sensitive to minimal variations:

> "[i]t can therefore be expected to be derailed by even smaller imprecisions [...] and/or tiny random outside influences. This is somewhat analogous to those pinball machine type puzzle where one is supposed to get a small

metal ball into a particular small region. You have to do
things just right to get it in but almost anything you do
gets it out into larger region." (Lebowitz, 1993, p. 9)

Although these arguments introduce elements that certainly
belong to the dynamics of the system, for the critics of STA, they
are too little and too late. A satisfactory causal story of equilibrium
must compellingly include Hamiltonians and a way to do that will
be discussed in the next section.

5.3.3. *The Combined Typicality Approach*

The STA claims that thermal systems exhibit thermodynamic-like
behaviors because, given the typicality of the equilibrium in the phase
space, exceptions to such behavior are negligible. We have seen that
this approach denies a significant role for the underlying dynamics,
which is enough for many physicists to reject it. However, there is
also a profound epistemological problems with the STA. Using the
scheme described in section 2.1, we can derive from (EQ) that, if x is
a phase point, then x is an equilibrium microstate, meaning it belongs
to M_E. But this conclusion falls short of what we want to explain.
Remember that our explanandum concerns the thermodynamic-like
behavior of physical systems with trajectories generated by Hamiltonian flows. This means that we are not interested in the properties of
a generic microstate, but only in the microstates contained in the subset T_H. We can claim to have an explanation of the thermodynamic-
like behavior only if equilibrium is typical in that subset. But we only
know (EQ) and therefore to conclude our explanation, we need somehow to ground a condition stating that the typicality of equilibrium
carries over into the trajectory space. If the typicality of equilibrium
(EQ) and this carry-over condition hold true, then we can conclude
that a system trajectory consists mainly of equilibrium points and
that it manifests a thermodynamic-like behavior. Now, if we want to
construct a typicality explanation, then the carrying-over condition
must also be a TC, that is, it must have the following form:

(CC) Typically, if equilibrium is typical in Σ, then it's typical
in T_H.

In other words, (CC) describes what happens for trajectories generated by typical Hamiltonian flows. I call the combination of (EQ) and (CC) the Combined Typicality Approach (CTA) to equilibrium.

It is important to appreciate a subtle epistemological point. The motivation behind the CTA is not the ultra-skeptical attitude of taking seriously even physically negligible exceptions. Even if we concede that atypical events do not happen in practice, there is still a meaningful distinction to drawn between the accessible phase space and the trajectory space. While the former summarizes what a system can do given very general constraints such as the total energy, the latter amounts to what a particular system actually does given its specific dynamics. By debunking the dynamics, the STA implicitly denies that the distinction between mathematical possibilities and their physical realizations is meaningless as far as equilibrium is concerned. By contrast, the CTA stresses that this distinction has epistemological and metaphysical juice in it or, to put it in slightly different terms, there are epistemological and metaphysical constraints that intervene in constructing the trajectory space and that must also be operative in producing equilibrium.

One of the most elaborated attempts at developing a CTA is due to Roman Frigg and Charlotte Werndl (Frigg & Werndl, 2012; Werndl, 2013; Werndl & Frigg, 2015). According to Frigg and Werndl, what grounds (CC) is on specific physical properties, ε-ergodicity. Roughly, ε-ergodicity is the property according to which the trajectory of the system roams fairly uniformly the accessible phase space. If this is the case, the system will spend, in each macrostate, a time proportional to the size of the macrostate and, as the equilibrium state is by far the largest, the trajectory will naturally be almost always in equilibrium. To put it in other terms, if a system is ε-ergodic, the ratios between the size of the macrostates in the phase space will be almost identical to the ratios of the corresponding macrostates in the trajectory subspace so that, as required, the dominance of equilibrium will carry over into the trajectory subspace. In addition, Frigg and Werndl manage to prove that the property of ε-ergodicity is typical among perturbed Lennard-Jones Hamiltonians, which are a common type of Hamiltonian. Frigg and Werndl understand the typicality of ε-ergodicity in MT terms, which is to say they show that the vast majority of perturbed Lennard-Jones Hamiltonians are ε-ergodic. This MT analysis introduces a characteristic difficulty because "function spaces, unlike phase spaces, do not come equipped with normalized measures" (Frigg & Werndl, 2012, p. 921), so it is not obvious how one can 'count' Hamiltonians. To solve the problem, Frigg and Werndl introduce the concept of comeagre set as the topological equivalent of measure-1 set, and seek a suitable topology with respect to which ε-ergodicity is typical for perturbed Lennard-Jones Hamiltonians. It turns out that the Whitney topology satisfies this condition.

Thus, according to the account by Frigg and Werndl, the typicality explanation of the equilibrium proceeds as following. First one has the TC:

(EQ)   The property of being a state of equilibrium is typical in Σ.

which rests on the structure of the phase space and the use of the Lebesgue measure. Next, one introduces the new TC concerning the typicality of ε-ergodicity:

(εE) ε-ergodicity is typical among perturbed Lennard-Jones Hamiltonians.

From (εE), one can derive the carry-over condition:

(CC)   Typically, if the equilibrium is typical in Σ, then it's typical in T_H.

Putting these three statements together, one can conclude that, typically, thermal systems will exhibit thermodynamic behavior.

On the side of the STA, Dustin Lazarovici and Paula Reichert have taken up the task to respond to Frigg's and Werndl's challenge. They insist that the whole point of statistical mechanics is precisely to explain thermodynamic-like behavior without invoking the microscopic dynamics, and that both ergodicity and ε-ergodicity are simply irrelevant for the purpose. Apart from the technical arguments produced against the Frigg-Werndl proposal, it is interesting to look at the philosophical rationale of Lazarovici and Reichert, because it shows where the crux of the debate lies and why it needs the epistemological reassessment I have tried to develop in this article.

According to Lazarovici and Reichert, what is lurking behind the complains about the neglected microdynamics is the worn-out deductive-nomological model of explanation (Lazarovici & Reichert, 2015, pp. 705–706). Critics such as Frigg and Werndl, so the argument goes, assume that to *explain* the thermodynamic-like behavior, one is required to derive it from dynamical laws and auxiliary assumptions, and in so doing they miss the point of Boltzmann's lesson: we cannot do such a derivation, we can only show that almost all initial conditions lead to equilibrium. Both camps, however, strike me as rather philosophically unsophisticated in this respect. As I have argued in section 2, typicality explanations can be carried out in two different fashions according to the circumstances of the problem at hand. Thus, Frigg's and Werndl's criticism that the STA unduly neglects dynamical considerations is a reasonable motivation to develop a CF-analysis, but it can hardly debunk the explanatory value of STA. Conversely, Lazarovici and Reichert seem to overlook the point that the search for dynamical factors is not a belated homage paid to the deductive-nomological model, but an approach wholly consistent with the epistemological structure of typicality explanations. As I have argued in section 2.1, typicality works explanatorily as an Hempelian surprise-remover, but the argumentative scheme it requires is not a deductive derivation. It seems, in other words, that the debate on the explanatory value of typicality in statistical mechanics is taking place before raising the key preliminary question of what is the explanatory value of typicality in general. The consequence is that both camps throw at each other fairly subtle physical arguments which, however, often fail to contribute to a well-defined philosophical project.

5.4. Conclusion

The previous discussion paints a highly fractured picture. Experts' opinions are fiercely split on the practice of the typicality explanation in statistical mechanics. On the one hand, the upholders of the STA claim that nothing more that the typicality of equilibrium in the accessible phase space is needed because the accessible exceptions are negligible. On the other hand, physicists and philosophers supporting the CTA reply that the underlying dynamics must play a substantial role and that a line must be drawn between the accessible phase space and what the system does in actual fact.

Underlying this clash is a more profound cleavage about the epistemological requirements of a satisfactory typicality explanation of equilibrium. Supporters of STA claim that the structure of the phase space is sufficient, possibly supplemented by considerations on the high number of degrees of freedom and trajectory instability. On the other side of the trenches, supporters of the CTA insist that a more robust causal account is required, which include claims on the typicality of certain physical properties among common Hamiltonians.

In this article, I have introduced a general epistemological framework and some useful notions to reconceptualize the debate on typicality in statistical mechanics. Now, I want to conclude on a slightly more ambitious note. One of the points established in section 2.2 is that typicality is a *pluralistic concept*. By saying that a property is typical, we might be referring to the whole of our experience of the commonality of that property, or to its being characteristic of a certain natural type. At times, typicality can be linked to the phenomenological act of counting or to the metaphysical investigation of tendencies. How to proceed in decoding a TC is largely a balance between contingent factors, such as our epistemological access to the system, and our explanatory goals. Choosing the MT analysis over the CF analysis is a matter that depends inherently on the problem at hand, and it is not a question that must have only one answer. In section 2.1. I have defended a very minimalistic conception of typicality explanation as the use of TC for explanatory purposes. This conception is justified precisely by the pluralistic nature of typicality. Apart from the inferential schema described in section 2.1, typicality explanations can differ substantially both epistemologically and metaphysically. From this perspective, the variety of positions in statistical mechanics is scarcely surprising and a pluralistic attitude toward the issue of typicality is high recommendable. The changing combination of measure-theoretical and causal factors is more of a plus than a minus; it reveals that a full picture of equilibrium requires the cooperation of multiple methods.

Interestingly enough, the inspiring figure of the typicality approach, Ludwig Boltzmann, understood this point very well. The supporters of the STA love to present their work as the truth continuation of Boltzmann's legacy. But beside the famous combinatorial argument, Boltzmann approached the problem of equilibrium in a variety of ways. For example, he made use of the ergodic hypothesis, which in turn he grounded in the high number of degrees of freedom. In another place, he argued that mechanical reversibility is practically impossible because of the instability of the atomic trajectory. He even mentioned the idea that a violation of the thermodynamic-like behavior requires a sort of conspiration:

> "If we choose the initial configuration on the basis of a previous calculation of the path of each molecule, so as to violate intentionally the laws of probability, then of course we can construct a persistent regularity." (Boltzmann, 1896, pp. I, 22)

In employing all these ideas, Boltzmann always maintained a pluralistic stance. He thought that a comprehension of equilibrium requires a careful combination of measure-theoretical considerations (what we know call) and mechanical theories. This is, I believe, a sensible attitude. Typicality explanation is not an all-or-nothing affair, but rather a business of balancing multiple epistemological constraints and metaphysical requirements to gain a total picture of the phenomenon of equilibrium.

References

Badino, M. (2011) Mechanistic Slumber vs. Statistical Insomnia: The Early History of Boltzmann's H-theorem, *European Physical Journal H*, **36**, 353–378.

Badino, M. (2015) *The Bumpy Road: Max Planck from Radiation Theory to the Quantum (1896-1906)* Springer, New York.

Boltzmann, L. (1877) Über die Beziehung zwischen dem zweiten Hauptsatze der mechanischen Wärmetheorie und der Wahrscheinlichkeitsrechnung respective den Sätzen über das Wärmegleichgewicht. *Sitzungsberichte der Akademie der Wissenschaften zu Wien*, **76**, 373–435.

Boltzmann, L. (1896) *Vorlesungen über Gastheorie.* Leipzig: Barth.

Dürr, D. (2001) Bohmian Mechanics. In J. Bricmont, D. Dürr, M. C. Galavotti, G. Ghirardi, F. Petruccione, & N. Zanghí (Eds.), *Chance in Physics: Foundations and Perspectives* (pp. 115–132), Springer, Berlin.

Dürr, D., Goldstein, S., Zanghí, N. (1992) Quantum Equilibrium and the Origin of Absolute Uncertainty, *Journal of Statistical Physics*, **67**, 843–907.

Frigg, R. (2007) Why Typicality Does Not Explain the Approach to Equilibrium. In M. Suarez (Ed.), *Probabilities, Causes and Propensities in Physics* (pp. 77–93), Springer, Berlin.

Frigg, R. (2009) Typicality and the Approach to Equilibrium in Boltzmannian Statistical Mechanics, *Philosophy of Science*, **76**(5), 997–1008.

Frigg, R. (2011) Why Typicality Does Not Explain the Approach to Equilibrium. In M. Suarez (Ed.), *Probabilities, Causes and Propensities in Physics* (pp. 77–93).

Frigg, R., Werndl, C. (2012) Demystifying Typicality, *Philosophy of Science*, **79**(5), 917–929.

Goldstein, S. (2001) Boltzmann's Approach to Statistical Mechanics. In J. Bricmont, D. Dürr, M. C. Galavotti, G. Ghirardi, F. Petruccione, & N. Zanghí (Eds.), *Chance in Physics: Foundations and Perspectives* (pp. 39–54), Springer, Berlin.

Goldstein, S. (2012) Typicality and Notions of Probability in Physics. In Y. Ben-Menahem & M. Hemmo (Eds.), *Probability in Physics* (pp. 59–71), Springer, Berlin.

Goldstein, S., Lebowitz, J. L. (2004) On the (Boltzmann) Entropy of Non-Equilibrium Systems, *Physica D*, **193**(53–66).

Hemmo, M., Shenker, O. R. (2012) *The Road to Maxwell's Demon. Conceptual Foundations of Statistical Mechanics*, Cambridge University Press, Cambridge.

Lazarovici, D., Reichert, P. (2015) Typicality, Irreversibility and the Status of Macroscopic Laws, *Erkenntnis*, **80**(4), 689–716. doi:10.1007/s10670-014-9668-z

Lazarovici, D., Reichert, P. (Unpublished) *Against (Epsilon-)Ergodicity.*

Lebowitz, J. L. (1993) Macroscopic Laws, Microscopic Dynamics, Time's Arrow and Boltzmann's Entropy, *Physica A*, **194**, 1–27.

Lebowitz, J. L. (1999) Statistical Mechanics: A Selective Review of Two Central Issues, *Reviews of Modern Physics*, **71**, 346–357.

Pitowsky, I. (2012) Typicality and the Role of the Lebesgue Measure in Statistical Mechanics. In Y. Ben-Menahem & M. Hemmo (Eds.), *Probability in Physics* (pp. 41–58), Springer, Berlin.

Volchan, S. B. (2007) Probability as Typicality, *Studies in History and Philosophy of Modern Physics*, **38**(4), 801–814. doi:10.1016/j.shpsb.2006.12.001

Werndl, C. (2013) Justifying Typicality Measures of Boltzmannian Statistical Mechanics and Dynamical Systems, *Studies in History and Philosophy of Modern Physics*, **44**(4), 470–479. doi:10.1016/j.shpsb.2013.08.006

Werndl, C., Frigg, R. (2015) Reconceptualising Equilibrium in Boltzmannian Statistical Mechanics and Characterising its Existence, *Studies in History and Philosophy of Modern Physics*, **49**, 19–31.

Woodward, J. (2000) Explanation and Invariance in the Special Sciences, *British Journal for the Philosophy of Science*, **51**, 197–254.

Woodward, J. (2003) *Making Things Happen. A Theory of Causal Explanation.* Oxford: Oxford University Press.

Zanghí, N. (2005) I fondamenti concettuali dell'approccio statistico in fisica. In V. Allori, M. Dorato, F. Laudisa, & N. Zanghí (Eds.), *La natura delle cose: Introduzione ai fondamenti e alla filosofia della fisica* (pp. 139–227). Rome: Carocci.

Chapter 6

The Logic of Typicality

Harry Crane and Isaac Wilhelm

The notion of typicality appears in scientific theories, philosophical arguments, mathematical inquiry, and everyday reasoning. Typicality is invoked in statistical mechanics to explain the behavior of gases. It is also invoked in quantum mechanics to explain the appearance of quantum probabilities. Typicality plays an implicit role in non-rigorous mathematical inquiry, as when a mathematician forms a conjecture based on personal experience of what seems typical in a given situation. Less formally, the language of typicality is a staple of the common parlance: we often claim that certain things are, or are not, typical. But despite the prominence of typicality in science, philosophy, mathematics, and everyday discourse, no formal logics for typicality have been proposed. In this paper, we propose two formal systems for reasoning about typicality. One system is based on propositional logic: it can be understood as formalizing objective facts about what is and is not typical. The other system is based on the logic of intuitionistic type theory: it can be understood as formalizing subjective judgments about typicality.

6.1. Introduction

Typically, gases in non-equilibrium macrostates evolve to the equilibrium macrostate relatively quickly. Not all gases do; in fact, the initial microstates of some gases prevent them from ever reaching equilibrium. But those initial microstates are unusual, or atypical. Nearly all initial microstates are not like that; nearly all initial microstates lead to equilibrium after a short while. This is a 'typicality fact': a fact about what is typical.

Typicality facts are studied in many areas of science, but they are particularly prominent in statistical mechanics and quantum mechanics. For instance, Boltzmann discusses a version of the typicality fact just mentioned: the overwhelming majority of initial conditions of a gas, he writes, reach equilibrium in a relatively short amount of time (1896/2003, p. 394). On the basis of his many-worlds interpretation of quantum mechanics, Everett argues that typically, the probabilistic predictions of the Born rule are valid (1956/2012, p. 123). In their analysis of Bohmian mechanics, Dürr, Goldstein, and Zanghì show that typically, initial configurations of the universe lead to empirical distributions that agree with the probabilistic predictions of the quantum formalism (1992, p. 846). Reimann (2007) shows that given certain generic conditions, pure quantum states typically yield more or less the same expectation values for sets of observables which are not too large. Kiessling (2011) shows that for N gravitationally-interacting bodies confined to the surface of a sphere, Boltzmann's H functional is minimized by states which are typical for those N-body systems (in the limit as N approaches infinity). Tasaki (2016) shows that pure quantum states in the microcanonical energy shell typically share a particular collection of properties associated with thermal equilibrium.[1]

These typicality facts are explanatory. The typicality fact about the initial conditions of gases, for example, explains their thermodynamic behavior (Goldstein, 2001, p. 52). Both the typicality fact discussed by Everett (2012) and the typicality fact discussed by Dürr, Goldstein, and Zanghì (1992) provide explanations of why the frequencies observed in quantum experiments conform to the probabilities predicted by the Born rule. There is, as Goldstein puts

---

[1]For examples of typicality results in mathematics, see Kesten (1980), Alon *et al.* (1998), and Ledoux (2001).

it, a "logic of appeal to typicality" in scientific explanation (2012, p. 70).

The logic of typicality extends beyond its role in explanation, however. Typicality is invoked in many different kinds of scientific reasoning: explanation, prediction, evaluation of hypotheses, and more. But what is the logic of that reasoning?[2]

Typicality is also invoked in non-rigorous mathematical reasoning. Take Goldbach's conjecture,[3] which has been shown to hold for all natural numbers less than 10^{18}.[4] The set of integers on which the conjecture has been verified is large but finite, and therefore does not pass the 'nearly all' threshold to be considered typical in the sense described above. But consider a number theorist whose past experience is such that when a mathematical claim has been verified on a similarly large sample of natural numbers, that claim has most often turned out to be true. Based on this experience, the number theorist may feel justified to conjecture that the claim holds in this specific case. Such judgments are based on assessments of typicality; they are based on reasoning to the effect that past conjectures which have been empirically tested to the same degree as Goldbach's conjecture have typically turned out to be true. This is a 'typicality judgment': a judgment about what is typical.

Typicality judgments are common in science, mathematical conjecture, and everyday reasoning. They help scientists arrive at hypotheses to test. They help mathematicians posit conjectures to

---

[2]The notion of typicality has historical roots in the writings of Bernoulli (1713) and Cournot (1843). Both Bernoulli and Cournot formulated principles of typicality reasoning using the notion of probability, not typicality. Roughly, they argued that events with very high probability are 'morally certain', and events with very small probability are 'morally impossible' (Shafer & Vovk, 2006, p. 72). These principles of reasoning may be different from the analogous principles that replace the notion of probability with the notion of typicality, since arguably, probability and typicality are distinct (Goldstein, 2012; Wilhelm, forthcoming). But regardless, these principles of reasoning are at least direct ancestors of similar principles based on the notion of typicality.

[3]Every even integer greater than 2 can be expressed as the sum of two primes.

[4]Empirical verification of Goldbach's conjecture is catalogued at http://sweet.ua.pt/tos/goldbach.html.

prove.[5] And they help guide reasoning in other domains as well. For example, consider a prosecutor who attempts to convince a jury of a defendant's guilt by linking blood at the crime scene to the defendant. This prosecutor appeals to typicality. For typically, the presence of blood indicates the defendant's involvement in the crime. It would be quite atypical, though not impossible, for the defendant's blood to be present if the defendant were in no way involved.

Like typicality facts, typicality judgments can figure in our explanations: they can justify conjectural claims which have not yet been proven, but which has sufficient evidence to support. Though Goldbach's conjecture has remained unresolved for over three centuries, its plausibility is undisputed. There is lots of empirical evidence in its favor, though a formal proof remains elusive. Even without a proof, belief in the conjecture seems to be based on sound reasoning. But what is the logic of that reasoning?

Despite the ubiquity of typicality reasoning, no formal systems for the logic of typicality have been proposed.[6] Most research on typicality either (i) proves results about what is typical, as in quantum mechanics and statistical mechanics; or (ii) explicates the notion of typicality — as in philosophy.[7] There is comparatively little research on the logical principles which govern reasoning that relies upon typicality facts. There are no rigorous formal languages designed to model claims about what is and is not typical, or to assess the soundness of typicality judgments. There is no detailed formal semantic theory and no detailed proof theory for reasoning about typicality. Of course, there are rigorous formal systems for other sorts of reasoning.

---

[5]Pólya (1954) discussed something like this 'non-rigorous' side of mathematics in depth in his two-volume treatise *Mathematics and Plausible Reasoning*. Mazur (2012) has also explored the role of plausibility in mathematical practice.

[6]That is, there are no formal systems for our notion of typicality, according to which something is typical when, roughly, *nearly all* things of a certain sort are a certain way. There are proposed formal systems for other typicality notions. One notion of typicality, for instance, is given by the notion of 'normal': something is typical just in case it is normal (relative to the entities in some class). See (Booth *et at.*, 2012) for a logic of this 'normalcy' notion of typicality.

[7]For recent work on the philosophical foundations of typicality, see Frigg (2011), Frigg & Werndl (2012), Werndl (2013), and Wilhelm (forthcoming).

Propositional logic and first-order logic are formal systems for deductive reasoning. Bayesian theory is a formal system for probabilistic reasoning. But there are no analogous formalisms for typicality reasoning.[8]

There are at least two other reasons to develop a logical system for typicality reasoning. First, such a system would unite and systematize the different ways of quantifying typicality: it would show that the different typicality measures employed by Everett (1956/2012), Dürr *et al.* (1992), Reimann (2007), and others are species of a common genus. In other words, a logical system for typicality reasoning would capture the formal unity of a wide variety of typicality results. It would reveal what many different approaches to typicality have in common. Second, and relatedly, a logical system for typicality reasoning would formulate the basic logical principles that seem to govern the intuitive notion of typicality. Here is an example of one such principle: if $p \wedge q$ is typical, then p is typical and q is typical, but if p is typical and q is typical, then it does not follow that $p \wedge q$ is typical. Here is another: if either p is typical or q is typical, then $p \vee q$ is typical, but if $p \vee q$ is typical, then it does not follow that either p is typical or q is typical. A logical system for typicality reasoning would be a rigorous theory of principles which, like these two, govern all rational reasoning that relies on claims about what is typical.

So in this paper, we present two logical systems for typicality. The first is propositional: it supplements the standard language of propositional logic with a new sentence operator 'Typ'. Intuitively, $Typ(p)$ says that p is typical. The second is type-theoretic: it supplements the standard language of intuitionistic Martin-Löf type theory (MLTT) (Martin-Löf, 1984) by introducing a new type former **Typ**. Intuitively, **Typ**(p) is a type corresponding to the proposition that p is typical, and each term of **Typ**(p) represents a justification for the judgment that p is typical.

In Section 6.2, we present the propositional formalism for typicality. We introduce the language, the semantics, and the proof theory for what we call *Typicality Propositional Logic* (TPL). We

---

[8]Steps towards a formalism are taken by Goldstein *et al.* (2010, pp. 3217–3220).

also establish some formal results. In Section 6.3, we present the type-theoretic formalism for typicality. We introduce the language of Martin-Löf type theory (MLTT), along with the semantics and proof theory for what we call *Typicality Intuitionistic Logic* (TIL).

6.2. Typicality propositional logic

In this section, we propose a formal logic for TPL. We introduce the language of that formalism — the basic vocabulary and the well-formed formulas — in Section 6.2.1. In Section 6.2.2, we propose a semantic theory for this language, and we prove some simple yet illuminating results. In Section 6.2.3, we propose a proof theory. In Section 6.2.4, we show that the proof theory is sound with respect to the semantic theory. Finally, in Section 6.2.5, we discuss some additional features of TPL.

6.2.1. *The language*

TPL is the language of standard propositional logic supplemented with a typicality operator 'Typ'. In particular, the logical vocabulary of TPL consists of three symbols: a binary sentence operator $\rightarrow$, a unary sentence operator $\neg$, and a unary sentence operator Typ. The non-logical vocabulary of TPL consists of infinitely many sentence letters and two bracket symbols: the sentence letters are p, q, and so on, and the bracket symbols are (and).

Well-formed formulas in the language of TPL are defined recursively, as follows.

(1) Each sentence letter is a well-formed formula.
(2) If ϕ is a well-formed formula, then $\neg\phi$ is a well-formed formula.
(3) If ϕ is a well-formed formula, then $Typ(\phi)$ is a well-formed formula.
(4) If ϕ and ψ are well-formed formulas, then $\phi \rightarrow \psi$ is a well-formed formula.
(5) Nothing else is a well-formed formula.

It follows that all well-formed formulas in the language of propositional logic are well-formed formulas in the language of TPL.

6.2.2. *The Semantics of TPL*

The models of the well-formed formulas of TPL are named 'TPL universes'. Each TPL universe is a pair $\langle \Gamma, \mathcal{V} \rangle$, where Γ is a large set and $\mathcal{V}$ is a set of truth functions from well-formed formulas of TPL to $\{0, 1\}$. Intuitively, Γ is a set of possible states or possible worlds, and $\mathcal{V}$ is a set of functions which assign a truth value to each sentence, with 0 representing 'false' and 1 representing 'true'. For each $w \in \Gamma$, there is exactly one function $f_w \in \mathcal{V}$. Intuitively, f_w expresses the facts about truth propositions at w: for well-formed formula ϕ, for $w \in \Gamma$, and for $f_w \in \mathcal{V}$, $f_w(\phi) = 1$ says that ϕ is true at world w.

The rigorous definition of the truth functions proceeds in two steps. First, for each $w \in \Gamma$, let g_w be a truth function defined over all well-formed formulas of the language PL, where PL is the standard language for propositional logic. So each g_w is just a truth function of propositional logic.

Second, for each $w \in \Gamma$, extend g_w to a truth function f_w defined over all of TPL. The extension is defined in terms of double recursion: at each step in the recursion, in addition to defining the truth function f_w for that step, a set must also be defined. Roughly put, the defined set is the set of all elements of Γ at which a certain well-formed formula is true.

More precisely, let $\mathcal{S}$ be a σ-algebra over Γ and let τ be a finite, non-zero measure over $\mathcal{S}$. Fix $\epsilon > 0$ such that $\epsilon \ll 1$. Then for each $w \in \Gamma$, the recursive definition of f_w is as follows.

(1) For each well-formed formula ϕ in the language of PL,

    (i) $f_w(\phi) = g_w(\phi)$, and
    (ii) $\Gamma_\phi = \{w' \in \Gamma \mid f_{w'}(\phi) = 1\}$.

(2) If ϕ is a well-formed formula in the language of TPL, then

    (i) $f_w(\neg\phi) = 1$ if and only if $f_w(\phi) = 0$, and
    (ii) $\Gamma_{\neg\phi} = \{w' \in \Gamma \mid f_{w'}(\phi) = 0\}$.

(3) If ϕ and ψ are well-formed formulas in the language of TPL, then

(i) $f_w(\phi \to \psi) = 1$ if and only if $f_w(\phi) = 0$ or $f_w(\psi) = 1$, and

(ii) $\Gamma_{\phi \to \psi} = \{w' \in \Gamma \mid f_{w'}(\phi) = 0 \text{ or } f_{w'}(\psi) = 1\}$.

(4) If ϕ is a well-formed formula in the language of TPL, then

(i) $f_w(Typ(\phi)) = 1$ if and only if there exists a set $X \in \mathcal{S}$ such that $X \subseteq \Gamma_\phi$ and $\frac{\tau(\Gamma \backslash X)}{\tau(\Gamma)} < \epsilon$, and

(ii) $\Gamma_{Typ(\phi)} = \{w' \in \Gamma \mid f_{w'}(Typ(\phi)) = 1\}$.

It follows from these clauses that for each well-formed formula ϕ, Γ_ϕ is the set of worlds in Γ at which ϕ is true; that is, $\Gamma_\phi = \{w' \in \Gamma \mid f_{w'}(\phi) = 1\}$. For the purposes of the proof theory in Section 6.2.3, say that $Y \subseteq \Gamma$ is a 'typical set' if and only if there is an $X \in \mathcal{S}$ such that $X \subseteq Y$ and $\frac{\tau(\Gamma \backslash X)}{\tau(\Gamma)} < \epsilon$.

Here is an informal description of what these clauses say. According to the first clause, f_w agrees with g_w on the formulas of PL. The second clause uses the value of f_w on unnegated formulas in TPL to define the value of f_w on negated formulas in TPL. The third clause uses the value of f_w on pairs of formulas in TPL to define the value of f_w on conditionals created out of those formulas in TPL. The fourth clause is more involved: it uses facts about the measures of sets to define the value of f_w on typicality statements in TPL. Intuitively, the fourth clause says that $Typ(\phi)$ is true at a world if and only if the set of worlds at which ϕ holds — that is, the set Γ_ϕ — contains a 'sufficiently large' set X, where X is 'sufficiently large' if and only if X is measurable and the size of the set of elements not in X (divided by the size of the set Γ) is very small.

Note that in condition 4(i), the sizes of sets are quantified using a measure. But there are other ways of quantifying the sizes of sets. For example, a cardinality-theoretic version of 4(i) can be used for the case where Γ is infinite:[9] $f_w(Typ(\phi)) = 1$ if and only if $|\Gamma \backslash \Gamma_\phi| < |\Gamma|$. In other words, ϕ is typical if and only if the set of $\neg \phi$ worlds

---

[9]We stipulate that Γ is infinite because if Γ were finite, then this version of 4(i) would imply that $f_w(Typ(\phi)) = 1$ if and only if Γ_ϕ is nonempty. But this means that ϕ is typical if and only if ϕ is true in at least one world — that is not how the intuitive notion of typicality works.

has strictly smaller cardinality than the set of ϕ worlds.[10] For a topological version of 4(i): $f_w(Typ(\phi)) = 1$ if and only if $\Gamma \setminus \Gamma_\phi$ is meager.[11] In other words, put roughly, ϕ is typical if and only if the set of $\neg\phi$ worlds is tightly packed, topologically, in Γ. And here is a version of 4(i) related to, but distinct from, the measure-theoretic version: given $\epsilon > 0$ such that $\epsilon \ll 1$, given a field of sets $\mathcal{F}$ whose members are subsets of Γ, and given a finite non-zero finitely additive measure μ on $\mathcal{F}$, $f_w(Typ(\phi)) = 1$ if and only if there exists a set $X \in \mathcal{F}$ such that $X \subseteq \Gamma_\phi$ and $\frac{\mu(\Gamma \setminus X)}{\mu(\Gamma)} < \epsilon$.

According to condition 4(i), the truth of a typicality statement depends on the value of ϵ. The parameter ϵ serves as a measure of smallness: very roughly, ϕ is typical if and only if ϕ is false at a sufficiently small proportion of worlds, where a proportion of worlds is 'sufficiently small' if and only if that proportion is smaller than ϵ. So different choices of ϵ yield different truth conditions for claims about what is typical.

There are roughly two different views of the relationship between ϵ and the truth conditions for typicality statements. According to one view — call it the 'context-dependent view' of ϵ — many different values for ϵ are permissible, and there is a different version of condition 4(i) for each such value. The correct value for ϵ — the value, that is, which correctly captures the truth conditions of typicality statements — is determined by context. But different contexts can determine different values for ϵ. For example, consider a context in which physicists are studying the entropic behavior of gases. These physicists make veridical claims like "Typically, gases with such-and-such an initial macrostate evolve to a higher-entropy macrostate in thus-and-so amount of time". In this context, the value of ϵ may be as low as 10^{-100}, because an absolutely massive number of gases exhibit the entropy-increasing behavior in question. In contrast, consider a context in which biologists are studying the behaviors of cells. These

---

[10] As discussed in Wilhelm (forthcoming), this quantification of typicality is only suitable for modeling some typicality claims.

[11] A set is meager if and only if it can be written as a countable union of nowhere dense sets. A set is nowhere dense if and only if its closure has empty interior.

biologists make veridical claims like "Typically, cells with sodium-potassium pumps transport such-and-such many sodium ions in thus-and-so amount of time". In this context, the value of ϵ may only be 10^{-4}, because the failure rate of the relevant transport processes is low but not as low as 10^{-100}.

According to another view — call it the 'context-independent view' of ϵ — only one value of ϵ is correct: only one version of condition 4(i) can be used in typicality reasoning. That value for ϵ is always the same; it does not vary with context. For example, the value of ϵ in the context of physicists studying gases is the same as the value of ϵ in the context of biologists studying cells.

We prefer the context-dependent view of ϵ. Typicality, in our view, is a context-dependent notion: whether or not something is typical varies with the standards of the context in question. A formal system for typicality should capture that contextual variability. This does not make TPL any less exact, however, nor does this make TPL imprecise. By invoking the parameter ϵ, TPL exactly and precisely captures the inexactness and imprecision which is inherent in the notion of typicality. A formal system which fails to capture the contextual variability of typicality is not any more exact or precise than TPL. Such a formal system is, in fact, not a formal system for typicality at all.

But in this paper, for the sake of brevity, we will not explore different ways of varying the parameter ϵ. So in the theorems to follow, we will assume that ϵ has been fixed to some particular value. We hope that future work will explore the consequences of allowing the value of ϵ to vary.[12]

We now present an account of truth at a world in a TPL universe, an account of truth in a TPL universe, and an account of logical truth. Let $M_0 = \langle \Gamma_0, \mathcal{V}_0 \rangle$ be a TPL universe, let $w_0 \in \Gamma_0$ be a world, and let ϕ be a well-formed formula in the language of TPL. First, ϕ is 'true in M_0 at w_0' if and only if $f_{w_0}(\phi) = 1$. In symbols: $\vDash_{M_0,w_0} \phi$. Second, ϕ is 'true in M_0' if and only if for every $w \in \Gamma_0$, ϕ is true in

---

[12]See footnote 13 for one example of how the present theory can be adapted to allow for contextual variable ϵ.

M_0 at w. In symbols: $\vDash_{M_0} \phi$. Third, ϕ is 'logically true' if and only if for every TPL universe M, ϕ is true in M. In symbols: $\vDash \phi$.

The definition of logical entailment in TPL is as follows. Let Σ be a set of well-formed formulas in the language of TPL. Let M be a TPL universe. Say that 'Σ logically entails ψ in M' if and only if the following holds: if ϕ is true in M for each $\phi \in \Sigma$, then ψ is true in M. In symbols, $\Sigma \vDash_M \psi$. And say that 'Σ logically entails ψ' if and only if for each TPL universe M, Σ logically entails ψ in M. In symbols, $\Sigma \vDash \psi$.[13]

A few simple results will help to clarify the nature of typicality in this semantic theory. The following lemma shows that typicality statements are 'all or nothing': either $Typ(\phi)$ is true at each world in the universe at issue or $Typ(\phi)$ is false at each world in the universe at issue.

Lemma 1. *Let* $\langle \Gamma, \mathcal{V} \rangle$ *be a TPL universe. For any well-formed formula* ϕ *in the language of TPL, either* $\Gamma_{Typ(\phi)} = \Gamma$ *or* $\Gamma_{Typ(\phi)} = \emptyset$.

Proof. Take any well-formed formula ϕ in the language of TPL. Suppose $\Gamma_{Typ(\phi)} \neq \Gamma$. Then, there is a world $w \in \Gamma$ such that $f_w(Typ(\phi)) = 0$. Recall that for any $w' \in \Gamma$, $f_{w'}(Typ(\phi)) = 1$ if and only if there exists a set $X \in \mathcal{S}$ such that $X \subseteq \Gamma_\phi$ and $\frac{\tau(\Gamma \backslash X)}{\tau(\Gamma)} < \epsilon$. Therefore, since $f_w(Typ(\phi)) = 0$, there is no set X which satisfies

---

[13]This definition of logical entailment depends upon a specific choice of $\epsilon > 0$ such that $\epsilon \ll 1$, since ϵ was fixed to a single value for the purposes of the recursive definitions of the functions f_w that assign truth values to sentences at worlds in TPL universes. But this definition of logical entailment can easily be adapted to allow for multiple values of ϵ. To do so, simply call this definition — the one which depends upon a fixed ϵ — an 'ϵ-relative' definition of logical entailment. Then, say that the set of sentences Σ logically entails the sentence ψ if and only if for all $\epsilon > 0$ such that $\epsilon < \frac{1}{2}$, Σ logically entails ψ according to the ϵ-relative definition of logical entailment. The choice of $\frac{1}{2}$ as an upper bound is not forced: a different number, for instance $\frac{1}{10^4}$ or $\frac{1}{10^{100}}$, could be used instead. But $\frac{1}{2}$ seems like the least arbitrary choice for this non-ϵ-relative definition of logical entailment. For any other choice, it seems reasonable to ask why ϵ cannot be just a little bit higher; and it is not obvious what the answer would be. But clearly, ϵ cannot be $\frac{1}{2}$ or greater: for if something is typical, then at the very least, it must be true more than half of the time.

those conditions. Therefore, for every $w' \in \Gamma$, $f_{w'}(Typ(\phi)) = 0$. And therefore, $\Gamma_{Typ(\phi)} = \emptyset$. So either $\Gamma_{Typ(\phi)} = \Gamma$ or $\Gamma_{Typ(\phi)} = \emptyset$. $\qquad\square$

Lemma 1 justifies the use of the phrase 'typicality facts' to refer to typicality statements in TPL. When a typicality statement holds at some world, it holds at all worlds. When a typicality statement fails to hold at some world, it fails to hold at all worlds. Therefore, the truth value of a typicality statement is constant over all worlds in a TPL universe, and is thus a *fact* about that universe.

The following theorem shows that iterated typicality claims do not change in truth value: that is, $Typ(Typ(\phi))$ holds if and only if $Typ(\phi)$ holds.

Theorem 1. *For any well-formed formula ϕ in the language of TPL, for any TPL universe $M = \langle \Gamma, \mathcal{V} \rangle$, and for any $w \in \Gamma$, $Typ(Typ(\phi))$ is true in M at w if and only if $Typ(\phi)$ is true in M at w.*

Proof. By definition, $Typ(Typ(\phi))$ is true in M at w if and only if $f_w(Typ(Typ(\phi))) = 1$. Similarly, $Typ(\phi)$ is true in M at w if and only if $f_w(Typ(\phi)) = 1$. So to establish the theorem, it suffices to show that $f_w(Typ(Typ(\phi))) = f_w(Typ(\phi))$.

By Lemma 1, either $\Gamma_{Typ(\phi)} = \emptyset$ or $\Gamma_{Typ(\phi)} = \Gamma$. To start, suppose that $\Gamma_{Typ(\phi)} = \emptyset$. Then for each set $X \in \mathcal{S}$ such that $X \subseteq \Gamma_{Typ(\phi)}$, $X = \emptyset$. Therefore, for such an X,

$$\frac{\tau(\Gamma \setminus X)}{\tau(\Gamma)} = \frac{\tau(\Gamma \setminus \emptyset)}{\tau(\Gamma)}$$

$$= \frac{\tau(\Gamma)}{\tau(\Gamma)}$$

$$= 1$$

$$\not< \epsilon.$$

And so by definition, $f_w(Typ(Typ(\phi))) = 0$. In addition, if $\Gamma_{Typ(\phi)} = \emptyset$, then $f_w(Typ(\phi)) = 0$ by definition. Therefore, if $\Gamma_{Typ(\phi)} = \emptyset$, then $f_w(Typ(Typ(\phi))) = f_w(Typ(\phi))$.

Now suppose that $\Gamma_{Typ(\phi)} = \Gamma$. Then there is a set $X \in \mathcal{S}$ such that $X \subseteq \Gamma_{Typ(\phi)}$ and $\frac{\tau(\Gamma \setminus X)}{\tau(\Gamma)} < \epsilon$; namely, the set $X = \Gamma$. So by definition, $f_w(Typ(Typ(\phi))) = 1$. In addition, if $\Gamma_{Typ(\phi)} = \Gamma$, then $f_w(Typ(\phi)) = 1$ by definition. So if $\Gamma_{Typ(\phi)} = \Gamma$, then $f_w(Typ(Typ(\phi))) = f_w(Typ(\phi))$.

Therefore, regardless of what $\Gamma_{Typ(\phi)}$ is, $f_w(Typ(Typ(\phi))) = f_w(Typ(\phi))$. □

The following corollary establishes the corresponding result for the sets $\Gamma_{Typ(Typ(\phi))}$ and $\Gamma_{Typ(\phi)}$.

Corollary 1. *For any well-formed formula ϕ in the language of TPL, and for any TPL universe $M = \langle \Gamma, \mathcal{V} \rangle$, $\Gamma_{Typ(Typ(\phi))} = \Gamma_{Typ(\phi)}$.*

Proof. By definition, $\Gamma_{Typ(Typ(\phi))} = \{w' \in \Gamma \mid f_{w'}(Typ(Typ(\phi))) = 1\}$ and $\Gamma_{Typ(\phi)} = \{w' \in \Gamma \mid f_{w'}(Typ(\phi)) = 1\}$. Theorem 1 implies that for each $w \in \Gamma$, $f_w(Typ(Typ(\phi))) = f_w(Typ(\phi))$. Therefore, $\Gamma_{Typ(Typ(\phi))} = \Gamma_{Typ(\phi)}$. □

6.2.3. *Tableaus for TPL*

In this subsection, we propose a proof theory for TPL based on tableaus; call it a 'TPL proof theory'.[14] The proof theory consists of a series of rules for decomposing well-formed formulas in the language of TPL. The decomposition resembles a tree-like structure; because of that, tableau proofs are often called 'proof trees'.

Each line in a proof tree consists of two parts: a formula and an index. The formula is just a well-formed formula in the language of TPL. The index is a syntactic parameter that keeps track of certain information about that well-formed formula. Intuitively, the index specifies the worlds at which the formula is true. There are three different types of indices: the 'all' index, denoted by the symbol a; 'nearly all' indices, denoted by symbols like n, n', and so on; and

---

[14] For discussion of tableau methods in propositional logic, first-order logic, and modal logic, see Smullyan (1968) and Priest (2008).

'world' indices, denoted by symbols like w, w', and so on. Intuitively, a line of the form 'A, a' says that A is true at all worlds. A line of the form 'A, n' says that there is a typical set such that A is true at each world in that set: think of n as that typical set's name. And a line of the form 'A, w' says that A is true at world w.

We now present the decomposition rules for the proof trees. In each rule, A and B are schematic letters for well-formed formulas of TPL. The first rule is the 'conditional' rule:

$$A \rightarrow B, w$$

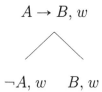

$$\neg A, w \qquad B, w$$

Intuitively, this rule is interpreted as saying that if $A \rightarrow B$ is true at world w, then either $\neg A$ is true at w or B is true at w. This rule is, however, purely syntactic. Later we will justify this interpretation of the syntax. Doing so is the first step towards showing that this TPL proof theory is sound with respect to the TPL semantic theory discussed in Section 6.2.2.

The next three rules are 'negated conditional' rules:

$$\neg(A \rightarrow B), a \qquad \neg(A \rightarrow B), n \qquad \neg(A \rightarrow B), w$$

$$A, a \qquad\qquad A, n \qquad\qquad A, w$$
$$\neg B, a \qquad\qquad \neg B, n \qquad\qquad \neg B, w$$

Intuitively, the rule with index 'a' is interpreted as saying the following: if $\neg(A \rightarrow B)$ is true at every world, then A and $\neg B$ are true at every world. The rule with index 'n' is interpreted as saying the following: if $\neg(A \rightarrow B)$ is true at every world in the typical set named by n, then A and $\neg B$ are true at every world in n. And the rule with index 'w' is interpreted as saying the following: if $\neg(A \rightarrow B)$ is true at world w, then A and $\neg B$ are true at w.

The next three rules are 'negation' rules:

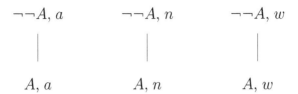

Intuitively, the rule with index 'a' is interpreted as saying the following: if $\neg\neg A$ is true at every world, then A is true at every world. The rule with index 'n' is interpreted as saying the following: if $\neg\neg A$ is true at every world in the typical set named by n, then A is true at every world in n. And the rule with index 'w' is interpreted as saying the following: if $\neg\neg A$ is true at world w, then A is true at w.

The next three rules are 'typicality' rules:

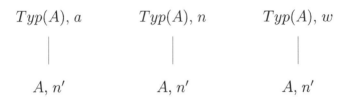

where in each rule, n' is a new 'nearly all' index, one not used earlier in the tableau. Think of n' as naming the set of worlds at which A holds. So intuitively, the first rule says that if A is typical at each world, then there is a typical set n' such that A holds at each world in n'. The second and third rules say something similar, except that they start with slightly different assumptions: the second starts with the assumption that A is typical at all worlds in some typical set n, and the third starts with the assumption that A is typical at some world w.

We require that n' be a *new* 'nearly all' index — we require, in other words, that n' not show up earlier in the proof tree — for the same reasons that in the tableau method for first-order logic, one always uses a new constant symbol '*c*' when decomposing a formula of the form $\exists x \phi(x)$ into $\phi(c)$.[15] $Typ(A)$ is, in a sense, an implicitly existential sort of formula: $Typ(A)$ holds in a TPL universe $\langle \Gamma, \mathcal{V} \rangle$ if and only if *there exists* a typical set $X \subseteq \Gamma$ such that A is true at each world in X. So, as with other such existential rules in other tableau methods, we require that the name for the item whose existence is asserted in the decomposition — the name '*n*', in this case, which denotes a typical set of worlds such that A is true at each world in that set — not be associated with any other properties.[16]

The next three rules are 'negated typicality' rules:

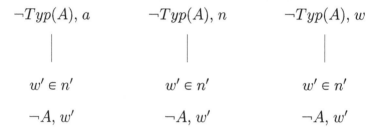

where n' is a 'nearly all' index, and w' is a 'world' index which was not invoked earlier in the proof. So intuitively, the first rule says that if it is not the case that A is typical at each world, then for any typical set n' there is a world in n' such that A is false at that world. The second and third rules say something similar, except that they start with slightly different assumptions: the second starts with the assumption that it is not the case that A is typical at all worlds in some typical set n, and the third starts with

---

[15]For a nice discussion, see Smullyan (1968, p. 54).

[16]If we did not require this, then the tableau method would not match the semantics of Section 6.2.2 in the right way. Proof trees would say more than is licensed by the semantics for TPL. For instance, suppose that some other typical set $n^\star$ is mentioned earlier in the proof tree, and B is true at each $n^\star$. We do not yet know whether or not A is also true at each world in $n^\star$. So if we decomposed the line '$Typ(A), a$' as '$A, n^\star$', we would be saying something stronger than, at this point in the tree, we legitimately can. We would be saying, for example, that A and B are both true at each world in $n^\star$. But we do not know that. We only know that A is typical and that B is typical: we do not know that their conjunction is typical (indeed, it may not be).

the assumption that it is not the case that A is typical in some world w.

Note that in this rule, the index n' may or may not have appeared earlier on. In fact, for various pragmatic reasons, this rule should often be applied whenever a new 'nearly all' index appears in the proof tree. For instance, if the tree features the three lines 'B, n_1', 'C, n_2', and '$\neg Typ(A), a$', then apply the first negated typicality rule twice: once for the case where $n' = n_1$, and once for the case where $n' = n_2$.[17]

In addition, note that the index w' must *not* have shown up earlier in the proof tree. The reason is that $\neg Typ(A)$ contains an implicit existential. $\neg Typ(A)$ holds in a TPL universe $\langle \Gamma, \mathcal{V} \rangle$ if and only if for each typical set $X \subseteq \Gamma$, *there exists* a world $w' \in X$ such that A is false at w'.[18] So, as with other such existential rules in other tableau methods, we require that the name for the item whose existence is asserted in the decomposition — the name 'w'', in this case, which denotes a world in Γ — not be associated with any other properties.

The remaining rules concern indices in particular. The first two are 'all-to-less' index rules:

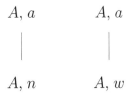

$$A, a \qquad\qquad A, a$$
$$| \qquad\qquad\qquad |$$
$$A, n \qquad\qquad A, w$$

[17]Whether or not this rule should *always* be applied for *each* new 'nearly all' index will depend, ultimately, on pragmatic considerations. Some tableau proofs, for some logical systems, can go on forever: see the discussion by Smullyan (1968). So, for pragmatic reasons, this rule should not be applied in a way that, in conjunction with the other rules, generates an infinite tree.

[18]For, if there were a typical set X in which no such world existed — that is, where A was true at each world in X — then $Typ(A)$ would be true in the TPL universe in question.

where n is a 'nearly all' index and w is a 'world' index. Intuitively, the first all-to-less index rule says that if A holds at every world, then for any typical set n, A holds at each world in n. The second all-to-less index rule says that if A holds at every world, then for any particular world w, A holds at w.

The next rule is an 'instantiation' index rule:

$$A, n$$

$$w \in n$$

$$A, w$$

Intuitively, the instantiation index rule says that if A is true at each world in the typical set n, and if w is in n, then A is true at w.

The final rule is the 'world introduction' rule:

$$w \in n$$

where n is a 'nearly all' index which may or may not have been invoked earlier in the proof, and w is a 'world' index which was not invoked earlier in the proof. Intuitively, the world introduction rule says that for any given typical set n, there is a world w in n.

It will be helpful to have some terminology for various features of proof trees. In general, proof trees have structures like the following:

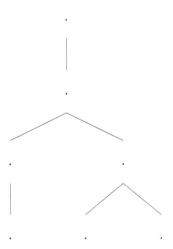

The dots in the above image, which in TPL proof trees consist of a collection of pairs of formulas and indices, are called 'nodes'. The 'initial' node is the one at the tree's top. A vertical line between two nodes indicates that one of the decomposition rules has been applied to the node at the top, yielding the node at the bottom. A 'branch' of a proof tree is a path from the initial node to a lower node. A 'closed' branch is a branch that contains two contradictory formulas with the same index: for instance, a branch that contains 'A, w' and '$\neg A, w$' is closed. An 'open' branch is a branch which is not closed. A 'closed' tree is a proof tree such that every branch is closed. An 'open' tree is a proof tree which is not closed. Finally, in what follows, proof trees are often called 'tableaus'.

Let us now define deductive TPL proofs. Let $\psi, \phi_1, \phi_2, \ldots, \phi_n$ be well-formed formulas in the language of TPL. Say that ψ is deducible from $\phi_1, \phi_2, \ldots, \phi_n$ if and only if there is a proof tree $\mathfrak{p}$ which satisfies the following conditions:

(1) The initial node of $\mathfrak{p}$ consists of the following collection of pairs of formulas and indices:

$$\phi_1, a$$
$$\phi_2, a$$
$$\vdots$$
$$\phi_n, a$$
$$\neg\psi, w$$

(2) The tree $\mathfrak{p}$ is closed.

Let $\phi_1, \phi_2, \ldots, \phi_n \vdash \psi$ denote that ψ is deducible from $\phi_1, \phi_2, \ldots, \phi_n$. More generally, if Σ is a finite set of well-formed formulas, let $\Sigma \vdash \psi$ denote that ϕ is deducible from the formulas in Σ. If ϕ is a well-formed formula in the language of TPL, then ϕ is a 'theorem' if and only if $\vdash \phi$; that is, ϕ is a theorem if and only if ϕ can be deduced without invoking any assumptions.

As an example, let us prove that $p \vdash Typ(p)$. Here is the proof tree. The $\times$ at the bottom indicates that the branch is closed.

$$p, a$$
$$\neg Typ(p), w$$
$$|$$
$$w' \in n$$
$$\neg p, w'$$
$$|$$
$$p, w'$$
$$\times$$

This tree was obtained by first applying the negated typicality rule for a world index to the second line, and then applying the all-to-less rule for a world index to the first line. The proof tree shows that if p is true at every world, and if $Typ(p)$ is false at some world, then a contradiction is obtained. And, intuitively, that is correct. For if something is true at every world, then that something must be typical at every world.

As another example, let us prove that $\{p \rightarrow q, Typ(p)\} \vdash Typ(q)$. Here is the proof tree.

$$p \rightarrow q,\ a$$

$$Typ(p),\ a$$

$$\neg Typ(q),\ w$$

$$|$$

$$p,\ n$$

$$p \rightarrow q,\ n$$

$$w' \in n$$

$$\neg q,\ w'$$

$$|$$

$$p,\ w'$$

$$p \rightarrow q,\ w'$$

$$\diagup\diagdown$$

$$\neg p,\ w' \qquad q,\ w'$$

$$\times \qquad\quad \times$$

This tree shows that if $p \rightarrow q$ is true at every world, if $Typ(p)$ is true at every world, and if $Typ(q)$ is false at some world, then a contradiction is reached. That is, if $p \rightarrow q$ is true at every world, and if $Typ(p)$ is true at every world, then $Typ(q)$ must also be true at every world.

As a final example, let us see why $\nvdash p \to Typ(p)$. Here is the relevant tree.

$$\neg(p \to Typ(p)), \ w$$

$$|$$

$$p, \ w$$

$$\neg Typ(p), \ w$$

$$|$$

$$w' \in n$$

$$\neg p, \ w'$$

The tree is not closed. Of course, more rules could still be applied. We could apply the negated typicality rule for a world index again, or we could apply the world introduction rule. But, clearly, repeated applications of these rules will never close the tree. So there is no proof of $p \to Typ(p)$. And that is intuitively the right result, since the formula '$p \to Typ(p)$' need not be true at every world in every TPL universe. Consider a TPL universe where p is false at all worlds but one. Then $p \to Typ(p)$ is false at that world: for at that world, p is true and $Typ(p)$ is false. Therefore, $p \to Typ(p)$ is not true in this TPL universe. And therefore, $p \to Typ(p)$ is not logically true.

6.2.4. *Soundness*

When explaining the intuitions motivating the decomposition rules, we often invoked semantic notions. In this section, we precisify the relationship between the proof theory of Section 6.2.3 and the semantic theory of Section 6.2.2. In particular, we show that the proof theory is sound with respect to the semantic theory.

To start, consider the following definition.

Definition 6.1 (Faithfulness). *Let* $M = \langle \Gamma, \mathcal{V} \rangle$ *be a TPL universe, and let* b *be any branch of a tableau. Say that* M *is 'faithful' to* b *if and only if there is a function* f *which takes each world index* w *on* b *to a world* $f(w) \in \Gamma$, *and takes each nearly all index* n *on* b *to a typical set* $f(n) \subseteq \Gamma$, *such that the following conditions hold:*

(1) For every node of the form 'A, a' on b, A is true at each world in Γ.

(2) For every node of the form 'A, n' on b, A is true at each world in $f(n)$.

(3) For every node of the form 'A, w' on b, A is true at $f(w)$.

(4) For every node of the form $w \in n$ on b, $f(w) \in f(n)$.

The following lemma will help to prove the soundness of the tableau method. It uses the notion of faithfulness to show that the decomposition rules of the proof theory preserve truth.

Lemma 2. *Let* b *be a branch of a tableau, and let* $M = \langle \Gamma, \mathcal{V} \rangle$ *be a TPL universe. If* M *is faithful to* b, *and a tableau rule is applied to* b, *then this tableau rule yields at least one extended branch to which* M *is faithful.*

Proof. Let f be the function which witnesses the fact that M is faithful to b. The proof of this lemma proceeds by checking every case: we show that for each tableau rule, an application of that tableau rule yields a branch b' to which M is faithful.

Suppose we apply the conditional rule for '$A \rightarrow B, w$' to b. Since '$A \rightarrow B, w$' is on b, $A \rightarrow B$ is true at $f(w)$. So either $\neg A$ is true at $f(w)$ or B is true at $f(w)$. So for at least one of these extensions of b — the one that extends to the left in the tree diagram, which has line '$\neg A, w$'; or the one that extends to the right in the tree diagram, which has line 'B, w' — M is faithful to that extension.

Suppose we apply the negated conditional rule for '$\neg(A \rightarrow B), a$' to b: the resulting branch b' has all the lines of b as well as the lines 'A, a' and '$\neg B, a$'. Since '$\neg(A \rightarrow B), a$' is on b and M is faithful to b, $\neg(A \rightarrow B)$ is true at each world in Γ. So A is true at each world in Γ and $\neg B$ is true at each world in Γ. Therefore, M is faithful to b'. Similarly, suppose we apply the negated conditional rule for

'$\neg(A \rightarrow B), n$' to b: the resulting branch b' has all the lines of b as well as the lines 'A, n' and '$\neg B, n$'. Since '$\neg(A \rightarrow B), n$' is on b and M is faithful to b, $\neg(A \rightarrow B)$ is true at each world in $f(n)$. So A is true at each world in $f(n)$ and $\neg B$ is true at each world in $f(n)$. Therefore, M is faithful to b'. Finally, suppose we apply the negated conditional rule for '$\neg(A \rightarrow B), w$' to b: the resulting branch b' has all the lines of b as well as the lines 'A, w' and '$\neg B, w$'. Since '$\neg(A \rightarrow B), w$' is on b and M is faithful to b, $\neg(A \rightarrow B)$ is true at $f(w)$. So A is true at $f(w)$ and $\neg B$ is true at $f(w)$. Therefore, M is faithful to b'.

Suppose we apply the negation rule for '$\neg\neg A, a$' to b: the resulting branch b' has all the lines of b as well as the line 'A, a'. Since '$\neg\neg A, a$' is on b and M is faithful to b, $\neg\neg A$ is true at each world in Γ. So A is true at each world in Γ. Therefore, M is faithful to b'. Similarly, suppose we apply the rule for '$\neg\neg A, n$' to b: the resulting branch b' has all the lines of b as well as the line 'A, n'. Since '$\neg\neg A, n$' is on b and M is faithful to b, $\neg\neg A$ is true at each world in $f(n)$. So A is true at each world in $f(n)$. Therefore, M is faithful to b'. Finally, suppose we apply the rule for '$\neg\neg A, w$' to b: the resulting branch b' has all the lines of b as well as the line 'A, w'. Since '$\neg\neg A, w$' is on b and M is faithful to b, $\neg\neg A$ is true at $f(w)$. So A is true at $f(w)$. Therefore, M is faithful to b'.

Suppose we apply the typicality rule for '$Typ(A), a$' to b: the resulting branch b' has all the lines of b as well as the line 'A, n''', where n' is a 'nearly all' index which does not appear in b. Since '$Typ(A), a$' is on b, $Typ(A)$ is true at each world in Γ. So there is a typical set $X \subseteq \Gamma$ such that A is true at each world in X. Extend f to a function f' as follows: for all w in the domain of f, $f'(w) = f(w)$; for all n in the domain of f, $f'(n) = f(n)$; and $f'(n') = X$.[19] By definition 6.1, M is faithful to b': that faithfulness is witnessed by f'.

---

[19]This last clause uses the fact that n' is a new 'nearly all' index, one which does not appear earlier on b'; that is, n' is a 'nearly all' index which does not show up on b. For if n' did show up on b, then n' would already be in the domain of f, so we would have to set $f'(n') = f(n')$; that is, we would not be free to set $f'(n') = X$.

Now suppose we apply the typicality rule for '$Typ(A), n$' to b: the resulting branch b' has all the lines of b as well as the line 'A, n''', where n' is a 'nearly all' index which does not appear in b. Since '$Typ(A), n$' is on b and M is faithful to b, $Typ(A)$ is true at each world in some typical set $Y \subseteq \Gamma$. Lemma 1, in conjunction with the fact that Y is nonempty, implies that $Typ(A)$ is true at each world in Γ. So by the exact same reasoning as before, M is faithful to b'.

Similarly, suppose we apply the typicality rule for '$Typ(A), w$' to b: the resulting branch b' has all the lines of b as well as the line 'A, n''', where n' is a 'nearly all' index which does not appear in b. Since $Typ(A), w$ is on b and M is faithful to b, $Typ(A)$ is true at world w. So once again, Lemma 1 implies that $Typ(A)$ is true at each world in Γ. And so by the exact same reasoning as before, M is faithful to b'.

Suppose we apply the negated typicality rule for '$\neg Typ(A), a$' to b: the resulting branch b' has all the lines of b as well as the two lines '$w' \in n''$' and '$\neg A, w''$', where w' is a 'world' index which does not show up on b and n' is a 'nearly all' index which may or may not show up on b. If n' shows up earlier on b, then $f(n')$ is already defined. Since '$\neg Typ(A), a$' is on b, A is false at some world q in $f(n')$. For if A were true at each world in $f(n')$, then by condition 4(i) of the recursive definition of truth functions in Section 6.2.2, and by Lemma 1, $Typ(A)$ would be true at each world in Γ. Extend f to a function f' as follows: for all n in the domain of f, $f'(n) = f(n)$; for all w in the domain of f, $f'(w) = f(w)$; and $f'(w') = q$. Then M is faithful to b', as witnessed by f'. Now suppose that n' does not show up earlier on b. Since '$\neg Typ(A), a$' is on b, Lemma 1 implies that $Typ(A)$ is false at each world in Γ. So take any typical set $X \subseteq \Gamma$. As before, since $Typ(A)$ is false at each world in Γ, there must be a world q in X at which A is false. Extend f to a function f' as follows: for all n in the domain of f, $f'(n) = f(n)$; for all w in the domain of f, $f'(w) = f(w)$; $f'(n') = X$; and $f'(w') = q$. Then M is faithful to b', as witnessed by f'.

Because of lemma 1, the same line of argument works for the negated typicality rules for '$\neg Typ(A), n$' and '$\neg Typ(A), w$'. So we do not present those arguments here. In each case, the conclusion is that M is faithful to the extension of b.

Suppose we apply the first all-to-less index rule for 'A, a' to b: the resulting branch b' has all the lines of b as well as the line 'A, n'. If n shows up on b, then $f(n)$ is already defined; $f(n) = Y$, say, where $Y \subseteq \Gamma$ is a typical set. Since 'A, a' is on b and M is faithful to b, A is true at each world in Γ. So A is true at each world in Y, and therefore, M is faithful to b'. If n does not show up on b, then take any typical set $X \subseteq \Gamma$. Extend f to a function f' as follows: for all n in the domain of f, $f'(n) = f(n)$; for all w in the domain of f, $f'(w) = f(w)$; and $f'(n') = X$. By definition 6.1, M is faithful to b': that faithfulness is witnessed by f'.

Similarly, suppose we apply the second all-to-less rule for 'A, a' to b: the resulting branch b' has all the lines of b as well as the line 'A, w'. If w shows up earlier on b, then $f(w)$ is already defined. So $f(w) = q$, say, where $q \in \Gamma$. And since 'A, a' is on b, A is true at each world in Γ. So A is true at q, and therefore, M is faithful to b'. If w does not show up earlier on b, then pick a world $q \in \Gamma$. Extend f to a function f' as follows: for all n in the domain of f, $f'(n) = f(n)$; for all w' in the domain of f, $f'(w') = f(w')$; and $f'(w) = q$. By definition 6.1, M is faithful to b': that faithfulness is witnessed by f'.

Suppose we apply the instantiation rule for 'A, n' and '$w \in n$' to b: the resulting branch b' has all the lines of b as well as the line 'A, w'. Since 'A, n' is true on b and M is faithful to b, A is true at each world in $f(n)$. Since f witnesses the fact that M is faithful to b, $f(w) \in f(n)$. Therefore, A is true at $f(w)$. So M is faithful to b'.

Finally, suppose we apply the world introduction rule to b: the resulting branch b' has all the lines of b as well as the line '$w \in n$', where n is a 'nearly all' index that may or may not appear earlier on b, and w is a 'world' index which does not show up earlier on b. Suppose that n appears earlier on b. Then, note that if $f(n)$ is empty, then $\frac{\tau(\Gamma \setminus f(n))}{\tau(\Gamma)} = 1 \not< \epsilon$, contradiction; thus, $f(n)$ is nonempty. So extend f to f' as follows: for all $n^\star$ in the domain of f, $f'(n^\star) = f(n^\star)$; for all w' in the domain of f, $f'(w') = f(w')$; and $f'(w)$ is some chosen element of $f(n)$. Therefore, M is faithful to b': that faithfulness is witnessed by f'. Now suppose that n does not appear earlier on b. Then, pick a typical set $X \subseteq \Gamma$, and extend f to f' as follows: for all $n^\star$ in the domain of f, $f'(n^\star) = f(n^\star)$; for all w' in the domain of f, $f'(w') = f(w')$; $f'(n) = X$; and $f'(w)$ is some chosen element of $f'(n)$. □

Now for the proof of soundness.

Theorem 2 (Soundness). *If* $\phi_1, \phi_2, \ldots, \phi_n \vdash \psi$, *then* $\phi_1, \phi_2, \ldots, \phi_n \vDash \psi$.

Proof. The proof establishes the contrapositve. So suppose that $\phi_1, \phi_2, \ldots, \phi_n \nvDash \psi$. Then, there is a TPL universe $M = \langle \Gamma, \mathcal{V} \rangle$ such that $\vDash_M \phi_1, \vDash_M \phi_2, \ldots, \vDash_M \phi_n$, but for some world $w \in \Gamma$, $\vDash_{M,w} \neg\psi$. Consider a tableau with $\phi_1, \phi_2, \ldots, \phi_n, \neg\psi$ at its start. Since M models $\phi_1, \phi_2, \ldots, \phi_n, \neg\psi$, M is faithful to this initial branch: the witnessing function f maps the world index for $\neg\psi$ in the initial node of this tableau to the world w. By lemma 2, every application of a tableau rule to this initial branch yields at least one extended branch to which M is faithful. So there is no finite sequence of applications of tableau rules to this initial branch which closes the whole tree. So $\phi_1, \phi_2, \ldots, \phi_n \nvdash \psi$. □

6.2.5. *Additional results*

In this subsection, we present a few more facts about TPL. In Section 6.2.5.1, we establish some simple results about the typicality of conjunctions and disjunctions. In Section 6.2.5.2, we establish two partial deduction theorems, one syntactic and one semantic. In Section 6.2.5.3, we outline an important difference between TPL and standard systems of modal logic.

6.2.5.1. *Conjunction and Disjunction*

In this subsection, we present some results concerning typicality statements about conjunction and disjunction. As usual, define $A \wedge B$

as $\neg(A \to \neg B)$, and define $A \vee B$ and $\neg A \to B$. It is straightforward to show that for any TPL universe $\langle \Gamma, \mathcal{V} \rangle$, any world $w \in \Gamma$, and any well-formed formulas A and B, $f_w(A \wedge B) = 1$ if and only if $f_w(A) = 1$ and $f_w(B) = 1$, and $f_w(A \vee B) = 1$ if and only if $f_w(A) = 1$ or $f_w(B) = 1$.

The typicality results about conjunction and disjunction are as follows.

Theorem 3. *Let ϕ and ψ be well-formed formulas of TPL. Then the following two conditions hold.*

(1) $Typ(\phi \wedge \psi) \vDash Typ(\phi) \wedge Typ(\psi)$.
(2) $Typ(\phi) \vee Typ(\psi) \vDash Typ(\phi \vee \psi)$.

Proof. To begin, let us establish condition 1. Let $M = \langle \Gamma, \mathcal{V} \rangle$ be a TPL universe, and suppose that $Typ(\phi \wedge \psi)$ is true in M. Take any $w \in \Gamma$. Since $Typ(\phi \wedge \psi)$ is true in M, $Typ(\phi \wedge \psi)$ is true at w. So by the recursive definition of truth in Section 6.2.2, there is a set $X \in \mathcal{S}$ such that $X \subseteq \Gamma_{\phi \wedge \psi}$ and $\frac{\tau(\Gamma \backslash X)}{\tau(\Gamma)} < \epsilon$. For each $w' \in \Gamma_{\phi \wedge \psi}$, $w' \in \Gamma_\phi$ and $w' \in \Gamma_\psi$, since if $\phi \wedge \psi$ is true at w', then ϕ is true at w' and ψ is true at w'. Therefore, $\Gamma_{\phi \wedge \psi} \subseteq \Gamma_\phi$ and $\Gamma_{\phi \wedge \psi} \subseteq \Gamma_\psi$. It follows that $X \subseteq \Gamma_\phi$ and $X \subseteq \Gamma_\psi$. Therefore, by the recursive definition of truth in Section 6.2.2, $Typ(\phi)$ is true at w and $Typ(\psi)$ is true at w. So Lemma 1 implies that both $Typ(\phi)$ and $Typ(\psi)$ are true at each world in M. Therefore, $Typ(\phi) \wedge Typ(\psi)$ is true in M. And therefore, $Typ(\phi \wedge \psi) \vDash Typ(\phi) \wedge Typ(\psi)$.

Now for condition 2. Let $M = \langle \Gamma, \mathcal{V} \rangle$ be a TPL universe, and suppose that $Typ(\phi) \vee Typ(\psi)$ is true in M. Take any $w \in \Gamma$. Since $Typ(\phi \vee \psi)$ is true in M, $Typ(\phi) \vee Typ(\psi)$ is true at w. So either $Typ(\phi)$ is true at w or $Typ(\psi)$ is true at w. Without loss of generality, suppose $Typ(\phi)$ is true at w. Then, by the recursive definition of truth in Section 6.2.2, there is a set $X \in \mathcal{S}$ such that $X \subseteq \Gamma_\phi$ and $\frac{\tau(\Gamma \backslash X)}{\tau(\Gamma)} < \epsilon$. For each $w' \in \Gamma_\phi$, $w' \in \Gamma_{\phi \vee \psi}$, since if ϕ is true at w', then $\phi \vee \psi$ is true at w'. Therefore, $\Gamma_\phi \subseteq \Gamma_{\phi \vee \psi}$. It follows that $X \subseteq \Gamma_{\phi \vee \psi}$. Therefore, by the recursive definition of truth in Section 6.2.2, $Typ(\phi \vee \psi)$ is true at w. So Lemma 1 implies that $Typ(\phi \vee \psi)$ is true at each world in M. Therefore, $Typ(\phi \vee \psi)$ is true in M. The same conclusion results if $Typ(\psi)$, rather than $Typ(\phi)$, is true at w. And therefore, $Typ(\phi) \vee Typ(\psi) \vDash Typ(\phi \vee \psi)$. $\qquad\qquad\square$

The reverse implications do not always hold: that is, $Typ(\phi) \wedge Typ(\psi) \nvDash Typ(\phi \wedge \psi)$, and $Typ(\phi \vee \psi) \nvDash Typ(\phi) \vee Typ(\psi)$. To see why $Typ(\phi) \wedge Typ(\psi)$ does not imply $Typ(\phi \wedge \psi)$, consider a TPL universe $M = \langle \Gamma, \mathcal{V} \rangle$ that satisfies the following four conditions. First, Γ contains 100 worlds. Second, $\epsilon = \frac{1}{10}$ and τ is the counting measure. Third, suppose that sentence letter p is true at 91 worlds. Fourth, suppose that sentence letter q is true at exactly 90 of the worlds at which p is true, and q is true at exactly one of the worlds at which p is false. Then $Typ(p)$ is true at each world in Γ: this follows from the fact that Γ_p contains exactly 91 worlds, and so $\frac{\tau(\Gamma \backslash \Gamma_p)}{\tau(\Gamma)} = \frac{9}{100} < \frac{1}{10} = \epsilon$. For analogous reasons, $Typ(q)$ is true at each world in Γ. Therefore, $Typ(p) \wedge Typ(q)$ is true in M. But $\Gamma_{p \wedge q}$ contains exactly 90 worlds, so $\frac{\tau(\Gamma \backslash \Gamma_{p \wedge q})}{\tau(\Gamma)} = \frac{10}{100} = \frac{1}{10} \not< \epsilon$. It follows that $Typ(p \wedge q)$ is not true at any world in M, and thus, that $Typ(p \wedge q)$ is false in M. Therefore, $Typ(\phi) \wedge Typ(\psi) \nvDash Typ(\phi \wedge \psi)$.

To see why $Typ(\phi \vee \psi)$ does not imply $Typ(\phi) \vee Typ(\psi)$, consider a TPL universe $M = \langle \Gamma, \mathcal{V} \rangle$ that satisfies the following four conditions. First, Γ contains 100 worlds. Second, $\epsilon = \frac{1}{10}$ and τ is the counting measure. Third, suppose that sentence letter p is true at 50 worlds. Fourth, suppose that sentence letter q is true at all, and only, the worlds at which p is false. Then $Typ(p \vee q)$ is true at each world in Γ: this follows from the fact that $\Gamma_{p \vee q}$ contains all 100 worlds, and so $\frac{\tau(\Gamma \backslash \Gamma_{p \vee q})}{\tau(\Gamma)} = \frac{0}{100} = 0 < \epsilon$. But Γ_p contains exactly 50 worlds, and so $\frac{\tau(\Gamma \backslash \Gamma_p)}{\tau(\Gamma)} = \frac{50}{100} = \frac{1}{2} \not< \epsilon$. It follows that $Typ(p)$ is false at each world in Γ. For analogous reasons, $Typ(q)$ is false at each world in Γ. Thus, $Typ(p) \vee Typ(q)$ is false at each world in Γ, so $Typ(p) \vee Typ(q)$ is false in M. And therefore, $Typ(\phi \vee \psi) \nvDash Typ(\phi) \vee Typ(\psi)$.

6.2.5.2. *Deduction theorems*

The deduction theorem of propositional logic says that for any finite set of well-formed formulas Σ, and for any well-formed formulas ϕ and ψ, $\Sigma \vdash \phi \rightarrow \psi$ if and only if $\Sigma, \phi \vdash \psi$. The semantic deduction theorem of propositional logic is similar: for any finite set of well-formed formulas Σ, and for any well-formed formulas ϕ and ψ, $\Sigma \vDash \phi \to \psi$ if and only if $\Sigma, \phi \vDash \psi$.

The left-to-right direction of each biconditional holds in TPL. The right-to-left direction of each biconditional, however, does not. Let us see why.

To start, consider the deduction theorem for $\vdash$.

Theorem 4 (Partial Deduction Theorem). *Let Σ be a finite set of well-formed formulas of TPL. Let ϕ and ψ be well-formed formulas of TPL. If $\Sigma \vdash \phi \to \psi$ then $\Sigma, \phi \vdash \psi$.*

Proof. Let $\Sigma = \{\sigma_1, \sigma_2, \ldots, \sigma_n\}$. Suppose $\Sigma \vdash \phi \to \psi$. Then, there is a closed tree whose initial node consists of the lines 'σ_i, a' ($1 \le i \le n$) along with the line '$\neg(\phi \to \psi), w$'; call this tree **p**.

Consider a tree **p'** whose initial node consists of the lines 'σ_i, a' ($1 \le i \le n$) along with the line 'ϕ, a' and the line '$\neg\psi, w$'. To complete the proof of this theorem, we explain how to decompose these initial lines of **p'** to close each branch. The decompositions depend on exactly how branches are closed in **p**. Put roughly, any way that **p** reaches a contradiction can be reached by **p'**, because the decomposition of the line '$\neg(\phi \to \psi), w$' is already 'built into' the initial node of **p'**. But to be completely rigorous, we must check all the possible cases: we must check all the ways that a branch of **p** might use the line '$\neg(\phi \to \psi), w$' to reach a contradiction.

Suppose **p** reaches a contradiction on a branch b without decomposing the line '$\neg(\phi \to \psi), w$'. Then, either **p** reaches that contradiction by producing a line of the form '$\phi \to \psi, w$' from the 'σ_i, a' lines, or **p** reaches that contradiction without using the line '$\neg(\phi \to \psi), w$' at all. In the latter case, the same contradiction can be reached on a corresponding branch b' in **p'** by applying the exact same rules in the exact same way. In the former case, it is possible to create a pair of branches in **p'** that correspond to b, both of which are closed, by doing the following: apply the exact same rules in the exact same way as in **p** to derive '$\phi \to \psi, w$'; use the conditional rule to create a '$\neg\phi, w$' branch b_1 and a 'ψ, w' branch b_2; use the second all-to-less index rule to decompose the line 'ϕ, a' in the initial node to 'ϕ, w' on b_1. Note that both of the newly-created branches contain a contradiction: b_1 contains lines 'ϕ, w' and '$\neg\phi, w$', and b_2 contains lines 'ψ, w' and '$\neg\psi, w$'.[20] So both branches close.

Alternatively, suppose $\mathfrak{p}'$ reaches a contradiction on a branch b by decomposing the line '$\neg(\phi \to \psi), w$' into the pair of lines 'ϕ, w' and '$\neg\psi, w$' via the 'world' index negated conditional rule. Then the same contradiction can be reached on a corresponding branch b' in $\mathfrak{p}'$, in the following way. First, apply the exact same rules in the exact same way as in $\mathfrak{p}$ — apart from that decomposition of initial line '$\neg(\phi \to \psi), w$' of course, since that initial line in $\mathfrak{p}$ is not an initial line in $\mathfrak{p}'$. Second, apply the second all-to-less index rule to decompose the line 'ϕ, a' in the initial node of $\mathfrak{p}'$ to get the line 'ϕ, w'. Then b' has both 'ϕ, w' and '$\neg\psi, w$' on it, just as b does. So the contradiction is reached on b' in just the same way that it is reached on b. $\qquad\square$

The other direction of this theorem does not hold: it is not the case that if $\Sigma, \phi \vdash \psi$ then $\Sigma \vdash \phi \to \psi$. A counterexample has already been given. As shown in Section 6.2.3, $p \vdash Typ(p)$, but $\not\vdash p \to Typ(p)$.

Let us now consider the deduction theorem for $\vDash$.

Theorem 5 (Partial Semantic Deduction Theorem). *Let σ be a finite set of well-formed formulas of TPL. Let ϕ and ψ be well-formed formulas of TPL. If $\Sigma \vDash \phi \to \psi$ then $\Sigma, \phi \vDash \psi$.*

Proof. Let $\Sigma = \{\sigma_1, \sigma_2, \ldots, \sigma_n\}$. Let $M = \langle \Gamma, \mathcal{V} \rangle$ be a TPL universe such that $\sigma_1, \sigma_2, \ldots, \sigma_n$, and ϕ are true in M. Suppose $\Sigma \vDash \phi \to \psi$. Then $\phi \to \psi$ is true at each world in Γ. Take $w \in \Gamma$. Then ϕ and $\phi \to \psi$ are both true at w. Therefore, ψ is true at w as well. Since this holds for each $w \in \Gamma$, ψ is true in M. And since this holds for arbitrary M, it follows that $\Sigma, \phi \vDash \psi$. $\qquad\square$

The other direction of this theorem does not hold: it is not the case that if $\Sigma, \phi \vDash \psi$ then $\Sigma \vDash \phi \to \psi$. For example, note that $p \vDash Typ(p)$:

---

[20]The line '$\neg\psi, w$' is in the initial node of $\mathfrak{p}'$.

this holds because for any TPL universe $M = \langle \Gamma, \mathcal{V} \rangle$, if $\vDash_M p$ — that is, if p is true at each $w \in \Gamma$ — then $\{w \in \Gamma \mid \vDash_{M,w} p\} = \Gamma$, and so $Typ(p)$ is true at each $w \in \Gamma$. But $\nvdash p \to Typ(p)$. To see why, let Γ be a large set, let $M = \langle \Gamma, \mathcal{V} \rangle$ be an TPL universe, and suppose p is true at just one world $w \in \Gamma$. Then, it is not the case that $Typ(p)$ is true in M. So $\vDash_{M,w} p$ but $\nvDash_{M,w} Typ(p)$. Therefore, $\nvDash_{M,w} p \to Typ(p)$, and so $\nvdash p \to Typ(p)$.

6.2.5.3. *Modal Logic and TPL*

In this subsection, we briefly discuss an important difference between TPL and modal logic. The typicality operator Typ is substantially different from the necessity operator $\Box$. Modal logics generally adopt the K-axiom schema: $\Box(A \to B) \to (\Box A \to \Box B)$. The corresponding schema for TPL would be: $Typ(A \to B) \to (Typ(A) \to Typ(B))$. But it can be shown that instances of this schema are false in some TPL universes. Thus, TPL is different in kind from the main systems of modal logic.

As an example of the falsity of schema $Typ(A \to B) \to (Typ(A) \to Typ(B))$ in some cases, consider the well-formed formula $Typ(p \to q) \to (Typ(p) \to Typ(q))$. And consider the following TPL universe $\langle \Gamma, \mathcal{V} \rangle$, where Γ contains 100 worlds. At 91 of those worlds, p is true. Of those 91 worlds, q is true at exactly 82 of them. In addition, q is false at each of the 9 worlds at which p is false. From all this, it follows that $p \to q$ is true at exactly 91 worlds: it is true at each of the 82 worlds at which q is true, it is true at each of the 9 worlds at which p is false, and it is false at the 9 worlds where p is true but q is false. Let $\epsilon = \frac{1}{10}$, and let τ be the counting measure. Then $\frac{\tau(\Gamma \backslash \Gamma_{p \to q})}{\tau(\Gamma)} = \frac{9}{100} < \frac{1}{10} = \epsilon$, so $Typ(p \to q)$ is true at each world in Γ. Similarly, $\frac{\tau(\Gamma \backslash \Gamma_p)}{\tau(\Gamma)} = \frac{9}{100} < \frac{1}{10} = \epsilon$, so $Typ(p)$ is true at each world in Γ. But $\frac{\tau(\Gamma \backslash \Gamma_q)}{\tau(\Gamma)} = \frac{18}{100} \not< \frac{1}{10} = \epsilon$, so $Typ(q)$ is false at each world in Γ. Therefore, $Typ(p \to q) \to (Typ(p) \to Typ(q))$ is false in $\langle \Gamma, \mathcal{V} \rangle$.

This is a feature of TPL, not a bug. The K-axiom makes intuitive sense in modal logic. Intuitively, it says that if $p \to q$ holds at each world, and if p holds at each world, then q holds at each world. But the corresponding axiom for typicality is overly strong, since typical statements need not hold everywhere. Indeed, typical statements are generally liable to exceptions: typicality is *nearly* all, not *absolutely* all. So if $p \rightarrow q$ is typical and p is typical, it follows that $p \rightarrow q$ is true at nearly all worlds, and it follows that p is true at nearly all worlds. But since $p \rightarrow q$ and p may not be true at all the same worlds, it does not follow that q is true at nearly all worlds. So there may not be enough q worlds for q to be typical.

6.3. Typicality Intuitionistic Logic

In this section, we describe an intuitionistic logic for typicality, which we call 'typicality intuitionistic logic' or TIL. To contrast TIL with TPL, we note that at the outset, there is a shift in perspective from the impersonal and objective interpretation of TPL using sets of possible worlds (Section 6.2) to the personal and subjective inter-pretation of TIL using sets of 'credal states' (below).[21] It is impor-tant not to confuse the credal states discussed in this section with the physical states (i.e., quantum states, microstates, or, generically, 'states of the world') discussed in the previous sections. Whereas the states of the previous section are objective properties, those of the current reflect subjective dispositions toward claims. These dispo-sitions guide an agent's judgment about what is true and what is typical.

Recall that in TPL, we associate each well-formed formula ϕ to the set of possible worlds Γ_ϕ at which ϕ holds. In this inter-pretation, when a statement holds at a world, it is a *fact* of that world, and the formalism of TPL makes precise how the notion of typicality can be integrated into such an outlook. In particular, if a statement holds at a sufficiently large subset of possible worlds,

---

[21]For our purposes, a credal state is a probability space $(\Omega, \mathcal{F}, \nu)$ with ν repre-senting the credences of a rational agent regarding the measurable subsets in $\mathcal{F}$. It is assumed, for the purposes of this discussion, that each subset in $\mathcal{F}$ represents a proposition in TPL.

then the typicality of that statement is a *fact* of every world in the universe.

In TIL, on the other hand, when a credence assigns maximal belief to a statement, an agent with that credence is justified in making a *judgment* that the claim holds, regardless of whether or not it actually does. Thus, in the alternative formalism given below, when a credence assigns a sufficiently large degree of belief to a statement, an agent with that credence is justified in make a *judgment* that the claim is typical. The validity of all such judgments is determined not by the actual states of the world, which are unknown to the agent making the judgments, but rather by the *context* within which the judgment is being made. In the proposed formalism, we interpret this notion of context as a constraint on the possible credal states of the agent making the judgment. The semantic difference between possible worlds and possible credal states in the two formalisms mirrors the philosophical distinction between 'typicality facts' and 'typicality judgments'.

6.3.1. *The language*

To define the language of TIL, we extend a fragment of the non-dependent version of intuitionistic Martin-Löf type theory (MLTT; see Martin-Löf, 1985) by specifying rules for an additional typicality type former **Typ**. The syntactic rules for the **Typ**-type specified below are a modification of an earlier development by Crane to append a probability type (denoted **Prob**) to the syntax of MLTT. See Crane (2018) for further details.

For clarity, we reserve capital Greek, lowercase Greek and lowercase Roman letters to represent the different primitive notions of context, types and terms, respectively. The basic components of the language are called judgments, each of which has the form of one of the following three primitive expressions:[22]

---

[22]The full syntax of MLTT has two additional primitive notions of 'judgmental equality' which play no substantive role in our treatment below and are thus omitted.

Formal	Natural language description
Δ **ctx**	Δ is a well-formed context
$\phi :$ **Type**	ϕ is a type
$a : \phi$	a is a term of type ϕ

Importantly, the syntax of MLTT does not permit 'untyped' statements, i.e., it is meaningless to refer to a term without reference to its type. In the preferred propositions-as-types interpretation we appeal to below, a judgment of the form $\phi :$ **Type** is interpreted to mean that ϕ is a proposition, and $a : \phi$ is interpreted to mean that a is a proof of the proposition ϕ (Curry and Feys, 1959; Howard, 1969). When understood in this light, the rules of MLTT presented below can be understood as an algorithmic prescription for how to prove compound propositions, as we illustrate below in the case of the product type.[23]

In MLTT, and thus in TIL, all judgments are made in a specific context and are expressed in the form

$$\text{context} \vdash \text{judgment}$$

$$\Delta \vdash \mathcal{J}, \tag{6.1}$$

where $\mathcal{J}$ has the form of one of the above primitive judgments and Δ is a finite list of the form

$$a_1 : \phi_1, \ldots, a_n : \phi_n.$$

The statement in (6.1) can be interpreted pre-formally to mean that the judgment $\mathcal{J}$ on the right is made in the context Δ on the left, where a context consists of a string of prior judgments about types $\phi_1, \ldots, \phi_n$.

From these primitive elements, the syntax of MLTT is built by specifying a collection of rules for how to derive new judgments from old. In the non-dependent version of MLTT featured here, these

---

[23]With this understanding of 'terms' as 'proofs', the meaninglessness of untyped statements — e.g., let 'a' be a term — becomes apparent; for if we interpret terms as proofs, the statement "let a be a proof" leaves ambiguous what it is that a is a proof of.

derived judgments correspond to the non-logical vocabulary for constructing the ×-type, +-type, and **0**-type. For example, the product (×) type is defined by the following rules:

$$\frac{\Delta \vdash \phi : \textbf{Type} \qquad \Delta \vdash \psi : \textbf{Type}}{\Delta \vdash \phi \times \psi : \textbf{Type}} \quad (\times\text{-form})$$

This formation rule says that given two types ϕ and ψ, a new type can be formed, denoted $\phi \times \psi$. This is akin to saying that $\phi \times \psi$ is a 'well-formed formula' in the language of MLTT or, more appropriately, that $\phi \times \psi$ is a 'well-formed type' in the context Δ. In the propositions-as-types interpretation, this rule corresponds to the conjunction formation rule in propositional logic, by which two well-formed formulas ϕ and ψ in PL can be combined to another well-formed formula $\phi \wedge \psi$ in PL.

$$\frac{\begin{array}{cc} \Delta \vdash \phi : \textbf{Type} & \Delta \vdash \psi : \textbf{Type} \\ \Delta \vdash a : \phi & \Delta \vdash b : \psi \end{array}}{\Delta \vdash (a,b) : \phi \times \psi} \quad (\times\text{-intro})$$

This introduction rule describes how terms of the product type $\phi \times \psi$ are constructed (or introduced) by combining terms $a : \phi$ and $b : \psi$ to form $(a,b) : \phi \times \psi$. In the corresponding propositions-as-types interpretation this rule establishes the truth conditions for $\phi \wedge \psi$. In particular, to verify that $\phi \wedge \psi$ is true, ϕ is true and ψ is true have to be verified individually. Combining these two individual verifications serves as a verification of their conjunction.

$$\frac{\begin{array}{c} \Delta, a : \phi, b : \psi \vdash \rho : \textbf{Type} \\ \Delta, a : \phi, b : \psi \vdash d(a,b) : \rho \end{array}}{\Delta, z : \phi \times \psi \vdash \texttt{split}_d(z) : \rho} \quad (\times\text{-elim})$$

This elimination rule asserts that to define a function out of $\phi \times \psi$, it is sufficient to define how the function acts on pairs of the form (a,b) for $a : \phi$ and $b : \psi$. Intuitively, though somewhat informally, this rule states implicitly that all terms of $\phi \times \psi$ consist of pairs of the form (a, b) for $a : \phi$ and $b : \psi$.[24,25] In the corresponding propositions-as-types interpretation, the elimination rule describes the conditions under which it is justified to deduce ρ on the basis of $\phi \wedge \psi$. In particular, if ρ holds whenever ϕ and ψ both hold, then ρ holds whenever $\phi \wedge \psi$ holds.

The coproduct + and-**0**-types are defined by analogous rules of formation, introduction, and elimination; see, e.g., Appendix A of Lumsdaine and Kupulkin (2014) or Tsementzis (2018) for a detailed description of those rules. We also list these rules explicitly in our soundness proof for TIL (Theorem 6).

6.3.2. *The Typicality Type*

To build a notion of typicality on top of the existing machinery of MLTT, we define the following rules for a new type **Typ** as follows.

The first formation rule (**Typ**-form) says that if ϕ is a well-formed type in context Δ, then $\mathbf{Typ}(\phi)$ is a well-formed type in that context. This operation is akin to the step taken when defining the well-formed formulas of TPL in Section 6.2, in which $Typ(\phi)$ is a well-formed formula whenever ϕ is a well-formed formula. Expressed formally,

---

[24]To understand the notation $d(a, b) : \rho$, it may be helpful to think of d as a function that takes terms $a : \phi$ and $b : \psi$ as input, and returns a term $d(a, b) : \rho$ as output. But, in general, this is simply notation that denotes a new term 'd' which is constructed in a way that depends on $a : \phi$ and $b : \psi$. Similarly, and informally, $\mathtt{split}_d$ in the conclusion can be thought of as a function that takes a term $z : \phi \times \psi$ of the product type as input and outputs $\mathtt{split}_d(z) : \rho$.

[25]In addition to the above three rules, the standard rules for MLTT specify the following computation rule:

$$\frac{\begin{array}{c} \Delta \vdash \phi : \mathbf{Type} \qquad \Delta \vdash \psi : \mathbf{Type} \\ \Delta, a : \phi, b : \psi \vdash \rho : \mathbf{Type} \\ \Delta, a : \phi, b : \psi \vdash d(a, b) : \rho \end{array}}{\Delta, (a, b) : \phi \times \psi \vdash \mathtt{split}_d((a, b)) \equiv d(a, b) : \rho} \ (\times\text{-comp})$$

The computation rule establishes coherence between $\mathtt{split}_d$ and the function d from which it is derived. In particular, it establishes that the introduction and elimination rules commute. Because we specialize here to the propositions-as-types case, in which all inhabited types have exactly one unique term up to identity, these computation rules play no material role and are therefore omitted.

the formation rule reads:

$$\frac{\Delta \vdash \phi : \mathbf{Type}}{\Delta \vdash \mathbf{Typ}(\phi) : \mathbf{Type}} \quad (\mathbf{Typ}\text{-form})$$

The next introduction rule ($\mathbf{Typ}$-intro) says that a judgment that a proposition ϕ is typical can be constructed from any proof of ϕ. In particular, from a proof $a : \phi$ that ϕ holds, we construct $\tau_\phi(a) : \mathbf{Typ}(\phi)$, where τ_ϕ is the constructor for the $\mathbf{Typ}(\phi)$-type.

$$\frac{\Delta \vdash \phi : \mathbf{Type}}{\Delta, a : \phi \vdash \tau_\phi(a) : \mathbf{Typ}(\phi)} \quad (\mathbf{Typ}\text{-intro})$$

The elimination rule ($\mathbf{Typ}$-elim) says that if ϕ implies ψ (i.e., if every proof of ϕ ($a : \phi$) can be turned into a proof of ψ ($d(a) : \psi$)), then $\mathbf{Typ}(\phi)$ implies $\mathbf{Typ}(\psi)$ (i.e., any justification that ϕ is typical — i.e., $x : \mathbf{Typ}(\phi)$ — can be used to construct a justification that ψ is typical ($\mathtt{imp}_d(x) : \mathbf{Typ}(\psi)$)).[26]

$$\frac{\Delta \vdash \phi : \mathbf{Type} \qquad \Delta \vdash \psi : \mathbf{Type}}{\Delta, a : \phi \vdash d(a) : \psi} \quad (\mathbf{Typ}\text{-elim})$$
$$\overline{\Delta, x : \mathbf{Typ}(\phi) \vdash \mathtt{imp}_d(x) : \mathbf{Typ}(\psi)}$$

Finally, the $\mathbf{Typ}$-0 rule states that a justification of the typicality of the empty type $\mathbf{0}$, i.e., $x : \mathbf{Typ}(\mathbf{0})$ can be used to construct an element of the empty type $\sigma(x) : \mathbf{0}$. In relating this rule to TPL, we interpret $\mathbf{0}$ as $\perp$, so that this rule can be translated to mean that $Typ(\perp)$ implies $\perp$ in any context.

$$\frac{\Delta \ \mathbf{ctx}}{\Delta, x : \mathbf{Typ}(\mathbf{0}) \vdash \sigma(x) : \mathbf{0}} \quad (\mathbf{Typ}\text{-0})$$

---

[26]In the $\mathbf{Typ}$-elimination rule, $\mathtt{imp}_d$ is a constructor for $\mathbf{Typ}(\psi)$ built from the constructor $d(-)$ in the premises. The main content of the elimination rule is its assertion of the validity of constructing $\mathtt{imp}_d : \mathbf{Typ}(\phi) \to \mathbf{Typ}(\psi)$ from $d : \phi \to \psi$.

6.3.3. *The Semantics of TIL*

We define a semantics for the above system of rules by associating judgments to subsets of probability spaces, where each probability space is understood as the credal state of a rational agent. First, let Π be the set of all sentences in PL and let Π^* consist of all sentences in TPL, obtained recursively by adding the unary predicate Typ to PL, as in Section 6.2.[27]

To formalize this, we let Ω be a set (i.e., the set of possible worlds) and associate each proposition $\phi \in \Pi^*$ to a subset of worlds at which ϕ holds, denoted $\tilde{\phi} \subseteq \Omega$. From this, we have the following:

(1) If ϕ and ψ are well-formed formulas, then

$$\widetilde{\phi \wedge \psi} = \{\omega \in \Omega \mid \omega \in \tilde{\phi} \text{ and } \omega \in \tilde{\psi}\} \equiv \tilde{\phi} \cap \tilde{\psi}.$$

(2) If ϕ and ψ are well-formed formulas, then

$$\widetilde{\phi \vee \psi} = \{\omega \in \Omega \mid \omega \in \tilde{\phi} \text{ or } \omega \in \tilde{\psi}\} \equiv \tilde{\phi} \cup \tilde{\psi}.$$

(3) If ϕ is a well-formed formula, then

$$\widetilde{\neg\phi} = \{\omega \in \Omega \mid \omega \notin \tilde{\phi}\} \equiv \tilde{\phi}^c.$$

(4) If ϕ is a well-formed formula, then we require $\widetilde{Typ(\phi)} \supseteq \tilde{\phi}$.
(5) Finally, we have $\tilde{\perp} = \emptyset$.

Altogether, a *credal state* is a probability space $(\Omega, \mathcal{F}, \nu)$ that assigns credence $\nu(\tilde{\phi})$ to every proposition $\phi \in \Pi^*$ whose representation satisfies $\tilde{\phi} \in \mathcal{F}$. Any ϕ in TPL for which $\phi \notin \mathcal{F}$ is a proposition for which the agent has no credence and thus suspends judgment. In total, the

---

[27]In Section 6.2, we defined the syntax of TPL using only the logical connectives $\neg$ and $\rightarrow$. We can define the connectives $\wedge$ and $\vee$ in the usual way and supplement TPL with these connectives.

credal state $(\Omega, \mathcal{F}, \nu)$ specifies the σ-algebra $\mathcal{F}$ of propositions about which an agent in this state has a credence along with a probability measure ν on $(\Omega, \mathcal{F})$ that specifies the agent's credences.

In what follows, we fix a set of possible worlds Ω along with a representation $\tilde{\phi} \subseteq \Omega$ of each $\phi \in \Pi^*$. For any fixed $\epsilon \in (1/2, 1)$, we write

$$\mathcal{P}(\Pi^*) = \{(\Omega, \mathcal{F}, \nu) \mid \forall \phi \in \Pi^* \; \tilde{\phi} \in \mathcal{F} \Rightarrow \widetilde{Typ(\phi)} \in \mathcal{F},$$

$$\forall \phi \in \Pi^* \; \nu(\widetilde{Typ(\phi)}) = 1 \Leftrightarrow \nu(\tilde{\phi}) \geq 1 - \epsilon\}$$

to denote the set of admissible credal states for a rational agent holding beliefs about propositions in Π^*. For every $\phi \in \Pi^*$ we define

$$\mathcal{S}_\phi = \{(\Omega, \mathcal{F}, \nu) \mid \tilde{\phi} \in \mathcal{F}\}$$
$$\mu_\phi = \{(\Omega, \mathcal{F}, \nu) \in \mathcal{P}(\Pi^*) \mid \tilde{\phi} \in \mathcal{F} \text{ and } \nu(\tilde{\phi}) = 1\}$$

We interpret the syntax from Section 6.3.1 and 6.3.2 into the semantics of TIL by regarding all type-theoretic judgments as set-theoretic statements. In particular, we have the following symbolic correspondence for symbols in type theory, propositional logic, and set theory:

MLTT	PL	set theory
Δ **ctx**	—	$\Delta \subseteq \mathcal{P}(\Pi^*)$
$\phi : \textbf{Type}$	ϕ well-formed formula	$\mathcal{S}_\phi$
$a : \phi$	ϕ	μ_ϕ
$\vdash$	$\vdash$	$\subseteq$
$\phi \times \psi$	$\phi \wedge \psi$	$\tilde{\phi} \cap \tilde{\psi}$
$\phi + \psi$	$\phi \vee \psi$	$\tilde{\phi} \cup \tilde{\psi}$
$\mathbf{0}$	$\perp$	$\varnothing$

In type-theory, on the left side of the turnstile, we interpret commas as $\cap$. With this translation, the basic judgments of MLTT are interpreted as:

Syntax	Semantics
Δ **ctx**	$\Delta \subseteq \mathcal{P}(\Pi^*)$
$\Delta \vdash \phi : \textbf{Type}$	$\Delta \subseteq \mathcal{S}_\phi$
$\Delta \vdash a : \phi$	$\Delta \subseteq \mu_\phi.$

Thus, the syntax Δ **ctx** translates to $\Delta \subseteq \mathcal{P}(\Pi^*)$, and hence justifying our interpretation of the context as a constraint on the credal states in which a judgment is made. The initial ('empty') context $\bullet$ is thus the one without any constraints on the credal state, namely $\bullet \equiv \mathcal{P}(\Pi^*)$. The syntax $\Delta \vdash \phi : \textbf{Type}$ translates to $\Delta \subseteq \mathcal{S}_\phi$, which imposes the constraint that the admissible credal states are those which assign some credence to ϕ. The syntax $\Delta \vdash a : \phi$ translates to $\Delta \subseteq \mu_\phi$, which imposes the constraint that the admissible credal states are those which assign maximal credence to ϕ.

For example, an agent in a specific credal state has a σ-algebra $\mathcal{F}$ corresponding to a subset of the propositions in Π^* and a probability measure ν that assigns a credence to each measurable subset of $\mathcal{F}$. If the subset $\tilde{\phi}$ corresponding to proposition ϕ is a measurable subset of $\mathcal{F}$, then the agent having this credence would make the judgment that 'ϕ holds' only if $\nu(\tilde{\phi}) = 1$, and that 'ϕ is typical' only if $\nu(\tilde{\phi}) \geq 1 - \epsilon$. The semantic interpretation of the rules of TIL does not require an agent's credal state to be pinned down to a single probability space in order for a judgment to be justified. The semantics only require that the agent's credal state lies in a subset of possible credal states that are consistent with the given judgment.

To illustrate the semantic translation of the syntax, a deduction of the form

$$
\begin{array}{c}
\Delta \ \textbf{ctx} \\[4pt]
\Delta \vdash \phi : \textbf{Type} \\[4pt]
\dfrac{\Delta, a : \phi \vdash \psi : \textbf{Type}}{\Delta, a : \phi \vdash b : \psi}
\end{array}
\qquad \text{translates to} \qquad
\begin{array}{c}
\Delta \subseteq \mathcal{P}(\Pi^*) \\[4pt]
\Delta \subseteq \mathcal{S}_\phi \\[4pt]
\dfrac{\Delta \cap \mu_\phi \subseteq \mathcal{S}_\psi}{\Delta \cap \mu_\phi \subseteq \mu_\psi.}
\end{array}
$$

6.3.4. *Soundness*

Theorem 6. *The syntax of TIL is sound with respect to the above interpretation.*

Proof. To prove soundness, we interpret each of the rules for the $\times$-, $+$-, **0**-, and **Typ**-types into the semantics and show that the rule holds. We begin by specifying the interpretation of the rules for contexts.

- **Structural rules, •-ctx:**

Syntax	Semantics
• **ctx**	$\mathcal{P}(\Pi^*) \subseteq \mathcal{P}(\Pi^*)$

Holds trivially: Every set is a subset of itself.[28]

- **Structural rules**, ext-**ctx**

Syntax		Semantics	
Δ **ctx**	$\Delta \vdash \phi : \textbf{Type}$	$\Delta \subseteq \mathcal{P}(\Pi^*)$	$\Delta \subseteq \mathcal{S}_\phi$
$\Delta, x : \phi$ **ctx**		$\Delta \cap \mu_\phi \subseteq \mathcal{P}(\Pi^*)$	

By assumption, $\Delta \subseteq \mathcal{P}(\Pi^*)$, and thus $\Delta \cap A \subseteq \Delta \subseteq \mathcal{P}(\Pi^*)$ for all other sets A. Instantiating $A = \mu_\phi$ gives the result.

- **Structural rules**, ax-**ctx**

Syntax	Semantics
$\Delta, a : \phi, \Xi$ **ctx**	$\Delta \cap \mu_\phi \cap \Xi \subseteq \mathcal{P}(\Pi^*)$
$\Delta, a : \phi, \Xi \vdash a : \phi$	$\Delta \cap \mu_\phi \cap \Xi \subseteq \mu_\phi$

By assumption, $\Delta \cap \mu_\phi \cap \Xi$ is a set, and for any sets A and B it is always the case that $A \cap B \subseteq A$, yielding the result.

It follows from these structural rules for contexts that every context is a finite list of judgments of the form

$$(a_1 : \phi_1, \ldots, a_n : \phi_n) \textbf{ ctx,}$$

which in our semantic interpretation translates to

$$\mu_{\phi_1} \cap \cdots \cap \mu_{\phi_n} \subseteq \mathcal{P}(\Pi^*).$$

Thus, in our semantic treatment, every context Δ can be expressed in the form

$$\Delta \equiv \mu_{\phi_1} \cap \cdots \cap \mu_{\phi_n} \tag{6.2}$$

---

[28]This rule states that there is an initial 'empty' context •. In the semantics, the context places constraints on an agent's credal states, and thus this initial 'empty' context corresponds to a context without constraints, i.e., $\Delta \equiv \mathcal{P}(\Pi^*)$.

for some finite list $\phi_1, \ldots, \phi_n \in \Pi^*$. This specific representation will become useful when we prove soundness for the coproduct and typicality types below.

We next prove soundness for the product type.

- **Product type**, formation rule:

Syntax	Semantics
$\Delta \vdash \phi : \textbf{Type}$	$\Delta \subseteq \mathcal{S}_\phi$
$\Delta \vdash \psi : \textbf{Type}$	$\Delta \subseteq \mathcal{S}_\psi$
$\Delta \vdash \phi \times \psi : \textbf{Type}$	$\Delta \subseteq \mathcal{S}_{\phi \wedge \psi}$

Let $(\Omega, \mathcal{F}, \nu) \in \Delta$ so that $\tilde{\phi} \in \mathcal{F}$ and $\tilde{\psi} \in \mathcal{F}$. Then, $\widetilde{\phi \wedge \psi} \equiv \tilde{\phi} \cap \tilde{\psi} \in \mathcal{F}$ because $\mathcal{F}$ is a σ-algebra on Ω and is closed under intersection. It follows that $\Delta \subseteq \mathcal{S}_{\phi \wedge \psi}$, as claimed.

- **Product type**, introduction rule:

Syntax		Semantics	
$\Delta \vdash \phi : \textbf{Type}$	$\Delta \vdash a : \phi$	$\Delta \subseteq \mathcal{S}_\phi$	$\Delta \subseteq \mu_\phi$
$\Delta \vdash \psi : \textbf{Type}$	$\Delta \vdash b : \psi$	$\Delta \subseteq \mathcal{S}_\psi$	$\Delta \subseteq \mu_\psi$
$\Delta \vdash (a, b) : \phi \times \psi$		$\Delta \subseteq \mu_{\phi \wedge \psi}$	

First note that $\mu_\phi \subseteq \mathcal{S}_\phi$ and $\mu_\psi \subseteq \mathcal{S}_\psi$, so that the premises together imply $\Delta \subseteq \mu_\phi \cap \mu_\psi$. Now take any $(\Omega, \mathcal{F}, \nu) \in \mu_\phi \cap \mu_\psi$. Since $\tilde{\phi} \in \mathcal{F}$ and $\tilde{\psi} \in \mathcal{F}$ by assumption, we must have $\tilde{\phi} \cap \tilde{\psi} \in \mathcal{F}$ because $\mathcal{F}$ is a σ-algebra. Furthermore, by the equivalence $\neg(\phi \wedge \psi) \equiv \neg\phi \vee \neg\psi$, we have $(\tilde{\phi} \cap \tilde{\psi})^c \equiv \tilde{\phi}^c \cup \tilde{\psi}^c$. The assumption that $\nu(\tilde{\phi}) = \nu(\tilde{\psi}) = 1$ implies $\nu(\tilde{\phi}^c) = \nu(\tilde{\psi}^c) = 0$, and thus

$$\nu(\tilde{\phi} \cap \tilde{\psi}) = 1 - \nu(\tilde{\phi}^c \cup \tilde{\psi}^c) \geq 1 - \nu(\tilde{\phi}^c) - \nu(\tilde{\psi}^c) = 1.$$

It follows that $(\Omega, \mathcal{F}, \nu) \in \mu_{\phi \wedge \psi}$, so that $\Delta \subseteq \mu_{\phi \wedge \psi}$, as claimed.

- **Product type**, elimination rule:

$$\begin{array}{cc}
\text{Syntax} & \text{Semantics} \\
\Delta \vdash \phi : \mathbf{Type} \quad \Delta \vdash \psi : \mathbf{Type} & \Delta \subseteq \mathcal{S}_\phi \quad \Delta \subseteq \mathcal{S}_\psi \\
\Delta, a : \phi, b : \psi \vdash \rho : \mathbf{Type} & \Delta \cap \mu_\phi \cap \mu_\psi \subseteq \mathcal{S}_\rho \\
\dfrac{\Delta, a : \phi, b : \psi \vdash d(a,b) : \rho}{\Delta, z : \phi \times \psi \vdash \mathtt{split}_d(z) : \rho} & \dfrac{\Delta \cap \mu_\phi \cap \mu_\psi \subseteq \mu_\rho}{\Delta \cap \mu_{\phi \wedge \psi} \subseteq \mu_\rho}
\end{array}$$

Let $(\Omega, \mathcal{F}, \nu) \in \Delta \cap \mu_{\phi \wedge \psi}$. Then, in particular, we must have $(\Omega, \mathcal{F}, \nu) \in \mu_{\phi \wedge \psi}$, from which it follows that $\min(\nu(\tilde{\phi}), \nu(\tilde{\psi})) \geq \nu(\tilde{\phi} \cap \tilde{\psi}) = 1$; whence $\nu(\tilde{\phi}) = \nu(\tilde{\psi}) = 1$ and $(\Omega, \mathcal{F}, \nu) \in \Delta \cap \mu_\phi \cap \mu_\psi$. Now, by assumption, we have $\Delta \cap \mu_\phi \cap \mu_\psi \subseteq \mu_\rho$ so that $(\Omega, \mathcal{F}, \nu) \in \mu_\rho$, and thus $\Delta \cap \mu_{\phi \wedge \psi} \subseteq \mu_\rho$, as claimed.

Before we move on to discuss the coproduct type, we can use the rules for product type to deduce that $\mu_\phi \cap \mu_\psi = \mu_{\phi \wedge \psi}$ for any $\phi, \psi \in \Pi^*$. From this and the representation of contexts in the form of (6.2), we can equivalently express any context as

$$\Delta \equiv \mu_{\phi_1 \wedge \cdots \wedge \phi_n}, \tag{6.3}$$

which can more compactly be written as

$$\Delta \equiv \mu_\Phi$$

for some $\Phi \in \Pi^*$, because $\tilde{\phi}_1 \cap \cdots \cap \tilde{\phi}_n \in \mathcal{F}$ whenever $\tilde{\phi}_1, \ldots, \tilde{\phi}_n \in \mathcal{F}$. This representation plays a role in our proof of soundness for the coproduct elimination rule below.

We next discuss the coproduct type.

- **Coproduct type**, formation rule:

$$\begin{array}{cc}
\text{Syntax} & \text{Semantics} \\
\Delta \vdash \phi : \mathbf{Type} & \Delta \subseteq \mathcal{S}_\phi \\
\dfrac{\Delta \vdash \psi : \mathbf{Type}}{\Delta \vdash \phi + \psi : \mathbf{Type}} & \dfrac{\Delta \subseteq \mathcal{S}_\psi}{\Delta \subseteq \mathcal{S}_{\phi \vee \psi}}
\end{array}$$

Let $(\Omega, \mathcal{F}, \nu) \in \mathcal{S}_\phi \cap \mathcal{S}_\psi$, then $\tilde{\phi}^c \in \mathcal{F}$, $\tilde{\psi}^c \in \mathcal{F}$, and $\tilde{\phi} \cap \tilde{\psi} \in \mathcal{F}$, because $\mathcal{F}$ is an algebra. Finally, by definition, we have $\widetilde{\phi \vee \psi} \equiv$

$\tilde{\phi} \cup \tilde{\psi} \equiv (\tilde{\phi}^c \cap \tilde{\psi}^c)^c \in \mathcal{F}$, because $\mathcal{F}$ is a σ-algebra and is closed under complementation and countable intersection. It follows that $(\Omega, \mathcal{F}, \nu) \in \mathcal{S}_{\phi \vee \psi}$.

- **Coproduct type**, introduction rule 1:

$$
\begin{array}{cc}
\text{Syntax} & \text{Semantics} \\
\Delta \vdash \phi : \mathbf{Type} & \Delta \subseteq \mathcal{S}_\phi \\
\Delta \vdash \psi : \mathbf{Type} & \Delta \subseteq \mathcal{S}_\psi \\
\dfrac{\Delta \vdash a : \phi}{\Delta \vdash \mathtt{inl}(a) : \phi + \psi} & \dfrac{\Delta \subseteq \mu_\phi}{\Delta \subseteq \mu_{\phi \vee \psi}}
\end{array}
$$

The three assumptions combine to imply $\Delta \subseteq \mu_\phi \cap \mathcal{S}_\psi$, so that any $(\Omega, \mathcal{F}, \nu) \in \Delta$ satisfies $\tilde{\phi}, \tilde{\psi} \in \mathcal{F}$ and $\nu(\tilde{\phi}) = 1$. Because $\mathcal{F}$ is an algebra (or, alternatively, by the preceding formation rule), we have $\tilde{\phi} \cup \tilde{\psi} \in \mathcal{F}$, and so ν assigns measure to it, and since probability measures are increasing, we must have $\nu(\tilde{\phi} \cup \tilde{\psi}) \geq \nu(\tilde{\phi}) = 1$; whence $\nu \in \mu_{\phi \vee \psi}$, as claimed.

- **Coproduct type**, introduction rule 2:

$$
\begin{array}{cc}
\text{Syntax} & \text{Semantics} \\
\Delta \vdash \phi : \mathbf{Type} & \Delta \subseteq \mathcal{S}_\phi \\
\Delta \vdash \psi : \mathbf{Type} & \Delta \subseteq \mathcal{S}_\psi \\
\dfrac{\Delta \vdash b : \psi}{\Delta \vdash \mathtt{inr}(b) : \phi + \psi} & \dfrac{\Delta \subseteq \mu_\psi}{\Delta \subseteq \mu_{\phi \vee \psi}}
\end{array}
$$

The three assumptions combine to imply $\Delta \subseteq \mu_\psi \cap \mathcal{S}_\phi$, so that any $(\Omega, \mathcal{F}, \nu) \in \Delta$ satisfies $\tilde{\phi}, \tilde{\psi} \in \mathcal{F}$ and $\nu(\tilde{\psi}) = 1$. Because $\mathcal{F}$ is an algebra (or, alternatively, by the preceding formation rule), we have $\tilde{\phi} \cup \tilde{\psi} \in \mathcal{F}$, and so ν assigns measure to it, and since probability measures are increasing, we must have $\nu(\tilde{\phi} \cup \tilde{\psi}) \geq \nu(\tilde{\psi}) = 1$; whence $\nu \in \mu_{\phi \vee \psi}$, as claimed.

- **Coproduct type**, elimination rule:

Syntax	Semantics
$\Delta \vdash \phi : \mathbf{Type} \qquad \Delta \vdash \psi : \mathbf{Type}$	$\Delta \subseteq \mathcal{S}_\phi \qquad \Delta \subseteq \mathcal{S}_\psi$
$\Delta, a : \phi, b : \psi \vdash \rho : \mathbf{Type}$	$\Delta \cap \mu_\phi \cap \mu_\psi \subseteq \mathcal{S}_\rho$
$\Delta, a : \phi \vdash d_l(a) : \rho$	$\Delta \cap \mu_\phi \subseteq \mu_\rho$
$\Delta, b : \psi \vdash d_r(b) : \rho$	$\Delta \cap \mu_\psi \subseteq \mu_\rho$

$$\Delta, z : \phi + \psi \vdash \mathsf{case}_{d_l, d_r}(z) : \rho \qquad\qquad \Delta \cap \mu_{\phi \vee \psi} \subseteq \mu_\rho$$

Here, we use the representation in (6.3) to express $\Delta \equiv \mu_\Phi$ for some $\Phi \subseteq \Pi^*$, so that the third and fourth assumptions and the conclusion on the righthand side, respectively, become

$$\mu_{\Phi \wedge \phi} \subseteq \mu_\rho,$$
$$\mu_{\Phi \wedge \psi} \subseteq \mu_\rho \quad \text{and}$$
$$\mu_{\Phi \wedge (\phi \vee \psi)} \subseteq \mu_\rho.$$

By assumption, we have $\mu_{\Phi \wedge \phi} \subseteq \mu_\rho$. Thus, any ν that satisfies $\nu(\tilde{\Phi} \cap \tilde{\phi}) = 1$ must also satisfy $\nu(\tilde{\rho}) = 1$, which is possible only if $\tilde{\Phi} \cap \tilde{\phi} \subseteq \tilde{\rho}$. Suppose that there is some $\omega \in \tilde{\Phi} \cap \tilde{\phi}$ for which $\omega \notin \tilde{\rho}$, then there is a measurable space $(\Omega, \mathcal{F}_\omega, \nu_\omega)$ with powerset σ-algebra $\mathcal{F}_\omega$ on Ω and ν_ω, the atomic measure, at $\{\omega\}$ (i.e., $\nu_\omega(\{\omega\}) = 1$). With $\omega \in \tilde{\Phi} \cap \tilde{\phi}$, it follows that $\nu_\omega(\tilde{\Phi} \cap \tilde{\phi}) \geq \nu_\omega(\{\omega\}) = 1$ and $\nu_\omega(\tilde{\rho}) = 0$, contradicting the assumption. By applying an analogous argument to the fourth assumption, we must have $\tilde{\Phi} \cap \tilde{\phi} \subseteq \tilde{\rho}$.
For the conclusion, note that $\tilde{\Phi} \cap (\tilde{\phi} \cup \tilde{\psi}) \equiv (\tilde{\Phi} \cap \tilde{\phi}) \cup (\tilde{\Phi} \cap \tilde{\psi})$, so that the conclusion reads

$$\mu_{(\Phi \wedge \phi) \vee (\Phi \wedge \psi)} \subseteq \mu_\rho.$$

Now, suppose $(\Omega, \mathcal{F}, \nu) \in \mu_{(\Phi \wedge \phi) \vee (\Phi \wedge \psi)}$ so that $\nu((\tilde{\Phi} \cap \tilde{\phi}) \cup (\tilde{\Phi} \cap \tilde{\psi})) = 1$. Then, by the preceding argument, we have $\tilde{\Phi} \cap \tilde{\phi} \subseteq \tilde{\rho}$

and $\tilde{\Phi} \cap \tilde{\psi} \subseteq \tilde{\rho}$, which implies

$$(\tilde{\Phi} \cap \tilde{\phi}) \cup (\tilde{\Phi} \cap \tilde{\psi}) \subseteq \tilde{\rho}.$$

It follows that

$$1 = \nu((\tilde{\Phi} \cap \tilde{\phi}) \cup (\tilde{\Phi} \cap \tilde{\psi})) \leq \nu(\tilde{\rho});$$

whence, $\nu(\tilde{\rho}) = 1$ and $(\Omega, \mathcal{F}, \nu) \in \mu_\rho$, as claimed.

We next discuss the **0**-type.

- **0-type**, formation rule:

Syntax	Semantics
$\dfrac{\Delta \text{ ctx}}{\Delta \vdash \mathbf{0} : \mathbf{Type}}$	$\dfrac{\Delta \subseteq \mathcal{P}(\Pi^*)}{\Delta \subseteq \mathcal{S}_\perp}$

By assumption, Δ is a subset of probability spaces $(\Omega, \mathcal{F}, \nu)$, with $\mathcal{F}$ a σ-algebra over Ω. As any σ-algebra contains $\emptyset$, it is immediate that $\mathcal{P}(\Pi^*) = \mathcal{S}_\perp$ and the conclusion follows.

- **0-type**, elimination rule:[29]

Syntax	Semantics
$\dfrac{\Delta \vdash \phi : \mathbf{Type}}{\Delta, x : \mathbf{0} \vdash \mathsf{efq}_\phi(x) : \phi}$	$\dfrac{\Delta \subseteq \mathcal{S}_\phi}{\Delta \cap \mu_\perp \subseteq \mu_\phi}$

The subset $\mu_\perp \subseteq \mathcal{P}(\Pi^*)$ consists of all probability spaces $(\Omega, \mathcal{F}, \nu)$ that assign measure 1 to $\emptyset$. By definition, any probability measure ν must satisfy $\nu(\emptyset) = 0$, so that $\mu_\perp = \emptyset$. Thus, the conclusion reads $\Delta \cap \mu_\perp \equiv \emptyset \subseteq \mu_\phi$, which holds trivially.

Finally, for the **Typ**-type.

---

[29] Here, efq stands for *ex falso quodlibet* ("from falsehood, anything follows"). Formally, this rule says that given any $\phi : \mathbf{Type}$ and a proof $x : \mathbf{0}$ of the contradiction, it is possible to construct a proof $\mathsf{efq}_\phi(x) : \phi$. This is a type-theoretic version of the principle of explosion in PL, $\perp \to \phi$.

- **Typicality-type**, formation rule:

Syntax	Semantics
$\dfrac{\Delta \vdash \phi : \textbf{Type}}{\Delta \vdash \textbf{Typ}(\phi) : \textbf{Type}}$	$\dfrac{\Delta \subseteq \mathcal{S}_{\phi_*}}{\Delta \subseteq \mathcal{S}_{Typ(\phi)}}$

By definition, we require that $\widetilde{Typ(\phi)}$ is a measurable set whenever $\tilde{\phi}$ is, so that the conclusion immediately follows by definition of $\mathcal{P}(\Pi^*)$.

- **Typicality-type**, introduction rule:

Syntax	Semantics
$\dfrac{\Delta \vdash \phi : \textbf{Type}}{\Delta, a : \phi \vdash \tau_\phi(a) : \textbf{Typ}(\phi)}$	$\dfrac{\Delta \subseteq \mathcal{S}_\phi}{\Delta \cap \mu_\phi \subseteq \mu_{Typ(\phi)}}$

As any $(\Omega, \mathcal{F}, \nu) \in \Delta \cap \mu_\phi$ must satisfy $\nu(\tilde{\phi}) = 1$ and it is required that $\widetilde{Typ(\phi)} \supseteq \tilde{\phi}$, we must have $\nu(\widetilde{Typ(\phi)}) \geq \nu(\tilde{\phi}) = 1$, and $(\Omega, \mathcal{F}, \nu) \in \mu_{Typ(\phi)}$, as claimed.

- **Typicality-type**, elimination rule:

Syntax	Semantics
$\dfrac{\Delta \vdash \phi : \textbf{Type} \qquad \Delta \vdash \psi : \textbf{Type} \qquad \Delta, a : \phi \vdash d(a) : \psi}{\Delta, x : \textbf{Typ}(\phi) \vdash \texttt{imp}_d(x) : \textbf{Typ}(\psi)}$	$\dfrac{\Delta \subseteq \mathcal{S}_\phi \qquad \Delta \subseteq \mathcal{S}_\psi \qquad \Delta \cap \mu_\phi \subseteq \mu_\psi}{\Delta \cap \mu_{Typ(\phi)} \subseteq \mu_{Typ(\psi)}}$

By (6.3), we can express the second assumption as $\mu_{\Phi \wedge \phi} \subseteq \mu_\psi$ for some $\Phi \in \Pi^*$, from which it follows that $\tilde{\Phi} \cap \tilde{\phi} \subseteq \tilde{\psi}$, by an argument already given above when proving the elimination rule for the coproduct type. Thus, we can rewrite the conclusion as

$$\mu_{\Phi \wedge Typ(\phi)} \subseteq \mu_{Typ(\psi)}.$$

Let $(\Omega, \mathcal{F}, \nu) \in \mu_{\Phi \wedge Typ(\phi)}$. Then, $\nu(\tilde{\Phi} \cap \widetilde{Typ(\phi)}) = 1$, and in particular $\nu(\tilde{\Phi}) = 1$ and $\nu(\widetilde{Typ(\phi)}) = 1$, implying that $\nu(\tilde{\phi}) \geq 1 - \epsilon$.

By definition we have

$$\nu(\tilde{\Phi} \cap \tilde{\phi}) = \nu(\tilde{\Phi}) + \nu(\tilde{\phi}) - \nu(\tilde{\Phi} \cup \tilde{\phi})$$
$$\geq \nu(\tilde{\Phi}) + (1 - \epsilon) - 1$$
$$= 1 - \epsilon.$$

By assumption, we have $\nu(\tilde{\psi}) \geq \nu(\tilde{\Phi} \cap \tilde{\phi}) \geq 1 - \epsilon$, and $\nu(\widetilde{Typ(\psi)}) = 1$ by definition.

- **Typicality-type, 0 rule:**

Syntax $\qquad\qquad$ Semantics

$$\dfrac{\Delta \;\textbf{ctx}}{\Delta, x : \textbf{Typ}(\textbf{0}) \vdash \sigma(x) : \textbf{0}} \qquad \dfrac{\Delta \subseteq \mathcal{P}(\Pi^*)}{\Delta \cap \mu_{Typ(\perp)} \subseteq \mu_\perp}$$

By definition, every probability measure is required to assign probability 0 to $\perp \equiv \varnothing$. Thus, there does not exist any probability measure ν for which $\nu(\varnothing) \geq 1 - \epsilon$, as is required to make the judgment that $\perp$ is typical. It follows that $\mu_{Typ(\perp)} \equiv \varnothing$; whence, $\Delta \cap \mu_{Typ(\perp)} \equiv \varnothing \subseteq \varnothing \equiv \mu_\perp$, as required.

This completes the proof of soundness.

$\square$

6.3.5. *Additional results*

The above rules for the **Typ**-type have several immediate consequences for derived inference rules involving typicality. For example, we have the following one-way hierarchy of typicality judgments involving conjunction and disjunction. With '$\Rightarrow$' understood informally as 'implies', we have

$$Typ(\phi \wedge \psi) \Rightarrow Typ(\phi) \wedge Typ(\psi) \Rightarrow Typ(\phi) \vee Typ(\psi) \Rightarrow Typ(\phi \vee \psi). \tag{6.4}$$

We prove this formally below.

Theorem 7. *The implication arrows in* (6.4) *correspond, respectively, to the following derived inference rules:*[30]

(1) $Typ(\phi \wedge \psi) \Rightarrow Typ(\phi) \wedge Typ(\psi)$:

$$\frac{\Delta \vdash \phi : \mathbf{Type} \qquad \Delta \vdash \psi : \mathbf{Type}}{\Delta, z : \mathbf{Typ}(\phi \times \psi) \vdash (\mathrm{imp}_{\mathrm{pr}_\phi}(z), \mathrm{imp}_{\mathrm{pr}_\psi}(z)) : \mathbf{Typ}(\phi) \times \mathbf{Typ}(\psi)}$$

(2) $Typ(\phi) \wedge Typ(\psi) \Rightarrow Typ(\phi) \vee Typ(\psi)$:[31]

$$\frac{\Delta \vdash \phi : \mathbf{Type} \qquad \Delta \vdash \psi : \mathbf{Type}}{\Delta, z : \mathbf{Typ}(\phi) \times \mathbf{Typ}(\psi) \vdash inl(\mathrm{pr}_{\mathbf{Typ}(\phi)}(z)) : \mathbf{Typ}(\phi) + \mathbf{Typ}(\psi)}$$

(3) $Typ(\phi) \vee Typ(\psi) \Rightarrow Typ(\phi \vee \psi)$:

$$\frac{\Delta \vdash \phi : \mathbf{Type} \qquad \Delta \vdash \psi : \mathbf{Type}}{\Delta, z : \mathbf{Typ}(\phi) + \mathbf{Typ}(\psi) \vdash \mathrm{case}_{\mathrm{imp}_{inl}, \mathrm{imp}_{inr}}(z) : \mathbf{Typ}(\phi + \psi)}$$

Proof. To prove the first implication, we observe first that

$$\Delta, x : \phi \times \psi \vdash \mathrm{pr}_\phi(x) : \phi \quad \text{and}$$
$$\Delta, x : \phi \times \psi \vdash \mathrm{pr}_\psi(x) : \psi$$

are both valid judgments in MLTT, which match the second line in the elimination rule for the **Typ**-type. It follows, from each of these

---

[30]Note we define below the projection operator pr_ϕ, which selects out the corresponding coordinate from a pair $(a, b) : \phi \times \psi$. For example, $\mathrm{pr}_\phi((a, b)) \equiv a : \phi$ and $\mathrm{pr}_\psi((a, b)) \equiv b : \psi$.

[31]An alternative formalization of this implication is given by

$$\frac{\Delta \vdash \phi : \mathbf{Type} \qquad \Delta \vdash \psi : \mathbf{Type}}{\Delta, z : \mathbf{Typ}(\phi) \times \mathbf{Typ}(\psi) \vdash \mathrm{inr}(\mathrm{pr}_{\mathbf{Typ}(\psi)}(z)) : \mathbf{Typ}(\phi) + \mathbf{Typ}(\psi)}$$

lines respectively, that

$$\Delta, y : \mathbf{Typ}(\phi \times \psi) \vdash \mathrm{imp}_{\mathrm{pr}_\phi}(y) : \mathbf{Typ}(\phi) \quad \text{and}$$

$$\Delta, y : \mathbf{Typ}(\phi \times \psi) \vdash \mathrm{imp}_{\mathrm{pr}_\psi}(y) : \mathbf{Typ}(\psi).$$

An application of the introduction rule for the product type gives

$$\Delta, y : \mathbf{Typ}(\phi \times \psi) \vdash (\mathrm{imp}_{\mathrm{pr}_\phi}(y), \mathrm{imp}_{\mathrm{pr}_\psi}(y)) : \mathbf{Typ}(\phi) \times \mathbf{Typ}(\psi).$$

To prove the second implication, it is enough to observe that $\Phi \times \Psi \Rightarrow \Phi + \Psi$ for all $\Phi, \Psi : \mathbf{Type}$ in any context. Formally, we have

$$\Delta, x : \mathbf{Typ}(\phi) \times \mathbf{Typ}(\psi) \vdash \mathrm{inl}(\mathrm{pr}_{\mathbf{Typ}(\phi)}(x)) : \mathbf{Typ}(\phi) + \mathbf{Typ}(\psi),$$

or alternatively

$$\Delta, x : \mathbf{Typ}(\phi) \times \mathbf{Typ}(\psi) \vdash \mathrm{inr}(\mathrm{pr}_{\mathbf{Typ}(\psi)}(x)) : \mathbf{Typ}(\phi) + \mathbf{Typ}(\psi).$$

To prove the third implication, we must apply the elimination rule for the co-product type. In this case, the elimination rule is applied by

$$\frac{\begin{array}{c} \Delta \vdash \mathbf{Typ}(\phi) : \mathbf{Type} \qquad \Delta \vdash \mathbf{Typ}(\psi) : \mathbf{Type} \\ \Delta, a : \mathbf{Typ}(\phi), b : \mathbf{Typ}(\psi) \vdash \mathbf{Typ}(\phi + \psi) : \mathbf{Type} \\ \Delta, a : \mathbf{Typ}(\phi) \vdash \mathrm{imp}_{\mathrm{inl}}(a) : \mathbf{Typ}(\phi + \psi) \\ \Delta, b : \mathbf{Typ}(\psi) \vdash \mathrm{imp}_{\mathrm{inr}}(b) : \mathbf{Typ}(\phi + \psi) \end{array}}{\Delta, z : \mathbf{Typ}(\phi) + \mathbf{Typ}(\psi) \vdash \mathrm{case}_{\mathrm{imp}_{\mathrm{inl}}, \mathrm{imp}_{\mathrm{inr}}}(z) : \mathbf{Typ}(\phi + \psi)} \qquad \square$$

We note that the arrows in (6.4) do not reverse in general. To see this, we can produce a semantic counterexample by giving a probability space $(\Omega, \mathcal{F}, \nu)$ for which the corresponding interpretation fails.

(1) $Typ(\phi \wedge \psi) \nLeftarrow Typ(\phi) \wedge Typ(\psi)$. To prove this, we instead show

$$\mu_{Typ(\phi \wedge \psi)} \nsupseteq \mu_{Typ(\phi)} \cap \mu_{Typ(\psi)}.$$

Let $\Omega = \{\omega_1, \omega_2, \omega_3\}$ have the power set σ-algebra, and define measurable sets corresponding to $\phi, \psi, Typ(\phi), Typ(\psi)$ and their conjunctions and disjunctions by

$$\tilde{\phi} = \{\omega_1, \omega_2\}$$
$$\tilde{\psi} = \{\omega_2, \omega_3\}$$
$$\widetilde{Typ(\phi)} = \{\omega_1, \omega_2, \omega_3\} = \Omega$$
$$\widetilde{Typ(\psi)} = \{\omega_1, \omega_2, \omega_3\} = \Omega$$
$$\widetilde{Typ(\phi \wedge \psi)} = \{\omega_2\}$$

Fix $\epsilon = 1/3$ and define measure ν by

$$\nu(\{\omega_1\}) = \nu(\{\omega_3\}) = 1/4 \quad \text{and} \quad \nu(\{\omega_2\}) = 1/2.$$

From this, we have

$$\nu(\tilde{\phi}) = \nu(\tilde{\psi}) = 3/4 \geq 1 - \epsilon,$$

so that ϕ and ψ are typical according to ν. (Also note that

$$\nu(\widetilde{Typ(\phi)}) = \nu(\widetilde{Typ(\psi)}) = \nu(\Omega) = 1,$$

as required to justify the judgments that ϕ and ψ are typical.) Thus, we have $Typ(\phi) \wedge Typ(\psi)$ in accordance with the righthand side, but

$$\nu(\tilde{\phi} \cap \tilde{\psi}) = \nu(\{\omega_2\}) = 1/2 < 1 - \epsilon,$$

so that $\phi \wedge \psi$ is not typical.

(2) $Typ(\phi) \wedge Typ(\psi) \not\models Typ(\phi) \vee Typ(\psi)$. To prove this, we instead show

$$\mu_{Typ(\phi)} \cap \mu_{Typ(\psi)} \not\supseteq \mu_{Typ(\phi)} \cup \mu_{Typ(\psi)}.$$

Let $\Omega = \{\omega_1, \omega_2, \omega_3\}$ have the power set σ-algebra, and define measurable sets corresponding to $\phi, \psi, Typ(\phi), Typ(\psi)$ and their conjunctions and disjunctions by

$$\tilde{\phi} = \{\omega_1, \omega_2\}$$
$$\tilde{\psi} = \{\omega_2, \omega_3\}$$
$$\widetilde{Typ(\phi)} = \{\omega_1, \omega_2, \omega_3\} = \Omega$$
$$\widetilde{Typ(\psi)} = \{\omega_2, \omega_3\}$$
$$\widetilde{Typ(\phi \wedge \psi)} = \{\omega_2\}$$

Fix $\epsilon = 1/3$ and define measure ν by

$$\nu(\{\omega_1\}) = 1 \quad \text{and} \quad \nu(\{\omega_2\}) = \nu(\{\omega_3\}) = 0.$$

Thus, $\nu(\tilde{\phi}) = 1$ justifies the judgment in $Typ(\phi)$, so that the righthand side holds. But $\widetilde{Typ(\phi)} \cap \widetilde{Typ(\psi)} = \{\omega_2, \omega_3\}$ has measure 0 under ν, so the lefthand side does not hold under ν.

(3) $Typ(\phi) \vee Typ(\psi) \not\models Typ(\phi \vee \psi)$. To prove this, we instead show

$$\mu_{Typ(\phi)} \cup \mu_{Typ(\psi)} \not\supseteq \mu_{Typ(\phi \vee \psi)}.$$

Let $\Omega = \{\omega_1, \omega_2, \omega_3\}$ have the power set σ-algebra, and define measurable sets corresponding to $\phi, \psi, Typ(\phi), Typ(\psi)$ and their conjunctions and disjunctions by

$$\tilde{\phi} = \{\omega_1, \omega_2\}$$
$$\tilde{\psi} = \{\omega_2, \omega_3\}$$
$$\widetilde{Typ(\phi)} = \{\omega_1, \omega_2\}$$
$$\widetilde{Typ(\psi)} = \{\omega_2, \omega_3\}$$
$$\widetilde{Typ(\phi \wedge \psi)} = \{\omega_2\}$$
$$\widetilde{Typ(\phi \vee \psi)} = \{\omega_1, \omega_2, \omega_3\} = \Omega$$

Fix $\epsilon = 1/10$ and define measure ν by

$$\nu(\{\omega_1\}) = \nu(\{\omega_3\}) = 1/4 \quad \text{and} \quad \nu(\{\omega_2\}) = 1/2.$$

In this case, we have

$$\nu(\tilde{\phi}) = \nu(\tilde{\psi}) = 0.75 < 1 - 1/10,$$

so that neither ϕ nor ψ is typical, and the lefthand side fails to hold. But the disjunction $\tilde{\phi} \cup \tilde{\psi} = \Omega$ has ν-measure 1, and is therefore typical, satisfying the righthand side.

6.4. Conclusion

Formally, TPL and TIL have a great deal in common. Their proof theories are sound with respect to their semantic theories. In both TPL and TIL, typicality distributes over conjunction in one direction but not in the other: if $p \wedge q$ is typical, then p is typical and q is typical; but if p is typical and q is typical, then it does not follow that $p \wedge q$ is typical. Similarly, in both TPL and TIL, typicality distributes over disjunction in one direction but not in the other: if p is typical or q is typical, then $p \vee q$ is typical; but if $p \vee q$ is typical, then it does not follow that p is typical or q is typical.

There are important formal differences between TPL and TIL, however. The logic underlying TPL is classical, whereas the logic underlying TIL is intuitionistic. And while TPL is formulated in the manner of propositional modal logic, TIL is formulated in the manner of type theory.

These formal differences correspond to conceptual and metaphysical differences between the notion of typicality captured by TPL and the notion of typicality captured by TIL. For instance, according to TPL, the notion of typicality conforms to the rules of classical logic. According to TIL, in contrast, the notion of typicality is more constructive than that, requiring that an agent make explicit judgments about the truth and typicality of propositions. And there are other conceptual and metaphysical differences between these two systems. Whereas the meanings of typicality statements in TPL are given by sets of possible worlds, the meanings of typicality statements in TIL are given by sets of possible credences. So TPL is perhaps best understood as formalizing objective typicality facts, while TIL is perhaps best understood as formalizing subjective typicality judgments. Given these differences, it is worth exploring — in future works — the circumstances in which one system may be preferable to the other.

Regardless, both systems offer rigorous regimentations of typicality reasoning. They formalize the logical structure of the notion of typicality: its grammar, its semantic content, its proof theory, and the sorts of valid inferences which it licenses. Because of that, TPL and TIL both offer formal frameworks for typicality reasoning on a par with the formal framework that first-order logic offers for quantificational reasoning, or the formal framework that Bayesian theory offers for reasoning in terms of credences. TPL and TIL limn the deep logical structure that is shared by many instances of typicality reasoning in the literature on quantum mechanics and statistical mechanics. These systems reveal, in other words, what is common to the diverse array of explanations, predictions, and other kinds of scientific reasoning, which invoke the notion of the typical. In short, TPL and TIL capture the logic of typicality.

References

Alon, N., Krivelevich, M., Sudakov, B. (1998) Finding a Large Hidden Clique in a Random Graph. *Random Structures & Algorithms*, **13**, (3–4), 457–466.

Bernoulli, J. (1713) *Ars Conjectandi*. Thurnisius, Basel.

Boltzmann, L. (2003) Reply to Zermelo's Remarks on the Theory of Heat. In S. G. Brush (Ed.), *The Kinetic Theory of Gases, Vol. 1* (pp. 392–402) London: Imperial College Press. (Original work published 1896)

Booth, R., Meyer, T., Varzinczak, I. (2012) PTL: A Propositional Typicality Logic. In L. F. del Cerro, A. Herzig, & J. Mengin (Eds.), *Logics in Artificial Intelligence* (pp. 107–119), Heidelberg: Springer.

Cournot, A. A. (1843) *Exposition de la théorie des chances et des probabilitiés*. Hachette, Paris.

Crane, H. (2018) Logic of probability and conjecture. *Researchers.One*. Accessed on February 5, 2019 at https://www.researchers.one/article/2018-08-5.

Curry, H., Feys, R. (1958) *Combinatory Logic. I*, North-Holland Publishing Company, Amsterdam.

Dürr, D., Goldstein, S., Zanghì, N. (1992) Quantum Equilibrium and the Origin of Absolute Uncertainty, *Journal of Statistical Physics*, **67**(5/6), 843–907.

Everett, H. (2012) The Theory of the Universal Wave Function. In J. Barrett &
P. Byrne (Eds.), *The Everett Interpretation of Quantum Mechanics* (pp. 72–
172). Princeton, NJ: Princeton University Press. (Original work published
1956)

Frigg, R. (2011) Why Typicality Does Not Explain the Approach to Equilibrium.
In M. Suárez (Ed.), *Probabilities, Causes and Propensities in Physics* (pp.
77–93), New York, NY: Springer.

Frigg, R., Werndl, C. (2012) Demystifying Typicality, *Philosophy of Science*, **79**,
917–929.

Goldstein, S. (2001) Boltzmann's Approach to Statistical Mechanics. In J. Bricmont, D. Dürr, M. C. Galavotti, G. Ghirardi, F. Petruccione, & N. Zanghì
(Eds.), *Chance in Physics* (pp. 39–54). Heidelberg: Springer.

Goldstein, S. (2012) Typicality and Notions of Probability in Physics. In Y. Ben-
Menahem & M. Hemmo (Eds.), *Probability in Physics* (pp. 59–71), New York,
NY: Springer-Verlag.

Goldstein, S., Lebowitz, J., Mastrodonato, C., Tumulka, R., Zanghì, N. (2010)
Normal typicality and von Neumann's quantum ergodic theorem. *Proceedings: Mathematical, Physical and Engineering Sciences*, **466**(2123), 3203–
3224.

Howard, W. A. (1969) The formulae-as-types notion of construction. In J. P.
Seldin, R. J. Hindley, & H. B. Curry: *Essays on Combinatory Logic, lambda
calculus and formalism* (pp. 479–490), Academic Press, New York, NY.

Kesten, H. (1980) The Critical Probability of Bond Percolation on the Square
Lattice Equals 1/2, *Communications in Mathematical Physics*, **74**, 41–59.

Kiessling, M. K-H. (2011) Typicality analysis for the Newtonian N-body problem
on $\mathbb{S}^2$ in the $N \rightarrow \infty$ limit, *Journal of Statistical Mechanics: Theory and
Experiment*, P01028.

Ledoux, M. (2001) *The Concentration of Measure Phenomenon*. Providence, RI:
American Mathematical Society.

Martin-Löf, P. (1984) *Intuitionistic type theory*. Sambin, Giovanni. Napoli: Bib-
liopolis.

Mazur, B. (2012) Is it plausible? Available at http://www.math.harvard.edu/
mazur/papers/Plausibility.Notes.3.pdf. Accessed on August 20, 2018.

Pólya, G. (1954) *Mathematics and Plauisble Reasoning*, Princeton, NY: Princeton
University Press.

Priest, G. (2008) *An Introduction to Non-Classical Logic*, New York, NY:
Cambridge University Press.

Reimann, P. (2007) Typicality for Generalized Microcanonical Ensembles, *Physical Review Letters*, **99**, 160404.

Shafer, G., Vovk, V. (2006) The Sources of Kolmogorov's "Grundbegriffe", *Statistical Science*, **21**(1), 70–98.

Smullyan, R. M. (1968) *First-Order Logic*, New York, NY: Springer-Verlag.

Tasaki, H. (2016) Typicality of Thermal Equilibrium and Thermalization in Isolated Macroscopic Quantum Systems, *Journal of Statistical Physics*, **163**,
937–997.

Tsementzis, D. (2018) A Meaning Explanation for HoTT. Available at http://rci.rutgers.edu/~dt506\\/A.Meaning.Explanation.for.HoTT.v3.5.public.pdf. Accessed on August 27, 2018.

Werndl, C. (2013) Justifying typicality measures of Boltzmannian statistical mechanics and dynamical systems, *Studies in History and Philosophy of Modern Physics*, **44**, 470–479.

Wilhelm, I. (forthcoming) Typical: A Theory of Typicality and Typicality Explanation. *The British Journal for the Philosophy of Science*.

Chapter 7

The Grammar of Typicality

Tim Maudlin

Department of Philosophy, New York University,
John Bell Institute for the Foundations of Physics
twm3@nyu.edu

It is easy to suspect that the appearance of probability measures —
or anything close to probability measures — in a physical theory
must ultimately trace back to the supposition of some probability
measure in the foundations of the theory. This latter could either
arise when specifying the dynamics of the theory (if the dynamics
is fundamentally stochastic) or else as a probability measure over
the space of initial states of the theory. The latter sort of postulate
is immediately puzzling: where should a *probability measure* over
initial conditions come from, and what are the truth conditions for
ascribing it to a system? This whole problematic is a conceptual
illusion. One can *derive* something close to a probability measure
from a deterministic theory without putting such a measure over
the space of initial states. The additional conceptual resource one
needs is the notion of *typicality*. I will formally exposit what the
concept of typicality is, why it is much less detailed than a probability measure. I will explain how to derive just what one wants
from it: the notion of typical empirical frequencies according to the
theory. The theory can then be tested by comparing the observed
empirical frequencies to the typical ones.

Statistical mechanics trades in probability measures because it trades
in statistics. If the fundamental physics involved is fundamentally
stochastic, such as the Ghirardi-Rimini-Weber (GRW) theory, then

it is no mystery where some of those probabilities come from: they are inherited from the fundamental probability structure of the dynamics. But if the fundamental dynamics is deterministic, the origin of the probability measures that arise in statistical mechanics are not so evident. Ultimately, of course, the data used to test the theory (and hence the data that needs to be accounted for by the theory) are empirical frequencies. Frequencies obey the axioms of a probability measure, so a probability measure must be the ultimate output of the statistical explanation machine. The question that confronts us is simple: if we do not get this probability measure as output by using some other probability measure as input, how can a deterministic dynamics yield a probability measure as a derived consequence? We need something more than just the deterministic dynamics to work with, but we also want the extra ingredient to be something less than a probability measure. That extra bit of information required for the derivation is provided by the notion of *typicality*.

I do not propose anything novel in this exposition; I just hope to anatomize the conceptual structure in enough explicit detail to lay to rest some possible confusions about how the trick is done.

The notion of typicality is, at the mathematical level, much less detailed than that of probability. Given a sample space, a probability measure assigns a real number between 0 and 1 to every measurable subset; that is, in an obvious sense, a tremendous amount of extremely detailed mathematical information. Indeed, it is so much highly detailed information that it is immediately clear that the role of probability measures for some theoretical purposes can only be that of an unrealistic idealization. For example, probability measures are commonly used to represent credal states — "degrees of belief" — in agents. But no actual human being has credal states that correspond to unique probability measures over propositions. For example, I believe that string theory will never pan out as a successful physical theory, and I can characterize that belief to some extent in terms of its strength. It is not very strongly held; it would not take much, in the way of a surprising successful empirical prediction coming out of the theory, to get me to change my mind. I could compare the strength of this belief with some other beliefs.

I believe that neutrinos have mass more strongly, but still not all that strongly. I believe that water is an H_2O molecule much, much, much more strongly. But these sorts of characterizations of the strength of my beliefs do not take us very far towards specifying a unique real number between 0 and 1 to attach to the proposition that string theory will never yield useful physics. Indeed, there is obviously no such unique number. For there is nothing in my cognitive architecture that would support the attribution of a precise number.

It is also clear how to model my credal state using probability measures. Just assign over all the propositions a probability measure that has this feature: when I believe one proposition more than I believe another, then the number assigned to the first is greater than the number assigned to the second. I can even imagine a theorem to the effect that such an association of degrees of belief with real numbers must be possible to avoid Dutch book. But even so, the constraints provided by my credal states are much, much, much too weak to implicitly single out a unique probability measure. If one measure does the job adequately, then an infinitude of others will do just as well. So the use of probability measures in the philosophy of mind and cognitive theory has the status of an unrealistic over-specification of a system that inherently has much less detailed structure. But whole approaches — e.g. Bayesianism — have this unrealistic modelling built deep into their set of technical tools, and would take considerable effort to wean from it.

A tempting first step in the weaning is an even higher abstraction: instead of representing the credal state with a single probability measure, use a set of probability measures. But, however practically useful that may be, it lies in the opposite direction from what we are seeking. Just as there is no sharply defined probability measure that corresponds uniquely to my credal state, so is there no sharply defined set of such measures. And if the only way to specify which probability measures are to go in the set — by demanding that certain relations among the assigned numbers be respected — then why not just use the set of constraints directly as the representative of the credal state? Why take such a roundabout route through idealized sets of probability measures?

The literature on this set of connected issues is vast, and I have neither space nor the expertise to sort through it. So let's just cut to the chase. The notion of a "typicality measure" that I want to exposit is something of an idealization as well. It is a more sharply defined sort of gadget than the underlying precise fundamental physics can specify. But it is still so much less detailed and specific than a probability measure that replacing probability measures with typicality measures makes the whole analysis much more realistic and satisfactory.

Rather than an assignment of real numbers to all the measurable subsets, a "measure of typicality" is a simple predicate on the set of subsets of the sample space. It defines what we may call a "big subset" of the sample space. That's it. From a formal point of view it is nothing but a set of elements of the power set of the sample space. No more and no less. Given that single predicate, we can immediately define two more: a subset of the sample space is a "small subset" iff it is the complement of a big subset, and a subset is "indeterminate" iff it is neither big nor small.

It must immediately be acknowledged that this formal account of a measure of typicality is also an unrealistic idealization. There is no sharp line dividing the big subsets from the indeterminate ones, or the indeterminate ones from the small. Indeed, it contradicts the very notion of a "big" subset to suggest that the loss of a single element would render it "indeterminate": big subsets are supposed to be so big that the addition or loss of a single element could not change their characters. In other words, the predicate "big" is a vague predicate, with all of the formal challenges that come with that feature. Trying to explicate vague predicates using the sharp tool of set theory raises a host of conceptual and technical challenges. But these are not new challenges — the Sorites paradox has a vast literature of its own — and we have every reason to expect that the idealization here is benign; such is often the case with vague predicates. The terms "shoes" and "boots" are vague; there is no sharp line dividing the one from the other. But in everyday practice, this makes no difference: the actual pieces of footware one encounters are either unproblematically shoes or unproblematically boots. Similarly, the concept of a "big" subset of the sample space is vague at the edges, but the edges never show up in practice, so it is fine to treat the predicate as precise even though it is not.

The technical challenge I want to address in this paper, then, is exactly how this conceptual architecture fits together. What we want is a scheme that takes a deterministic dynamics + a "measure of typicality" as input, and yields a relevant sort of probability measure — or something close to a probability measure — as output.

The basic move is simple to specify, so I'll just break the suspense and give away the answer first, and fill in the details later. Given a deterministic dynamics and a "measure of typicality", it is easy to define the *typical empirical frequencies* that are implied by the theory and the measure of typicality. And since frequencies must — as a matter of conceptual necessity — be specified by a mathematical probability measure, it comes as no surprise that one gets something like a probability measure as output. But one does not get out a perfectly precise probability measure, analogous to the probability measure postulated in the GRW theory, for example. Rather, one gets out a vague or approximate probability measure. All of this falls straight out of the analysis, so let's begin.

What is typical?

In everyday talk it is perfectly acceptable to characterize a particular individual as "typical". Peoria, Illinois, for example, might be cited as a "typical Midwestern town". But in the usage I insist on, this locution cannot be correct. Individuals are never typical. The only things that can be typical in the strict sense are *generic features or properties* of individuals. And the condition for a generic feature to be typical is simple: the set of individuals that have that generic feature must be a big set according to the measure of typicality. So, for example, it may be a typical feature of Midwestern towns that they have a town hall, but it cannot possibly be a typical feature of Midwestern towns that they be Peoria, Illinois since the extension of the latter property is a unit set in the sample space and no unit set is a big set. "Peoria, Illinois is a typical Midwestern town" can at best

be parsed as "Peoria, Illinois has all (or most) of the typical features
of Midwestern towns", but this is a secondary, or derivative locution.
It certainly cannot mean "all of the features of Peoria, Illinois are
typical features of Midwestern towns" because among the features of
Peoria, Illinois is being in Illinois, and that is not a typical feature
of Midwestern towns.

Just to make sure this is clear, being white may be a typical
feature of being a swan, and that explains our tendency to call black
swans atypical, but a particular black swan may be atypical in its
color and typical in its having two legs. And once we get to very, very
large populations with many, many relevant features, it is perfectly
possible that no individual at all have *all* the typical features of
the population of individuals. In short, it may well be typical to be
atypical, as paradoxical as that may sound.

Indeed, once the logic of the locution becomes clear it may seem
inevitable that it be typical to be atypical in at least one respect. Here
is the argument: every individual has the property of being that very
individual, but that property cannot possibly be a typical property
because its extension is a unit set. Therefore, every individual in
the sample space has at least one atypical property. But the set
that includes the whole sample space is certainly a big set, so being
atypical is necessarily typical. QED.

The preceding argument may be tempting, but I want to reject
it. It relies on a contentious use of the term "property". David Lewis
famously made a distinction between an "abundant" sense of the
term "property" and a "natural" sense of the term (Lewis, 1983). In
the abundant sense, every subset of a set counts as a property, and
in the natural sense the set of properties is much, much more restrictive. In the abundant sense, of course, there is a property of, e.g.
being Peoria, Illinois, because there is a unit set whose only member
is Peoria, Illinois. But, as a matter of fact, the term "property" has
never in the history of philosophy been used in Lewis's abundant
sense. In the abundant sense, every pair of individuals have an infinite number of properties in common, because there are an infinite
number of sets of which they are both members. But this just fails to
correspond to any actual use of the term "property" in the history of

mankind. If I assert that I have a property in common with a particular solar neutrino, and someone asks me what property that is, the answer "The set whose members are myself and the solar neutrino" is non-responsive. No one, ever, would accept that answer, in any context. If I respond "The property of having mass", then that is acceptable. But that is a natural property in Lewis's sense. So the conclusion that it is typical to be atypical cannot be secured by the quick argument above.

Nonetheless, there are acceptable arguments in many cases that atypicality is typical. In the class of human beings, for example, a particular genetic structure is certainly a genuine property and one that is certainly not typical (even when the extension is not a unit set, as with identical twins). So, in realistic cases, we still get the conclusion that "X is a typical F" cannot be interpreted to mean "all of Xs properties are typical for Fs". And since the locution "X is a typical F" plays no essential role in our explication, we will from now on ignore and avoid that locution entirely. We will only predicate typicality of generic features, not of individuals. And once again, a generic feature is typical if and only if the extension of the feature is a big set according to the "measure of typicality" being used.

From a purely logical point of view, then, we postulate a predicate — "big set" — and use it to define a second-order predicate (i.e. a predicate of predicates) "typical". None of this requires anything like a probability measure over any set. But then how does it help us with our original task of deriving probability talk from a deterministic dynamics together with the "big set" predicate?

The answer runs through empirical frequencies. To cut to the chase again, probability talk within deterministic theories is really talk about typical empirical frequencies. An example can make this clear.

Suppose we have a perfectly symmetrical coin in this sense: the mass distribution is symmetric around a plane that geometrically bisects the coin. What we want to conclude from the perfect symmetry is that the coin is fair, and what we want to conclude from the fairness is that the coin has a 50% chance of coming heads and a 50% chance of coming tails if properly flipped. There is a very quick argument to the 50-50 conclusion that runs through a principle of indifference: since there is nothing about one side that distinguishes it from the other, we have to treat the two sides the same, and the only way to treat them probabilistically the same is to assign them the same probability, which must therefore be 50%.

But this symmetry argument is clearly broken-backed. First of all, the symmetry between the two sides must not be exact. In order to distinguish "heads" from "tails", there must be some qualitative difference between the sides. In order to keep the exact symmetry of the mass distribution, let's stipulate that the sides are different colors: one red and the other blue. But given this difference, it is possible that the coin will not in fact be fair; that depends on the physical circumstances in which it is flipped. If there is, for example, an extremely strong beam of light coming from some direction, the differential absorption and reflection of the light by the two sides could create a bias. So we have to somehow establish that the physical difference between the sides will not have any effect on the trajectory of the coin once it is flipped. This is a substantial physical claim, and cannot be settled otherwise than by physical analysis. That physical analysis would show that the spatial trajectory of a coin placed red-side-up and flipped in some precise way would be identical to that of the coin placed blue-side-up and flipped. Let's presume that this physical argument has been made. We are still not done.

Given a deterministic dynamics and a precisely specified method of flipping and environment, the dynamics will provide a map from the details of the flipping procedure to one of the alternatives {same side up, opposite side up}. That is, the dynamics will either imply that if the coin starts out placed red-side-up it will land red-side-up, or it will imply that if the coin starts out red-side-up it will land blue-side-up. And given the irrelevance of the color to the trajectory, the same map will apply to a coin initially placed blue-side-up, mutatis mutandis. What the perfect mass-distribution symmetry of the coin buys us is the identity of these two maps. Nothing more. But how can we parlay this result into a probability attribution of 50-50?

One unhelpful suggestion is to appeal to a 50% chance that the coin starts out red-side-up as opposed to blue-side-up. That just pushes the whole problem back to a probability distribution over these two possible initial conditions, and is no help at all.

Indeed, it is obvious that we can make no progress at all on the problem if we only consider a single, precise method of flipping. If exactly the same method of flipping is used every time the coin is flipped, then the whole flipping procedure is just a physically complicated way to implement one of the two maps: either the initial upward-facing color of the coin will be the final upward-facing color or else the opposite color will be the final one. If we happen to know which of the two maps is implemented by the upcoming flip then the method of coin flipping will no longer be considered a way to produce a 50% chance of each outcome. Rather, the whole issue will turn on the initial orientation of the coin before it is flipped. So to move us forward we evidently have to consider not just a single precise method of flipping but a whole range of possibilities. The range will include flips with different initial linear momenta and angular momenta of the coin, different heights from the floor etc. Each particular precise flipping method will yield either the same-side or the opposite-side initial-to-final dynamical map, but different precise flippings will yield different maps. That diversity appears to provide us some resource that can be leveraged to get us to 50-50. But how can that resource be deployed?

Once again, there is a temptation to appeal to a probability distribution over the various possible precise flipping procedures. If we are handed such a probability distribution then the dynamics will induce a probability distribution over the pair of outcomes {same side up, opposite side up}. And if that probability distribution happens to be 50-50, then we will have vindicated assigning a 50% chance to each outcome. But on reflection this whole line of reasoning reveals itself to be a waste of time. Our original problem was how to get a probability distribution — or at least something akin to a probability distribution — out of a deterministic dynamics plus something logically weaker than a probability distribution. Getting a probability distribution out of a deterministic dynamics plus a probability distribution over initial conditions is no great trick and solves no conceptual problems. It just pushes the problems back onto the probability distribution over the initial conditions. Where did *that* come from?

I conclude that this method of attack is essentially hopeless. There is no way to derive single-case probabilities (other than 0 or 1) from a deterministic dynamics in an explanatorily useful way. You can push the lump in the rug around from here to there, but you just can't get rid of it. *Non-trivial objective single-case probabilities cannot be conjured out of a deterministic dynamics no matter what you try.* This same intuition was expressed by Lewis when he asserted that a deterministic dynamics is just flatly incompatible with objective chance (Lewis, 1986, p.118). I'm with Lewis here. Objective non-trivial single-case probabilities in a deterministic world are, strictly speaking, incoherent. We will revisit this stark conclusion at the end, and try to ameliorate it a bit, but at a foundational level it is essentially correct.

Empirical statistics

What's so be done? Well, in a deterministic world, if you want to get out a probability like 50% from an analysis, you had better have something there can be 50% of. Given a single coin flip, there just is nothing that is vaguely 50%-ish about the experiment or its outcome: the outcome is either red-side-up or blue-side-up. Period.

But if I flip a coin 100 times, or 1,000 times, or 1,000,000 times, or if I flip 1,000,000 different coins all at once, then there is *something* that can be 50%! It is possible (albeit not physically necessary) that 50% of the outcomes are red-side-up and 50% are blue-side-up. Looking not at single outcomes but at collections of outcomes offers a ray of hope to find a way forward. Let's follow that ray and see where it leads.

Flipping the coin in *exactly* the same way all 1,000,000 times is evidently no improvement: in the deterministic setting we either get 1,000,000 red-side-ups or 1,000,000 blue-side-ups. Still nothing like 50% in sight. But if the 1,000,000 coin flips are done in 1,000,000 even slightly different ways then we have something to work with. The mere fact that the underlying dynamics is deterministic leads to no conclusion at all. The percentage of red-side-up outcomes could be anything between 0 and 100.

Since the outcome is a function not only of the precise flipping method but also of the initial orientation of the coin, let's take that latter factor out of the analysis by stipulating that the coin always start out red-side-up. Our initial intuition is that if the coin is fair *and the flipping process itself is somehow fair*, then the 50% chance over the different outcomes should be obtained even if you always start with the same side up.

So we now are working with a large class of possible individual-coin-flip procedures. Each particular flip in our collection of 1,000,000 flips will utilize one of these particular flipping mechanisms with the coin starting red-side-up. And the dynamics will map every such possible 1,000,000 flip collective into an *empirical distribution*: a precise percentage of red-side-up outcomes. Every possible percentage will be achieved by some collection of flipping procedures or other. Now what?

Once again, what we want to avoid is some explicit or tacit appeal to a probability distribution over these various possible 1,000,000-flip sequences. That would just be shoving our conceptual problem back on to the initial conditions. But now, at last, the weaker notion of typicality can come to the rescue.

Recall: we are looking for a way to somehow ascribe something like the probability of 50% to each possible outcome of the flip of a fair coin. And we have now found something — the empirical distribution — that can display a 50% outcome. But a moment's thought shows that, intuitively, the chance of getting *exactly* 500,000 red-side-up outcomes in a sequence of 1,000,000 flips is tiny. We expect the outcome to be *about* 50%, but do not at all expect the outcome to be *exactly* 50%. And the more we flip, the lower the chance of getting exactly 50%. If you want a precisely 50% outcome your best bet is to only flip the coin twice. Flipping more than that just *lowers* the chance of success. But as the chance of an *exactly* 50% outcome goes down, the chance of a *nearly* 50% outcome goes up, for any choice of what to mean by "nearly". This is just the Law of Large Numbers in action. And this observation provides the final link in

T. Maudlin

the conceptual chain leading from the deterministic dynamics to the probability. For once we get to a collective containing enough coin flips, outcomes that are *approximately* 50% heads can become typical. In other words, given a measure of typicality over the sample space of all possible precise-flipping-sequences, there can be a tolerance ϵ such that the outcome displaying an empirical distribution of 50%±ϵ is a typical outcome. The set of possible flipping sequences leading to the outcome 50%±ϵ can be a big set. Intuitively, *overwhelmingly most* of the ways one could flip the coin 1,000,000 times (or flip 1,000,000 different coins all at the same time) yield an empirical distribution of about 50%. Furthermore, for longer and longer sequences of flips, the tolerance ϵ gets smaller and smaller. In the limit as the number of flips goes to infinity, ϵ tends to 0. (But, of course, in the limit as the number of flips goes to infinity, the chance of getting *exactly* 50% as the outcome also goes to zero! So don't cheat and say that in the limit, the typical outcome is exactly 50%. That is just not true. Invocation of the tolerance ϵ is essential to the analysis, and does not disappear even in the "infinite limit".)

And now we have finally got a result we can use. Given as input nothing but a deterministic dynamics, a specification of the physical structure of the coin, a physical specification of a class of precise flipping methods *plus a measure of typicality over the precise flipping methods*, we can in principle derive the result that a certain percentage of red-side-up outcomes ±ϵ is typical. This is not exactly the thing we were originally after — namely the attribution of the precise probability 50% to the symmetric coin — but is close enough for all practical purposes. And it has the tremendous conceptual advantage that the symmetry of the coin is nowhere required for that analysis. Every actual coin is not precisely symmetrical in its mass distribution. Is the coin still fair, or at least fair enough FAPP? Well, there is no easy shortcut to answering that interesting question. The only way to tell is to carry out the long and laborious calculation of exactly what outcome the deterministic dynamics yields for each and every possible way that the coin might be flipped. As Detlef Dürr once remarked in a similar situation, we are infinitely far from being able to actually carry out such an analysis for an actual coin. Is the flipping of an actual US quarter, with its precise matter distribution, perfectly fair? That is very doubtful. Indeed, it would be almost incredible if it were true. But in order that the slight bias of the coin show up in the analysis, in order that the range of typical results to actually *exclude* 50%, the number of flips must presumably be taken to be astronomically high. Such a coin is therefore fair FAPP.

Note that at no point in this entire analysis has any appeal been made to anyone's state of information about the coin or about the flipping process. We nowhere invoked the internally incoherent command to treat processes as probabilistically identical if our information about them is the same. That command requires us to ascribe the same probabilities to all events of which we have zero information, and that is not only bad and unjustified advice, it is not even logically possible to follow. Rather than being rooted in *ignorance*, our derivation of typical empirical frequencies from deterministic dynamics + a measure of typicality over precise initial conditions is rooted in *knowledge*: precise knowledge of the dynamics, of the physical makeup of the coin and of the various possible flipping mechanisms. If the coin does display some *physical* symmetry that may provide a means to a quick physical analysis, or at least to a quick plausibility argument for what the proper physical analysis would yield — but even if such a symmetry is *convenient* for analytical purposes, it is not in the least *necessary* — the conceptual apparatus applies just as well to coins that display no physical symmetries at all. It is an all-purpose almost-probability-generating mechanism. They are almost-probabilities rather than probabilities on account of the tolerance $\pm\epsilon$, which cannot be dispensed with. Without the tolerance, no empirical frequency will be typical. If I flip a coin 1,000,000 times, no *precise* outcome frequency is typical: only a range of output frequencies can be. With respect to precise empirical frequencies, it is typical to be atypical.

The single case

In some practical circumstances, I am not interested in empirical frequencies in a large collective, I am rather interested in a single, discrete outcome. If I am going to bet a large sum of money on a single throw of the dice, for example, then what I care about is what the outcome of that particular throw is going to be, not what the long-term frequency of outcomes over a long sequence of throws will be. It is this single-case probability that I will use to calculate an expected utility, on the basis of which I may make the practical decision about whether or not to make the bet. Of course, this sort of betting-on-a-single-outcome is not something that is part of the scientific practice we are primarily concerned with. That practice is not one of practical reasoning but rather of theoretical reasoning: we want to be able to predict empirical frequencies in large collectives, so we can use the actual outcomes of experiments on large collectives as evidence for or against various theories. For that purpose we already have what we want. If, according to a theory, a certain range of frequencies $\pm\epsilon$ is typical for a large collection of outcomes, then doing an experiment and finding a frequency outside that range counts as strong evidence against the theory. Finding a frequency that lies within the range may not be so strong an evidence in favor of the theory; that depends on the existence or absence of rival theories that can make equally acceptable predictions.

So asking for single-case probabilities is something that properly lies outside the scope of our inquiry. A single result is an *anecdote*, not *data*. Nonetheless, our results do suggest a practical piece of advice. If one is faced with a single case — one single coin flip — and wants to ascribe a single-case probability to it, then a reasonable way to proceed is to see if a long sequence of flips would yield a typical frequency $F\pm\epsilon$ in the long term. If so, then ascribe the probability F to the single case. One can cook up cases where this seems like bad advice (for example, where the long-term frequency settles down to a value, but the initial members of the sequence are generated in a completely different way, so the long-term frequency is clearly irrelevant to the one-off initial result one is betting on), but as a default position it seems quite reasonable. In this way, we can derive non-trivial single-case probabilities from a deterministic dynamics (+ measure of typicality), Lewis's skepticism notwithstanding.

The advice in the previous paragraph is conditional: *if* the dynamics yields typical (nontrivial) approximate frequencies in the long run, use those as (approximate) single-case probabilities. But there is no logical guarantee that the dynamics will yield what we want. A sufficiently chaotic dynamics, for example, may simply fail to generate any useful typical frequency ranges, even as the set of flips gets large. There may or may not be useful generalizations about the sorts of dynamics that yield good results and those that don't, but that strikes me as a mathematically extremely difficult question, and one that is certainly above my pay grade.

One question that arises is what one wants these single-case probabilities *for* in the first place, and reflection on that question yields some interesting results. One of the main uses for single-case probabilities is in practical decision-making. And the standard advice there is to use the single-case probabilities to calculate the expected utilities of the various options available, and choose the option with the highest expected utility. But this bit of advice is both somewhat conceptually opaque and also yields what appear to be some dreadful pieces of advice. The conceptual worry is this: why should I choose the option with the highest expected utility in the one-off case, since what I care about is not the expected utility (which I will certainly *not* receive) but the actual utility that I will receive. Having maximized my expected utility is cold comfort if I have had a disaster with the actual utility of the outcome.

It is this fact that fuels the St. Petersburg Paradox. Recall, the paradox involves a bet with this payoff structure: a coin will be flipped as long as it keeps landing heads. If it lands tails on the first flip, the payoff is zero; if the first tails occurs on the second flip, the payoff is $1; on the third flip, the payoff is $2; on the fourth $4; on the fifth $8 etc. How much should one be willing to pay for such a bet? The expected utility of the bet is infinite: $(1/2 \times 0) + (1/4 \times 1) + (1/8 \times 2) + (1/16 \times 4) + (1/32 \times 8) \ldots = 0 + 1/4 + 1/4 + 1/4 + 1/4 + \ldots$, which grows unboundedly. So according to the expected utility rule, one should be willing to pay any finite amount at all for the bet. And even worse, the expected utility remains infinite even if the first billion payoffs are all reduced to zero. Accepting such a bet for some huge finite stake in the one-off case will maximize my expected utility while at the same time being almost certain to vastly reduce my actual utility! This strikes many people as completely insane, and I am with them.

We are now in a position to solve this difficulty by proposing a different decision rule. That rule is simple: Expect the Typical. If I expect what is typical, then if I have both the resources and the opportunity to play the St. Petersburg game enough times, I can expect to *eventually* make money, even if it will typically take a very long time to turn a profit. But if it is a one-off, then given a high enough stake, the typical outcome is to suffer a loss, and that is a bad outcome. So Expect the Typical does a better job of providing rational practical advice than Choose the Highest Expected Utility does. Of course, if you plan to repeat the same decision strategy over and over for many runs, then Expect the Typical yields the same advice as Choose the Highest Expected Utility, since the typical long-term outcome after a large number N plays is (N x Expected Utility) $\pm\epsilon$. So, in the case of repeated plays, maximizing the *expected* utility typically maximizes the *actual* utility, and the standard advice is good iff Expect the Typical is. That is why it is rational (and profitable) to *run* a casino. But in the one-off case, the two decision rules yield different advice, and the advice given by Expect the Typical seems intuitively more rational than the advice of Choose the Highest Expected Utility. That's what the St Petersburg paradox is pointing out and the same holds for everyday lotteries. The *typical* result of playing a powerball lottery once is to be a dollar poorer the next day, even when the payoffs are high enough to drive the expected utility positive. So we have both vindicated the standard advice in the circumstances where it does well, and explained its failure in situations where it fails.

Expect the Typical is essentially a version of Cournot's Principle, couched in terms of measures of typicality rather than in terms of probability measures. Cournot's principle, in its unrefined form, advises that once the probability of an event gets high enough (above, say, .99999999), just act as if you are certain it will occur. In our terms, we can translate that into typicality talk as Expect the Typical. As we have seen, this addresses both problems in confirmation theory and problems in practical reasoning at one fell swoop. That is a nice result.

Expect the Typical is of no practical use where nothing is typical. And in many one-off or few-off cases, there are no typical results at all. So if you want an all-purpose decision rule, Expect the Typical will not satisfy your desires while Choose the Highest Expected Utility will (assuming you always have single-case probabilities to plug into the expected utility calculation). One might tout that as a virtue of the standard approach. But if the advice given by the more extensive rule happens to be dreadfully bad in some cases (such as St. Petersburg), it is hard to see the extensiveness *per se* as a virtue.

Typicality From Probability

So far, our task has been to define something probability-like, given only a deterministic dynamics and something logically much weaker than a probability measure as input. We have seen that this can be done, in the appropriate circumstances, by appeal to typical approximate long-term frequencies. But suppose one actually starts with something stronger to begin with. What if we start already with a probability measure in hand? Can we then use that probability measure to define a measure of typicality?

This may seem to be of merely academic interest, but there are a couple of reasons to be interested in the answer. One is that such a definition allows for a sort of consistency check on the theory. The probability measure can be used directly to address a problem, such as making a practical decision, or it can be employed indirectly by first using it to define a measure of typicality and then using that to confront the problem. We cannot expect these two approaches to yield identical results in all cases — the step from a probability measure to a measure of typicality loses information, after all — but the results ought to be very, very similar over a wide range of cases.

The other reason for generating a measure of typicality from a probability measure is that it creates equivalence classes of probability measures: all the probability measures that generate the same measure of typicality can be grouped together. This makes formally precise the sense in which two probability measures are the same

FAPP. Insofar as one can use a given measure of typicality to generate quasi-probabilities, the quasi-probabilities can be thought of as arising from any of the probabilities in the equivalence class. This undercuts the suspicion that the whole procedure is just — in the words of Detlef Dürr — Garbage In, Garbage Out. That is, the suspicion is that one is only getting a particular quasi-probability out at the end because one put a corresponding probability measure in as input. Even more concretely, in the case of quantum theory, we want an explanation of why quantum systems display the empirical frequencies predicted by Born's Rule, namely ψ-squared frequencies. (Note: the ψ in the previous sentence is the wavefunction of a *subsystem* of the universe of interest. There can be many, many such subsystems with the same wavefunction ψ.) If one appeals to typicality as an explanation of this, and then uses the *universal* Ψ-squared probability measure to generate the measure of typicality, then it might appear as if you are only getting out the desired complex-squared result by using a complex-squared input probability measure to generate it. But if the only role that the initial universal probability measure plays is as generating a measure of typicality, then this criticism of being question-begging falls by the wayside. To be even more concrete, if using Ψ-squared to generate a measure of typicality yields Born statistics as typical, then using Ψ-fourth to generate a measure of typicalty *will also yield Born statistics*, because Ψ-squared and Ψ-fourth generate the very same measure of typicality FAPP. So one is not getting ψ-squared out as typical due to using Ψ-squared input, one is rather getting ψ-squared out *as a consequence of the nature of the fundamental dynamics*. The quasi-probability really is a consequence of the dynamics, not of some precise probability measure over universal initial conditions.[1]

---

[1] One might wonder: if using Ψ-squared to define a measure of typicality gives the same result FAPP as using Ψ-fourth, why is Ψ-squared always used and Ψ-fourth never used? The answer is practical: the Ψ-squared measure is conserved by the dynamics of the theory (it is *equivariant*) while Ψ-fourth isn't. So calculations using Ψ-squared are much easier to do than the equivalent ones using Ψ-fourth.

Using a probability measure to define a measure of typicality is trivial: all one needs to do is to choose a threshold near 1, such as .99999999 or .999999999999. The threshold should be close enough to 1 to render Cournot's Principle valid: events of interest with that level of certainty always actually occur. They are, as the old-time nomenclature has it, morally certain to occur even if not physically guaranteed. If I flip a coin and let it fall to a flat stainless-steel floor, then it is physically possible that it ends up standing on edge rather than falling red-side-up or blue-side up. But — as a practical matter — that event never actually occurs. No matter how many times you flip, the outcome is either red-side-up or blue-side-up. Making provisions in case the coin lands on edge is wasted effort, just as making contingency plans in case the Second Law of Thermodynamics fails would be. Yes, such a phenomenon is physically possible, but as a matter of fact, it just isn't going to happen.

Having chosen a threshold near 1, the rest is child's play. Every measurable set with a measure above the threshold counts as a "big" set. That defines a measure of typicality. Done.

Now, clearly the precise extension of "big" is a function of the threshold: different thresholds yield different extensions. Call a set that counts as "big" under the .99999999 threshold but not under the .999999999999 threshold as a "borderline big" set. Just as the theoretical existence of borderline footware that is no more shoes than boots need not cause any trouble if there is no such actual footware, so too the existence of borderline big sets causes no practical problem if none of the predictions or explanations we want to make involve borderline big sets. In exactly the same way, using Ψ-squared to generate a measure of typicality (given a fixed threshold) yields a slightly different outcome than using Ψ-fourth with the same threshold, and there will be borderline big sets that are big according to one measure but fail to be big according to the other. But so long as none of the questions we are actually interested in involve these borderline cases, the resulting vagueness or ambiguity or arbitrariness (in using one measure rather than the other) is benign. It makes no practical difference at all.

Here is an analogy that may help explain the situation. Suppose I am about to flip a coin 1,000,000,000,000 times, and there is a set of possible precise flipping mechanisms that might be used. The Cartesian product of that set with itself 1,000,000,000,000 times forms the configuration space for the experiment: the actual sequence of flips will be represented by a point in this space. Now, for $\epsilon = .01$, take every point that yields an empirical frequency of $.5 \pm \epsilon$ for red-side-up and paint that point green. The intuition is that *overwhelmingly most* of the points in the configuration space will be green. Overwhelmingly most by what measure? Well, by any measure of typicality which you might hit on by making some reasonable choice for the measure of typicality; by any such measure, the set of green points in configuration space will be a big set. As an analogy, consider a wall that is painted almost entirely green with microscopic regions of red dotted here and then. Such a wall will appear to be overwhelmingly green. To who? Well, to anyone who looks at it, even if different people have different eyesight, or even if some of the people are looking through distorting lenses. In order for it not to appear overwhelmingly green, one would have to look at it through a powerful microscopic trained on one or another precise region of the wall. And the choice to do that cannot be motivated by any desire to explain anything.

On the one hand, statistical explanations are in some sense probabilistic explanations. As Maxwell pointed out with his demon example, the laws of thermodynamics are really based in probabilities rather than in strict physical necessity. But, ironically, the Second Law of Thermodynamics is also characterized as an obstacle that simply cannot be circumvented, even if our views on what the fundamental laws are should change. Its validity somehow transcends even the precise fundamental law, because it would continue to hold even were the fundamental law to be somewhat different. The way to reconcile these seemingly contradictory claims is via typicality. Typical behavior is not guaranteed by the fundamental dynamics, but also many changes to the fundamental dynamics of, e.g., the molecules in a box of gas will not change their typical behavior.

The challenge of deriving probabilities — or quasi-probabilities, probabilities with tolerances — from an underlying deterministic dynamics can be met. Typicality is the conceptual tool by which the trick is done.

References

Lewis, D. (1983) "New Work for a Theory of Universals", *Australasian Journal of Philosophy*, **61**(4), 343–377

Lewis, D. (1986) *Philosophical Papers, Volume 1*, Princeton University Press, Princeton.

Part III

The Past Hypothesis, the Arrows of Time, and Cosmology

Chapter 8

Can the Past Hypothesis Explain the Psychological Arrow of Time?

Meir Hemmo[†] and Orly Shenker[‡]

[†]*Philosophy Department, University of Haifa*
meir@research.haifa.ac.il
[‡]*Edelstein Center for Philosophy of Science,*
The Hebrew University of Jerusalem
orly.shenker@mail.huji.ac.il

Can the second law of thermodynamics explain our mental experience of the direction of time? By this experience, we mean our psychological feel of distinction between the past and the future; we call this experience the 'psychological arrow of time'. An influential approach (for example, by Albert and Loewer) argues that the second law gives rise to the psychological temporal arrow. On this approach, accounting for the temporal directionality of the second law in terms of fundamental physics (which is time-symmetric) requires introducing the past hypothesis of low entropy. The idea is that the directionality of the universal increase of entropy explains how the psychological arrow comes about. We argue that there are two necessary conditions on the workings of the brain that any account of the psychological arrow of time must satisfy, and we show that the past hypothesis of a universal low entropy state does not satisfy these conditions. We propose a new reductive (physical) account of the psychological arrow compatible with time-symmetric physics, according to which our two necessary conditions are sufficient. Our proposal has some radical implications, for example, that the psychological arrow is fundamental, whereas the temporal direction of entropy increase in the past hypothesis is derived from it, rather than the other way around.

8.1. Introduction

The question we address in this paper is whether and how contemporary physics may explain the psychological arrow of time. This question has various aspects, such as: What does one mean by the psychological arrow of time? Another aspect is as follows: Contemporary physics is symmetric under the reversal of the direction of time,[1] so the explanation of the direction of time should involve the breaking of this symmetry in some way. In this paper, we focus on a certain approach to the second aspect, namely the attempts to explain the psychological arrow on the basis of the past hypothesis of low entropy in statistical mechanics (see Albert, 2000, 2014; Loewer, 2012; we expand on this hypothesis below). As is well known, the past hypothesis is introduced in statistical mechanics over and above the time-symmetry of the underlying fundamental laws of motion to account for the asymmetric temporal direction of the universe's entropy increase. Albert and Loewer (*ibid*), for example, attempt to explain the psychological arrow of time on the basis of the past hypothesis of low entropy and an additional hypothesis (we call the 'ready-state hypothesis'), according to which the low entropy initial state of the universe is *also* a universal 'ready-to-measure' state. In this paper, we examine this approach, criticize it and propose an alternative.

The paper is structured as follows: In Section 2, we set up the phenomenon that we seek to explain by physics, namely the psychological arrow of time. In Section 3, we start by describing some necessary conditions that a physical explanation of the psychological arrow of time should satisfy. In Section 4, we describe what we take to be the best case for Albert's (2000, 2014) and Loewer's (2012) approach, according to which the psychological arrow is to be

---

[1]The exact meaning and nature of time reversal invariance is currently under debate; see Allori (2015) and references therein. Asymmetries like the charge-parity and time asymmetry in the electroweak quantum theory, as well as (for example) the electromagnetic radiation asymmetry and the thermodynamic asymmetry in classical and quantum statistical mechanics, are effective: they are compatible with the time-symmetry of the fundamental laws but they require special conditions.

explained in fundamental physics by the past hypothesis of universal low entropy. In Section 5, we argue that this approach fails. In Section 6, we propose an alternative approach to explaining the psychological arrow of time on the basis of *local* and non-temporal symmetry breaking in the brain (yet to be discovered), and examine its implications.

8.2. The psychological arrow of time

Let us start by setting up the stage for the explanandum we seek in this paper: It is the *fact* that we have a *psychological arrow of time*. We distinguish between the direction we call past and the direction we call future — for example, a memory that we have at the present feels different than expectation of the same content! We don't address the general question of how to explain mental states in terms of physics. We are only after necessary conditions on the physical explanation of this arrow of experience. And we don't address here other aspects of the experience of time such as duration, flow, passage, the notion of the present and becoming, etc.[2] Einstein, for example, seems to have worried about explaining the present (or flow).[3] Here is Reichenbach on this issue:

> "Our observations of physical things, our feelings and emotions, and our thinking processes extend through time and cannot escape the steady current that flows unhaltingly from the past by way of the present to the future" (Reichenbach 1956, p. 1).

Our question is this: Can contemporary time-symmetric physics explain the phenomenon of the psychological direction of time? What kind of physical facts, as described by contemporary theories of

---

[2]There is an enormous body of literature on these notions and distinctions, but we only focus here on the physical explanation of the psychological arrow of time. Recent literature on the passage (or flow) of time includes Ismael (2010, 2017), Maudlin (2007, Ch. 4), Dieks (2016); see Le Poidevin (2015), Markosian (2016), Phillips (2017) for overviews and references; see Block (2014) on temporal experience as studied in contemporary psychology and cognitive science.
[3]See Carnap (1963, pp. 37–38).

physics, could bring about (or be correlated with) the psychological arrow of time?

Note that when we say physical 'explanation' of the psychological arrow of time, all we mean is that the arrow is *correlated* with a *directed* sequence of physical states. We do not go into the more general question (which is part of the mind-body problem) of what might be counted as an explanation of our psychology by physics. All we need and all we are seeking is some (robust) *correlation* that satisfies supervenience (logical or nomological) of the (kind of) sequence of psychological states, in which we have the feeling of a direction of time and a corresponding sequence of physical states. We do not go into the question of the origin and what kind of correlation it is — e.g., whether it is accidental, lawlike, or metaphysically necessary (in whatever sense) — since our arguments in this paper do not depend on any specific answer to this question; they apply to any view which requires such correlations.[4] In what follows, our phrasing might seem to presuppose some sort of a physicalist approach to psychology, or sometimes even phrases expressing a psycho-physical identity, but nothing hinges on this. It is just for the sake of brevity.

Finally, there are two general approaches to time (and the direction of time) in the literature: A non-reductive view according to which there is a primitive fundamental arrow of time (see, e.g., Earman, 1974; Maudlin, 2007), which is a fact above and beyond the facts described by the time-symmetric theories of physics and entropy increases relative to this arrow. It is a good question, what explains this alleged alignment on this approach, but we set this entire approach aside in this paper. By contrast, according to the second approach (e.g., Boltzmann's), which we support[5] and which is the target of this paper, the direction of time is considered to be reduced to some non-temporal asymmetry (for current reductive approaches, see for example, Albert (2000, 2014), Loewer (2012),

---

[4]We do not go into the question of what such correlation may be taken to imply with respect to hypotheses about realization, grounding, or the identity of psychological states and physical states. The arguments to follow are independent of the open issues concerning these ideas.

[5]We do not argue for a reductive approach here, but see the last section.

Smith (2014), Mlodinow and Brun (2014)). Indeed, on the alternative approach we propose in Section 7, the psychological arrow of time is meant to be reduced without remainder to some non-temporal asymmetry in the brain.

8.3. Some constraints on the physical explanation of the psychological arrow of time

In his brief discussion of the psychological arrow of time, Boltzmann (1896–1898; see a discussion in Sklar 1981) proposed a brilliant analogy between the experience of the arrow of time and the sensation of the gravitational arrow:

> "For the universe, the two directions of time are indistinguishable, just as in space there is no up and down. However, just as at a particular place on the earth's surface we call "down" the direction toward the center of the earth, so will a living being in a particular time interval of such a single world distinguish the direction of time toward the less probable state from the opposite direction (the former toward the past, the latter toward the future)." (Boltzmann 1896–98, pp. 446–7)

However, while the local gravitational field is measured by the vestibular system, which then sends electrochemical signals to the brain, leading to the feeling we call "up-down"[6], Boltzmann does not worry about the way the local "entropy field" can be measured or sensed by us to bring about our feeling of temporal directionality, which is the psychological arrow of time. In this sense, Boltzmann's approach to the psychological arrow here is only partial.

Smith (2014) and, perhaps, also Mlodinow and Brun (2014) attempted to complete one of the missing links in Boltzmann's argument by pointing out that memory formation in the human brain involves processes of diffusion which are locally entropy-increasing.[7]

---

[6]We do not address the so-called "hard problem" (Chalmers, 1996) of how mental states are brought about by the physics of the brain.

[7]Hawking's (1988) attempt to explain the psychological arrow of time by the thermodynamics of local memory systems is flawed, since it relies on mistaken ideas about the relation between entropy and computation (see Hemmo and Shenker, 2012, 2013, 2016).

Smith conjectures that these processes may be responsible for the psychological arrow of time. But other relevant brain processes may be entropy-decreasing as the brain (which is an open system) becomes more complex with time, and they too could, in principle, account for the psychological arrow.[8] It is thus a question of fact, which of these kinds of processes actually gives rise to the experience of a temporal arrow. However, according to this proposal, the relevant entropy gradient is in the brain, regardless of whether or not this gradient is correlated with the entropy in the environment — and *ipso facto* in the rest of the universe — by some measurement or otherwise, so that *prima facie* it is possible that if we had the right physical processes in the brain, we could have a psychological arrow of time regardless of what is happening in the environment.[9]

In this paper, we take it that all the facts concerning temporal experience (and perhaps all other facts of experience) are, as a matter of principle, proper subject matters for physical investigation, even if it is unclear at the moment how to go about providing some of these physical accounts (or, for that matter, whether contemporary physics is true at all[10]). In other words, we assume as a working hypothesis that the mental supervenes on the physical (either logically or nomologically, although we do not argue here for this view). By 'physical', we mean the principles of fundament physics, which include not only the fundamental laws of physics, but also boundary conditions, and everything that follows from them.

Here, we are only looking for necessary conditions for physical explanations of the psychological arrow. The idea is this: Some local degrees of freedom in the brain at time t (equally: at a time interval), perhaps, together with some other parts of the body — for simplicity call all of them "the brain" — should explain the psychological

---

[8]We thank Ronen Golan for discussions of relevant brain processes.
[9]Paul (2015) addresses the question of whether the psychological arrow of time can be illusory in some sense. This is relevant for our proposal; see the last section.
[10]Here, Hempel's dilemma concerning physicalism may arise; see Ney (2008) on this topic.

arrow. These local degrees of freedom must have the following two properties:

MENTAL: Some degrees of freedom in the brain, their physical state at t, or a sequence of such states, should give rise to the mental experience of temporal arrow at t or along the sequence of states.

ASYMMETRY: There must be some physical symmetry-breaking in the degrees of freedom that instantiate MENTAL.

The ASYMMETRY requirement is necessary, since there cannot be any symmetry-breaking in the mental experience unless there is some symmetry-breaking in the physics, if the mental is to be explained by the physical.[11] This requirement involves the ontology. From an epistemological perspective, this could mean that, ideally, by looking at the state of the brain at t and identifying the relevant asymmetry, one could tell on which "side" of t the psychological past is.

In the MENTAL condition, we are looking for something in the physics of the brain that will give rise to the psychological experience of time. Presumably, not every physical system has experiences, but only systems with certain kinds of physical features. Human beings are such systems and there may be other such systems, provided they have the right sort of physical features. Here is a case in which the MENTAL condition is not satisfied. Albert (2014, p. 66) considers a pendulum clock and takes it to be a "device which has no other business than distinguishing between what it has just done and what it is to do next — the paradigmatic distinguisher, the distinguisher par excellence, between what it has just done and what it is to do next." Presumably, pendulum clocks do not have any experience whatsoever, let alone the experience of temporal direction, and therefore they do not satisfy the MENTAL condition.[12] Whether or not we have a mental experience of anything, and in particular of temporal

---

[11] Arguing for this point in all its generality is beyond the scope of this paper; see for example, Van Fraassen (1989).

[12] It seems to us that a physical realization of a Turing machine is no different in this respect. But this question goes beyond the scope of this paper, so we set it aside.

direction, upon observing a pendulum clock depends on the physical features of our brains, not the features of the pendulum.

8.4. The past hypothesis and the psychological arrow

Some influential approaches to the psychological arrow of time attempt to show that the direction of the psychological arrow of time is aligned with (or even fixed by) the temporal direction of universal entropy increase described by the statistical mechanical counterpart of the second law of thermodynamics. The direction of entropy increase is, in turn, determined by the so-called past hypothesis (of low entropy), which is introduced into statistical mechanics for the following reason.

Contemporary fundamental physics is time-symmetric, and therefore a proof that entropy is likely to increase towards the future entails that entropy is equally likely to increase towards the past. This entails that our memories of past events (and other records, described and generalized by the second law of thermodynamics), as well as our belief that the second law holds at all times (so that entropy increases to the future but decreases to the past), are highly likely to be false (this is called sometimes the 'local minimum entropy problem' or the 'retrodiction problem'). To solve this problem, Feynman (1965, p. 116) proposed to "add to the physical laws the hypothesis that in the past the universe was more ordered, in the technical sense, than it is today."[13] There are recent attempts at showing that

---

[13]For various versions of this solution, see Feynman (1965, p. 116), Sklar (1973), Albert (2000, 2014), Loewer (2001, 2012), Callender (2010), Wallace (2017), Hemmo and Shenker (2012). Price (1996) criticizes the postulation of temporal partiality. Earman (2006, p. 410) discusses Boltzmann's remarks about this problem and criticizes Feynman's (1965) dismissal of some of Boltzmann's ideas; see also Goldstein, Tumulka and Zanghi (2016) on this issue. Whether the past hypothesis can be derived from first principles is an open question, and there are ongoing attempts to ground it in considerations of cosmology; see for example, Caroll and Chen (2004); Barbour, Koslowski and Merkati (2014); Goldstein, Tumulka and Zanghi (2016); see also criticism by Winsberg (2004); Earman (2006); Hemmo and Shenker (2012, Ch. 10). Whether or not the past hypothesis needs explanation at all is discussed in Price (2010), Callender (2004).

the hypothesis is unnecessary, since fundamental physics entails the empirical asymmetry (e.g., along the lines of Carroll and Chen, 2004; Carroll, 2010).

Albert (2000) formulates the hypothesis of low entropy, which he calls the *past hypothesis*, as follows[14]:

> "[T]he world first came into being in whatever particular low-entropy highly condensed big-bang sort of macrocondition it is that the normal inferential procedures of cosmology will eventually present to us." (Albert, 2000, p. 96)

This is in fact a conjunction of two independent hypotheses: the first one is what we call the 'universal low entropy hypothesis', according to which the macroscopic state of the universe next to the big bang is of low entropy; and the second one is what we call the 'ready-state hypothesis', according to which this low entropy state is of some particular sort of state of the universe, the details of which will be presented to us by cosmology.

According to Albert and Loewer (*ibid*) these two hypotheses, if true, provide a sensible physical explanation of the psychological arrow of time. We will later argue that neither of these hypotheses is necessary or sufficient for explaining the psychological arrow of time, and we shall propose a new approach that seems to us more plausible as an account of how the psychological arrow of time comes about, and with it all the other temporal arrows in physics. But let us start by outlining how the explanation of the psychological temporal arrow from the past hypothesis should go about, and what is the division of labor between these two hypotheses in Albert and Loewer's approach. The explanation should go along the following lines:

(I) The psychological arrow is to be explained by a *local entropy gradient in the brain*, thus satisfying the MENTAL and ASYMMETRY requirements, as for example in Smith (2014) or Mlodinow and Brun (2014).

---

[14] Albert (2000, Ch. 4) adds some more details to the past hypothesis concerning the dynamics and the probabilistic algorithm of statistical mechanics; see also Hemmo and Shenker (2012, Ch. 10, 2014).

The next task is to show that the "mental arrow of time is aligned with the thermodynamic arrow" (Goldstein, Tumulka and Zanghi (2016, p. 5); that is, that our experience reflects a fact about the external world.[15] (Towards the end of this paper, we challenge this task.) Here is how this is supposed to be done:

(II) The local entropy gradient in the brain *reliably reflects* a *universal* entropy increase (in either the entire universe or its relevant part). By reducing the notion of a temporal arrow to the notion of an entropy arrow, it can be said that the experience is one *of* an arrow of time in the universe. Here, we would need a proof, based on the standard statistical-mechanical probabilistic algorithm and other considerations, that the universal increase of entropy is such that subsystems of the universe (such as the brain) are likely to increase their entropy. (We later consider the proposal of Albert, 2000, 2014; and Loewer, 2012 on this point).

(III) The universal entropy increase (which is reflected in the brain) is to be explained, as usual, from first principles (that is, without resorting to our *experience* of second law behavior) by a combination of (a) a proof that the entropy of the universe is highly likely to increase towards the future;[16] and (b) a proof of the past hypothesis concerning universal low entropy in the remote past that comes from fundamental physics (e.g., along the lines of Carroll, 2010; and Carrol and Chen, 2014).

In our reconstruction so far, the ready-state hypothesis played no role in determining the psychological arrow of time. The explanation

---

[15]Whether or not the past hypothesis of low entropy prevents a causal influence about the past is an open question; see Albert (2000, 2014), Elga (2000), Frisch (2010). We don't address this question here.

[16]For overviews of such attempted proofs in the classical case, see Uffink (2007), Frigg (2008), Werndl and Frigg (2015). In our view, there can be no general proofs of this statement, since Maxwell's Demon is compatible with the principles of statistical mechanics, unless one *postulates* that Demons are impossible and *conjectures* that the Hamiltonian of the universe precludes them (see Hemmo and Shenker, 2010, 2012, 2016, 2017). There are, however, some interesting results concerning entropy increase in special cases, such as Lanford's theorem, but these are subject to the minimum problem (see Uffink and Valente, 2010, 2015). We do not expand on this point here.

of the directionality rests entirely on the entropy gradient implied by the past hypothesis and the second law of thermodynamics. So, what is the role of the ready-state hypothesis in Albert and Loewer's approach? The starting point for this hypothesis is the retrodiction problem we mentioned above, which follows from the time-symmetry of the statistical mechanical probabilistic algorithm. The problem is, essentially, *epistemic*. Since the statistical mechanical algorithm implies that our memories and all other records that appear in our inferences about the past are very likely unreliable, what is the *justification* for our reliance on them?[17] The ready-state hypothesis is introduced by Albert (2000, 2014) to answer this question (together with the second law of thermodynamics and the past hypothesis of low entropy).

Consider the following example of an *object* that you would intuitively take to be a photograph of a scene from your fifth birthday. Usually, we treat this object as a reliable record and infer, from observing it now (say, at time t_0), that the birthday event took place at time t_{-1}. If we carry out predictions regarding the future state of this object, relying on the dynamical and probabilistic hypotheses of statistical mechanics (which are compatible with the second law of thermodynamics), then we infer that at the future time t_{+1}, the entropy of the object will be greater than it is now. For example, the photograph may be less clear and, depending on the details of its material and the circumstances of its storage, may even decay altogether, so that we would no longer consider that its decayed elements constitute a photograph. It follows from the minimum theorem that a retrodiction concerning the state of this object at time t_{-1} entails that at t_{-1} — at which your fifth birthday presumably took place — the object was not a photograph either. And the conclusion is that what *appears* (at present) to be a record of your fifth birthday is,

---

[17] Albert (2000) presents the problem as leading to a *skeptical catastrophe*, since it casts doubt on the *reliability* of the contents of our memories and all other records, on which we rely in everything we do, and in fact on the very empirical statements that led us to endorse mechanics — and hence statistical mechanics, in which the retrodiction problem arises — to begin with.

actually, *not* a record of anything: the object at the present is more likely to be a rare fluctuation out of chaos that came to existence a very short time ago (see also the *asymmetry problem* in Earman, 2006, p. 401).

According to Albert, when we read the measurement outcome imprinted on a measuring device we in fact make an inference pertaining to *three* states of the universe, the present one being the latest one of them. We assume.

(1) That in the past (*prior* to t_{-1} in the above example), the measuring device was in some state called a '*ready state*', which is (either explicitly or implicitly) *known* to us. In our example, the measuring device is the photographic paper, and we *know* that *prior* to t_{-1}, it was blank and not, for example, filled with ink marks similar to those it has now.

(2) That at some intermediate time (t_{-1} in our example), this device underwent a suitable interaction with the measured system (in our example, it is the state of affairs at your fifth birthday), in such a way that the state of the measuring device (in our example, the photographic paper) changed (at t_{-1}) to some "outcome state" (in which the paper is covered with the ink marks it now has). We must know (or assume) the general nature of this interaction, at least the general relations between its inputs and outputs, in order to make inferences concerning the significance of the "outcome state" and its relation to the event being recorded.

(3) And that this outcome state remained effectively unchanged until now, when we are observing the device (that is, looking at the photograph at t_0).

When we say that the photograph is a record of your fifth birthday, we are implicitly referring to this entire complex (3-stage) chain of events.

Obviously, there is a vicious regress here: How do we know that this 3-stage chain of events actually took place, for example, that the measuring device actually was in the ready state at t_{-1} (e.g., that the photographic paper at t_{-1} was blank)? We can know this only if we have some other (3-stage) record of this ready state in the form of some other measurement outcome which took place earlier in the past. But this obviously leads to a vicious regress, which can be halted only if we are in a position to assume that in the remote past, the universe was in a ready state, *without* having any record of it. According to Albert, the only way to stop this regress is to introduce a hypothesis to the effect that the universal low entropy state near the big bang is *also* a ready state of the entire universe, the nature of which will be discovered by cosmology.[18] Albert calls this state, the "mother (as it were) of all ready conditions" (Albert, 2000, p. 118; 2014, p. 38).[19] This assumption that the low entropy initial macrostate of the universe is also a universal ready state is the ready-state hypothesis.

What source of knowledge concerning the ready-state hypothesis is available to us? Surely, an oracle is out of the question. Albert's proposal is that the universal ready state is hard-wired into our cognitive apparatus, by natural selection and everyday experience, possibly amended and expanded by explicit scientific practice (Albert, 2014, p. 39).[20] (Perhaps, what is hard-wired into us is not only the ready state, but also the interaction and evolution in the second and third stages of the complex assumption).

---

[18]But, if (for example) the photograph is to be treated as a record of the past event of your fifth birthday, we need to know details about the universal ready state that go *far* beyond what is given to us by the low-entropy hypothesis, and far beyond what is given to us by current cosmology. But we set aside this issue here.

[19]According to mechanics, if we know the present microstate of the device, we can calculate its microscopic ready state.

[20]According to Albert (2014, p. 17, 39), evolutionary arguments support this idea. We have proposed elsewhere (Hemmo and Shenker, 2012, Ch. 10) to replace the ready-state hypothesis with a local *reliability* hypothesis about human beings (and other surviving animals), according to which our memories and what we take to be local records of the past (given the psychological temporal direction that is fixed, say by some entropy gradient in our brains; see below) just are — as a matter of contingent fact — reliable states of affairs with respect to the past. Evolutionary arguments seem to us to support this *reliability* hypothesis.

What is the connection between the two hypotheses, of low entropy and of the ready state? Assuming that the second law of thermodynamics holds towards the future of the ready state, the complex assumption of records — involving the above three stages — justifies inferring that the record is *reliable*; that is, that the event in question indeed took place; and that it is not, for instance, a random fluctuation out of chaos. In other words, in Albert's approach, our records and memories are *reliable* (with high probability) if in addition to the ready-state hypothesis we have, (i) the past hypothesis of low entropy, in the *same* side of the sequence of states at which the ready state is placed; and (ii) the second law of thermodynamics, holds. Therefore, the role of the ready-state hypothesis is to explain (given that these two conditions hold) how it is that we have a reliable epistemic access to the past by means of measurements and records.

There seems to be two additional roles played by the ready-state hypothesis, according to Albert and Loewer, in the context of the psychological arrow of time, as follows:

(IV) The psychological arrow of time includes not only the feeling of a direction, but also a feeling that the past is qualitatively different from the future in the sense that, for example, the past is (felt as) fixed and the future is (felt as) open. According to Albert (2014, p. 56), this fact is explained by the ready-state hypothesis. We make inferences concerning matters of fact in the past by reliance on records based on the ready-state hypothesis, and inferences of matters of fact in the future by the usual probabilistic algorithm of statistical mechanics. "And this distinction seems to me to make obvious and immediate sense of the everyday phenomenological feel of the difference between the past and the future." (Albert 2014, p. 56) This requires some notion of "present", but we do not address this point here. Note that if this approach is to work, there must be some physical labels in the brain indicating the way in which our inferences are made, either by measurement or by using the standard probabilistic algorithm of statistical mechanics.

(V) Another crucial role that the ready-state hypothesis may have is the following: Although the local entropy gradient in the brain (which gives rise to the psychological arrow of time) may be correlated with an entropy gradient in the universe as a whole, this correlation may not be analytic and may fail to obtain. Therefore, assuming that our psychology is such that we feel a temporal direction due to the local entropy gradient in the brain, it is still meaningful to ask at which side of the sequence of states of the universe the entropy of the universe is low. To answer this question, one needs to rely on records, and to do so reliably, one needs a ready-state hypothesis. Now, *if* the ready state is on the same side of the sequence of states as the low entropy past — that is, if the ready-state hypothesis and the low entropy hypothesis pertain to states on the same side of the sequence of states (note that this is not analytic), *then* the answer will be the one we now take to be empirically supported. This is to say that the low entropy of the universe was on the side that we feel (due to the local entropy gradient in the brain) as "past". Evolutionary arguments may suggest that the alignment of the psychological arrow with the universal thermodynamic arrow of entropy has a survival advantage: creatures for which the psychological arrow is reversed relative to the external thermodynamic arrow (i.e. creatures which take the ready-state hypothesis to hold on the side of the sequence of states opposite to that of the past hypothesis) would not survive, since the predictions of such creatures would be wrong and their memories unreliable.

To sum up: according to our analysis of Albert and Loewer's approach, the psychological arrow of time is to be explained by the hypothesis of universal low entropy; namely, the assumption that near the big bang, the universe was in a low entropy macrostate. The ready-state hypothesis is formulated against the background of this arrow of entropy increase, it is introduced to explain our epistemic access to the past, given the arrow of entropy. It seems that accepting the low entropy hypothesis and the ready-state hypothesis provides a unified picture in which, on the one hand, the psychological arrow of time is determined by the universal arrow of entropy, while on the other hand, the ready-state hypothesis makes it likely that our memories and other mental states reliably track this entropic arrow. However, unfortunately, this picture does *not* work, for reasons we now turn to describing.

8.5. Why the past hypothesis fails to explain the psychological arrow of time

In the literature, there are two lines of thinking that attempt to underpin the psychological arrow of time in the first principles of fundamental physics, which also ground the second law of thermodynamics. We now describe these two approaches, and argue that the past hypothesis (in particular, the hypothesis of low entropy) is insufficient for grounding the psychological arrow of time. One line of thinking attempts to connect the past hypothesis of low entropy with the local gradient of entropy in the brain by typicality arguments. The other line of thinking attempts to connect the past hypothesis of low entropy with the local gradient of entropy in the brain by dynamical or causal considerations. We shall start with the typicality line of thinking. Then, we will argue that the past hypothesis of low entropy is not necessary for explaining the psychological arrow.

(I) **The typicality line of thinking** is that the physical explanation of the psychological arrow must involve two stages. Stage 1 connects the hypothesis of low entropy in the remote past of the *universe* — that is, the hypothesis of a universal entropy gradient — with a local entropy gradient in the *present*, in our *brain*. Stage 2 connects the local entropy gradient in our brain at present with the psychological arrow at present. The two stages together should establish that the "mental arrow of time is aligned with the thermodynamic arrow" (Goldstein, Tumulka and Zanghi, 2016, p. 5).

Let us begin with Stage 1. An example of an argument for Stage 1 is by Loewer (2012) and Albert (2014), which is roughly as follows: The initial conditions of the universe, which are compatible with the past hypothesis, form two sets. One set gives rise to trajectories in which the entropy of the universe increases, effectively and to a good approximation in accordance with the second law, while the other gives rise to all sorts of other trajectories. The typicality assumption[21] is that according to "standard Boltzmannian arguments" (Albert, 2014, p. 65), the former set is much larger (by some appropriate measure). Its trajectories are, in this sense, "typical". This set of typical entropy-increasing trajectories, in turn, has two subsets. In the first subset, *most* [quasi-isolated] subsystems are entropy-increasing, and in the second subset, this is not the case (for instance, in the second subset, entropy may increase substantially in only a few subsystems and yet, due to the additivity of entropy, the global entropy of the universe would increase). The first subset — in which *most* [quasi-isolated] subsystems are entropy-increasing (so the argument goes) — is much larger (by some appropriate measure). In this sense, its members are "typical". Two levels of typicality are involved here: globally entropy-increasing universes are typical and, within them, universes in which *most* [quasi-isolated] subsystems are entropy increasing are typical. *If* our universe belongs to the typical set and typical subset, then — so the argument goes — this entails the empirically verified prediction that (and thereby also explains why) quasi-isolated subsystems (such as our brains) are entropy-increasing; and this would become part of the physical explanation of the psychological arrow of time.

The question is, however, why we should accept the latter "*if*" clause. This question arises with respect to all of the typicality arguments in the foundations of statistical mechanics, and this is not our topic here, so we only briefly describe the line of thinking that might be behind accepting the abovementioned "if" clause (according to which our universe belongs to the typical set and its typical subset). Two reasons are presented in the literature for accepting the "if" clause.

The first is as follows: (1) The spatiotemporal part of the universe that we observe obeys the second law of thermodynamics; this

---

[21]For a critical discussion of the typicality approach, see for example, Hemmo and Shenker (2012, 2014).

statement has ample empirical support. (2) There is no reason to think that the spatiotemporal part of the universe that we observe is special, that is, *atypical* (see the abovementioned debate on this matter between Feynman, 1965 and Earman, 2006); hence, we can assume that it is typical, in the two abovementioned levels. (3) Therefore, the rest of the universe is highly likely to be the same as the spatiotemporal part of the universe that we observe. (4) It follows that our universe is thermodynamic, and most of its subsystems obey the second law of thermodynamics. (5) It is tempting to continue by saying that statement (4) explains why the spatiotemporal part of the universe that we observe is thermodynamic, but this would be circular, given the above argument. We started off from the actual case that we observe and inferred that typical cases are like it; and therefore, since the actual dictates the typical, the typical cannot explain the actual. Without this idea, that the actual dictates the typical (sometimes called a Copernican argument), the fact that we *observe* entropy increase in our vicinity does not imply that entropy increase is *typical* of most sub-systems of our universe. Without a Copernican argument, it may well be that our entropy-increasing universe is very special. In particular, in statistical mechanics, the measure used to determine the size of the abovementioned sets is selected so as to provide results that will satisfy the Copernican argument, and therefore this measure is not explanatory.

The second line of reasoning in the literature for accepting the "if" clause (which states that our universe belongs to the typical set and its typical subset) is that the measure used to determine "typicality" in the above two senses is dictated by the dynamics of the universe (see, e.g. Dürr, 2001; Goldstein, 2012). We have argued elsewhere in the context of both classical and quantum statistical mechanics (see Hemmo and Shenker, 2012, 2014; Shenker, 2017a,b, 2018) that none of the dynamical arguments that have been presented in the literature are compelling.

Perhaps, Loewer (2012) is acknowledging the weakness of typicality arguments when he writes: "But this is as should be. The job is to get the second law ... *in so far as* the second law is correct". Our point here is that to complete Stage 1 of the first line of thinking, one must accept the typicality argument in at least one of its versions sketched above. If one wishes to avoid appealing to typicality, one option is to make use of the dynamics of the actual universe (see Winsberg, 2004; Frisch 2010). Here, there is no appeal to counterfactual universes, but rather to the dynamics of our actual universe. Such a line of thinking is, indeed, a way to derive the entropy gradient in the brain from first principles. However, in this way of thinking, the stipulated dynamics does *all* the work; therefore, the past hypothesis by itself is neither necessary nor sufficient for generating a local entropy gradient in the brain.

Once Stage 1 is completed (contra the above arguments), the task is to proceed to Stage 2 and connect the local entropy gradient in our brain with the psychological arrow. This task is to be completed by arguments such as those of Mlodinow and Brun (2014) and Smith (2014), who attempt to satisfy the ASYMMETRY and MENTAL conditions. One point that arises with respect to Stage 2 is the following:

The entropy gradient (at a given point of time t) is *not* one of the macrovariables described by statistical mechanics, since all the thermodynamic macrovariables are time-symmetric (e.g. average kinetic energy, which is the counterpart of temperature of an ideal gas in equilibrium; coarse-grained distribution of positions, which is the counterpart of volume; and the degree of uniformity of the distribution of temperature, or pressure). For example, the set of microstates that share the same average kinetic energy (at some time t) contains microstates that are present both on trajectories that are entropy-increasing (at t) and those that are entropy-decreasing (at t). In this sense, the set of microstates that share the same entropy gradient is a subset of the set of microstates that share, for example, the same temperature (at t). In the approaches we are considering here, the psychological arrow of time is brought about by this 'finer-grained' property, and *not* by the thermodynamic properties that appear in statistical mechanics and thermodynamics. This point is also relevant to the second approach we now consider.

(II) **The second proposal** for explaining the psychological arrow of time appeals to some sort of causal considerations. Albert (2014,

p. 16) suggests that an appropriate correlation between our brain state at the present and the low-entropy past "will have been hard-wired into us as far back as when we were fish." This proposal is not committed to the idea that it is entropy gradient in the brain that satisfies the ASYMMETRY and MENTAL conditions described above. It could be some other asymmetry, provided that the latter is causally correlated with the entropy gradient induced by the past hypothesis (see also the next section). Another crucial point is that this proposal needs to be completed with important details similar to the ones missing in Boltzmann's proposal aforementioned. The "hard-wiring" hypothesis assumes that at some point — namely, in the remote past when the hard-wiring developed — there was (in us or somewhere in our environment) an *entropy-gradient sensor* that measured the entropy gradient of its environment (which, by assumption, reflected the entropy gradient of the universe at that time), such that the outcome of this ancient measurement is (possibly following a chain of intermediate stages) registered in precisely those degrees of freedom in the brain that also bring about the psychological temporal experience (as required in the MENTAL condition).

Two crucial elements are missing in this approach. (a) Assuming an entropy gradient in the environment (either at present or in the remote past when the hard-wiring took place), we need to find a way to measure this gradient so that the measurement outcome will be reflected in some aspect of the physical state of the brain (this is the way in which the past hypothesis should be shown to ground our ASYMMETYRY condition). (b) The aspect of the physical state of the brain must give rise to the psychological experience that we call temporal direction (this is our MENTAL condition). As we already said, ways of providing this last element have been sketched, for example, by Mlodinow and Brun (2014) and Smith (2014). Let us focus on (a) for now.

(1) **The external field:** What is the external field that causally brings about the temporal arrow, in analogy to the local external gravitational gradient? There are two options.

(i) In a non-reductive view, time is a primitive aspect of ontology that we perceive when we experience the temporal arrow. We have not address this view of time in this paper, but it is worth mentioning that it cannot be ruled out, especially because of the hard questions concerning the physical explanation of the present and the flow or passage of time. Indeed, this seems to be Einstein's worry in the quote cited in the introduction to this paper. But we are setting aside the non-reductive view now.

(ii) In a reductive view, the field in question is not time itself, but something else, associated with some spatial or other degrees of freedom. Boltzmann thought that the relevant field in this context should be the local entropy gradient. Importantly, *entropy is not inherently temporally oriented*, and the gradient of entropy (global or local) is not in any intrinsic way correlated with a direction of time (i.e., one side of the sequence of states as opposed to the other) more than, say, the distribution of particles in space is. To see this, consider the efforts invested in *proving* the second law. If entropy were inherently temporal, say its gradient would be temporally directed, this proof would be analytic and trivial. Entropy is a function of spatial[22] degrees of freedom that may have a variety of temporal behaviors; fundamental mechanics is compatible with a world in which entropy (both universal and local) zigzags. Thus, there may be other quantities, other external fields, other symmetry breakings, that could give rise to the temporal arrow (in a reductive view). In this paper, we focus on the second law and the past hypothesis, and therefore we do not examine other options for such external fields.

(2) **The sensor**: How is the external field sensed? We know that the vestibular system is the sensor for the external gravitational field; what could measure the field that leads to the psychological arrow of time? Suppose that the relevant external field is an entropy gradient (and let us assume that this gradient satisfies the second law of thermodynamics). Since entropy is a function of the measure over the state space — that is, of a continuous set of microstates, where only one of which is actual while the rest are counterfactual at every moment — entropy is a number without units that is a function of the macrovariables (state functions) of sets of microstates. Therefore,

---

[22] And in quantum mechanics possibly of other degrees of freedom, e.g., spin.

an entropy sensor should (a) be sensitive to all the macrovariables
of which entropy is a function; and (b) carry out the calculation
of this function.[23] The fact that we don't have "entropy meters",
despite the importance of this quantity, raises the *suspicion* that our
physiology does not contain such a sensor, in which case an entropy
gradient (either at present or in the remote past) cannot account
for the psychological arrow of time. This is only a suspicion, however. Perhaps, there is an entropy sensor — perhaps, nature came
up with an entropy gradient sensor that we, at the moment, cannot
understand or imitate.[24]

Nevertheless, this suspicion brings to mind a third option. Perhaps
there is no sensor, and it is not needed at all, because the asymmetry
required for the psychological arrow of time is generated within the
brain and is not correlated with an external field outside it. This
option has some radical consequences, which we consider in the last
section.

So far, we have argued that the past hypothesis is insufficient for
explaining the psychological arrow of time. We now turn to explain
why it is unnecessary. It would be very elegant if the past hypothesis,
which involves an entropic asymmetry in the history of the universe,
could also explain the psychological arrow of time. However, elegance
considerations are not compelling, and the main claim in this section
is that there can be a *physical* explanation for the psychological arrow
of time that is not based on the past hypothesis (in either of the
senses in (I) or (II) above).

Obviously, the second law of thermodynamics, which states that
entropy is unlikely to decrease with time, generates a tight connection
between entropy and the direction of time. But this connection is
not of a conceptual or logical kind, and the very fact that the second
law requires a proof in statistical mechanics shows that it is not

---

[23]There is an open debate in the literature on the status of the measure needed
to determine the size of sets of microstates. It is not analytic that the function
for calculating entropy should be the Lebesgue measure; see Hemmo and Shenker
(2012, Ch. 7) on this point.
[24]The special nature of entropy in this respect was brought to our attention in
discussions with Yemima Ben-Menahem.

analytic.[25] This is the case because entropy (and its gradient) are functions of the microstate of the system, and this function may change over time in different ways depending on the initial conditions and the dynamics, so that there is *no a priori* correlation between the direction of time and the behavior of entropy.

Once we realize this, it becomes clear that there are other functions of the system's microstate that evolve in a variety of ways; and these other state functions provide other possibilities for relevant local asymmetries in the brain that may satisfy the ASYMMETRY requirement and could — for all we know — also satisfy the MENTAL requirement. Any effective local symmetry-breaking compatible with the time-symmetry of fundamental physics could, as a matter of principle, give rise to the psychological arrow of time; the question of which of them also satisfies the MENTAL condition is to be settled by brain research. For all we know, the function of the microstate called an entropy gradient could be as good (for this purpose) as the function of the microstate of the distance of a certain molecule from our left ear. (It is known today, for example, that memories and expectations are stored in two distinct brain regions).[26] Of course effective asymmetries like the violation of charge-parity asymmetry and time asymmetry in the quantum electroweak interaction,[27] as well as (for example) the electromagnetic radiation asymmetry or even of the local gravitational field

---

[25]The unsettled question of Maxwell's Demon supports this point.

[26]Obviously, this correlation can be explained in many ways. It could be, for example, that an event occurred "in the past" or is expected to happen "in the future" makes it the case that it is stored in these regions. But it could also be the other way around; the fact that an event is stored in these regions makes it the case that it is felt to be "in the past" or "in the future". So any conclusion from these empirical findings should be made with caution.

[27]The charge-parity and time asymmetries, which have been observed in the electroweak interaction, are often ruled out as irrelevant for the psychological arrow of time due to the levels of energy involved. But it has been conjectured (see Atkinson 2006) that these asymmetries result from the low entropy state assumed by the past hypothesis. If this conjecture is true, our argument that the past hypothesis is neither insufficient nor necessary for the psychological time arrow will indirectly rule out also these asymmetries.

asymmetry (as in Boltzmann's remarks concerning spatial direc- tions) could — for all we know — give rise to the psychological experience of the arrow of time. Another alternative may be the time-asymmetric collapse of the wavefunction in quantum theories, as in the spontaneous localization theory of Ghirardi, Rimini and Weber (1986). However, von Neumann stated: "It is desirable to uti- lize the thermodynamic method of analysis, because it alone makes it possible for us to understand correctly the difference between [Schrödinger's unitary transformation] and [the measurement trans- formation], into which reversibility questions obviously enter." (Von Neumann, 1955, p. 358).

8.6. A new proposal for explaining the psychological arrow

There is a fundamental difference in the order of explanation between approaches based on a *local* entropy gradient (e.g., Smith, 2014) or some other physical asymmetry in the brain and approaches based on the past hypothesis (e.g., Albert, 2014; and Loewer, 2012) or some other *global* physical asymmetry. Both approaches should agree that it is the *local* entropy gradient (or some other asymmetry) in the brain that *directly* gives rise to the psychological arrow of time. In Albert and Loewer's approach, for example, the idea is that the low- entropy past hypothesis is the basic assumption and the psychological arrow is explained by it, via the claim that the global past hypothesis confers high probability on a local entropy gradient in the brain, or via the claim that the global entropy gradient is causally wired into our brains. By contrast, in approaches such as that of Smith (2014), the order of explanation is reversed. The local entropy gradient is the basic empirical datum, which we learn from observing the brain (and not from attempted proofs of a global second law) and *this* local fact breaks the symmetry. Once we have this local asymmetry, we can then say that the low entropy assumed by the past hypothesis is on the "side" of the sequence of states that we *feel* to be "past" due to the local entropy gradient. However, the trouble with Smith's pro- posal is that it is not explicitly based on fundamental physics, since the second law itself and its directionality is assumed here rather than explained.

Our proposal for explaining the arrow is different from both approaches. We propose that the psychological arrow of time is grounded in a local asymmetry in the brain, where this asymmetry is an effective feature of fundamental physics (and it may or may not have anything to do with the second law of thermodynamics). This proposal is different from Albert and Loewer's approach in that it explains the arrow locally and regardless of the past hypothesis. It reverses the order of explanation in that the temporal direction of the past hypothesis is explained by the local facts about the brain. Our proposal shares with Albert and Loewer's approach the reductive nature of the explanation of the arrow and that the explanation is given in terms of fundamental physics. In this last feature, our proposal differs from Smith's (2014) with whom we share the locality of the explanation.

So far, we argued that the past hypothesis (of low entropy of the universe in one side of the sequence of states) and the ready-state hypothesis (of a ready state of the universe in the remote past, where the side of the past is fixed by the past hypothesis) *cannot* explain the psychological arrow. What, then, can provide such an explanation? As we said earlier, we do not purport to provide a complete account, since contemporary science (including brain science, cognitive psychology, and cognitive science) is far from explaining how mental experience comes about (although the empirical findings in brain research seem to support our proposal). What we are after, instead, are some *necessary* conditions for a *physical* account of the psychological arrow of time. Given that the fundamental laws are time-*symmetric*, what in the first principles of fundamental physics can (as a matter of fact) explain the psychological arrow of time? Here we follow Boltzmann's insight, which we mentioned before, and then go back to the proposal by Smith (2014) and some of its possible consequences.

Boltzmann draws an analogy between the gravitational asymmetry and the temporal asymmetry (see Sklar, 1981). He begins by noticing that the up-down sensation is brought about causally by the local gradient of the gravitational field on the face of Earth. But — applying our ASYMMETRY and MENTAL conditions — the up-down *experience* is to be explained by an asymmetry in *the brain itself.* In taking the case of the gravitational arrow as analogous to the case of the temporal arrow, whether or not the asymmetry in the brain corresponds (causally or otherwise) to an asymmetry in an external "field," it is crucial to note the following points:

(I) Whether the psychological arrow is causally (or otherwise) correlated with an external field, or whether it is generated within the brain, in both cases the two conditions — ASYMMETRY and MENTAL — need be satisfied by the *brain's* degrees of freedom. The ASYMMETRY and MENTAL conditions are satisfied if, ideally, by looking at the state of the brain at t, one can ascertain at which "side" of t is the psychological past.[28]

(II) The identification of the degrees of freedom that satisfy the MENTAL requirement is an issue for brain research. The only insight into this matter that current physics can offer concerns the ASYMMETRY requirement. Here, lacking an account of the condition MENTAL, *any* effective local symmetry-breaking in the brain that is compatible with fundamental physics should be considered. A local entropy gradient *in the brain* (not in the universe now nor in the past), the violation of parity and time symmetries (in the electroweak interaction), the electromagnetic radiation arrow, or even the distance of a certain molecule from our left ear, are all examples of effective local symmetry breaking that could, for all we know, explain the psychological arrow of time.[29]

(III) The statement concerning the local entropy gradient (whether or not it reflects the global one) is a statement concerning a matter of fact. (If it were the conclusion of a proof that there is a local entropy gradient, the so-called minimum entropy problem

---

[28] Assuming, of course, that supervenience (logical or nomological) of the mental on the physical holds. Also, see Arntzenius (2000) for the question of instantaneous physical states and the Zeno paradox.

[29] If there is a fundamental collapse of the quantum state (as in the GRW theory), then physics is not time-reversal-invariant, in which case the symmetry-breaking that might explain the psychological arrow of time need not be merely effective.

would arise). The claim is that the matter-of-fact local entropy gradient brings about the psychological arrow of time. It is totally irrelevant for our point whether or not there is some proof that this local entropy gradient is likely. Such likelihood would be important only as part of a *justification* for a belief that this local entropy gradient obtains, in the absence of any empirical evidence to that effect (see Section 4).

However, if the brain's degrees of freedom do not reflect some external field (as we propose), if the psychological arrow is *generated* by degrees of freedom in the brain that satisfy the ASYMMETRY and MENTAL conditions, this has a number of interesting and quite radical implications which we now consider briefly.

(IV) One may object to Boltzmann's analogy between the case of gravity and the case of time. It is uncontroversial that the degrees of freedom in the brain that give rise to the up-down experience need not have anything to do with gravity. But there is a debate about; that is, whether the degrees of freedom in the brain must share some temporal properties with what is represented and whether the brain represents at all (see LePoidevin, 2015; Phillips, 2014; Dennett and Kinsbourne, 1992). We do not presuppose any relation of representation between brain states and what they represent (mental states or external-world states). In our reductive approach, the asymmetry in the brain that is the psychological arrow of time may be spatial or any other non-temporal asymmetry precisely because it does not represent anything. Here, it is also crucial to realize (as we said in the previous section, point (ii)) that entropy gradient (local or global) is not intrinsically future-directed, or more generally, it is not temporally oriented at all. The fact that there are mechanical trajectories along which entropy changes in almost anyway one likes — increases, decreases, zigzags — supports this point. If entropy were inherently temporally oriented, proofs of the second law from the underlying mechanical theory (classical or quantum) would be analytic and trivial.[30] We take this to mean that to explain the psychological arrow

---

[30] One might want to argue that although proofs of the second law of thermodynamics in statistical mechanics are not analytic, the second law is a fact in our

of time, entropy gradient (local or global) is conceptually on a par with any other spatial asymmetry, so it has *no a priori* status in this respect.

(V) If the psychological arrow is generated by some asymmetry in the brain, as we propose, this has implications concerning the origin of the *direction* in *all* other so-called 'temporal' arrows (in and outside of physics). In particular, if we consider the direction of entropy increase in the second law of thermodynamics, the "side" on which we place the past hypothesis is *derivative* and determined by the degrees of freedom in the brain that satisfy the ASYMMETRY and MENTAL conditions. If it is, in fact, the case that some local asymmetry gives rise to the psychological arrow of time, then we have a direction of time *before* postulating the past hypothesis and regardless of its truth. The past hypothesis is placed on the "side" of the sequence of states which we already experience as "past", and the explanation of the latter fact may have nothing to do with entropy at all, but rather may be based on other effective local asymmetry that is compatible with the time-symmetry of fundamental physics. This means that the direction of the psychological arrow of time determines the temporal *direction* of the past hypothesis, rather than the other way around.

There are two points to note here. (i) Consider, for example, some other 'arrows of time' such as the cosmological arrow, according to which the big bang occurred at the 'initial' time and at *later* times, the universe expands. Or consider the thermodynamic arrow, according to which (as we said) the low entropy macrostate occurs at the *beginning* of time and, by the second law, the entropy of the universe increases since *then*. On our proposal, the *directions* of these two arrows are fixed by the direction of the psychological arrow, so that the side of the sequence of states which we feel as earlier (or

---

world that metaphysically grounds (or explains) the direction of time (local or global). But since there are always trajectories of any isolated system along which entropy decreases or zigzags, it seems to us that proofs of the second law from the underlying mechanical theory (classical or quantum) cannot establish any metaphysical connection, which therefore (if it exists) must come from somewhere else.

past) is fixed by the asymmetry in our brains. It turns out that, as a matter of fact, the side that (we feel) is earlier (or past) according to the cosmological arrow is the one which is closer to the big bang. Likewise, it turns out as a matter of fact that the side of the same sequence of states which (we feel) is earlier (or past), according to the thermodynamic arrow, is the side at which the entropy is low. And similarly, the same occurs with respect to *all* other arrows such as the quantum mechanical arrow manifested in the violation of the charge-parity symmetry; the electromagnetic radiation asymmetry; and even the arrows of evolutionary processes and causation.

But it does *not* follow from this that (for example) the cosmological arrow and the thermodynamic arrow should be *aligned* relative to each other, such that what is earlier according to the thermodynamic arrow is *also earlier* according to the cosmological arrow. So, although — in our proposal — the direction of the psychological arrow completely fixes the *directions* in all the other arrows, how these arrows are aligned relative to each other is a different question. And with respect to this question, it might be that one or another of these arrows (perhaps the thermodynamic arrow) is more fundamental than the others; perhaps in the sense that it brings about, or grounds, the others. This is another task of Albert (2000, 2014) and Loewer's (2012) approach which we did not address here.

(ii) Another implication of our proposal is as follows: It turns out that if there were no creatures like us that have the local asymmetry that feels like the direction of time, the sequence of states of the universe would, of course, still remain the same (as we just saw in the previou point), although (given time-symmetric physics) it would be *directionless*. By this, we mean the following: Suppose, for example, that the universe is in a low entropy state at one temporal end (say near the big bang), and in a high entropy state at the other. And suppose that the entropy of the universe unfolds from one end to the other in accordance with the equations of motion. Of course, the two ends of the sequence of states in this scenario are distinguishable. One may be tempted to say that the second law holds in this sequence of states in the sense that the entropy of the universe increases as one moves from the low entropy end to the high entropy end. But this last phrase presupposes a direction of time, which we argue is given by some asymmetry in our brains. Therefore, if there are no creatures with some in-brain asymmetry that feels like the direction of time, one could equally and mistakenly say that the entropy of the universe along this sequence of states decreases as one moves from the high entropy end to the low entropy end. If the direction of time is fixed by some local asymmetry in the brain, and if this asymmetry ceases to exist, the sequence of states along which the universe evolves is *directionless* in the sense that there is no fact of the matter as to which side of this sequence is earlier and which is later.

It seems to us that this conclusion cannot be avoided as long as we are seeking an explanation of the psychological arrow of time within current physics. If this sounds disappointing, then one should either look for a new physics that would *not* be time-symmetric, or else adopt some non-reductive philosophy in which the direction of time would remain unexplained and even unreflected in physics.

Acknowledgements

We thank two anonymous reviewers for helpful comments on an earlier draft of this paper. This research was supported by the Israel Science Foundation, grant number 1148/18.

References

Albert, D. (2000) *Time and Chance*, Harvard University Press.

Albert, D. (2014) *After Physics*, Harvard University Press.

Allori, V. (2015) "Maxwell's Paradox: The Metaphysics of Classical Electrodynamics and its Time-Reversal Invariance," *Analytica*, **1**, 1–19.

Atkinson, D. (2006) "Does Quantum Electrodynamics Have an Arrow of Time?", *Studies in History and Philosophy of Modern Physics*, **37**, 528–54.

Arntzenius, F. (2000) "Are There Really Instantaneous Velocities?" *The Monist*, **83**(2), 187–208.

Barbour, J., Koslowski, T. Mercati, F. (2014) "Identification of a gravitational arrow of time," *Physical Review Letters*, **113**: 181101 (http://arxiv.org/abs/1409. 0917)

Block, R. A. (2014) *Cognitive Models of Psychological Time*, Psychology Press, New York.

Boltzmann, L. (1896–1898) *Lectures on Gas Theory* (Trans. by S. Brush (Berkeley: University of California Press, 1964).

Callender, C. (2004) "Measures, Explanation and the Past: Should 'Special' Initial Conditions be Explained?" *British Journal for the Philosophy of Science*, **55**, 195–217.

Callender, C. (2010) "The Past Hypothesis Meets Gravity," in G. Ernst and A. Hüttemann (eds.), *Time, Chance, and Reduction: Philosophical Aspects of Statistical Mechanics*, Cambridge University Press, pp. 34–58.

Carnap, R. (1963) "Carnap's Intellectual Biography", in *The Philosophy of Rudolf Carnap*, P. A. Schilpp (ed.), La Salle, IL: Open Court, pp. 3–84.

Carroll, S. (2010) *From Eternity to Here*, New York, Dutton.

Carroll, S., Chen, J. (2004) "Spontaneous Inflation and the Origin of the Arrow of Time," arXiv:hep-th/0410270.

Dieks, D. (2016) "Physical Time and Experienced Time," in Y. Dolev and M. Roubach (eds.), *Cosmological and Psychological Time*, Boston Studies in Philosophy of Science, Vol. 285, Springer, pp. 3–20.

Dürr, D. (2001) "Bohmain Mechanics," in J. Bricmont, D. Dürr, M. C. Galavotti, G. Ghirardi, F. Petruccione, N. Zanghi (eds.), *Chance in Physics: Foundations and Perspectives*, Lecture Notes in Physics, Springer-Verlag, pp. 115–132.

Earman, John (1974) "An Attempt to Add a Little Direction to "The Problem of the Direction of Time"," *Philosophy of Science*, **41**(1), 15–47.

Earman, J. (2006) "The Past Hypothesis: Not Even False," *Studies in History and Philosophy of Modern Physics*, **37**, 399–430.

Einstein, Albert (1919) "Time, Space, and Gravitation," Times London, 28 November 1919, 13–14. Reprinted as "What is the Theory of Relativity?" In Einstein, *Ideas and Opinions*, 227–232, New York, Bonanza Books, 1954.

Elga, A. (2001) "Statistical Mechanics and the Asymmetry of Counterfactual Dependence," *Philosophy of Science*, **68**, No. S3, S313–S324.

Feynman, R. (1965) *The Character of Physical Law*, Cambridge, MIT Press.

Frigg, R. (2008) "A Field Guide to Recent Work on the Foundations of Statistical Mechanics," in D. Rickles (ed.), *The Ashgate Companion to Contemporary Philosophy of Physics*, London, Ashgate, pp. 99–196.

Frisch, M. (2010) "Does a Low-Entropy Constraint Prevent Us From Influencing the Past?" In G. Ernst and A. Hüttemann (eds.), *Time, Chance, and Reduction: Philosophical Aspects of Statistical Mechanics*, Cambridge University Press, pp. 13–33.

Ghirardi, G., Rimini, A., Weber, T. (1986) "Unified Dynamics for Microscopic and Macroscopic Systems," *Physical Review, D*, **34**, 470–479.

Goldstein, S. (2012) "Typicality and Notions of Probability in Physics," in Y. Ben-Menahem and M. Hemmo (eds.), *Probability in Physics*. The Frontiers Collection, Berlin Heidelberg: Springer-Verlag, pp. 59–72.

Goldstein, S., Huse, D. A., Lebowitz, J. L., Tumulka R. (2016) "Macroscopic and Microscopic Thermal Equilibrium," Unpublished manuscript. Available at arXiv:1610.02312v1 [quant-ph].

Goldstein, S., Tumulka, R., Zanghi, N. (2016) "Is the Hypothesis About a Low Entropy Initial State of the Universe Necessary for Explaining the Arrow of Time?" (unpublished manuscript).

Hawking, S. W. (1988)*A Brief History of Time*, London, Bentam Press.

Hemmo, M., Shenker, O. (2010) "Maxwell's Demon," *The Journal of Philosophy*, **107**, 389–411.

Hemmo, M., Shenker, O. (2011) "Szilard's Perpetuum Mobile," *Philosophy of Science*, **78**, 264–283.

Hemmo, M., Shenker, O. (2012) *The Road to Maxwell's Demon*, Cambridge University Press.

Hemmo, M., Shenker, O. (2013) "Entropy and Computation: The Landauer-Bennett Thesis Reexamined," *Entropy* 2013, **15**, 3297–3311.

Hemmo M., Shenker O. (2014) "Probability and Typicality in Deterministic Physics," *Erkenntnis*, ISSN 0165-0106 DOI 10.1007/s10670-014-9683-0.

Hemmo, M., Shenker, O. (2016), "Maxwell's Demon," *Oxford Handbooks Online*. (Oxford University Press). http://www.oxfordhandbooks.com/view/10.1093/oxfordhb/9780199935314.001.0001/oxfordhb-9780199935314-e-63?rskey$=$plUl7T{\&}result$=$1

Hemmo, M., Shenker, O. (2018) "The Past Hypothesis and the Psychological Arrow of Time". Forthcoming.

Ismael, J. (2010) "Temporal Experience," in C. Callender (ed.) *Oxford Handbook on Time*, Oxford University Press.

Ismael, J. (2017) "Passage, Flow and the Logic of Temporal Perspectives," in C. Bouton and P. Huneman (eds.), *Time of Nature and the Nature of Time*, *Boston Studies in the Philosophy and History of Science*, Springer, pp. 23–38.

Le Poidevin, R. (2015) "The Experience and Perception of Time," *The Stanford Encyclopedia of Philosophy*, Edward N. Zalta (ed.), URL =<http://plato.stanford.edu/archives/sum2015/entries/time-experience/.>

Linden, N., Popescu, S., Short, A.J., Winter, A. (2009) "Quantum Mechanical Evolution Towards Thermal Equilibrium," *Physical Review E*, 79, 061103-1-12.

Loewer, B. (2001) "Determinism and Chance," *Studies in History and Philosophy of Modern Physics*, **32**, 609–620.

Loewer, B. (2012) "Two Accounts of Laws and Time," *Philosophical Studies*, **160** (1), 115–137.

Markosian, N. (2016) "Time", *The Stanford Encyclopedia of Philosophy* (Fall 2016 Edition), Edward N. Zalta (ed.), URL = <https://plato.stanford.edu/archives/fall2016/entries/time/.>

Maudlin, T. (2007) *The Metaphysics Within Physics*, Oxford University Press.

Mlodinow, L., Brun, T. (2014) "On the Relation Between the Psychological and the Thermodynamic Arrows of Time," *Physical Review E* 89, 052102. Also in arXiv:1310.1095.

Ney, A. (2008) "Physicalism as an Attitude," *Philosophical Studies* **138**, 1–15.

Paul, L. A. (2015) "Experience and the Arrow," in A. Wilson (ed.), *Chance and Temporal Asymmetry,* Oxford University Press.

Phillips, Ian, (2014) "Experience Of and In Time," *Philosophy Compass,* **9/2**, 131–144.

Phillips, Ian (2017) *The Routledge Handbook of Philosophy of Temporal Experience,* Routledge.

Price, H. (1996) *Time's Arrow and Archimedes' Point,* Oxford University Press.

Price, H. (2010) "Time's Arrow and Eddington's Challenge," *Seminaire Poincare* XV, 115–140.

Reichenbach, H. (1956) *The Direction of Time,* Berkeley, University of California Press.

Shenker, O. (2017a) "Foundations of Statistical Mechanics: Mechanics by Itself," *Philosophy Compass* (Oxford University Press). Forthcoming.

Shenker, O. (2017b) "Foundations of Statistical Mechanics: The Auxiliary Hypotheses," *Philosophy Compass* (Oxford University Press). Forthcoming.

Shenker, O. (2018) "Foundations of Quantum Statistical Mechanics." In: E. Knox and A. Wilson (eds.): *Routledge Companion to the Philosophy of Physics,* Oxford: Routledge. Forthcoming.

Sklar, L. (1973) "Statistical Explanation and Ergodic Theory," *Philosophy of Science,* **40**, 194–212.

Sklar, L. (1981) "Up and Down, Right and Left, Past and Future," *Nous,* **15**(2), 111–129.

Sklar, L. (1993) *Physics and Chance,* Cambridge, Cambridge University Press.

Smith, R. (2014) "Do Brains Have an Arrow of Time?," *Philosophy of Science,* **81**(2), 265–275.

Uffink, J. (2007) "Compendium to the Foundations of Classical Statistical Physics," in J. Butterfield and J. Earman (eds.), *Handbook for the Philosophy of Physics, Part B,* pp. 923–1074.

Uffink, J. Valente, G. (2010) "Time's Arrow and Lanford's Theorem," *Seminaire Poincare* XV, 141–173.

Uffink, J. Valente, G. (2015) "Lanford's Theorem and the Emergence of Irreversibility," *Foundations of Physics,* **45**, 404–438.

Van Fraassen, B. (1989) *Laws and Symmetry,* Oxford: Clarendon.

Von Neumann, J. (1932) *Mathematical Foundations of Quantum Mechanics,* Eng. Trans. by R. T. Beyer 1955, Princeton: Princeton University Press.

Wallace, D. (2017) "The Logic of the Past Hypothesis," in B. Loewer, B. Weslake and A. Winsberg (eds.), *Time's Arrow and the Origin of the Universe: Reflections on Time and Chance: Essays in Honor of David Albert's Work,* Cambridge, Ma: Harvard University Press. Forthcoming.

Werndl, C., Frigg, R. (2015) "Rethinking Boltzmannian Equilibrium," *Philosophy of Science,* **82** (5), 1224–1235.

Winsberg, E. (2004) "Can Conditioning on the 'Past Hypothesis' Militate Against the Reversibility Objections?", *Philosophy of Science,* **71**, 489–504.

Chapter 9

Eternal Recurrence Worlds and the Best System Account of Laws

Ryan A. Olsen and Christopher J. G. Meacham

University of Massachusetts Amherst

In this paper, we apply the popular Best System Account of laws to typical eternal recurrence worlds — both classical worlds and worlds of the kind posited by popular contemporary cosmological theories. Our first thesis is that, according to the Best System Account, such worlds will have no laws that meaningfully constrain boundary conditions.

It is generally thought that lawful constraints on boundary conditions are required to avoid skeptical arguments. Thus, the lack of such laws given the Best System Account may seem like a severe problem for the view. Our second thesis is that at eternal recurrence worlds, lawful constraints on boundary conditions do little to help fend off worries. So with respect to handling these worries, the proponent of the Best System Account is no worse off than their competitors.

9.1. Introduction

One of the most popular accounts of laws is the Best System Account (BSA).[1] On this account, the laws are roughly simple and informative descriptions of what the world is like. This account is popular for a number of reasons: its deflationary nature avoids uncomfortable metaphysical commitments; it lines up well with the methodological

---

[1] Prominent proponents include Lewis (1994), Loewer (2001), Hoefer (2005), and Albert (2012).

virtues scientists often espouse; and, most importantly, it seems to yield the kinds of laws that scientists have suggested would hold. For example, given a typical classical world, the BSA seems to yield something like the laws of classical statistical mechanics; given the kind of cosmology physicists envision, the BSA seems to yield something like the inflationary theories physicists have offered, and so on.[2]

In this paper we will argue that in a wide range of cases this last claim is mistaken: the BSA won't yield the kinds of laws that physicists suggest. In particular, some prominent physical theories have been proposed that arguably require lawful constraints on boundary conditions.[3] And we will argue, for at least some of these theories, that if the world eternally recurs, the BSA won't yield laws that constrain boundary conditions in the ways these theories suggest.

It's generally assumed that these lawful constraints on boundary conditions are required in order to avoid skeptical consequences — viz. that we should be near certain that our evidence about the past is highly misleading.[4] So the failure to yield such constraints might seem like a devastating blow to the BSA. But we will argue that, surprisingly, this turns out not to be the case. When one works through the details of how these skeptical arguments are supposed to go at eternal recurrence worlds, one finds that the theories which don't posit lawful constraints on boundary conditions are, in fact, no more

---

[2]We use the term "typical" here in its colloquial sense, that (in some good sense) the vast majority the things we are talking about (viz. worlds of a certain kind) are as we describe. That said, we take our use of the term "typical" to line up closely with the formal notions of typicality used in the typicality literature; see for example, Maudlin (2007b) and Frigg and Werndl (2012). One example is as follows: our descriptions are true of (near) measure 1 of classical one-way eternal recurrence worlds, using the Liouville measure on phase space.

[3]Interestingly, those working on these issues have taken different attitudes towards these constraints in different contexts. For example, in classical statistical mechanics, many have taken the requirement for such constraints to be an unproblematic feature of the theory (e.g., see Feynman (1965) and Albert (2000)). By contrast, in eternal inflation theories, most of those who have thought that such requirements might be needed have taken this to be a demerit of the theory (e.g., see Steinhardt (2011)).

[4]For example, see Albert (2000) and Carroll (forthcoming b).

susceptible to skeptical worries than the theories which do. Thus, at the end of the day, it's not clear that the BSA's failure to yield these lawful constraints on boundary conditions at eternal recurrence worlds is something that should bother proponents of the BSA.

This paper will proceed as follows: In section 9.2, we will lay out some background. In section 9.3, we will argue that at typical classical eternal recurrence worlds, the BSA won't yield meaningful constraints on boundary conditions. In section 9.4, we will argue that at typical eternal inflation worlds, the BSA won't yield meaningful constraints on boundary conditions. In section 9.5, we will assess the skeptical consequences of these results. In section 9.6, we will briefly summarize our results.

9.2. Background

9.2.1. *Classical statistical mechanics*

Statistical mechanics (SM) aims to predict thermodynamic phenomena, like milk diffusing into tea, ice cubes melting in warm water, and a raw egg's absorbing a hot pan's heat. In part, then, SM aims to predict entropy's tendency to increase over time in (approximately) isolated systems that aren't already at their maximum entropy.[5] For now, we restrict our focus to Newtonian worlds, fundamentally describable in terms of the motion of point particles, and so to *classical* statistical mechanics (CSM). We begin by briefly summarizing Albert's (2000) formulation of CSM and the justification for its inclusion of a nondynamical law, the past hypothesis.

CSM's predictions are probabilistic. For example, the CSM probability that (approximately) isolated, non-maximum entropy systems evolve in entropy-increasing ways is overwhelmingly high.[6] Hence, we

---

[5]SM doesn't always predict entropy increase. Here are two examples. First, SM predicts that a system in its maximum entropy (equilibrium) state will tend to remain in that state. Second, SM predicts infrequent entropic decreases.

[6]This is essentially the probabilistic version of the second law of thermodynamics. That said, it should be noted that this claim is not entirely uncontentious; see Shenker and Hemmo (2012).

reasonably expect such evolutions. The CSM apparatus for generating these probabilities has several components.

First, it assumes the laws of classical mechanics (CM).

Second, it employs a *phase space*. The classical phase space for a system of n particles is a $6n$-dimensional space representing both the position and momentum of each particle. Every point of phase space thus represents a complete specification of an instantaneous state the classical system could be in. Call these the *microstates* of such a system, and say that a system is *located at* the point in phase space which represents its current microstate. Given the determinism of classical mechanics,[7] a system's being in a microstate at a time determines what microstates it was and will be in. Likewise, a system's current location in phase space determines where it has been and where it will be located in that space.

Third, it distinguishes microstates from *macrostates*. Macrostates are macroscopic ways things can be, sequences of which comprise the thermodynamic evolutions that we directly observe. Many different microstates are macroscopically indistiguishable from each other, so that any of these microstates would yield (underlie, realize, etc.) the very same macrostate. In other words, such microstates are *compatible with* the same macrostate. Just as microstates are represented by points of phase space, macrostates are represented by *regions* of that space — viz. the region of points representing microstates that are compatible with that macrostate. When there is little danger of confusion, we conflate macrostates with the regions of phase space representing them.

Fourth, it employs a *probability distribution* over phase space. In particular, CSM uses a flat distribution defined over the standard Liouville measure of phase space regions. Higher-entropy macrostates are larger on this measure than lower-entropy ones, and a macrostate preserves its measure through time. Call this probability distribution

---

[7]Strictly speaking, classical mechanics is not deterministic (for various kinds of counterexamples, see Earman (1986), Xia (1992), and Norton (2008)). But these counterexamples are widely believed to be of measure of zero, and so for our purposes can be safely ignored.

the *Statistical Postulate* (SP). This yields a general definition of the CSM probability that macrostate A obtains, given that macrostate B does. In other words, relative to background propositions K, like CM and all lawful boundary conditions, the following holds true:

$$Prob_K(A|B) = \frac{m(A \cap B \cap K)}{m(B \cap K)}$$

In other words, the probability of A, given B (and given background conditions K), is a matter of *how much* of B (intersected with K) is taken up by A. Any same-measure part of B is equally likely, so the probability that the actual microstate is in A, given its being located in B, is a matter of the size of their intersection.

To demonstrate, suppose B is the macrostate that some ice cubes are floating in a cup of hot water at time t, and A is the macrostate of there being at t-plus-five-minutes only tepid water in that cup. Then, it turns out that the vast majority of B is taken up by $A \cap B$: nearly all the microstates in B (on the standard measure) are such that time evolving them forward by five minutes yields microstates wherein the cup contains only tepid water. That is, the probability of this evolution is extremely high. It will be useful to have a name for the apparatus described thus far, namely, the combination of CM and SP. Call this CSM⁻.

More is needed. For just as B is overwhelmingly comprised (on the standard measure) of microstates that evolve forward to be compatible with A, something analogous is true for microstates evolved *backward*: the vast majority of microstates in B had their ice melted in the past, too. In other words, CSM⁻ wrongly has it that ice spontaneously materialized from warm water.

Moreover, if we compare the probability CSM⁻ assigns to the world having been like what we think it was a minute ago, and the probability that our memories are false and we spontaneously fluctuated out of a higher entropy state, we will find that CSM⁻ assigns a vastly higher probability to the latter. And thus, if our beliefs should line up with the chances (cf. section 9.2.4), then it seems we are rationally required (conditional on CSM⁻ being true) to disbelieve our memories. As Albert (2000) has argued, this seems to make CSM⁻

self-undermining. For if we disbelieve our memories, then we lose our reasons for believing something like CSM⁻ in the first place.

The canonical solution is to add to the background propositions K a *Past Hypothesis* (PH), to the effect that the world was initially in a very simple, low-entropy macrostate M of the sort that cosmology presumably aims to discover. Nearly all microstates in B that increase in entropy toward the past, rather than decrease, are incompatible with the PH. So by adding the PH, the vast majority of remaining microstates evolved in ways compatible with what we remember, and so the probability that the world evolved in this way is high. CSM is the result of adding the PH to CSM⁻.

To summarize, Albert's formulation of CSM is the conjunction of the following three theses:

(CM) The laws of classical mechanics.

(SP) The statistical postulate, i.e. a flat probability distribution over the Liouville measure of phase space regions.

(PH) The past hypothesis, i.e. a statement to the effect that the universe was in a very simple, low-entropy, globally initial macrostate M of the sort we might expect cosmology to discover.

Lastly, we will focus on classical worlds of eternal recurrence, by which we mean eternal worlds such that the Poincaré recurrence theorem holds. This theorem states that if a classical system is restricted to a finite region of phase space, and is in macrostate M, then given enough time the system will (with probability 1) evolve into macrostate M again. And given infinite time, the system will return to macrostate M infinitely many times. That is, M will (with probability 1) eternally recur.[8]

---

[8]For an accessible presentation of the recurrence theorem's proof, see Albert (2000, 73–75). One might question how realistic the restriction to a finite region of phase space is. But for our purposes, such as issues are beside the point. These classical examples highlight a straightforward consequence of eternal recurrence for the BSA. And as we will argue, similar consequences arise in more realistic cosmological models, like that of eternal inflation.

9.2.2. *Eternal inflation*

Eternal inflation is a cosmological model of considerable popularity and interest. It is best introduced by way of the more familiar Big Bang model, for inflationary cosmology aims to explain facts the Big Bang model cannot. Eternal inflation is then thought to be a consequence of the mechanism that drives the hallmark of inflationary cosmology: rapid spatial expansion in the early universe.[9]

According to Big Bang cosmology, the early universe was comprised of an extremely hot, dense plasma which expanded and cooled. This cooling plasma then synthesized into lighter chemical elements which, under the attraction of gravity, coalesced to form stars and galaxies. This, in the barest of outlines, is the standard cosmological model. But there are facts this model cannot explain, and inflationary cosmology promises to fill the gaps.

One putative gap is the flatness of space. The curvature of the early universe must have been extremely flat; otherwise, the universe now would be far more curved than it in fact is. But why was space so flat to begin with? Another gap is that soon after the Big Bang, the universe was remarkably homogeneous. We know this by observing the cosmic microwave background (CMB), the remnant radiation left by the cooling plasma. Everywhere across the sky, this radiation is very evenly distributed, to about one part in 100,000, indicating that the plasma that produced the CMB was also highly uniform.[10] But why was this plasma so homogeneous? A third gap is that the early universe wasn't perfectly thermally uniform, and the way in which it deviated from uniformity is also interesting. Specifically, the inhomogeneities found in the CMB are very nearly *scale-invariant*; that is, their magnitudes are largely the same whether you look at smaller or larger regions of the CMB. But why should the early universe have been inhomogeneous in just this way?[11]

---

[9]For an accessible discussion of eternal inflation, see Guth (2001, 2007), and Steinhardt (2011).
[10]Guth (2001, p. 75).
[11]To explain the apparently remarkable flatness of the early universe is to solve the so-called *flatness problem*. Conventionally, solving this problem helps motivate

Roughly, inflationary cosmology explains these puzzling facts as follows. Prior to the formation of this hot, dense plasma, the universe quickly underwent enormous spatial expansion, growing within a mere 10^{-30} seconds by a factor of at least 10^{25} — i.e., from about one quadrillionth of the size of an atom to around that of a dime.[12] Such expansion stretched any prior spatial curvature to a tiny fraction of what it had been, leaving a virtually flat arena for the hot plasma that we indirectly observe. Inflation fills the other two gaps by way of its hypothesized mechanism. This is a scalar field, called the *inflaton*, whose energy density varies across space. The field's density in a region determines its potential energy, which is repulsive rather than attractive; call this *inflationary energy*. So where there is high inflationary energy, space undergoes rapid, exponential expansion; and when and where that energy fizzles out, inflation ends.

Like other scalar fields, the inflaton classically moves from higher potential energies to its minimum, like a ball rolling down a hill. And when the inflaton's potential energy drops, inflationary energy converts into ordinary energy, thereby "thermalizing" the newly expanded space and creating the sort of plasma that gave rise to the CMB. Moreover, rapid expansion results in a highly uniform distribution of this energy in the expanded region, the decay of which yields a largely uniform thermalization. And quantum fluctuations in the inflaton field give rise to scale invariant inhomogeneities in the resulting thermalization: earlier fluctuations are magnified by expansion to cosmic scales, later fluctuations undergo less magnification, and so fluctuations over the course of the inflationary period yield similar thermal nonuniformities on scales from small to cosmic.

---

inflationary cosmology (cf. Guth (2007), p. 6813). However, see section 9.4 of Carroll (forthcoming a) for an argument that the flatness of the early universe is typical, given the canonical measure on possible spacetime trajectories the universe might take. While this may undercut the flatness problem, Carroll argues that the homogeneity of the early universe is extremely atypical on this canonical measure, lending value to explanations of that homogeneity.

[12]Steinhardt (2011, p. 40).

Eternal inflation — the thesis that there is always some region that has a high inflationary energy and so undergoes rapid exponential expansion — seems to be a consequence of the picture just described. Like the half-life of radioactive decay, inflationary energy decreases only with some probability.[13] So there is a non-zero chance that this energy remains high in a region. And because this potential energy doesn't dilute with expansion, even small regions where the inflaton hasn't decayed quickly grow to dominate space. If the chance that inflationary energy remains high is not too small — and the exponential expansion caused means even a small chance is sufficient — then this pattern of growth and decay and subsequent growth elsewhere continues *ad infinitum.*

Eternal inflation leads to a fractal spacetime structure. As an initial region of high inflationary energy evolves, some thermalizes and some does not; the space that does not thermalize likewise expands, some of which thermalizes and some of which does not; and so on. A helpful way to picture this is in terms of a simplified nested structure, illustrated in Figure 9.1.

Each elipse in Figure 9.1 represents the beginning of a new region of rapid expansion, and the white-to-gray gradient indicates the transition from high inflaton potential energy to thermalization. So the outermost elipse thermalizes around a single region of rapid expansion, which expands and thermalizes around a second region of rapid expansion, and so on. The illustration is simplified in several ways. First, we shouldn't expect a single region of continuing rapid expansion; rather, expanding regions are likely scattered in a sea of thermalizing space. They then grow to dominate the spatial volume, surrounding and cutting off pockets of thermalization from one another. Second, the diagram is not to scale. It does roughly

---

[13]These probabilities result from the combination of the inflaton's classical movement (toward its minimum potential energy) with chancy, quantum fluctuations that "bump" the inflaton either lower or higher on its potential. These fluctuations were mentioned earlier in explaining the scale invariant inhomogeneity in the CMB.

High Inflaton Potential Energy, Rapid Expansion

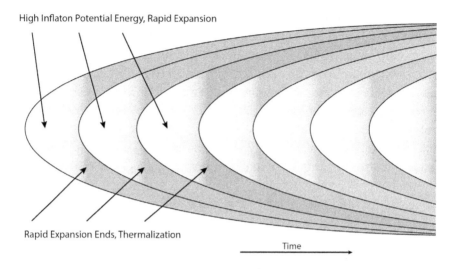

Rapid Expansion Ends, Thermalization

Time

Fig. 9.1.   Simplified Eternal Inflation.

represent, however, the way that expanding space comes to dominate the spatial volume: as one progresses farther to the right, less
and less of the diagram's height is comprised of thermalized space
from earlier phases of expansion. And third, thermalized space can
beget its own "child" regions of high inflationary energy and subsequent rapid expansion, namely, by quantum tunneling to a high
inflationary energy (though this is extremely unlikely). These child
regions are not represented.

We assume that this eternal inflation world began globally with a
largely homogeneous distribution of high inflationary energy, which
then sets off the infinite cascade of thermalization and further inflation. This assumption is plausible. As Brandenberger (2017) summarizes, recent modeling suggests that large inhomogeneities in the
initial conditions would stop inflation from getting started. Since the
world we are considering is such that there was initial inflation, we
assume that it was not initially inhomogeneous in that way.

Observe then that essentially the same macrocondition, call it M,
arises non-initially: those regions that do not thermalize expand to
have largely homogeneous distributions of high inflationary energy,

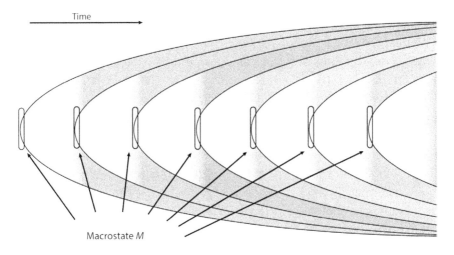

Fig. 9.2.   Infinite Recurrence of Macrostate M.

of effectively the same size and shape as the initial conditions. This occurs infinitely many times, and so in addition to the initial occurrence of M, there are infinitely many occurrences of M at particular local non-initial regions, as shown in Figure 9.2.[14]

The laws that eternal inflation theories appeal to are not as clear cut as those of CSM, and there's some variation between different versions of eternal inflation regarding what the dynamical laws are and whether it needs to include a constraint on initial conditions. That said, it will be convenient to have a name for the laws that eternal inflation theories propose. In what follows, we will take COS to refer to these laws, when understood as including a lawful constraint on initial conditions, and COS$^-$ to refer to the laws one gets by removing this constraint on initial conditions.

---

[14]Might the initial and later macrostates of high inflationary energy be different? One clear difference is that the initial macrostate occurs globally, whereas later macrostates occur only locally. Another difference is that while exponential expansion leaves non-initial macroconditions of high inflationary energy quite uniform, the initial condition may be less so. We return to these points in section 9.4, but neither consideration plausibly affects our argument.

9.2.3. *The Best System Account*

The Best System Account (BSA) of laws is a version of the regularity account of laws. Like other regularity accounts, it offers a deflationary account of what the laws are: according to the BSA, the laws are nothing more than a simple and informative description of what the world is like.

The BSA is popular for several reasons. One reason is that the BSA seems to closely adhere to the kind of methodology employed by the sciences, and to yield the kinds of laws our scientific community would endorse. A second reason is that by taking the laws to merely be a certain kind of description of what the world is like, it avoids the spookiness of primitive laws, or laws grounded in non-occurrent facts, like counterfactuals or necessitation relations between universals.[15] A third reason is that the BSA is compatible with the popular thesis of Humean Supervenience — the claim that all of the qualitative features of the actual world supervene on the qualities instantiated at points, and the spatiotemporal relations between them.

The classic formulation of the Best Systems Account (BSA) of laws comes from Lewis (1994). On this account, we can determine the laws of a world w in the following way. First, consider a language whose only predicates are: (a) predicates corresponding to perfectly natural properties and relations; and (b) a chance predicate. Second, consider every set of sentences in this language, and remove any set containing sentences that are false at w, or chance assignments that aren't probabilistic. Third, evaluate the remaining sets of sentences according to three desiderata: simplicity, informativeness, and "fit", where fit is a measure of how high a chance these sentences assign to the history of w. If one of these sets of sentences is robustly best with respect to these desiderata — i.e., it does substantially better

---

[15]For proponents of primitive laws, see Carroll (1994) and Maudlin (2007a); for an account that analyzes the laws in terms of counterfactuals, see Lange (2009); for an account that analyzes the laws in terms of necessitation relations between universals, see Armstrong (1983). That said, not all proponents of the BSA are moved by this kind of consideration; see Demarest (2017) for a proponent of the BSA that also accepts these kinds of intangible non-occurrent facts.

than any other candidate — then it's the *best system* at w. And any regularity entailed by the best system is a law.

What if there are several sets of sentences that are (roughly) tied, and so no set of sentences that's robustly better than any of the others? Lewis vacillated about how to best handle these cases. One approach, suggested by Lewis (1986), is to take the laws in this case to be only those regularities that are entailed by all of the systems that are in contention. Another approach, advocated by Lewis (1994), is to take there to be no laws in such cases. We will take both of these approaches to be live possibilities in what follows; as we will see, our arguments will go through either way.[16]

Although this classical formulation of the BSA is attractive in many respects, a number of issues arise regarding this formulation, especially when viewed through the lens of statistical mechanics.[17]

One issue is that the classical BSA restricts laws to *regularities*. As a result, it rules out lawful constraints on boundary conditions by fiat. And this makes it unable to recover popular formulations of statistical mechanics that appeal to lawful constraints on initial conditions, like Albert's (2000) Past Hypothesis. In light of this, we will join most modern proponents of the BSA in discarding the constraint

---

[16]Glynn *et al.* (2014) suggest that, with respect to ties between systems that differ only in their chance assignments, we should adopt a third option: taking the chances to not be sharp. In such cases, we should take the chances to be given by the set of probability functions the roughly tied candidates assign. It's worth noting that this is not, in fact, a third way to treat ties. For this treatment of chances is entailed by Lewis's (1986) suggestion to take the propositions entailed by all of the viable candidates to be laws. In particular, suppose that there are two candidate systems, S_1 and S_2, that disagree only about whether the chances should be given by ch_1 or ch_2. Both of these systems will entail disjunctions of the form "The chance of A is x [= $ch_1(A)$] or y [= $ch_2(A)$]". Thus, on Lewis's (1986) proposal, these disjunctive claims about chances would be lawful, even though each of its disjuncts would not (since neither disjunct is entailed by both systems). Thus, the proposal of Glynn *et al.* (2014) in fact follows from Lewis's first suggestion regarding how to handle ties.

[17]In addition to the three issues discussed below, a number of other worries have been raised about using the BSA to recover the laws of statistical mechanics; for further discussion, see Meacham (2010) and the references therein.

to regularities, allowing any proposition entailed by the best system to be a law.[18]

Another issue is that the classical BSA requires the language we use to evaluate candidate systems for simplicity to be one in which all of the predicates (except the chance predicate) refer to perfectly natural properties and relations. But if this is how we are evaluating candidate systems, it's unclear how something like the Past Hypothesis could be one of the laws. For describing this macrostate in the language of perfectly natural properties and relations (e.g., mass, charge, spatiotemporal relations) would yield something extraordinarily complicated. And thus any system entailing such a proposition would presumably itself be very complicated, and thus not a plausible candidate for being the best system.[19]

In light of this, we will join many modern proponents of the BSA in modifying the account in order to allow for laws like the Past Hypothesis. There are a number of different variants of the BSA in the literature which do this; the particular variant we will adopt draws from Hoefer (2005) and Loewer (2007).[20] On this approach, we allow candidate systems to be formulated in *any* language. And then we evaluate candidate systems for simplicity, informativeness and fit, where these criteria are now understood in terms of simplicity and informativeness for subjects like us, with scientific communities like ours.[21] Given this way of understanding the BSA desiderata, it's no longer implausible that something like the Past Hypothesis could be a law.

A third issue is that the notion of fit Lewis employs seems incapable of recovering the chances of statistical mechanics. On Lewis's

---

[18]For example, see Loewer (2001), Hoefer (2005), Ismael (2009), and Albert (2012). It's worth noting that, at times, Lewis himself seemed amenable to this suggestion; see Lewis (1986, p. 123).

[19]For versions of this worry, see Schaffer (2007) and Winsberg (2008).

[20]For a survey of these variants of the BSA, and a discussion of their relative merits, see Eddon and Meacham (2015) and the references therein.

[21]One might worry about the objectivity of appealing to what is simple and informative for such communities. For discussion of this point, see Loewer (2007, p. 325 and 327), and Eddon and Meacham (2015, sections 2.3 and 3.5).

conception, the fit of a system at w is determined by how high a chance the system assigns to w's history. But statistical mechanics will assign a chance of 0 to any particular history coming about. And this would still be true if we changed the chances statistical mechanics assigned to individual events — e.g., by making the chance of heads $2/3$ instead of $1/2$. Thus, Lewis's notion of fit won't discriminate between statistical mechanics and alternatives to statistical mechanics that assign different chances to individual events, because all of these theories assign the same chance to w's history — 0. And so, given this notion of fit, it's unclear how the BSA could recover statistical mechanics, since statistical mechanics will do no better with respect to fit than alternative theories which assign different chances to individual events.

In response to this problem, we will follow Elga (2004b) in modifying the notion of fit the BSA employs. Elga's proposal is to assess the fit of systems by looking at the chances they assign to a restricted set of test propositions. Then, we compare the fit of different systems by comparing the chances they assign to the true test propositions, with higher chances indicating better fit.

This proposal requires a way of picking out a restricted set of test propositions. Elga's suggestion is to take the test propositions to be those that are simply expressible in a privileged language; presumably the privileged language Lewis proposed to use when formulating candidate systems. Since, following Loewer, we don't appeal to Lewis's privileged language, we will adopt a slightly different approach: we will take the test propositions to be those that are simply expressible for subjects like us, and scientific communities like ours.[22]

---

[22]Although Elga's proposal is perhaps the most sophisticated proposal for how to assess fit on offer, one might have worries regarding how to spell out the details in a satisfactory way. For example, one might worry that given the kind of privileged language Lewis proposes, there will be infinitely many true simple sentences expressible in that language. (E.g., if we have names for spacetime points, and a fundamental exact occupation relation, then there will be uncountably many simple true sentences asserting that something does/does not occupy a given spacetime point.) And one might worry that the project of spelling out how to

9.2.4. *Credence and chance*

In this section, we lay out the standard framework for the relationship between *de se* credences and chances, namely, in terms of centered worlds and Lewis's Principle Principle. For such a framework makes clear the self-undermining nature of theories like CSM⁻. Given that we are rationally required to match our credences to the chances, then conditional on CSM⁻, we should be virtually certain that our memories are false. And yet if so, we are left with little reason to believe CSM⁻ in the first place.

Let a *centered world* be an ordered triple consisting of a world w, a time t, and an individual i. Let a *de dicto proposition* be a proposition that is true at a centered world $\langle w, t, i \rangle$ *iff* it's true at every other centered world at w. Let an *irreducibly de se proposition* be a proposition that doesn't satisfy this clause — a proposition such that, for some w, it's true at some centered worlds at w and false at others. Intuitively, *de dicto* propositions correspond to claims that are entirely about what the world is like, while irreducibly *de se* propositions correspond to claims that are also about one's location within a world.[23]

For any irreducibly *de se* proposition A, there will be a corresponding *de dicto* proposition that is true at all of the centered worlds A is true at, and also true at any centered world located at the same world as one of the centered worlds where A is true. This is the proposition that there exists some individual at some time for whom that *de se* proposition is true. For convenience, we will use the following notation to flag this relationship: if A is an irreducibly *de se* proposition, then $\hat{A}$ is the *de dicto* proposition that one gets by "filling in" A in the manner just described.

---

weigh and agglomerate the chances assigned to infinitely many propositions in a satisfactory way will raise many of the same worries that arose for Lewis's proposal regarding how to assess fit. Due to space constraints, we won't try to resolve these issues here.

[23]This framework borrows heavily from Lewis (1979), though he uses slightly different terminology (e.g., Lewis calls *de se* propositions "properties").

Let a subject's *credence function* (cr) be a function which assigns values to propositions between 0 and 1, representing the subject's confidence in those propositions, with 0 representing maximal confidence that the proposition is false and 1 representing maximal confidence that the proposition is true. It's generally assumed that rational subjects will have probabilistic credence functions. In what follows, we will restrict our attention to credence functions that are probabilistic.

For concreteness, we will assume Lewis's (1996) conception of evidence. In this picture, your total evidence E corresponds to the set of centered worlds $\langle w, t, i \rangle$ such that at world w, at time t, individual i has the same perceptual experiences and memories as you do. (Following Lewis, we are understanding "memories" here in a non-veridical sense; e.g., having the memory that it rained on your tenth birthday does not entail that it actually did.)

If a subject with total evidence E satisfies the Bayesian updating rule (Conditionalization), then we can express their current credences as a function of their initial credences (or "priors") ic and their total evidence E as follows:[24]

IC-Conditionalization: $cr_E(A) = ic(A \mid E)$, if defined.

We take there to be some constraints on rational priors (such as the Principal Principle, described below); that said, we will adopt a

---

[24]Two caveats. First, while IC-Conditionalization is largely uncontroversial if A and E are *de dicto* propositions, a number of tricky questions arise if A and E are *de se* propositions. (See Titelbaum (2012) and the references therein.) To skirt these issues, in what follows we will only rely on arguments that appeal to IC-Conditionalization (and this equation) when A and E are *de dicto* propositions. Second, this equality only holds if a subject doesn't lose evidence. If a subject loses evidence, and so E is strictly weaker than the conjunction of their evidence, then this equation won't express what Conditionalization prescribes. In what follows, we will assume that the subjects we are considering don't lose (*de dicto*) evidence. (For discussion of this issue, see Meacham (2016) and the references therein.)

broadly permissive approach which takes a number of different prior
functions to be rationally permissible.[25]

Let a *chance function* $ch_{T,K}(A)$ be the chance assigned to A by
chance theory T, given background K. As usually understood, A,
T and K must all be *de dicto* propositions.[26] It's widely held that
one's beliefs about the chances place constraints on what it's rational
for one to believe. A typical formulation of this constraint is Lewis's
(1980) Principal Principle, which holds that a rational subject's initial credence function should be such that:[27]

Principal Principle: $ic(A \mid T \wedge K) = ch_{T,K}(A)$, if defined.

It follows from IC-Conditionalization and the Principal Principle
that for rational subjects, $cr_{T \wedge K}(A) = ch_{T,K}(A)$, when defined. But
subjects like us will virtually never have total evidence of the form
$T \wedge K$ needed to yield well-defined chance assignments. After all,
our evidence is compatible with many different chance theories, and
many different backgrounds given those chance theories. Nonetheless, the Principal Principle will impose strong constraints on us. In
particular, it will require our credence in A to be equal to the average
of the chances assigned to A by the different $T \wedge K$s compatible with
our evidence, weighted by our credence that those $T \wedge K$s obtain:

$$cr_E(A) = \sum_i cr_E(T \wedge K_i) \cdot ch_{T,K_i}(A), \text{ if defined,}$$

---

[25]We make this assumption largely for concreteness; most of our points can be
tweaked to go through, given a picture on which there's only one rational prior,
given that this rational prior lines up with priors like ours (cf. section 9.5). The
question of whether there are many rational priors or only one is a contentious
issue; for a classic attack on permissivist stances on priors, see White (2005), for
some defense see Urbach and Howson (2005), and Meacham (2013).

[26]See Lewis (1980).

[27]Lewis required K to be a complete history up to a time; like many other authors,
we drop this constraint in order to allow for statistical mechanical chances (see
Loewer (2001), Meacham (2005), Hoefer (2005), Winsberg (2008), and Ismael
(2009)). Lewis (1994) himself later followed Hall (1994) in endorsing a more complicated principle — the "New Principle" — in order to address certain worries
regarding the compatibility of the Principal Principle and the BSA. However, in
the cases we are concerned with, the two principles will yield largely the same
prescriptions. So we will employ the simpler Principal Principle in what follows.

where i ranges over the elements of some partition of E into $T \wedge K$s compatible with E.[28]

9.3. The classical case

In this section, we argue that, by the BSA's lights, typical examples of classical eternal recurrence worlds — worlds that globally satisfy the Poincaré recurrence theorem — have no lawful boundary conditions. Or, at the very least, our argument shows that lawful boundary conditions those worlds do have are so weak, that they impose essentially no constraints on the chances.

The argument, in outline, is as follows. Consider again PH, the thesis that the world was initially in a very simple, low-entropy macrostate M. Call the PH a *boundary proposition* (BP), as it claims that a certain boundary condition obtains. Given the version of the BSA we have assumed, it is plausible that the PH earns its keep in a best system and so, other things being equal, it is a lawful boundary condition. After all, the PH is relatively simple to state, in a language that is salient to us. And together with SP and CM, the PH plausibly underwrites a vast increase in the fit of systems including it, by way of assigning high SM probabilities to test propositions, like those concerning the evolution of everyday thermodynamic systems.[29]

Say then that the PH is *highly eligible* to be a member of a best system, given its balance of complexity and fit. Observe then that in the worlds we are considering — where there is one-way eternal recurrence of the initial low-entropy macrostate M — there is not

---

[28]This is assuming we are restricting our attention to cases where Hall's (1994) New Principle and Lewis's (1980) Principal Principle yield approximately the same descriptions (cf. footnote 26). In the fully general case proponents of the BSA will need to adopt a more sophisticated account of the relation between one's credences in a proposition and one's credences in different chance assignments to that proposition. (For discussion of some problems that arise in the general case for the formula given in the text, see Pettigrew (2013). For an argument that we should still accept this formula even if we do adopt the BSA, see Ismael (2008).)

[29]For discussions of how the BSA could yield the laws of CSM, see Loewer (2001), Hoefer (2005), Winsberg (2008), and Albert (2012).

just one highly eligible BP, but plausibly an infinite number of them. That is, in the same way that PH seems highly eligible in such worlds, so do later *middling hypotheses* (MHs), which are BPs which state that M occurs at specific, non-initial times — viz. those at which M does in fact occur. In such worlds, PH and infinitely many MHs are more-or-less on a par with respect to the BSA's desiderata of simplicity, informativeness, and fit. And we argue that, given the BSA, this result — plus the plausible claim that there are no other, more eligible BPs — vitiates against any one of these BPs being a lawful boundary condition.

9.3.1. *Parity between boundary propositions*

Different collections of PH and/or MHs are axioms of different systems; on the BSA, the best of these is such that its theorems (axioms included) are the laws. Of course, there are other BPs, not all of which are made equal; plausibly, only PH and MHs could *prima facie* earn their keep in a best system. Setting aside those that are less eligible, we will use 'BPs' below to refer only to the PH and MHs, and 'recurrences' to refer only to occurrences of M. In this section, we observe that in a typical classical one-way eternal recurrence world w, infinitely many BPs are roughly on a par with respect to the BSA's desiderata of simplicity, informativeness, and fit.

It is easy to see that this claim is true for simplicity and informativeness. There are infinitely many MHs in addition to the PH, and all of these BPs state that the simple, low-entropy macrocondition M occurs. One way a BP p could be more informative than another BP p' is if p entails, but is not entailed by p'. So note then that no BP entails any other: each is such that some microstates compatible with it are not compatible with the other BPs.[30] Moreover, the only difference between these BPs is at which time condition M is said to occur. Stating that M occurs at t or at t', however, makes little

---

difference to either the relative complexity or the informativeness of the BPs in question. Hence,

> (1) All BPs in w are roughly on a par with respect to informativeness and simplicity.

To see that infinitely many BPs are on a par for fit, our focus on a typical recurrence world becomes important. At typical worlds of this kind, we should expect the world to come to have macrostate M infinitely many times. (It is nomologically possible, but extraordinarily unlikely, for such worlds to have macrostate M only finitely many times). Thus, at typical worlds, there will be infinitely many BPs that are roughly equal in the fit they confer on systems containing them.

We make two assumptions about what a typical recurrence world w is like, concerning the typicality of recurrences that take place within it. The first assumption:

> (2) In w, a typical recurrence r and its corresponding BP p are such that r's surrounding evolution is what is highly likely to occur, conditional on p.

Consider the recurrence of macrostate M at time t. Leading up to and following M-at-t are various events and thermodynamic phenomena, comprising the broad reduction of entropy resulting in M-at-t, and the more-or-less steady entropic increase thereafter. These events comprise the *surrounding evolution* for that recurrence, and so we also say that these events *surround* that recurrence. We can sharpen (2) somewhat. Focus on a certain kind of event, viz. the obtaining of macroconditions which are simple and salient: such as milk's being diffused in coffee and an ice cube's having not yet melted. Then, (2) entails the following:

> (3) In w, a typical recurrence r and its corresponding BP p are such that a large majority of the simple, salient events surrounding r each have a very high SM probability of occurring, conditional on p.

The restriction to simple, salient events here is important, as is allowing typical recurrences to have only a "large majority" of highly likely events surrounding it. That is because arbitrary compounds of simple events have arbitrarily low SM probabilities, conditional on p; and the SM probability (conditional on p) that some unlikely, simple and salient events will infrequently occur is high.[31] A further point: there must be a limit to how temporally distant events can be from M-at-t to be counted as surrounding it. Go too far and events occur which are very unlikely conditional on M-at-t, e.g., earlier and later recurrences of M. For definiteness, say that M-at-t's surrounding evolution occurs during the following interval: conditional on M-at-t, its beginning marks the most likely last point of entropic maximum (equilibrium) earlier than t, and its end marks the most likely first point of entropic maximum later than t. In other words, a recurrence's surrounding evolution takes place during the interval around t of plus or minus its *relaxation time*.[32]

We now turn to our second assumption about typical BPs:

> (4) For each typical recurrence r in w, there are infinitely many others such that they arise and evolve in largely the same way as r.

No doubt there are many highly likely ways for a typical recurrence's surrounding evolution to go. But quite plausibly in typical classical recurrence worlds, for each of these ways, there are infinitely many recurrences that evolve quite similarly. Even unlikely things happen infinitely many times in eternal recurrence worlds, so surely *likely* things happen infinitely many times, too. And similarly-evolving typical recurrences are among the highly likely ways that a surrounding evolution could go.

(4) entails a relevant lemma about test propositions. Recall that in the context of SM, test propositions must include statements stating

---

[31]The focus on simplicity and salience here connects with the BSA framework we have assumed, viz. one uses Elga's "test propositions" to measure fit and the notion of what is salient for creatures like us in measuring simplicity and informativeness. See section 9.2.3.

[32]Recall that we focus on classical worlds that have a bounded phase space, with the result that they have an equilibrium state.

that salient macroscopic events occur. Given this, (4) entails that:

> (5) For each typical recurrence r in w, there are infinitely many others such that they are largely similar to r concerning which of their surrounding events correspond to test propositions.

If infinitely many recurrences are largely similar in their surrounding events, then whichever events are test propositions in one surrounding evolution will largely correspond to test propositions in the others. To extend our usage, call the test propositions corresponding to events surrounding a recurrence r the *surrounding test propositions* for both r and the BP corresponding to it.

Combining (5) with (3) and our definition of fit yields:

> (6) In w, for each BP p corresponding to a typical recurrence, there are infinitely many other BPs such that each of these, p', purchases roughly equal additional fit as p, concerning the surrounding test propositions of p and p', respectively.

In other words, if infinitely many typical recurrences have largely similar sets of test propositions surrounding them, and the BPs corresponding to each assign roughly the same (high) SM probabilities to their respective surrounding events, then those BPs are roughly the same with respect to the fit that each purchases from their respective test propositions.

Of course, (6) alone is not sufficient to show that BPs corresponding to typical recurrences are roughly on a par for fit, since the fit they purchase overall concerns more than the probabilities they assign to their surrounding test propositions. What of their non-surrounding test propositions? These concern events that occur beyond a recurrence r's surrounding evolution, namely, the events occurring beyond the interval of plus-or-minus r's relaxation time. In fact, it is a consequence of SM that no two BPs differ in the probabilities they assign to non-surrounding events. That is because beyond the relaxation time — either forward or backward in time — the SM probabilities, conditional on the recurrence, invariably assign overwhelmingly high probabilities to the equilibrium state. (This is related to SM's prediction that a system at equilibrium stays at equilibrium.) Therefore, any two BPs make essentially the same statistical predictions for events falling outside their respective surrounding evolutions. In other words:

> (7) In w, any two BPs p and p' corresponding to typical recurrences are such that they purchase equal fit with respect to test propositions that surround neither p nor p'.

Moreover, consider any two BPs p and p' that are roughly on a par for the fit they earn from their respective surrounding test propositions. Each will do as poorly as the other for the probabilities they assign to the other's surrounding test propositions. That is, they will each predict that equilibrium occurs during the other's surrounding evolution, rather than the interesting low-entropy events that in fact take place. And so:

> (8) In w, any two typical BPs p and p', which earn the roughly the same fit for their respective surrounding events, are such that they earn equal fit for surrounding events of the other.

Together, (6), (7), and (8) concern the fit that may be earned from all test propositions. Hence, the roughly equal fit earned between p and p' for all of these entails:

> (9) In w, for each BP p corresponding to a typical recurrence, there are infinitely many other BPs such that each, p', purchases roughly equal fit for all particular macroscopic events as p, and so p and p' are roughly equal in fit overall.

Finally, we combine the rough parity of simplicity, informativeness, and fit claimed by (1) and (9) to get:

> (10) For each typical BP p, there are infinitely many others that are roughly equal to p with respect to all three of the BSA's desiderata.

For a world like w — that is, a typical, classical world that (i) begins in a simple, low-entropy macrostate M; and (ii) has a bounded phase space, so that it undergoes infinite recurrence — any BP is such that it is either bested by or is roughly on a par with infinitely many other BPs.

9.3.2. *Three salient possibilities*

Which of these infinitely many BPs is part of a best system for such worlds? We consider three salient answers to this question:

> (I)   Zero BPs are part of a best system for such worlds.
> (II) A finite, positive number of BPs are part of a best system for this world.
> (III) Infinitely many BPs are part of a best system for this world.

We argue that none plausibly leads to there being (meaningful) lawful BPs, given the BSA. Clearly (I) leads straightforwardly to our conclusion. So our arguments in the subsections below focus on (II) and (III). In outline, we argue that (II) results in infinitely many effectively tied-for-best systems, where the different ways BSA could treat such ties each yield our conclusion. Our response to (III), on the other hand, is different: we argue that this putative possibility is untenable for the BSA.

9.3.2.1. *From (II) to effective ties*

Suppose that a best system S includes n BPs, where $n > 0$, and consider an abitrary BP p in S. Given the conclusion of the previous section, there are infinitely many BPs such that each one, p', is roughly on a par with p overall. So consider a system S' which differs from S only in that it contains p' instead of p. Since p' is roughly on a par with p overall, then the same is true of S' and S.[33] And since

---

[33]This assumes that the contribution a BP makes to a system containing it is invariant from system to system, and so viz. that a BP's contribution of fit is invariant. This assumption seems right, for it is plausible that BPs are statistically independent of one another.

on the BSA, systems that are roughly tied are effectively tied — that
is, a uniquely best system must be *robustly* best — it follows that
S is effectively tied with S'. The reasoning generalizes to any BP in
S, and thus to any putative best system containing n many BPs, for
any $n > 0$.

What should the BSA say the laws are, when there are effectively
tied-for-best systems? Recall the two main approaches discussed in
section 9.2.3. According to the first, if there is no robustly best system, then no theorems of any system deserve to be called 'laws', so
there are none. Furthermore, if there are no laws, *a fortiori* there are
no lawful boundary conditions, and so our conclusion follows. Accod-
ing to the second approach, when there are effectively tied-for-best
systems, the laws are the theorems shared between all such systems.
For the latter, there *are* BSA-lawful boundary conditions at w, of a
sort. Since the disjunction of all the BPs found among the tied-for-
best systems will be among the theorems for each tied system, this
disjunction will be a law. Given the conclusion of the previous section, there are infinitely many such BPs, and so the lawful boundary
condition in question here is an infinite disjunction — that the world
is in state M at t_1 *or* state M at t_2 *or* ... for infinitely many ts. This is
an extraordinarily weak constraint — adding such a constraint would
yield chances virtually indistinguishable from those of CSM^-, and
would be far too weak to help with the kinds of reversibility objections raised against CSM^-. Such lawful boundary conditions are
hardly meaningful. So on either standard way of dealing with ties,
our conclusion follows.

9.3.2.2. *Against (III)*

If a best system for w contained any finite number of BPs, our conclusion follows. But what of systems with infinitely many BPs —
might one of those be best? In this section, we argue that this claim
is untenable. In support of that conclusion, we offer three considerations, from specific to general. First, in a classical world, on one
natural way of understanding the probabilities in question, the claim
that a best system has infinitely many BPs is incoherent. Second, a

best system with infinitely many BPs would undermine one of the main motivations for the BSA. And third, such a system's being best violates the very spirit of the BSA.

9.3.2.3. *Incoherence*

If one adopts a standard understanding of the probabilities involved, one can argue that it's incoherent to suppose that a system S with infinitely many BPs is best in a classical world w. The reason is that infinitely many BPs cannot all earn their keep in such a system, and yet they must do so if that system is best.

In more detail, recall that BPs earn their keep in a best system by way of significantly increasing that system's fit, such that the fit they purchase outweighs their cost in simplicity. Otherwise, a system S^- lacking those BPs that do not significantly increase S's fit would be better than S. S^- would forgo the cost in simplicity of adding those BPs, with little sacrifice in fit. This yields the following necessary condition:

> (N1) A system S containing some BPs is best only if each
> BP in it individually and significantly increases the
> fit of S.

Moreover, in a classical world a BP increases fit by increasing the SM probabilities of test propositions. In part, a BP increases SM probabilities by being added to the background propositions K in the SP:

$$Pr_K(A|B) = \frac{m(A \cap B \cap K)}{m(B \cap K)}$$

Adding a BP p to K increases fit when it increases the above ratio, and A is a test proposition. In general, adding p to K reduces both the numerator and denominator in the above ratio — call these the *SP numerator* and *denominator*. To significantly increase the SM probability of $(A|B)$, adding p to background propositions K must reduce the SP denominator by an amount significantly less than the SP numerator. This can only be true if adding p to background propositions K reduces the SP denominator significantly full stop.

We have, then, a necessary condition on a BP's significantly increasing the SM probability of test propositions, and so the fit of a system containing it is as follows:

> (N2) A BP p increases the fit of a system S only if adding
> p to K significantly decreases the SP denominator.

Finally, the argument. Suppose, for reductio, that S — which contains infinitely many BPs — is best in a classical recurrence world w. By (N1), it follows that each such BP significantly increases the fit of S. And by (N2), adding each of these BPs to the background propositions K must significantly decrease the SP denominator. Take any sequence of all the BPs in S, and add them one by one to K. Each addition reduces the SP denominator by a significant amount and, in the limit of adding them all, the SP denominator goes to zero. (Since the reductions in the denominator must be significant, they must be non-vanishing.) Standardly, however, this would mean that the SM probabilities underwritten by S are all undefined, including those assigned to test propositions. Such a system earns no fit; hence, its BPs do not earn their keep, and thus that system cannot be best.

Now there are various ways in which one might resist this argument (by adopting non-standard probabilities, for example), but we won't explore this issue further. For there are more general reasons to think that, given the BSA, an infinitary system like S cannot be best, we turn to these next.

9.3.2.4. *Undermining the motivation and spirit of the BSA*

There are two kinds of infinite systems and one has to distinguish them. We'll call an infinite system *robustly infinite* if there's no salient, intelligible and straightforward way to express it in finitary terms. By contrast, we will call an infinite system *non-robustly infinite* if there is a salient, intelligible and straightforward way to express it in finitary terms.

For example, consider the laws of CM, whose axioms plausibly include something like Newton's Second Law, standardly formulated as $F = ma$. Read literally, this formulation seems to suggest that the determinable property *force* is identical to the product of two other determinables, *mass* and *acceleration*. But that can't be right: a product of two determinables seems incoherent. Rather, the law expresses systematic relations between the determinate properties falling under those determinables. If axioms are to be formulated in terms of determinate properties — particular quantities of mass, acceleration, and force — then an infinitely large family of laws results, describing the relationship between triples of these determinates: $F_1 = m_1 a_1$, $F_2 = m_2 a_2$, $F_3 = m_3 a_3$, etc. We take this to be a non-robustly infinite system, since we can intelligibly and straightforwardly formulate these laws in a finite way, such as by quantifying over the determinate properties and the systematic relationships that hold between them.[34]

By contrast, a system S that includes infinitely many BPs is robustly infinite. A finite axiomatization of such a system would require finitely expressing each of the infinitely many times that M recurs, as specified by S's constituent BPs. There is no intelligible and straightforward way of doing this. And so S must include an infinitely long list of times at which M recurs.

As we are understanding the BSA, if a system is non-robustly infinite, this need not be a significant mark against its simplicity. For there's a salient way of formulating it in a finite manner that we have no trouble understanding. Robustly infinite systems, on the other hand, are deeply problematic — they do extraordinarily poorly with respect to the notion of simplicity relevant to the BSA. Since the system with infinitely many BPs we are considering is robustly infinite, it does poorly with respect to the BSA's criteria.

Moreover, the robustly infinite nature of such a system undermines a main motivation for the BSA, viz. that it mirrors actual scientific practice. Indeed, many have noted that the BSA is a kind of idealization of theory choice in science, using, as an abductive base, not our empirical evidence but all the particular matters of fact over the course of the world's history. Woodward's (2013) characterization

---

[34]Cf. Hawthorne (2006, pp. 236–7), and Eddon (2013, pp. 96 7).

of this motivation is a good example:

> "A substantial part of the appeal of the BSA is that it is
> supposed to correspond (in a very idealized form) to how
> abductive inference and theory choice in science work —
> the HSB [the Humean Supervenience Base] represents the
> most extensive body of inductive evidence we could possi-
> bly possess, and (it is contended) simplicity and strength
> are the criteria scientists actually employ in choosing the-
> ories and laws on the basis of this evidence."[35]

The idea that the standards of simplicity and informativeness used
in the BSA are those actually employed in science should be taken
seriously.[36] Then, observe our straightforward rejection of proposals
of laws that are infinitely complex. For instance, a revision of the
standard model that included an infinite variety of fundamental particles (and an infinite variety that couldn't be described in a salient,
intelligible and straightforward way) wouldn't be seriously considered. If the world had an infinite variety of particles, some finite
systematization of them would be called for. And if no such systematization were possible, then such a world might well be deemed too
chaotic to be lawful.

Our rejection of robustly infinite accounts of laws suggests that
our standards of simplicity and informativeness do not allow such systems to be best; that is, their cost in simplicity is insurmountable.
If that is indeed true then any version of the BSA which allowed a
robustly infinite system to be best would not accord with the standards of simplicity and informativeness employed by our scientific
community, and would fail to line up with actual scientific practice.

The claim that a robustly infinite system like S is best clashes with
the BSA in another way, for it violates the spirit of that view. In an
early description of it, Lewis likens the BSA to "a *Concise Encyclopedia of Unified Science*" provided to humankind by God.[37] Concision
is required, of course; God's List of All Truths would be useless for

---

[35]Woodward (2013, p. 49).
[36]See Eddon and Meacham (2015, sections 3.5 and 3.6.)
[37]Lewis (1973, p. 74).

finite creatures like us. But for the same reason that God's List would be too complex for us to grasp, the same is true for a robustly infinite system like S. Supposing S to be best, then, would be to forsake this core idea of the BSA: that it provides the best systematization of truths about the world that is intelligible for creatures like us.

9.4. The cosmological case

In this section, we consider typical eternal inflation worlds, w', of the sort described in section 9.2.2: w' begins in a largely homogeneous macrostate M of high inflationary energy, and it goes on to evolve in a way that M infinitely recurs.[38] Analogous to the previous section, we argue that this recurrence vitiates any (meaningful) lawful boundary conditions at w', given the BSA. Like before, we will restrict ourselves to what are plausibly the most eligible BPs: the PH and infinitely many MHs. And by 'recurrences', we will narrowly mean recurrences of M.

---

[38] It is worth noting that our argument does not actually require that w' begin in the highly uniform state M. That is because, even given the other viable initial conditions that w' might have, the resulting BPs are not more eligible than MHs stating that M holds at later times. To see this, first note that the initial conditions at w' cannot be too inhomogeneous. (See section 9.2.2 for discussion of this point.) And the effects of not too large initial inhomogeneities will be swamped by the ensuing spatial expansion and subsequent thermalization. So a BP's specifying those initial inhomogeneities — and paying the corresponding cost in simplicity — purchases only small increases in fit. At best then, those costs and benefits balance out, meaning a BP stating that a less homogeneous initial macrocondition holds in w' is still roughly on a par with BPs stating that M holds at w' at later times. The possibility thus does not affect our argument, and so to simplify we assume in the main text that w' begins in M. Similar remarks apply to inflationary cosmological models that aim to explain later local instances of M without positing an initial, global instance of high inflationary energy. An example is the model proposed by Carroll and Chen (2004). On that model, instances of M eventually arise via extremely unlikely quantum fluctuations from an otherwise largely static vacuum state, VAC. But a BP stating that VAC holds at a time surely earns no more in fit compared to BPs stating when and where M occurs, nor does it seem simpler to express. Hence the VAC BP seems at best no more eligible to be part of the best system than do the MHs stating M occurs. Thanks to an anonymous referee for raising the question of models like Carroll and Chen's.

It is again plausible that each BP incurs roughly the same cost in simplicity. One notable difference concerns the fact that M's initial occurrence is global whereas its later occurrences are merely local. This means the PH and MHs do differ in complexity: the former is surely simpler, since it is easier to state (where does M occur? Everywhere!). But this difference is small. The main simplicity cost comes with having to specify a local region at all. But specifying different local regions requires little change in complexity. Different choices of coordinate systems make it arbitrarily easier (or harder) to specify different local regions, and simplicity ought to be invariant between such choices. But the mere difference between having to specify a local region and not having to do so seems to involve no great jump in complexity. If so, then BPs stating that M occurs locally are close enough to be roughly equal in simplicity to the PH. So it is plausible that:

> (1′) In w', each BP is roughly on a par with every other with respect to simplicity.

Turning to informativeness, it is important to observe a certain ambivalence in our judgments about which BPs are more or less informative. On the one hand, all BPs involve the same phenomenon occurring in regions of the same size, suggesting that they are equally informative. On the other hand, cosmic expansion means later recurrences are situated in larger spatial "arenas", so to speak, suggesting that later BPs tell us less about what's going on therein. Consider an analogy: You have a list of the books in your library's science fiction collection. The library then expands, adding new books but no new Sci-Fi titles. Post expansion, your list is just as accurate as it was, suggesting it is equally informative. But your list also now tells you less about the library's collection overall, suggesting it is less informative. One might call these *positive* and *negative* conceptions of informativeness: the former gauges a statement's informativeness by what it tells you; the latter gauges informativeness by how much it hasn't told you.

Both conceptions are intuitively compelling. Yet, if informativeness is univocal — an implicit assumption of the BSA — they are incompatible. In what follows, we will assume the former "positive" conception. This conception of informativeness yields the following:

> (2′) In w', each BP is roughly on a par for informativeness.

Like before, we unpack the typicality of w' by way of specific assumptions about what typical recurrences are like in such a world. The first:

> (3′) In w', each typical recurrence r is such that its subsequent evolution is highly likely, conditional on the BP corresponding to r.

In the context of SM, we focussed on a recurrence's surrounding evolution; that is, the macroscopic events occurring before and after it, within a certain interval. For eternal inflation, we focus instead on a recurrence's *subsequent evolution*, namely, the later events over which the recurrence has causal influence. In other words, these are the events in the recurrence's future light cone.[39] Whatever chances the correct theory of eternal inflation trades in, typical recurrences are such that their forward light cones contain events that are highly likely (for chances of that theory), conditional on the relevant BP.

The second assumption:

> (4′) In w', each typical recurrence r is such that there are infinitely many other recurrences whose subsequent evolutions develop in largely similar ways to r.

Like before, (4′) plausibly entails a relevant lemma concerning test propositions:

> (5′) In w', each typical recurrence r is such that infinitely many other recurrences have subsequent evolutions which include roughly similar constellations of test propositions to r.

---

[39] Note that, given exponential expansion ensuing from that recurrence, the size of its light cone also grows exponentially.

The following is a consequence of (5′) and (3′), plus our conception of fit:

> (6′) In w' and for each typical BP p, there are infinitely many other BPs p' such that p and p' earn roughly as much fit for the events in their respective subsequent evolutions.

What does (6′) tell us overall about the fit between such BPs? To better appreciate the fit a BP earns, it is helpful to return to our simplified diagram; Figure 9.3 illustrates the fit earned by the PH.

Recall that the chance of high inflationary energy in a region not decaying is low. Hence, conditional on the PH, a low chance is assigned to there being high inflationary energy in the region R where M next recurs. Likewise, the events comprising the subsequent evolution of M-in-R are also assigned lower chances conditional on PH. And the same is true for each next expanding region in the infinite sequence of such expansions. The result: the PH earns less and less fit for events in subsequent regions of expansion.

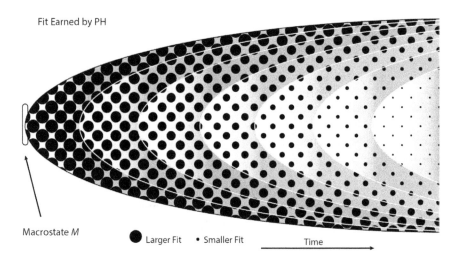

Fig. 9.3. Fit Earned by PH

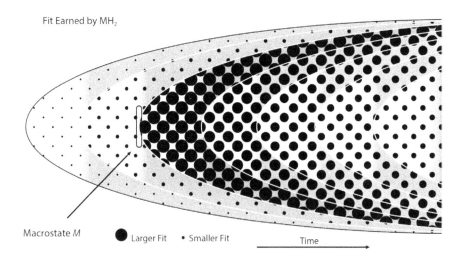

Fit Earned by MH₂

Macrostate *M* ● Larger Fit • Smaller Fit Time

Fig. 9.4. Fit Earned by MH₂

The subsequent evolution of the PH comprises all the events of w', and so all opportunities for the PH to earn fit. Now, recall what (6′) says — that typical BPs earn roughly the same fit for events in their respective subsequent evolutions. Thus, the question of how the MHs stand fit-wise to the PH (and to each other) comes down to what fit they earn from events that aren't in its subsequent evolution, i.e., events outside their future light cones. This, of course, depends on what probabilities a MH underwrites for those events. It is plausible that these events have some low but non-zero probability, conditional on the relevant MH.[40]Figure 9.4 is such a possible example.

The smaller the probabilities a MH underwrites for events outside its future light cone, the less additional fit it earns from those events, and the less difference this extra fit makes to a MH's fit overall. Assuming that these probabilities are small enough, and given (6′),

---

[40]Some versions of quantum mechanics rule this out. In particular, GRW theory has the consequence that later BPs underwrite no probabilities to events lying outside their future light cones; see Albert (2000, chapter 7). In that case, (6′) alone secures the result that the relevant BPs are roughly on a par for fit.

we have the following:[41]

> (7′) In w' and for each typical BP p, there are infinitely
> many other BPs that are roughly equal in fit to p,
> overall.

Finally, combining (7′) with (1′) and (2′) results in:

> (8′) In w' and for each typical BP p, there are infinitely
> many other BPs that are roughly equal to p with
> respect to simplicity, informativeness and fit.

Given (8′), there are infinitely many systems that are effectively tied
for best in w: for any system S with finitely many BPs, there are
infinitely many other BPs that could be traded out to produce a
system S' that is roughly just as good. (Recall from section 9.3.2.2
that it is untenable that a best system have infinitely many BPs.)
And as argued in section 9.3.2.1, on either way that the BSA might
approach ties, the result is that no (meaningful) boundary conditions
are lawful.

---

[41]We assume this for simplicity (and because we take this to be plausible), but
our conclusion follows regardless of the magnitudes of these probabilities. There
are a few salient possibilities.

First, the probabilities a MH underwrites for events outside its future light
cone might decrease the further they are (spatiotemporally) from the recurrence
corresponding to the MH, at a rate fast enough to yield a limit on how much extra
fit a MH earns beyond the fit it derives from events in its subsequent evolution.
(Figure 9.4 represents this.) (1) If this limit is low, then although subsequent MHs
have higher and higher fit, the differences are relatively small, and all of the MHs
(and the PH) will be roughly on a par with respect to fit. (This is the possibility
assumed in the text.) (2) Alternatively, if this limit is high, then earlier MHs (and
the PH) may fall far enough behind with respect to fit to be eliminated as viable
candidates for a best system. But there will still be infinitely many later MHs
that are roughly on a par for fit, which is all we need to derive our conclusion.

(3) A different possibility is that the probabilities a MH underwrites for events
outside its future light cone do not decrease, or don't decrease fast enough to
yield a limit on how much extra fit a MH earns beyond the fit it derives from its
subsequent evolution. If so, then for any BP, there is another BP that is arbitrarily
better with respect to fit. This will yield an infinite sequence of better and better
systems — since one can always trade worse BPs for better ones — so there will
be no system that is best, and thus no BSA laws. A fortiriori, there will be no
lawful constraints on boundary conditions, and our conclusion follows.

The foregoing argument is somewhat speculative. There's no established view on how one should conceive of informativeness on the BSA, nor is there a consensus with respect to how to conceive of fit. And, of course, the cosmological theory itself is somewhat speculative. But the assumptions made above are plausible, and we argue that the conclusion inherits this plausibility. If ours is a typical world of eternal inflation, and if the BSA is true, we have good reason to think that there are no (meaningful) lawful boundary conditions.

9.5. Skeptical arguments

In section 9.2.1, we noted that it's standardly held that CSM⁻ (the standard classical statistical mechanical laws minus the Past Hypothesis) is self-undermining — the chances it assigns seem to rationally require subjects like us, with evidence like ours, to believe that CSM⁻ doesn't hold. This, in turn, suggests that subjects like us, with evidence like ours, should believe that CSM⁻ is false. Thus, only theories like CSM, which posit lawful boundary conditions, should be serious contenders for belief.

In a similar vein, it's been suggested that the corresponding versions of cosmological theories which don't posit any lawful constraints on boundary conditions (COS⁻) also seem to be self-undermining.[42] This again suggests that subjects like us, with evidence like ours, should believe COS⁻ to be false, and that only cosmological theories like COS, which posit lawful boundary conditions, should be serious contenders for belief.

Combined with the results from sections 9.3 and 9.4, this suggests trouble for the BSA. That is because that the BSA entails that, at typical eternal recurrence worlds, there won't be any (or only very weak) lawful constraints on boundary conditions. And if subjects like us shouldn't believe such theories, then the BSA leads to a skeptical problem: subjects like us at such worlds shouldn't believe that the laws are what they actually are.

---

[42]See Carroll (forthcoming b).

In this section, we will show that this conclusion is mistaken. Starting with the classical case (in section 9.5.1), we will spell out in detail the standard argument for the claim that subjects like us should not believe CSM^-. Doing so will allow us to see (in section 9.5.2) that the standard argument is too quick — some of its premises become implausible when we consider eternal recurrence worlds. Focusing on eternal recurrence worlds, we will then show (in section 9.5.3) that given certain assumptions, one can repair the argument against believing CSM^-. But we will also show that these assumptions allow one to construct a similar argument against believing CSM. Thus, at typical classical eternal recurrence worlds — the worlds where the BSA yields CSM^- — CSM^- laws are no more subjected to skeptical worries than CSM laws.

We will then turn to theories of eternal inflation, i.e., COS and COS^-. Here, we will show (in section 9.5.4) that a similar dialectic obtains. In particular, we will show how given certain assumptions eternal inflation theories, both with and without lawful boundary conditions, are subject to skeptical worries to the effect that we should not believe such theories are true. And we will show how given a different set of assumptions, eternal inflation theories, both with and without lawful boundary conditions, are not subject to skeptical worries. Thus again we find that at typical eternal inflation worlds — the worlds where the BSA yields COS^- — COS^- laws are no more subject to skeptical worries than COS laws.

9.5.1. *The Argument Against* CSM^-

In this section we will spell out the standard argument for why subjects like us shouldn't believe CSM^-. To simplify things, we will start by introducing some notation.

Let E be our total evidence. It corresponds to the set of centered worlds that are centered on possible individuals who have the same perceptual experiences and memories as we do. Then, $\hat{E}$ corresponds to the set of centered worlds located at worlds at which there exists an individual who has the same perceptual experiences and memories as we do.

Let V be the proposition that we have total evidence E, and that our evidence about the past is largely veridical, that is, not extraordinarily misleading. Then, $\hat{V}$ corresponds to the set of centered worlds located at a world at which there exists an individual who has our total evidence E and whose evidence about the past is largely veridical.

Consider the CSM$^-$ worlds compatible with $\hat{E}$. Let the *canonical partition*$_{\hat{E}}$ of these $\hat{E} \wedge \text{CSM}^-$ worlds be the partition of these worlds into the coarsest background propositions $\hat{E}_i$ such that $\hat{E}_i$ fixes the non-dynamical properties of the world (e.g., the number of particles, the spatiotemporal extension, etc.). Therefore, the canonical partition$_{\hat{E}}$ is the coarsest way of carving up the $\hat{E} \wedge \text{CSM}^-$ worlds into propositions that yield well-defined chance assignments to the dynamical properties of the system (the particle positions and velocities).

The standard argument for the conclusion that subjects with priors and evidence like ours should believe CSM$^-$ is false requires three premises.[43]

The first premise of the argument is that for evidence E like ours, every element $\hat{E}_i$ of the canonical partition$_{\hat{E}}$ will be such that $ch_{\text{CSM}^-,\hat{E}_i}(\hat{V}) \approx 0$. That is, given any way things might be such that there exists a subject with our evidence, the chance, according to CSM$^-$, of there existing a subject with our evidence, whose evidence about the past is largely veridical, is approximately 0. This seems plausible because according to CSM$^-$ it seems much more likely that an individual with our evidence is the recent result of a spontaneous fluctuation from a higher entropy state — and so their evidence about the past is extraordinarily misleading — than it is that they came from the kind of low entropy past which would make their evidence about the past largely veridical.[44]

---

[43]See Meacham (forthcoming) for a (differently formatted) version of this argument.

[44]Of course, there will be some special backgrounds K for which the CSM$^-$ chance of $\hat{V}$ is reasonably high — e.g., Ks that specify that everything evolved from a (non-lawfully required) very low-entropy initial condition. But given evidence E like ours, such Ks won't be members of the canonical partition$_{\hat{E}}$.

The second premise of the argument, usually left implicit, is that for subjects with priors and evidence like ours, $cr_{\hat{E}}(\hat{V} \mid \text{CSM}^-) \approx cr_E(V \mid \text{CSM}^-)$. Roughly, the idea is that for subjects with priors like ours, changing our evidence from E to $\hat{E}$ and the object of our credence from V to $\hat{V}$ shouldn't really change what our credences are. In other words, whether we work with irreducibly *de se* propositions or their *de dicto* counterparts shouldn't really bear on our credences regarding subjects having veridical evidence like ours.

The third premise of the argument is that for subjects with priors and evidence like ours, $cr_E(\text{CSM}^- \mid \neg V) \approx 0$. That is, for subjects with priors and evidence like ours, our credence in CSM^-, on our evidence and the past being extraordinarily misleading should be very low. This is because our only reasons for believing that, say, something like classical mechanics holds is that our evidence suggests strong agreement between what classical mechanics tells us and how the world has behaved so far. And if our evidence about the past turned out to be extraordinarily misleading, then we'd lose our reason for thinking there is such agreement, and thus lose our reason for believing classical mechanics holds.

Given these premises, the standard argument against believing CSM^- goes as follows:[45]

The Argument Against CSM⁻:

P1. For evidence E like ours, every member $\hat{E}_i$ of canonical partition$_{\hat{E}}$ will be such that $ch_{\text{CSM}^-, \hat{E}_i}(\hat{V}) \approx 0$.

P2. For subjects with priors and evidence like ours, $cr_{\hat{E}}(\hat{V} \mid \text{CSM}^-) \approx cr_E(V \mid \text{CSM}^-)$.

P3. For subjects with priors and evidence like ours, $cr_E(\text{CSM}^- \mid \neg V) \approx 0$.

---

[45]To get L1, note that for evidence E like ours, P1 and the Principal Principle entail that rational subjects will be such that, for every $\hat{E}_i$ in the canonical partition$_{\hat{E}}$, $ic(\hat{V} \mid \hat{E}_i \wedge \text{CSM}^-) \approx 0$. Since these $\hat{E}_i$s form a partition of $\hat{E} \wedge \text{CSM}^-$, the probability axioms entail that $ic(\hat{V} \mid \hat{E} \wedge \text{CSM}^-) \approx 0$; that and IC-Conditionalization then entail that $cr_{\hat{E}}(\hat{V} \mid \text{CSM}^-) \approx 0$.

To get L2, note that L1 and P2 entail that, for rational subjects with priors and evidence like ours, $cr_E(V \mid \text{CSM}^-) \approx 0$.

L1. For rational subjects with priors and evidence E like ours, $cr_{\hat{E}}(\hat{V} \mid \text{CSM}^-) \approx 0$. [From P1, the Principal Principle and IC-Conditionalization.]

L2. For rational subjects with priors and evidence like ours, $cr_E(V \mid \text{CSM}^-) \approx 0$. [From P2, L1.]

L3. For rational subjects with priors and evidence like ours, $cr_E(V \wedge \text{CSM}^-) \approx 0$. [From L2.]

L4. For subjects with priors and evidence like ours, $cr_E(\neg V \wedge \text{CSM}^-) \approx 0$. [From P3.]

C. For rational subjects with priors and evidence like ours, $cr_E(\text{CSM}^-) = cr_E(V \wedge \text{CSM}^-) + cr_E(\neg V \wedge \text{CSM}^-) \approx 0$. [From L3, L4.]

Note that a similar argument doesn't seem to work against CSM, because the analog of P1 is implausible. The reason being that unlike CSM$^-$, the CSM chance of there being an E-having subject with veridical evidence, given that there exists an E-having subject, is reasonably high.

9.5.2. *The Argument Against* CSM$^-$ *and Eternal Recurrence Worlds*

The argument presented in section 9.5.1 is valid. However, once we take the possibility of eternal recurrence worlds into account, worries arise regarding whether it's sound.

To get L3, note that L2 entails that $cr_E(V \wedge \text{CSM}^-) \approx 0$, since it can only be the case that $cr_E(V \mid \text{CSM}^-) \approx 0$ if the numerator of the conditional probability is much smaller than the denominator, which entails that the numerator of the conditional probability is very small.

To get L4, note that P3 entails that $cr_E(\text{CSM}^- \wedge \neg V) \approx 0$, since it can only be the case that $cr_E(\text{CSM}^- \mid \neg V) \approx 0$ if the numerator of the conditional probability is much smaller than the denominator, which entails that the numerator of the conditional probability is very small.

To get C, note that the probability axioms entail that $cr_E(\text{CSM}^-) = cr_E(V \wedge \text{CSM}^-) + cr_E(\neg V \wedge \text{CSM}^-)$. L3 entails that one of these terms is approximately zero, and L4 entails that the other term is also approximately zero. Thus, for rational subjects with priors and evidence like ours, $cr_E(\text{CSM}^-) \approx 0$, which gives us our conclusion.

Consider P1, the claim that for evidence E like ours, every element $\hat{E}_i$ of the canonical partition$_{\hat{E}}$ is such that $ch_{\text{CSM}^-,\hat{E}_i}(\hat{V}) \approx 0$. Once we notice that there will be elements of the canonical partition$_{\hat{E}}$ that pick out eternal recurrence worlds, this claim seems false, since there will be some $\hat{E}_i$ such that $ch_{\text{CSM}^-,\hat{E}_i}(\hat{V}) \not\approx 0$. To see this, consider an $\hat{E}_i$ which specifies that the temporal extension of the world is infinite, and consider the chance of such a world coalescing into the kind of very low entropy macrostate that the Past Hypothesis refers to. This will be extraordinarily unlikely to happen in any time scale we are familiar with, but given enough time, the chance goes to 1. Likewise, the chance of the world coalescing into that low-entropy state and then evolving (over the next several billion years) to give rise to a subject with evidence E that's largely veridical about the past, is extraordinarily low over any time scale we are familiar with. But again, given enough time, the chance goes to 1. Since $\hat{V}$ is the proposition that there exists an E-having subject whose evidence about the past is largely veridical, this entails that $ch_{\text{CSM}^-,\hat{E}_i}(\hat{V}) \approx 1$, not 0. Thus, we have a counterexample to P1.

Likewise, consider P2, the claim that for subjects with priors and evidence like ours, $cr_{\hat{E}}(\hat{V} \mid \text{CSM}^-) \approx cr_E(V \mid \text{CSM}^-)$. But once we take eternal recurrence worlds into consideration, this claim is implausible. To see this, consider a subject whose priors assign most of their credence in CSM^- to an element $\hat{E}_i$ of the canonical partition$_{\hat{E}}$ which entails that the world eternally recurs.[46] As we have just seen, for such $\hat{E}_i$, $ch_{\text{CSM}^-,\hat{E}_i}(\hat{V}) \approx 1$. It follows from the Principal Principle that $ic(\hat{V} \mid \text{CSM}^- \wedge \hat{E}_i) \approx 1$, and thus (given our stipulation) that $ic(\hat{V} \mid \text{CSM}^- \wedge \hat{E}) \approx 1$. Given IC-Conditionalization, it then follows that $cr_{\hat{E}}(\hat{V} \mid \text{CSM}^-) \approx 1$.[47] But while holding such beliefs seems plausible for such a subject (since at

---

[46]This assumption simplifies the argument, but is stronger than required. All that is needed is the assumption that a subject with priors like ours might assign a non-trivial amount of their credence in CSM^- to element(s) $\hat{E}_i$ that entail the world eternally recurs.

[47]Assuming, of course, that the subject is approximately rational, at least in these respects.

a CSM$^-$ world of eternal recurrence, the chance of *some* subject with evidence E having veridical evidence about the past is ≈ 1), holding that $cr_E(V \mid \text{CSM}^-) \approx 1$ (as P2 would require) does not. After all, that would entail that, conditional on CSM$^-$ obtaining, they're virtually certain that their evidence is largely veridical, despite being confident that there are infinitely many subjects with the same evidence for whom this evidence is extraordinarily misleading! Thus, at least sometimes, subjects with priors and evidence like ours are such that $cr_{\hat{E}}(\hat{V} \mid \text{CSM}^-) \napprox cr_E(V \mid \text{CSM}^-)$.

Here's another way to see why P2 is implausible. Suppose you know you're in a CSM$^-$ world, that you have evidence E, and that there are 10^{100} E-having subjects. P2 requires your credence that *someone* has veridical evidence E ($\hat{V}$) to be the same as your credence that *you* have veridical evidence E (V). And your credence that someone has veridical evidence E will be the same (≈ 1) whether you know that only one of the 10^{100} E-having subjects has veridical evidence or whether you know that all of the 10^{100} E-having subjects have veridical evidence. But many subjects like us would not be virtually certain that their evidence E is veridical if they knew that only one of the 10^{100} subjects with evidence E is situated such that their evidence is veridical.

9.5.3. *Repairing the Argument Against* CSM$^-$

Once we take eternal recurrence worlds into consideration, the argument for disbelieving CSM$^-$ no longer goes through. But it's natural to think that one could repair the argument against CSM$^-$ so that it does work at eternal recurrence worlds. Let's look at how one might do this.

Since our primary concern is eternal recurrence worlds, we can simplify the dialectic by restricting our attention to typical eternal recurrence worlds. Let "CSM$^-_\infty$" stand for the conjunction of CSM$^-$ and the claim that the world eternally recurs. Similarly, let "CSM$_\infty$" stand for the conjunction of CSM and the claim that the world eternally recurs. In what follows, we will focus our attention on whether one can construct skeptical arguments against CSM$^-_\infty$ and CSM$_\infty$.

In the Argument Against CSM^-, P1 and P2 serve the role of allowing us to derive L2: that $cr_E(V \mid \text{CSM}^-) \approx 0$. To avoid the worries facing P1 and P2, we need to provide a rationale for the eternal recurrence version of L2 that does not rely on those suspect premises. It follows from the probability axioms that:[48]

$$cr_E(V \mid \text{CSM}_\infty^-) = \sum_{w \in \text{CSM}_\infty^-} \frac{cr_E(w)cr_E(V \mid w)}{cr_E(\text{CSM}_\infty^-)}$$

One natural thought is that for subjects like us, $cr_E(V \mid w)$ will be generally equal to the proportion of E-having subjects at w whose evidence is veridical. After all, if there are two E-having subjects at a world, and that evidence is veridical for one of them, then adopting a credence that your evidence is veridical (given w) that's greater than $1/2$ might seem overly optimistic, and a credence lower than $1/2$ unduly pessimistic. Call this assumption about the priors of subjects like us the Indifference Assumption.[49] At typical CSM_∞^- worlds, the vast majority of E-having subjects are situated such that their evidence about the past is extraordinarily misleading, since it's much more likely (according to CSM_∞^-) for a subject with E to have fluctuated into existence from a higher entropy state than it is for them to have correctly remembered that they came from an even lower entropy state. Thus, given the Indifference Assumption, and the assumption that for subjects like us most of the credence in CSM_∞^- is assigned to CSM_∞^- worlds that are typical, it follows that one's credence of one's evidence about the past is largely veridical given CSM_∞^- should be very low; i.e., $cr_E(V \mid \text{CSM}_\infty^-) \approx 0$.

However, this argument runs into difficulties when one realizes the infinite numbers involved. This argument relies on the intuitively plausible claim that the vast majority of E-having subjects don't

---

[48]For simplicity, we assume here that there are countably many CSM_∞^- worlds. One can drop this assumption in the usual way.

[49]This constraint is entailed by the restricted Indifference Principle proposed by Elga (2004a). Elga's Indifference Principle is contentious (see Weatherson (2005)). But the argument under consideration doesn't need to endorse Elga's Indifference Principle, it merely needs to assume that subjects like us generally have priors that line up with the prescriptions that Elga's Indifference Principle makes.

have veridical evidence about the past. And this claim is hard to make sense of at typical CSM_∞^- worlds, since such worlds will contain both infinitely many E-having subjects with veridical evidence and infinitely many E-having subjects with non-veridical evidence. (This is an instance of what cosmologists call "the measure problem".)[50]

In order to circumvent these difficulties, we need a more sophisticated way of measuring ratios between E-having subjects with and without veridical evidence about the past. Here is a schema for how to provide such an account.[51] Specify a way of picking out a spatiotemporal region at a world containing a finite number of centered worlds that are subjectively indistinguishable from your own. Assess the proportion of these centered worlds at which A is true. Then, specify a way of sequentially expanding this region, and assess the proportion of centered worlds at which A is true for these larger and larger regions. If the proportion of centered worlds at which A is true converges to x in the limit, then (according to the account) x is the correct proportion of these centered worlds at which A is true.[52]

Of course, this is just a schema, and there are a number of ways to fill in the details. But for a wide range of plausible ways of filling in these details, the proportion of E-having subjects with veridical evidence will converge to ≈ 0. And if subjects like us generally have priors corresponding to such a proposal, and most of our credence in CSM_∞^- worlds is assigned to typical CSM_∞^- worlds, then it follows that subjects like us should generally be such that $cr_E(V \mid \mathrm{CSM}_\infty^-) \approx 0$.[53]

---

[50]See Carroll (forthcoming b), and the references therein.

[51]See Arntzenius & Dorr (2017), and Carroll (forthcoming b).

[52]The relevant notion of "correct" here is correct with respect to the Indifference Assumption; i.e., a correct description of the priors of subjects like us.

[53]It is perhaps worth flagging that, as we note in section 9.2.4, we are assuming a somewhat permissivist account of rational priors here. Thus, the justification for holding that subjects like us will have priors that yield $cr_E(V \mid \mathrm{CSM}_\infty^-) \approx 0$ is simply that it's plausible that subjects like us have priors which roughly satisfy the condition we have described. And if someone had rational priors that didn't satisfy these conditions, then this skeptical argument regarding CSM_∞^- wouldn't apply to them. (We thank a referee for encouraging us to address this question.)

Thus, given the Indifference Assumption and a plausible way of applying it to the infinite case, we can modify the Argument Against CSM$^-$ so that it works, given eternal recurrence worlds, as follows:

The Modified Argument Against CSM$_\infty^-$:

P1. For rational subjects with priors and evidence like ours, $cr_E(V \mid \text{CSM}_\infty^-) \approx 0$.

P2. For rational subjects with priors and evidence like ours, $cr_E(\text{CSM}_\infty^- \mid \neg V) \approx 0$.

L1. For rational subjects with priors and evidence like ours, $cr_E(V \wedge \text{CSM}_\infty^-) \approx 0$. [From P1.]

L2. For rational subjects with priors and evidence like ours, $cr_E(\neg V \wedge \text{CSM}_\infty^-) \approx 0$. [From P2.]

  C. For rational subjects with priors and evidence like ours, $cr_E(\text{CSM}_\infty^-) = cr_E(V \wedge \text{CSM}_\infty^-) + cr_E(\neg V \wedge \text{CSM}_\infty^-) \approx 0$. [From L1, L2.]

The conclusion of this argument is that rational subjects with priors and evidence like ours should be virtually certain that CSM$_\infty^-$ is false. This leaves open the question of whether the same is true for CSM$_\infty$. Can we construct a similar argument against believing CSM$_\infty$?

Let's consider how the argument looks if we replace CSM$_\infty^-$ with CSM$_\infty$ throughout. The argument will remain valid, so the only question is whether it's sound.

The CSM$_\infty$ version of P2 (roughly, that for subjects like us, $cr_E(\text{CSM}_\infty \mid \neg V) \approx 0$) is just as plausible as the CSM$_\infty^-$ version. For in both cases, it's *prima facie* plausible that if our evidence about the past is largely misleading, then our credence in the kinds of theories suggested by this evidence — e.g., theories to the effect that the world behaves according to the laws of classical mechanics — would drop.

The CSM$_\infty$ version of P1 (roughly, that for subjects like us, $cr_E(V \mid \text{CSM}_\infty) \approx 0$) is also plausible. Let Δ_r be the interval of time that starts at the initial state, and ends when the world first reaches equilibrium. It's true that at CSM$_\infty$ worlds, the ratio of E-having subjects with veridical evidence will generally be greater than at CSM_∞^- worlds during Δ_r. But in intervals of time following Δ_r, the expected proportion of E-having subjects with veridical evidence will be exactly the same at CSM_∞ and CSM_∞^- worlds — namely, very small. And as we take more and more of the time following Δ_r into account, the impact of the post-Δ_r period will continue to increase, eventually swamping the contributions of the Δ_r period. Thus, in the limit, the expected proportion of E-having subjects with veridical evidence will be the same at CSM_∞ worlds and CSM_∞^- worlds. And so, the claim that $cr_E(V \mid \mathrm{CSM}_\infty) \approx 0$ is just as plausible as the claim that $cr_E(V \mid \mathrm{CSM}_\infty^-) \approx 0.$[54]

So, once we restrict our attention to eternal recurrence worlds, we find that this kind of skeptical argument can be raised against both CSM_∞^- and CSM_∞. In either case, given certain assumptions (i.e., the Indifference Assumption, a plausible way of applying it to the infinite case, and the assumption that most of our credence in CSM_∞^- is assigned to typical CSM_∞^- worlds), one can argue that subjects like us should be virtually certain these theories are false. Thus, at typical classical eternal recurrence worlds — the worlds where the BSA yields CSM^- laws — CSM^- laws are no more subject to skeptical worries than CSM laws. And thus, the fact that the BSA yields CSM^- laws at such worlds is clearly not a mark against it.

9.5.4. *The Argument Against* COS_∞^-

Let "COS_∞^-" stand for the conjunction of COS^- and the claim that the world eternally recurs, and let "COS_∞" stand for the conjunction of COS and the claim that the world eternally recurs. In the previous section, we looked at arguments for why subjects like us shouldn't believe CSM_∞^-. And we assessed whether such arguments give us a reason to favor CSM_∞ over CSM_∞^-. In this section, we will consider

---

[54] Again, we are assuming here that it's plausible that subjects like us have priors which roughly satisfy the condition we have described (cf. footnote 53). For subjects with priors that don't satisfy these conditions, this argument isn't rationally compelling. (We thank a referee for encouraging us to address this question.)

arguments for why subjects like us shouldn't believe COS_∞^-, and whether such arguments give us a reason to favor COS_∞ over COS_∞^-.

In section 9.5.3, we presented a valid argument for the conclusion that subjects like us shouldn't believe CSM_∞^-. By replacing CSM_∞^- with COS_∞^-, we can provide a similar valid argument for the conclusion that subjects like us shouldn't believe COS_∞^-:

The Argument Against COS_∞^-:

P1. For rational subjects with priors and evidence like ours, $cr_E(V \mid \text{COS}_\infty^-) \approx 0$.

P2. For rational subjects with priors and evidence like ours, $cr_E(\text{COS}_\infty^- \mid \neg V) \approx 0$.

L1. For rational subjects with priors and evidence like ours, $cr_E(V \wedge \text{COS}_\infty^-) \approx 0$. [From P1.]

L2. For rational subjects with priors and evidence like ours, $cr_E(\neg V \wedge \text{COS}_\infty^-) \approx 0$. [From P2.]

  C. For rational subjects with priors and evidence like ours, $cr_E(\text{COS}_\infty^-) = cr_E(V \wedge \text{COS}_\infty^-) + cr_E(\neg V \wedge \text{COS}_\infty^-) \approx 0$. [From L1, L2.]

The second premise of this argument is plausible for the same reasons as before. If our evidence about the past is extraordinarily misleading, then we had have little reason to be confident that something like COS_∞^- — an account largely motivated by past empirical evidence — is true. So the soundness of this argument hangs on the first premise. Is it plausible that rational subjects like us will be virtually certain that our evidence isn't veridical given COS_∞^-?

The typical COS_∞^- worlds will be eternal inflation worlds with both infinitely many E-having subjects whose evidence is veridical and infinitely many E-having subjects whose evidence is not veridical. So if one assumes that our credence that our evidence is veridical generally mirrors the expected proportion of E-having subjects whose evidence is veridical — that is, if one adopts the Indifference Assumption — then one will run into the same kinds of comparing infinities worries that came up in section 9.5.3.

We can address this problem in the same way as before: by adopting an account of how to correctly measure proportions in these infinite cases, and using that to apply the Indifference Assumption. For eternally recurring CSM$^-$ worlds, it seemed like most reasonable ways of measuring proportions yield the result that vanishingly few E-having subjects are positioned so that their evidence is veridical. The same is not true at eternal inflation worlds, since the fractal structure of typical eternal inflation worlds allows for different reasonable measures to yield wildly diverging results.

For example, some reasonable ways of measuring proportions will yield the result that the vast majority of E-having subjects were spontaneously created near the beginning of the local big bang of some emerging bubble universe, and so they have evidence about the past that's extraordinarily misleading. Other reasonable ways of measuring proportions will yield the result that the vast majority of E-having subjects were spontaneously created in empty space at times long after the local big bang, and have evidence about the past that's extraordinarily misleading. And yet a third class of reasonable ways of measuring proportions will yield the result that some non-trivial number of E-having subjects — perhaps the majority of them — came about in a manner similar to how we did, and have evidence about the past that's largely veridical.[55]

So the plausibility of the first premise of the Argument Against COS$_\infty^-$ hangs on what we take the correct way to measure these proportions to be. Given some reasonable measures, P1 will be plausible; for others, it will not.

Now, let's assess the merits of a similar argument against COS$_\infty$. This argument will be the same as the Argument Against COS$_\infty^-$, but with COS$_\infty^-$ replaced by COS$_\infty$.

As before, the argument will be valid, and P2 will be plausible. So the plausibility of the argument against COS$_\infty$ hangs on the plausibility of P1 — the claim that rational subjects like us will be virtually certain our evidence isn't veridical given COS$_\infty$.

---

[55]See De Simone *et al.* (2010) for examples of all three kinds of measures.

As before, if one takes the Indifference Assumption to be plausible, then it's natural to address this question by adopting an account of how to correctly measure proportions in infinite cases, and using that to apply the Indifference Assumption. And, again, different ways of measuring proportions will yield different results. Some ways of measuring proportions will yield the result that, despite the lawful constraints on initial conditions imposed by COS_∞, the vast majority of E-having subjects will have evidence about the past which is extraordinarily misleading. While other ways of measuring proportions will yield the result that a significant proportion — perhaps even the majority of — E-having subjects will have evidence about the past which is largely veridical.

So, as in the classical case, at eternal recurrence worlds this kind of skeptical argument can be raised against both COS_∞^- and COS_∞. Given some ways of measuring proportions, both theories will succumb to such arguments; given some other ways of measuring proportions, both theories will escape such arguments. Either way, these considerations give us little reason to think that COS^--laws are any more subject to skeptical worries than COS laws. Thus, the fact that the BSA yields COS^- laws at typical eternal inflation worlds is not clearly a problem for the BSA.

9.6. Conclusion

In this paper, we have looked at two prominent theories that take there to be lawful constraints on boundary conditions. And we have argued that at typical eternal recurrence worlds of the kind these laws describe, the BSA won't take these constraints to be laws.

It's generally thought that, for certain theories, the lawful constraints on boundary conditions allow us to avoid skeptical results. Without them, it seems, we should believe that it's extremely likely that we are the result of spontaneous fluctuations out of the void with highly misleading evidence about the past. So, at first glance, the conclusion that the BSA won't yield such lawful BPs at eternal recurrence worlds seems like a threat to the tenability of the BSA.

But we have argued that, surprisingly, at eternal recurrence worlds the versions of these theories without lawful BPs are no worse off than the versions which include lawful BPs. In broad strokes, the reason is that the kinds of lawful constraints on initial conditions some versions of these theories impose only have a finite effect on the ratio of E-having subjects with veridical versus non-veridical evidence. And as the duration of the universe increases, the impact of such lawful initial conditions becomes less and less meaningful, and in the limit, has no effect on the proportion of E-having subjects with veridical versus non-veridical evidence. Thus, at eternal recurrence worlds, we end up in the same epistemic situation regardless of whether we believe theories which posit such lawful constraints on initial conditions or not.

Therefore, at the end of the day, we suggest that this is not really a problem for the BSA. Although the BSA's deviations from these prominent theories is initially surprising, it's not clear that it gives us any reason to be concerned about the BSA.

References

Albert, D. Z. (2012) Physics and chance. In Y. Ben-Menahem, & M. Hemmo (Eds.) *Probability in Physics,* (pp. 17–40). Springer.

Albert, D. Z. (2000) *Time and Chance.* Harvard University Press.

Armstrong, D. M. (1983) *What is a Law of Nature?.* Cambridge University Press.

Arntzenius, F., Dorr, C. (2017) Self-locating priors and cosmological measures. In K. Chamcham, J. Barrow, S. Saunders, & J. Silk (Eds.) *The Philosophy of Cosmology,* (pp. 396–428), Cambridge University Press, Cambridge.

Brandenberger, R. (2017) Initial conditions for inflation — a short review, *International Journal of Modern Physics D,* **26**(01), 1740002.

Carroll, J. W. (1994) *Laws of Nature.* Cambridge University Press.

Carroll, S. M. (forthcoming a) In what sense is the early universe fine-tuned? In B. Loewer, B. Weslake, and E. Winsberg (Eds.) *Time's Arrows and the Probability Structure of the World.* Harvard University Press.

Carroll, S. M. (forthcoming b) Why boltzmann brains are bad. In S. Dasgupta, and B. Weslake (Eds.) *Current Controversies in the Philosophy of Science.* Routledge.

Carroll, S. M., Chen, J. (2004) Spontaneous inflation and the origin of the arrow of time. URL: https://arxiv.org/abs/hepth/0410270

De Simone, A., Guth, A. H., Linde, A., Noorbala, M., Salem, M. P., Vilenkin, A. (2010) Boltzmann brains and the scale-factor cutoff measure of the multiverse, *Physical Review D,* **82**(6), 063520.

Demarest, H. (2017) Powerful properties, powerless laws. In J. Jacobs (Eds.) *Causal Powers,* (pp. 38–53). Oxford University Press.

Earman, J. (1986) *A Primer on Determinism.* D. Reidel.

Eddon, M. (2013) Fundamental properties of fundamental properties. In K. Bennett and D. Zimmerman (Eds.) *Oxford Studies in Metaphysics, Volume 8*, (pp. 78–104).

Eddon, M., Meacham, C. J. G. (2015) No work for a theory of universals. In J. Schaffer, and B. Loewer (Eds.) *A Companion to David Lewis*, (pp. 116–137). Wiley-Blackwell.

Elga, A. (2004a) Defeating Dr. Evil with self-locating belief, *Philosophy and Phenomenological Research*, **69**(2), 383–396.

Elga, A. (2004b) Infinitesimal chances and the laws of nature, *Australasian Journal of Philosophy*, **82**(1), 67–76.

Feynman, R. (1965) *The character of physical law*. MIT press.

Frigg, R., Werndl, C. (2012) Demystifying typicality, *Philosophy of Science*, **79**, 917–929.

Glynn, L., Dardashti, R., Thebault, K. P. Y., Frisch, M. (2014) Unsharp humean chances in statistical physics: A reply to Beisbart. In M. C. Galavotti (Ed.) *New Directions in the Philosophy of Science*, (pp. 531–542) Springer.

Guth, A. H. (2001) Eternal inflation, *Annals of the New York Academy of Sciences*, **950**(1), 66–82.

Guth, A. H. (2007) Eternal inflation and its implications, *Journal of Physics A: Mathematical and Theoretical*, **40**(25), 6811.

Hall, N. (1994) Correcting the guide to objective chance, *Mind*, **103**(412), 505–518.

Hawthorne, J. (2006) Quantity in Lewisian metaphysics. In J. Hawthorne (Ed.) *Metaphysical Essays*, (pp. 229–237). Oxford University Press.

Hoefer, C. (2005) The third way on objective probability: A sceptic's guide to objective chance, *Mind*, **116**(463), 549–596.

Ismael, J. T. (2008) Raid! Dissolving the big, bad bug. *Noûs*, **42**(2), 292–307.

Ismael, J. T. (2009) Probability in deterministic physics, *Journal of Philosophy*, **106**(2), 89–108.

Lange, M. (2009) *Laws and Lawmakers: Science, Metaphysics, and the Laws of Nature*. Oxford University Press.

Lewis, D. (1979) Attitudes de dicto and de se, *Philosophical Review*, **88**(4), 513–543.

Lewis, D. (1980) A subjectivist's guide to objective chance. In R. C. Jeffrey (Ed.) *Studies in Inductive Logic and Probability, Volume II*, (pp. 263–293). Berkeley: University of California Press.

Lewis, D. (1986) A subjectivist's guide to objective chance, postscript. In *Philosophical Papers Vol. II*, (pp. 114–132). Oxford University Press.

Lewis, D. (1994) Humean supervenience debugged, *Mind*, **103**(412), 473–490.

Lewis, D. (1996) Elusive knowledge, *Australasian Journal of Philosophy*, **74**(4), 549–567.

Lewis, D. (1973) *Counterfactuals*. Blackwell.

Loewer, B. (2001) Determinism and chance, *Studies in History and Philosophy of Science Part B: Studies in History and Philosophy of Modern Physics*, **32**(4), 609–620.

Loewer, B. (2007) Laws and natural properties, *Philosophical Topics*, **35**(1/2), 313–328.

Maudlin, T. (2007a) *The Metaphysics Within Physics*. Oxford University Press.

Maudlin, T. (2007b) What could be objective about probabilities? *Studies in History and Philosophy of Modern Physics*, **38**, 275–291.

Meacham, C. J. G. (2005) Three proposals regarding a theory of chance, *Philosophical Perspectives*, **19**(1), 281–307.

Meacham, C. J. G. (2010) Contemporary approaches to statistical mechanical probabilities: A critical commentary — part ii: The regularity approach, *Philosophy Compass,* **5**(12), 1127–1136.

Meacham, C. J. G. (2013) Impermissive bayesianism, *Erkenntnis*, (S6), 1–33.

Meacham, C. J. G. (2016) Ur-priors, conditionalization, and ur-prior conditionalization, *Ergo: An Open Access Journal of Philosophy*, **3**.

Meacham, C. J. G. (forthcoming) The meta-reversibility objection. In B. Loewer, B. Weslake, and E. Winsberg (Eds.) *Time's Arrow and the Probability Structure of the World*.

Norton, J. D. (2008) The dome: An unexpectedly simple failure of determinism. *Philosophy of Science*, **75**(5), 786–798.

Pettigrew, R. G. (2013) What chance-credence norms should not be. *Noûs*, **47**(3), 177–196.

Schaffer, J. (2007) Deterministic chance? *British Journal for the Philosophy of Science*, **58**(2), 113–140.

Shenker, O., Hemmo, M. (2012) *The Road to Maxwell's Demon: Conceptual Foundations of Statistical Mechanics*. Cambridge University Press.

Steinhardt, P. J. (2011) The inflation debate, *Scientific American* **304**(4), 36–43.

Titelbaum, M. G. (2012) *Quitting Certainties: A Bayesian Framework Modeling Degrees of Belief*. Oxford University Press.

Urbach, P., Howson, C. (2005) *Scientific Reasoning: The Bayesian Approach*. Open Court, 3 ed.

Weatherson, B. (2005) Should we respond to evil with indifference? *Philosophy and Phenomenological Research*, **70**(3), 613–635.

White, R. (2005) Epistemic permissiveness, *Philosophical Perspectives*, **19**(1), 445–459.

Winsberg, E. (2008) Laws and chances in statistical mechanics, *Studies in History and Philosophy of Science Part B: Studies in History and Philosophy of Modern Physics*, **39**(4), 872–888.

Woodward, J. (2013) Laws, causes, and invariance. In S. Mumford, and M. Tugby (Eds.) *Metaphysics and Science*. Oxford University Press.

Xia, Z. (1992) The existence of noncollision singularities in the n-body problem, *Annals of Mathematics*, **135**, 411–468.

Chapter 10

Arrow(s) of Time without a Past Hypothesis

Dustin Lazarovici[*,‡] and Paula Reichert[†,§]

Université de Lausanne, Section de Philosophie
†LMU München, Mathematisches Institut
‡dustin.lazarovici@live.com
§reichert@math.lmu.de

The paper discusses recent proposals by Carroll and Chen, as well as Barbour, Koslowski, and Mercati to explain the (thermodynamic) arrow of time without a Past Hypothesis, that is, the assumption of a special (low-entropy) initial state of the universe. After discussing the role of the Past Hypothesis and the controversy about its status, we explain why Carroll's model — which establishes an arrow of time as typical — can ground sensible predictions and retrodictions without assuming something akin to a Past Hypothesis. We then propose a definition of a Boltzmann entropy for a classical N-particle system with gravity, suggesting that a Newtonian gravitating universe might provide a relevant example of Carroll's entropy model. This invites comparison with the work of Barbour, Koslowski, and Mercati that identifies typical arrows of time in a relational formulation of classical gravity on shape space. We clarify the difference between this gravitational arrow in terms of shape complexity and the entropic arrow in absolute spacetime, and work out the key advantages of the relationalist theory. We end by pointing out why the entropy concept relies on absolute scales and is thus not relational.

10.1. The easy and the hard problem of irreversibility

What is the difference between past and future? Why do so many physical processes occur in only one time direction, despite the fact that they are governed or described, on the fundamental level, by time-symmetric microscopic laws? These questions are intimately linked to the notion of *entropy* and the second law of thermodynamics. From the point of view of fundamental physics, it is the second law of thermodynamics that accounts for phenomena such as gases expand rather than contract, glasses break but don't spontaneously reassemble, heat flows from hotter to colder bodies, a car slows down and doesn't accelerate once you stop hitting the gas. All these are examples of irreversible processes, associated with an increase of entropy in the relevant physical systems.

Goldstein (2001) — possibly inspired by Chalmers' discussion of the mind-body problem (Chalmers, 1995) — distinguishes between the *easy part* and the *hard part* of the problem of irreversibility. The easy part of the problem is: *Why do isolated systems in a state of low entropy typically evolve into states of higher entropy (but not the other way round)?* The answer to this question was provided by Ludwig Boltzmann, who reduced the second law of thermodynamics to the statistical mechanics of point particles. He thereby developed insights and concepts whose relevance goes beyond the confines of any particular microscopic theory.

The first crucial concept of Boltzmann's statistical mechanics is the distinction between micro- and macrostates. Whereas the microstate $X(t)$ of a system is given by the complete specification of its microscopic degrees of freedom, its macrostate $M(t)$ is specified by (approximate) values of "observables" that characterize the system on macroscopic scales (typical examples are volume, pressure, temperature, magnetization, and so on). The macroscopic state of a system is completely determined by its microscopic configuration, that is $M(t) = M(X(t))$, but one and the same macrostate can be realized by a large (in general infinite) number of different microstates, all of which "look macroscopically the same". Partitioning the microscopic state space into sets corresponding to macroscopically distinct configurations is therefore called *coarse-graining*.

Turning to the phase space picture of Hamiltonian mechanics for an N-particle system, a microstate corresponds to one point $X = (q, p)$ in phase space $\Omega \cong \mathbb{R}^{3N} \times \mathbb{R}^{3N}$, $q = (\mathbf{q}_1, \ldots, \mathbf{q}_N)$ being the position- and $p = (\mathbf{p}_1, \ldots, \mathbf{p}_N)$ the momentum-coordinates of the particles, whereas a macrostate M corresponds to an entire region $\Gamma_M \subseteq \Omega$ of phase space ("macroregion"), namely, the set of all microstates coarse-graining to M. Boltzmann then realized that for macroscopic systems — that is, systems with a very large numbers of microscopic degrees of freedom — different macrostates will in general correspond to macroregions of vastly different size, as measured by the pertinent stationary phase space measure (the Lebesgue- or Liouville-measure in case of a classical Hamiltonian system), with the equilibrium state corresponding to the macroregion of by far the largest measure, exhausting almost the entire phase space volume. The Boltzmann entropy of a system is now defined as the logarithm of the phase space volume covered by its current macroregion (times a dimensional constant called Boltzmann constant):

$$S(X(t)) = k_B \, \log |\Gamma_{M(X(t))}|. \tag{10.1}$$

Since the entropy is an extensive macrovariable (proportional to N, the number of microscopic constituents), we see that the ratio of phase space volume corresponding to macroregions of significantly different entropy values is of order $\exp(N)$, where $N \sim 10^{24}$, even for "small" macroscopic systems (the relevant order of magnitude is given by Avogadro's constant). We thus understand why, under the chaotic microdynamics of a many-particle system, a microstate starting in a small (low-entropy) region of phase space will typically evolve into larger and larger macroregions, corresponding to higher and higher entropy, until it reaches the equilibrium state where it will spend almost the entire remainder of its history (that is, apart from rare fluctuations into lower-entropy states). Taken with enough grains of salt, we can summarize that an irreversible (i.e., entropy increasing) process corresponds to an evolution from less likely into more likely macrostates.

Notably, the time-reversal invariance of the microscopic laws implies that it cannot be true that *all* microstates in a low-entropy macroregion Γ_{M_1} evolve into states of higher entropy. But microconditions leading to an entropy-decreasing evolution are *atypical* — they form a subset of extremely small measure — while nearly all microstates in Γ_{M_1} evolve into states of higher entropy, i.e., entropy increase is *typical*.

The easy problem of irreversibility can be arbitrarily hard from a technical point of view if one seeks to obtain rigorous mathematical results about the convergence to equilibrium in realistic physical models. It is easy in the sense that, conceptually, Boltzmann's account is well understood and successfully applied in physics and mathematics — despite ongoing (but largely unnecessary) controversies and misconceptions in the philosophical literature (see Bricmont (1999), and Lazarovici and Reichert (2015) for a more detailed discussion and responses to common objections).

The hard problem begins with the question: *Why do we find systems in low-entropy states to begin with if these states are so unlikely?* Often the answer is that *we* prepared them, creating low-entropy subsystems for the price of increasing the entropy in their environment. But why then is the entropy of this environment so low — most strikingly in the sense that it allows *us* to exist? If one follows this rationale to the end, one comes to the conclusion that the universe as a whole is in a state of low entropy (that is, globally, in a spatial sense; we don't just find ourselves in a low-entropy pocket in an otherwise chaotic universe) and that this state must have evolved from a state of even lower entropy in the distant past. The latter assumption is necessary to avoid the absurd conclusion that our present macrostate — which includes all our memories and records of the past — is much more likely the product of a fluctuation out of equilibrium than of the low-entropy past that our memories and records actually record. In other words: only with this assumption does Boltzmann's account "make it plausible not only that the paper will be yellower and ice cubes more melted and people more aged and smoke more dispersed in the future, but that they were less so (just as our experience tells us) in the past." (Albert (2015, p. 5); for a good discussion of this issue, see also Feynman (1967, Ch. 5), Carroll (2010).)

In summary, the hard part of the problem of irreversibility is to explain the existence of a *thermodynamic arrow of time in our universe*, given the fact that the universe is governed, on the fundamental level, by reversible microscopic laws. And the standard account today involves the postulate of a very special (since very low-entropy) initial macrostate of the universe. Albert (2001) coined for this postulate the now famous term *Past Hypothesis* (PH). But the status of the Past Hypothesis is highly controversial. Isn't the very low-entropy beginning of the universe itself a mystery in need of scientific explanation?

In the next section, we will briefly recall this controversy and the various attitudes taken towards the status of the PH. Section 10.3 then introduces recent ideas due to Sean Carroll and Julian Barbour to explain the arrow of time *without* a Past Hypothesis, namely as a feature of *typical* universes. In Section 10.4, we discuss if the Carroll model can indeed ground sensible inferences about the past and future of our universe without assuming something akin to the PH. Section 10.5 will propose a Boltzmann entropy for Newtonian gravity, suggesting that a universe of N gravitating point particles provides a relevant realization of Carroll's entropy model. Section 10.6, finally, introduces the results of Barbour, Koslowski, and Mercati that establishes an arrow of time in a relational formulation of Newtonian gravity on "shape space". We will work out potential advantages of the relational theory and clarify the differences between the "gravitational arrow" of Barbour *et al.* and the entropic arrow identified in the preceding section.

10.2. The controversy over the Past Hypothesis

The standard response to the hard part of the problem of irreversibility involves the postulate of a very low-entropy beginning of our universe; but, what is the status of this Past Hypothesis? In the literature, by and large, three different views have been taken towards this issue.

1. The low-entropy beginning of the universe requires an explanation.
2. The low-entropy beginning of the universe does not require, or allow, any further explanation.
3. The Past Hypothesis is a law of nature (and therefore does not require or allow any further explanation).

The first point of view is largely motivated by the fact that our explanation of the thermodynamic arrow is based on a *typicality reasoning* (e.g., Lebowitz (1993), Goldstein (2001), Lazarovici and Reichert (2015)). Assuming a low-entropy initial macrostate of the universe, Boltzmann's analysis allows us to conclude that *typical* microstates relative to this macrostate will lead to a thermodynamic evolution of increasing entropy. It is then not a good question to ask why the actual initial conditions of the universe were among the typical ones. Once a fact about our universe — such as the existence of a thermodynamic arrow — turns out to be typical, given the fundamental laws and the relevant boundary conditions, there is nothing left to explain (except, possibly, for the boundary conditions). On the flipside, atypical facts are usually the kind of facts that cry out for further explanation (cf. the contribution of Maudlin to this volume). And to accept the PH is precisely to assume that the initial state of our universe was atypical, relative to all possible microstates, in that it belonged to an extremely small (i.e. very low-entropy) macroregion. Penrose (1999) estimates the measure of this macroregion relative to the available phase space volume to be at most $1 : 10^{10^{123}}$ — a mind-bogglingly small number. Notably, the explanatory pressure is mitigated by the fact that the PH entails only a special initial macrostate rather than a microscopic fine-tuning. In the case of a gas in a box, this would be the difference between atypical conditions that one can create with a piston and atypical conditions that one could create only by controlling the exact position and momentum of every single particle in the system. The point here is not that this makes it easier for a hypothetical creator of the universe, but that only the latter (microscopic) kind of fine-tuning gives rise to the worry that — given the huge number of microscopic degrees of

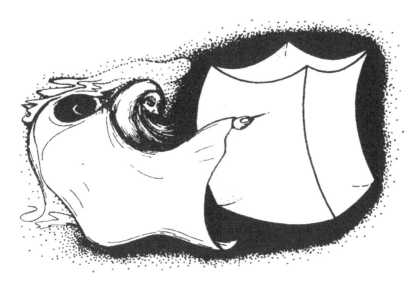

Fig. 10.1. God picking out the special (low-entropy) initial conditions of our universe. Penrose (1999).

freedom and the sensitivity of the evolution to variations of the initial data — atypical initial conditions could explain *anything* (and thus explain nothing; cf. Lazarovici and Reichert (2015)). Nonetheless, the necessity of a PH implies that our universe looks very different from a typical model of the fundamental laws of nature — and this is a fact that one can be legitimately worried about.

The second point of view was in particular defended by Callender (2004). While Callender is also sympathetic to the third option (regarding the PH as a law), he makes the broader case that a) there is no single feature of facts — such as being atypical — that makes them require explanation; and b) the conceivable explanations of the Past Hypothesis aren't much more satisfying than accepting it as a brute and basic fact. Notably, Ludwig Boltzmann himself eventually arrived at a similar conclusion:

> "The second law of thermodynamics can be proven from
> the mechanical theory if one assumes that the present state
> of the universe, or at least that part which surrounds us,
> started to evolve from an improbable state and is still in a

> relatively improbable state. This is a reasonable assump-
> tion to make, since it enables us to explain the facts of
> experience, and one should not expect to be able to deduce
> it from anything more fundamental." (Boltzmann, 1897)

The third option, finally, is most prominently advocated by Albert (2001) and Loewer (2007) in the context of the Humean *Best System Account* of laws. Upon their view, the laws of nature consist in a) the microscopic dynamical laws, b) the PH, and c) a probability (or typicality) measure on the initial macroregion. This package has been dubbed the "mentaculus" (Loewer, 2012). It is supposed to correspond to the best-system-laws because it strikes the optimal balance between being simple and being informative about the history of our universe (the "Humean mosaic"). In particular, adding b) and c) to the microscopic laws comes at relatively little cost in terms of simplicity but makes the system much more informative, precisely, because it accounts for the thermodynamic arrow of time and allows for probabilistic inferences. In addition, Albert (2001) and Loewer (2007) employ the mentaculus in a sophisticated analysis of records, counterfactuals, and more, the discussion of which goes beyond the scope of this paper. Instead, it is important to note that the proposition which Albert wants to grant the status of a law is not that the universe started in *any* low-entropy state. The PH, in its current form, is rather a placeholder for "the macrocondition ... that the normal inferential procedures of cosmology will eventually present to us" (Albert, 2001, p. 96). Ideally (we suppose), physics will one day provide us with a nice, and simple, and informative characterization of the initial boundary conditions of the universe — maybe something along the lines of Penrose's Weyl Curvate Conjecture (Penrose, 1999) — that would strike us as "law-like". But this is also what many advocates of option 1 seem to hope for as an "explanation" of the PH. So while option 3 sounds like the most clear-cut conclusion about the status of the PH, it is debatable to what extent it settles the issue. The more we have to rely on future physics to fill in the details, the less is already accomplished by calling the Past Hypothesis a law of nature.

10.3. Thermodynamic arrow without a Past Hypothesis

In recent years, Sean Carroll together with Jennifer Chen (2004; see also Carroll (2010)) and Julian Barbour together with Tim Koslowski and Flavio Mercati (2013, 2014, 2015) independently put forward audacious proposals to explain the arrow of time *without* a Past Hypothesis. While Barbour's arrow of time is not, strictly speaking, an *entropic* arrow (but rather connected to a certain notion of complexity), Carroll's account is largely based on the Boltzmannian framework, although with a crucial twist. For this reason, we shall focus on the Carroll account first, before comparing it to the theory of Barbour *et al.* in Section 10.6.

The crucial assumption of Carroll and Chen is that the relevant stationary measure on the phase space of the universe is unbounded, allowing for macrostates of *arbitrarily high entropy*. Hence, *every* macrostate is a non-equilibrium state from which the entropy can typically increase in both time directions, defining a thermodynamic arrow — or rather two opposite ones — on either side of the entropy minimum. A typical entropy curve (one hopes) would thus be roughly parabolic or "U-shaped", attaining its global minimum at some moment in time and growing monotonously (modulo small fluctuations) in both directions of this vertex (Fig. 10.2). Barbour *et al.* (2015) describe such a profile as "one-past-two-futures", the idea being that observers on each branch of the curve would identify the direction of the entropy minimum — which the authors name the *Janus point* — as their past. In other words, we would have two future-eternal episodes making up the total history of the universe, with the respective arrows of time pointing in opposite directions.

The Carroll model is intriguing because it is based on the bold, yet plausible assumption that the universe has no equilibrium state — a crucial departure from the "gas in the box" paradigm that is still guiding most discussions about the thermodynamic history of the universe. And it is particularly intriguing for anybody worried about the status of the Past Hypothesis, because it seeks to establish the existence of a thermodynamic arrow in the universe as *typical*. This is

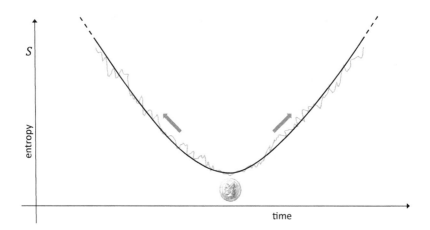

Fig. 10.2. Typical entropy curve (with fluctuations and interpolated) for a
Carroll universe. The arrows indicate the arrow(s) of time on both sides of the
"Janus point" (entropy minimum).

in notable contrast to the standard account, in which we saw that an
entropy gradient is typical only under the assumption of atypical —
and time-asymmetric — boundary conditions.

Prima facie, it seems plausible that an eternal universe with
unbounded entropy would exhibit the U-shaped entropy profile
shown in Fig. 10.2. Since if we start in *any* macrostate, the usual
Boltzmannian arguments seem to suggest that typical microstates
in the corresponding macroregion lead to a continuous increase of
entropy in both time directions (since there are always vastly larger
and larger macroregions, corresponding to higher and higher entropy
values, that the microstate can evolve into). And then, any sensible
regularization of the phase space measure would allow us to conclude that a U-shaped entropy profile is typical *tout court*, that is,
with respect to all possible micro-histories.

However, if we assume, with Carroll, a non-normalizable
measure — that assigns an infinite volume to the total phase space
and thus allows for an unbounded entropy — the details of the
dynamics and the phase space partition must play a greater role
than usual in the Boltzmannian account. For instance, the measure of low-entropy macroregions could sum up to arbitrarily (even

infinitely) large values, exceeding those of the high-entropy regions. Or the high-entropy macroregions could be arbitrarily far away in phase space, so that the dynamics do not carry low-entropy configurations into high-entropy regions on relevant time-scales. The first important question we should ask is therefore:

> Are there any interesting and realistic dynamics that give rise to typical macro-histories as envisioned by Carroll and Chen?

The original idea of Carroll and Chen (2004) is as fascinating as it is speculative. The authors propose a model of eternal spontaneous inflation in which new baby universes (or "pocket universes") are repeatedly branching off existing ones. The birth of a new universe would then increase the overall entropy of the multiverse, while the baby universes themselves, growing from a very specific pre-existing state (a fluctuation of the inflaton field in a patch of empty de-Sitter space), would typically start in an inflationary state that has much lower entropy than a standard big bang universe. This means, in particular, that our observed universe can look like a low-entropy universe, with an even lower-entropy beginning, even when the state of the multiverse as a whole is arbitrarily high up the entropy curve. The details of this proposal are beyond the scope of our paper and do not (yet) include concrete dynamics or a precise definition of the entropy.

In more recent talks, Carroll discusses a simple toy model — essentially an ideal gas without a box — in which a system of N free particles can expand freely in empty space. The only macrovariable considered is the moment of inertia, $I = \sum_{i=1}^{N} \mathbf{q}_i^2$, providing a measure for the expansion of the system. It is then easy to see that I will attain a minimal value at some moment in time t_0, from which it grows to infinity in both time directions (cf. equation (10.8) below). The same will hold for the associated entropy since a macroregion, corresponding to a fixed value of I, is just a sphere of radius $\sqrt{I}$ in the position coordinates (while all momenta are constant). The entropy curve will thus have the suggested U-shape with vertex at $t = t_0$.

A detailed discussion of this toy model can be found in Reichert (2012), as well as Goldstein *et al.* (2016).

In this paper, we will not discuss these two models in any more detail. Instead, we are going to argue in Section 10.5 that there exists a dynamical theory fitting Carroll's entropy model that is much less speculative than baby universes and much more interesting, physically, than a freely expanding system of point particles. This theory is *Newtonian gravity*. It will also allow us to draw interesting comparisons between the ideas of Carroll and Chen and those of Barbour, Koslowski, and Mercati.

First, however, we want to address the question, whether this entropy model would even succeed in explaining away the Past Hypothesis. Are typical macro-histories as envisioned by Carroll and sketched in Fig. 10.2 sufficient to ground sensible inferences about our past and future? Or would we still require — if not the PH itself, then a close variant — an equally problematic assumption about the specialness of the observed universe?

10.4. The (dispensible) role of the Past Hypothesis

The question to be addressed in this section is thus the following:

> Can Carroll's entropy model ground sensible statistical inferences about the thermodynamic history of our universe without assuming (something akin to) a Past Hypothesis?

To approach this issue, and clarify the role of the PH in the standard account, we have to disentangle two questions that are often confounded in such discussions:

i) Given the fundamental laws of nature, what do typical macro-histories of the universe look like? In particular, is the existence of a thermodynamic arrow typical?

ii) Given our knowledge about the present state of the universe, what can we reasonably infer about its past and future?

The answer to question i) will, in general, depend on the dynamical laws as well as cosmological considerations. If we have infinite time and a finite maximal entropy, a typical macro-history will be in thermodynamic equilibrium almost all the time, but also exhibit arbitrarily deep fluctuations into low-entropy states, leading to periods with a distinct entropy gradient, i.e., a local thermodynamic arrow. This *fluctuation scenario* was in fact Boltzmann's initial response to the hard problem of irreversibility (Boltzmann, 1896).

However, to assume a fluctuation as the origin of our thermodynamic arrow is highly unsatisfying, Feynman (1967, p. 115) even calls it "ridiculous". The reason is that fluctuations which are just deep enough to account for our present macrostate are much more likely (and would thus occur much more frequently[1]) than fluctuations producing an even lower-entropy past from which the current state could have evolved in accordance with the second law. We would thus have to conclude that we are currently experiencing the local entropy minimum, and that our present state — including all our records and memories — is, in fact, the product of a random fluctuation rather than a lower-entropy past. Feynman makes the further case that the fluctuation scenario leads not only to absurd conclusions about the past, but to wrong ones about the present state of the universe, as it compels us to assume that our current fluctuation is not any deeper — and hence more unlikely — than necessary to explain the evidence we already have: If we dig in the ground and find a dinosaur bone, we should not expect to find other bones nearby. If we stumble upon a book about Napoleon, we should not expect to find other books containing the same information about a guy called Napoleon. The most extreme form of this reductio ad absurdum is the *Boltzmann brain problem* (see e.g., Carroll (2010) for a recent discussion): a fluctuation that is just deep enough to account for your empirical evidence (many people claim) would produce only your brain, floating in space, with the rest of the universe at equilibrium. You should thus conclude that this is, by far, the most likely state of the universe you currently experience.

The only possible escape in such a fluctuation scenario is to involve the additional postulate — a form of Past Hypothesis — that

---

[1]e.g. in the sense, $\limsup_{T \to +\infty} \frac{1}{T}$ (number of fluctuations with entropy minimum $\approx S$ in the time-interval $[-T, T]$).

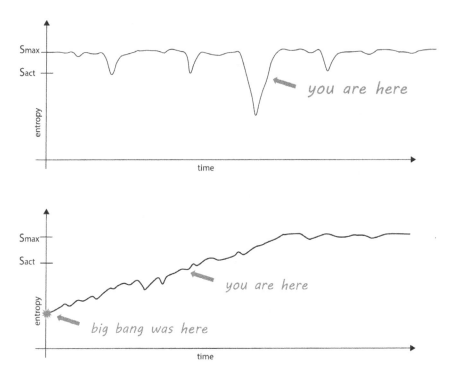

Fig. 10.3. Self-location hypothesis in the fluctuation scenario (upper image) and big bang scenario (lower image) with bounded entropy. Time-scales in the upper image are much larger than below, and periods of equilibrium are much longer than depicted.

the present macrostate is not the bottom of the fluctuation, but has been preceded by a sufficiently long period of entropy increase from a state of much lower entropy, still. In this context, the PH would thus serve a *self-locating* function, taking the form of an indexical proposition that locates our present state on the upwards-slope of a particularly deep fluctuation (Fig. 10.3).

The now standard account assumes a bounded entropy and a relatively young universe — about 13.8 billion years according to estimates from modern big bang cosmology. In this setting (we interpret the big bang as the actual beginning of time), a typical history would not have any thermodynamic arrow at all (the time-scale of $\sim 10^{10}$ years is too short for significant entropy fluctuations on cosmological scales). Thus, we need the PH to account for the existence of a thermodynamic arrow in the first place by postulating a low-entropy boundary condition at the big bang. A self-locating proposition is still crucial and hidden in the assumption of a young universe. Winsberg (2012) makes it explicit in what he calls the "Near Past Hypothesis" (NPH), which is that our present state lies between the low-entropy beginning of the universe and the time of first relaxation into equilibrium. Without such an assumption — and assuming that the universe is eternal in the future time direction — we would essentially be back in a fluctuation scenario with all its Boltzmann-brain-like absurdities. In a future-eternal universe with bounded entropy, there are still arbitrarily many entropy fluctuations that are just deep enough to account for our present evidence (but not much deeper). And we would still have to conclude that we are much more likely in one of these fluctuations than on the initial upwards slope originating in the very low-entropy big bang (cf. also the contribution by Loewer in this volume).

The self-locating role of the PH (which we take to include the NPH — for what would be the point otherwise) is thus indispensible. And it is, in fact, the indexical proposition involved, rather than the non-dynamical boundary condition, that we would be surprised to find amongst the fundamental laws of nature as we consider this option for the status of the Past Hypothesis.

Carroll's model, finally, postulates an eternal universe and unbounded entropy, suggesting that typical macro-histories will have the U-shaped entropy profile depicted in Fig. 10.2. If this works out — and we will argue that it does, at least we see no reason why it couldn't — the existence of a thermodynamic arrow (respectively two opposite ones) will be *typical*. (For completeness, we could also discuss the option of a temporally finite universe and unbounded entropy, but this model does not seem advantageous and goes beyond the scope of the paper.) In the upshot, the Carroll model can indeed explain the existence of a thermodynamic arrow without invoking a PH as a fundamental postulate over and above the microscopic laws and the pertinent typicality measure. It may still turn out that the theory requires a PH for its self-locating function *if* it would otherwise imply that our current macrostate is the global entropy minimum, i.e., has not evolved from a lower-entropy past. The relevant version of the PH may then take the form of an indexical clause — stating that our present state is high up the entropy curve — or be a characterization of the entropy minimum (Janus point) of our universe. (In the first case, the PH would first and foremost locate the present moment within the history of an eternal universe; in the latter, it would first and foremost locate the actual universe within the set of possible ones.) But it is not obvious why the Carroll model would lead to the conclusion that we are currently at (or near) the entropy minimum, and the issue actually belongs to our second question — how to make inferences about the past — to which we shall now turn.

10.4.1. *Predictions and retrodictions*

The most straightforward response to question ii) — how to make inferences about the past or future — is the following method of statistical reasoning: Observe the current state of the universe (respectively, a suitably isolated subsystem), restrict the pertinent probability (more correctly, typicality) measure to the corresponding macroregion in phase space, and use the conditional measure to make probabilistic inferences about the history of the system. We shall call this *naive evidential reasoning* (reviving a terminology introduced in an unpublished 2011 draft of Goldstein *et al.*, 2016). The negative connotation is warranted because we know that while this kind of reasoning works well for *predictions* — inferences about the future — it leads to absurd, if not self-refuting, conclusions when applied for *retrodictions*, that is, inferences about the past.

The now standard move to avoid this predicament is to employ the PH to block naive evidential reasoning in the time direction of the low-entropy boundary condition. For sensible retrodictions, we learn, one must conditionalize on the low-entropy initial state in addition to the observed present state. It is rarely, if ever, noted that an appeal to a PH may be sufficient but not necessary at this point. The key is to appreciate that the second question — how to

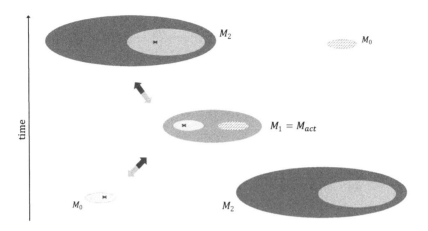

Fig. 10.4. Typical microstates in the intermediate macroregion $M_1 = M_{act}$ evolve into a higher-entropy region M_2 in both time directions. Only a small subset of microstates (light grey area) have evolved from the lower-entropy state M_0 in the past; an equally small subset (shaded area) will evolve into M_0 in the future. The actual microstate (cross) has evolved from the lower-entropy state in the past; only its future time-evolution corresponds to the typical one relative to the macrostate M_{act}.

make inferences about the past and future of the universe — must be addressed subsequently to the first, whether a thermodynamic arrow in the universe is typical. Since if we have good reasons to believe that we live in a universe with a thermodynamic arrow of time, this fact alone is sufficient to conclude the irrationality of retrodicting by conditionalizing the phase space measure on the present macrostate. More precisely, it follows from the Boltzmannian analysis that, in a system with a thermodynamic arrow, the evolution towards the future (the direction of entropy increase) looks like a *typical* one relative to any intermediate macrostate, while the actual microstate is necessarily atypical with respect to its evolution towards the entropic past (see Fig. 10.4). This is essentially the reversal of the familiar paradox that entropy increase in *both* time directions comes out as typical relative to any non-equilibrium macrostate.

In the upshot, the fact that naive evidential reasoning doesn't work towards the entropic past can be inferred from the existence of a thermodynamic arrow; it does not have to be inferred from the assumption of a special initial state. The explanation of the thermodynamic arrow, in turn, may or may not require a special initial state, but this was a different issue, as discussed above.

If the relevant physical theory tells us that a thermodynamic arrow is typical — that is, exists in almost all possible universes — we have a very strong theoretical justification for believing that we actually live in a universe with a thermodynamic arrow. And if we believe that we live in a universe with a thermodynamic arrow, a rational method for making inferences about the past is not naive evidential reasoning but *inference to the best explanation*. Rather than asking "what past state (or states) would be typical given our present macrostate?", we should ask "what past state (or states) would make our present macrostate typical?". In more technical terms, rather than adjusting our credence that the past macrostate was M_0 according to $\mathbb{P}(M_0 \mid M_{act})$, where M_{act} is our present macrostate, we should "bet" on macrostates M_0 that maximize $\mathbb{P}(M_{act} \mid M_0)$. If we find a dinosaur bone, we should infer a past state containing a dinosaur. If we find history books with information about Napoleon, we should infer a past state containing a French emperor by the name of Napoleon. We do not claim that this amounts to a full-fledged analysis of the epistemic asymmetry, but it is a large enough part, at least, to uphold the reliability of records and get a more fine-grained analysis going. In particular, considering the universe as a whole, the fact that it has evolved from a lower-entropy state in the past is *inferred*, rather than assumed, by this kind of abductive reasoning.

By now it should be clear that the debate is not about whether some version of the PH is *true*, but about whether it is an *axiom*. And the upshot of our discussion is that if the existence of a thermodynamic arrow in the universe turns out to be typical, we can consider our knowledge of the low-entropy past to be reasonably grounded in empirical evidence and our best theory of the microscopic dynamics (as any knowledge about our place in the history of the universe arguably should).

Another way to phrase the above analysis goes as follows: Naive evidential reasoning applied to both time directions will always lead to the conclusion that the current macrostate is the (local) entropy minimum. However, if we know that we observe a universe (or any other system) with a thermodynamic arrow, we also know that this conclusion would be wrong *almost all the time*. More precisely, it would be wrong unless we happened to observe *a very special period* in the history of the universe in which it is close to its entropy minimum.

Goldstein, Tumulka, and Zanghì (2016) provide a mathematical analysis of this issue in the context of Carroll's toy model of freely expanding particles. Their discussion shows that the two opposing ways of reasoning — typical microstates within a given macroregion versus typical time-periods in a history characterized by a U-shaped entropy curve — come down to different ways of regularizing the unbounded phase space measure by choosing an appropriate cut-off. Goldstein *et al.* then argue against the first option, corresponding to naive evidential reasoning, by saying that certain facts about the past amount to "pre-theoretical" knowledge. In contrast, our arguments for the same conclusion are based on a theoretical (Boltzmannian) analysis. Nonetheless, from a formal point of view, a certain ambiguity remains. In Section 10.6, we will discuss how the relational framework of Barbour *et al.* is able to improve upon this situation.

10.4.2. *The mystery of our low-entropy universe*

Another possible objection to the Carroll model (disregarding baby universes) goes as follows: Doesn't the fact that the entropy of the universe could be arbitrarily high make its present very low value — and the even lower value at the Janus point — only more mysterious? In other words, doesn't the fact that the entropy could have been arbitrarily high just increase the explanatory pressure to account for the specialness of the observed universe? While we cannot completely deny the legitimacy of this worry, our intuition is that the Carroll model precludes any *a priori* expectation of what the entropy of the universe *should* be. If it can be arbitrarily (but not infinitely) large, any possible value could be considered "mysteriously low" by skeptics.

Sidney Morgenbesser famously responded to the ontological question *Why is there something rather than nothing*: "If there was nothing, you'd be still complaining!" In the same spirit (though not quite as witty), our reaction to the question *Why is the entropy of the universe so low?* would be: "If it was any higher, you'd be still complaining!"

We concede, nonetheless, that divergent intuitions about this question are possible. In fact, the ambiguity is once again paralleled by mathematical issues arising from the non-normalizability of the phase space measure. When Penrose (1999) estimates that the entropy of the universe near the big bang could have been about 10^{123} times higher, the common worry is not that the actual value was so low in comparison, but that a 10^{123} times higher entropy would seem $10^{10^{123}}$ times *as likely*. While this conclusion is questionable even in the standard Boltzmannian framework (with a finite phase space volume), the interpretation of a non-normalizable phase space measure as a *probability* measure is problematic, to say the least, leading in particular to the paradox that any finite range of entropy values has probability zero. Again, we'll have to leave it at that, as far as the discussion of the Carroll model is concerned, explaining instead in the last section how the shape space theory of Barbour *et al.* is able to resolve the issue. First, though, we owe the reader some evidence that we have not been talking about the empty set, but that Newtonian gravity might actually provide a relevant example of a Carroll universe.

10.5. Entropy of a classical gravitating system

There is a lot of confusion and controversy about the statistical mechanics of classical gravitating systems, despite the fact that statistical methods are commonly and successfully used in areas of astrophysics that are essentially dealing with the Newtonian N-body problem (see, for example, Heggie and Hut (2003)). (An excellent paper clearing up much of the confusion is Wallace (2010);

see Callender (2009) for some problematic aspects of the statistical mechanics of gravitating systems, and Padmanabhan (1990) for a mathematical treatment.) Some examples of common claims are:

a) Boltzmann's statistical mechanics is not applicable to systems in which gravity is the dominant force.
b) The Boltzmann entropy of a classical gravitating system is ill-defined or infinite.
c) An entropy increasing evolution for a gravitating system is exactly opposite to that of an ideal gas. While the tendency of the latter is to expand into a uniform configuration, the tendency of the former is to clump into one big cluster.

We believe that the first two propositions are simply false, while the third is at least misleading. However, rather than arguing against these claims in the abstract, we will provide a demonstration to the contrary by proposing an analysis of a classical gravitating system in the framework of Boltzmann's statistical mechanics.

We start by looking at the naive calculation, along the lines of the standard textbook computation for an ideal gas, that finds the Boltzmann entropy of the classical gravitating system to be infinite (see, for example, Kiessling (2001)). For simplicity, we always assume N particles of equal mass m. We have

$$S(E, N, V) := k_B \log|\Gamma(E, N, V)|$$

$$= k_B \log \left[\frac{1}{h^{3N} N!} \int_{V^N} \int_{\mathbb{R}^{3N}} \delta(H - E) \, \mathrm{d}^{3N} q \, \mathrm{d}^{3N} p \right],$$

(10.2)

with

$$H^{(N)}(q, p) = \sum_{i=1}^{N} \frac{\mathbf{p}_i^2}{2m} - \sum_{1 \leq i < j \leq N} \frac{Gm^2}{|\mathbf{q}_i - \mathbf{q}_j|}$$

(10.3)

and

$$\int_{V^N} \int_{\mathbb{R}^{3N}} \delta(H - E) \mathrm{d}^{3N}p \, \mathrm{d}^{3N}q$$

$$= C \int_{V^N} \left(E + \sum_{i<j} \frac{Gm^2}{|\mathbf{q}_i - \mathbf{q}_j|} \right)^{\frac{3N-2}{2}} \mathrm{d}^{3N}q = +\infty. \quad (10.4)$$

For $N > 2$, the integral (10.4) diverges due to the singularity of the gravitational potential at the origin. Note that in the present context, there are no physical boundaries (i.e. no box) confining the particles to a given volume. The macrovariable V rather describes a volume (e.g., the smallest sphere, or cuboid, or convex set) enclosing the particle configuration in empty space.

There is nothing mathematically wrong with the above calculation, it just doesn't actually compute what it's supposed to. One problem is that as we integrate over V^N, we sum over all possible configurations of N particles (with total energy E) within the volume V. This includes configurations in which the particles are homogeneously distributed, but also configurations in which most particles are concentrated in a small subset of V (Fig. 10.5). In the case of the ideal gas in a box, the contribution of the latter is negligible since almost the entire phase space volume is concentrated on spatially homogeneous configurations. It is the entropy (or phase space volume) of this equilibrium state that we actually want to compute, and the mistake we make by including non-equilibrium configurations (in which the particles are concentrated in one half, or one quarter or one third, etc., of the volume) is so small that it is hardly ever mentioned.

In the case of a gravitating system, the situation is distinctly different, since the spatial configuration is correlated with the kinetic

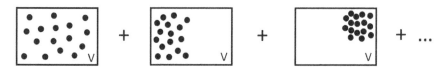

Fig. 10.5. Performing the volume integral in (10.4), we sum over *all* possible configurations of the particles within the given volume V.

energy or, in other words, with the possible momentum configurations of the system. Simply put, for an attractive potential and constant energy, a more concentrated spatial configuration corresponds to higher kinetic energy and thus larger phase space volume in the momentum variables. The "total volume" V is thus not a good macrovariable to describe a system with gravity. In particular, if we want to know whether the entropy of a gravitating system is increasing as the configuration clusters, we have to consider a macroscopic variable that actually distinguishes between more and less clustered configurations (as those shown in Fig. 10.5).

We thus propose to describe a system of N gravitating point particles by the following set of macrovariables:

- $E(p,q) = \frac{p^2}{2m} + V(q) = \sum_{i=1}^{N} \frac{\mathbf{p}_i^2}{2m} - \sum_{1 \leq i < j \leq N} \frac{Gm^2}{|\mathbf{q}_i - \mathbf{q}_j|}$ is the total energy of the system, which is a constant of motion.
- $I(q) = \sum_{i=1}^{N} m(\mathbf{q}_i - \sum_{j=1}^{N} \mathbf{q}_j)^2$ is the moment of inertia that will quantify how much the particles are spread out over space. In the center of mass frame — which we can and will use without loss of generality — it simplifies to $I(q) = mq^2 = \sum_{i=1}^{N} m\mathbf{q}_i^2$. Recall that the system is not confined by physical boundaries but can expand arbitrarily in empty space (I can grow arbitrarily).

The moment of inertia alone is still too coarse to differentiate between, let's say, a uniform configuration and a concentrated cluster with few residual particles far away. To distinguish between more and less clustered configurations, we thus have to introduce a further macrovariable. We choose:

- $U(q) := -V(q) = \sum_{1 \leq i < j \leq N} \frac{Gm^2}{|\mathbf{q}_i - \mathbf{q}_j|}$, which is just the absolute value of the potential energy. Since the total energy is $E(q,p) = T(p) + V(q)$, specifying the value of E and U is equivalent to specifying E and the kinetic energy $T(p) = \sum_{i=1}^{N} \frac{\mathbf{p}_i^2}{2m}$. An increase of $U(q)$ thus signifies both clustering and heating of the system.

Note that defining macrostates in terms of U (respectively, the potential energy) automatically takes care of the *ultraviolet divergence* in the computation of the associated entropy, since the minimal particle distance r is bounded as $r \geq \frac{Gm^2}{U}$.

Obviously, we do not claim that the moment of inertia or the gravitational poential energy of the universe can be precisely measured. What makes them macrovariables is, first and foremost, the fact that they are coarse-graining: many different micro-configurations of an N-particle universe realize the same values of I and U. Moreover, in the next subsection, it will become clearer that the evolution of these macrovariables does indeed provide relevant information about the large-scale structure of a gravitating universe.

Now, to determine the entropy of the respective macrostates, we have to compute the phase space volume corresponding to a macroregion $\Gamma(E, I, U)$, that is

$$|\Gamma(E,I,U)| = \iint_{\mathbb{R}^{3N} \times \mathbb{R}^{3N}} \delta\left(\frac{p^2}{2m} + V(q) - E\right)$$
$$\times \delta(V(q) + U)\,\delta(mq^2 - I)\,\mathrm{d}^{3N}q\,\mathrm{d}^{3N}p \quad (10.5)$$

for fixed values of E, I, and U. Unfortunately, we weren't able to solve this integral analytically (and maybe this is, in fact, impossible). However, if we replace the sharp values of the macrovariables with a small interval $I(q) \in (I - \Delta I, I + \Delta I)$, $|V(q)| \in (U - \Delta U, U + \Delta U)$ with, e.g., $\Delta I = \frac{I}{\sqrt{N}}$, $\Delta U = \frac{U}{\sqrt{N}}$ (roughly a standard deviation), we can obtain the bounds:

$$C(E + U)^{\frac{3N-2}{2}} I^{\frac{3N-3}{2}} \left(\frac{Gm^2}{U}\right)^3 \mathrm{e}^{-4N} \leq |\Gamma(E, U \pm \Delta U, I \pm \Delta I)|$$
$$\leq C(E + U)^{\frac{3N-2}{2}} I^{\frac{3N}{2}} \mathrm{e}^{3\sqrt{N}},$$

for sufficiently large values of I and U (more precisely, of the dimensionless quantity $\frac{\sqrt{I}U}{Gm^{5/2}}$) and $E \geq 0$, where C is a positive constant depending only on N and m. A precise statement and proof (valid for any E) is given in the appendix. Thus, we have

$$|\Gamma(E, U, I)| \approx const. \cdot (I(E+U))^{\frac{3N}{2}}, \qquad (10.6)$$

and, ignoring an additive constant,

$$S(E, I, U) \approx \frac{3N}{2}(\log(E+U) + \log(I)). \qquad (10.7)$$

10.5.1. *Typical evolutions*

We now provide a discussion of this result.

1. With our choice of macrovariables, the associated Boltzmann entropy of a gravitating system is well-defined and finite. We also see that the entropy can grow without bounds, either due to continuous expansion of the system ($I \to +\infty$) and/or due to continuous clustering and self-heating ($U \to +\infty$).
2. While common wisdom says that the typical evolution of a gravitating system is one of clumping and clustering, our computation shows that clustering and expansion (as quantified by the macrovariable U and I, respectively) can contribute equally to an increase of entropy. This fits well with the observed processes of gravithermal collapse that are known to show a "core-halo" pattern (see, for example, Heggie and Hut (2003, Ch. 23)): the configuration of masses splits into a core that collapses and heats up (increase of U), and a collection of particles on the outskirts that are blown away (increase of I).

   On even larger (cosmological) scales, a gravitating system in a homogeneous configuration can increase its entropy along both "dimensions" by forming many local clusters ("galaxies") that disperse away from each other — a process that would look very much like structure formation!

   Hence, it seems to be precisely the interplay between the opposing tendencies of clustering and expansion that makes classical gravity much more interesting, from a thermodynamic point of view, than often assumed.

3. Analytical and numerical results support the conclusion that the typical evolution of a gravitating system is one in which the entropy in (10.7) increases from a minimum value in both time directions, giving rise to the U-shaped entropy curves proposed by Carroll and Chen. The first analytical result is the classical *Lagrange-Jacobi equation* for the gravitational potential:

$$\ddot{I} = 4E - 2V. \qquad (10.8)$$

From this equation, which is a standard result in analytical mechanics, it follows immediately that if $E \geq 0$, the second time derivative of the moment of inertia is strictly positive (note that V is negative), meaning that $I(t)$ is a strictly convex (upwards curving) function. Together with the fact that $I \to \infty$ as $t \to \pm\infty$ (Pollard, 1967), we can conclude that the graph of I has precisely the kind of U-shape that we expect for the entropy.

Thanks to the results of Saari (1971), and Marchal and Saari (1974), we have an even more precise picture of the asymptotic behavior of the Newtonian gravitational N-particle system. Their work studies the inter-particle distances $|\mathbf{q}_i - \mathbf{q}_j|$, as well as the dispersion from the center of mass, for $t \to \infty$, independent of the total energy. It is found that either the minimal particle distance goes to zero

$$\lim_{t\to\infty} r(t) := \lim_{t\to\infty} \min_{i\neq j} |\mathbf{q}_i(t) - \mathbf{q}_j(t)| = 0,$$

while the greatest particle distance goes to infinity faster than t

$$\lim_{t\to\infty} \frac{R(t)}{t} := \lim_{t\to\infty} t^{-1} \max_{i\neq j} |\mathbf{q}_i(t) - \mathbf{q}_j(t)| = \infty,$$

or the asymptotic behavior in the center-of-mass frame is characterized by

$$\mathbf{q}_i(t) = \mathbf{A}_i t + \mathcal{O}(t^{2/3}) \quad \forall i = 1, \ldots, N \quad \text{and} \quad \limsup_{t\to\infty} r > 0,$$

$$(10.9)$$

where $\mathbf{A}_i \in \mathbb{R}^3$ are constant vectors (possibly the zero vector). Note that since the dynamics are time-reversal invariant, the results hold for $t \to -\infty$, as well.

The first case describes the so-called "super-hyperbolic escape". This scenario is consistent with an increase of our gravitational entropy (10.7), implying both $I \to \infty$ and $U \to \infty$, as $t \to \infty$. It also includes, however, the pathological cases in which solutions diverge in finite time. It is the second case (when super-hyperbolic escape is excluded), in which the Newtonian N-body system is much more interesting and generally well-behaved. More precisely, we see that if (10.9) holds, all inter-particle distances fall into one of the following three classes (see Saari (1971), Cor. 1.1, together with Marchal and Saari (1974), Cor. 6):

$$|\mathbf{q}_i - \mathbf{q}_j| = L_{ij}t + \mathcal{O}(t^{2/3}), \tag{10.10}$$

or

$$|\mathbf{q}_i - \mathbf{q}_j| = \mathcal{O}(t^{2/3}) \tag{10.11}$$

or

$$|\mathbf{q}_i - \mathbf{q}_j| \le L \tag{10.12}$$

for large t and positive constants L, L_{ij}.

The result can be summarized as follows (cf. Saari (1971, p. 227)): On sufficiently large time-scales, the system forms clusters, consisting of particles whose mutual distances remain bounded. These clusters form subsystems (clusters of clusters) that are reasonably well isolated (energy and angular momentum are asymptotically conserved in each one of them separately), the distance between their centers of mass growing proportional to t. Finally, within each of these subsystems, the clusters separate approximately as $t^{2/3}$. In other words, the long-term behavior of such a Newtonian universe looks very much like *structure formation*, with local clumping into "galaxies", and global expansion due to galaxies and galaxy clusters receding from each other. Even quantitatively, the result fits reasonably well with the behavior of our actual universe: identifying a "scale factor" $a(t)$ with the typical inter-galaxy distance, we find $H(t) = \frac{\dot{a}}{a} \sim t^{-1}$ for large t, consistent with Hubble's law for a universe dominated by matter and/or radiation.

In regard to entropic considerations, i.e., equation (10.7), we note that the moment of inertia will grow asymptotically like $I(t) \sim t^2$, while the macrovariable $U(t)$ is at least bounded from below by some multiple of $\frac{N}{L^2}$ (assuming that the number of particles in a cluster is of order N). What happens at intermediate times? Assuming, henceforth, a non-negative total energy, we already know that $I(t)$ is strictly convex. Together with its quadratic growth for $t \to \pm\infty$, we can conclude that it has a unique global minimum, let's say at $t = \tau$, from which it increases in both time directions. $U(t)$ will in general fluctuate, but if we exclude particle collisions and "near particle collisions" (very close encounters), it will remain bounded and not fluctuate too quickly ($\dot{U}$ remains bounded, as well). Hence, we expect that the graph of $(E + U(t))I(t)$ (the logarithm of which is proportional to our gravitational entropy) looks qualitatively like that of $I(t)$, namely by and large parabolic. Indeed, numerical simulations by Barbour et al. (2013, 2015) for the $E = 0$ universe (with $N = 1000$ and random initial data) support the claim that the evolution of $I \cdot U$ is well interpolated by a parabola of the form $\alpha(t - \tau)^2 + \beta$ with $\alpha, \beta > 0$. All these suggest the desired U-shaped evolution of the entropy $S(E, I, U) \approx \frac{3N}{2}(\log(E+U)+\log(I))$ as a function of time for a Newtonian gravitating universe with non-negative energy. (Actually, on large time-scales, the shape looks less like a U and more like Y — how some children draw birds on the horizon — since $S(t)$ grows only logarithmically as $|t - \tau| \to \infty$.)

We conclude that a Newtonian gravitating universe is indeed a "Carroll universe" which has no equilibrium state and for which entropy increase (in opposite directions from a global minimum) is typical. This entropy increase is, moreover, consistent with structure formation. It does not merely lead to one big boring clump of matter.

10.6. Gravity and typicality from a relational point of view

Starting from Machian/Leibnizian principles, Barbour, Koslowski, and Mercati (2013, 2014, 2015) discuss the Newtonian gravitational system from a relationalist perspective. According to the relational framework that Julian Barbour has championed over the past decades, all physical degrees of freedom are described on *shape space S*, which is obtained from Newtonian configuration space by factoring out absolute rotations, translations, and scale, leaving us with a $3N - 7$ dimensional space for an N-particle system. The configuration of N point particles is then characterized by the angles and ratios between their (Euclidean) distance vectors — or, in other words, by its *shape* — independent of extrinsic scales. The lowest-dimensional shape space is that of $N = 3$ particles. In this case, the shapes are those of triangles — specified by 2 angles or the ratios between 3 distances — and the topology of shape space is that of a 2-dimensional (projective) sphere.

Considering standard Newtonian gravity on absolute space and trying to extract, so to speak, its relational essence, we have to eliminate all dependencies on extrinsic spatio-temporal structures. To this end, we restrict ourselves to models with vanishing total momentum, $\mathbf{P} = \sum_{i=1}^{N} \mathbf{p}_i \equiv 0$, and angular momentum, $\mathbf{L} = \sum_{i=1}^{N} \mathbf{q}_i \times \mathbf{p}_i \equiv 0$, excluding rotating universes and propagations of the center of mass, respectively.[2] Furthermore, the rejection of absolute time-scales leads to considering only universes with zero total energy ($E \equiv 0$), since this is the only value invariant under a rescaling of time-units.

The problematic issue when it comes to Newtonian gravity is its lack of scale-invariance. Newtonian gravity has models that do

---

[2]Arbitrary solutions of Newtonian mechanics can be projected onto shape space, but the total angular momentum (let's say) cannot be captured by relational initial data. It corresponds to a particular choice of absolute spatio-temporal reference frame ("gauge") in the shape space theory; cf. Dürr *et al.* (2019). $\mathbf{L} = 0$ is then the only canonical choice, and the only one suggested by Machian principles.

not rotate ($L \equiv 0$) and models that do not propagate ($P \equiv 0$), but it does not have models that do not expand ($D := \frac{1}{2}\dot{I} = \sum_{i=1}^{N} \mathbf{q}_i \cdot \mathbf{p}_i \equiv 0$; Barbour calls D the *dilatational momentum*). The characteristic size of an N-particle system is given by $\sigma = \sqrt{\frac{I}{m}}$, where $I = m \sum_{i=1}^{N} \mathbf{q}_i^2$ is the center-of-mass moment of inertia, and we have already seen that I can never be constant for non-negative energy (equation (10.8)) but is roughly parabolic as a function of time. In other words, an N-particle universe interacting by Newton's law of gravity always changes in size.[3] Of course, we can (and will) insist that this is meaningless from a relational point of view — since absolute distance is meaningless — but the process has nonetheless dynamical (and thus empirical) consequences in Newtonian theory. Simply put, for constant energy, a gravitating system that expands also slows down.[4] Conversely, if we eliminate scale by hand, namely by a time-dependent coordinate transformation $q \to \frac{q}{\sigma(t)}$, the resulting dynamics can be formulated on shape space, but will no longer have the standard Newtonian form. Instead, the dynamics become non-autonomous (time-dependent) with scale acting essentially like friction (Barbour *et al.*, 2014).

How to capture this time-dependence without reference to external time? Barbour *et al.* make use of the fact that the dilatational momentum $D = \frac{1}{2}\dot{I}$ is monotonously increasing ($\ddot{I} > 0$ by equation (10.8)) and can thus be used as an internal time-parameter, a kind of universal clock. In particular, we observe that $D = 0$ precisely when I reaches its global minimum. This *central time* thus marks the midpoint between a period of contraction and a period of expansion, or

---

[3]In fact, the general Lagrange-Jacobi identity shows that $E = 0$ and $I \equiv const.$ is possible only if the potential is homogeneous of degree -2. This had motivated the alternative, scale-invariant theory of gravitation proposed in Barbour (2003). Here, we discuss the relational formulation of Newtonian gravity with the familiar $\frac{1}{r}$-potential.

[4]Compare this with the following: absolute rotations are meaningless on shape space, but centrifugal forces would be observable through their effect on the particle motions; cf. Maudlin (2012) for an excellent discussion of this issue going back to the famous Leibniz-Clarke debate.

better (though this remains to be justified): the Janus point between two periods of expansion with respect to opposite arrows of time. It provides, in particular, a natural reference point for parametrizing solutions of the shape space theory in terms of *mid-point data* on the shape phase space T^*S.[5]

There is one last redundancy from the relational point of view that Barbour *et al.* (2015) called *dynamical similarity*. It comes from the invariance of the equations of motion under a simultaneous rescaling of internal time D and shape momenta. More simply put, two solution trajectories are physically identical if they correspond to the same geometric curve in shape space, the same sequence of shapes, even if that curve is run through at different "speeds". Thus, factoring out the absolute magnitude of the shape momenta at central time, we reduce the relevant phase space (that parametrizes solutions) by one further dimension. The resulting space PT^*S (mathematically, this is the projective cotangent bundle of shape space S) is *compact*, which means, in particular, that it has a *finite total volume* according to the uniform volume measure. And this is where the relational formulation, i.e., the elimination of absolute degrees of freedom, really starts to pay off. Since the uniform measure on PT^*S — that Barbour *et al.* take to be the natural typicality measure, following Laplace's principle of indifference, — is normalizable, it allows for a statistical analysis that avoids the ambiguities resulting from the infinite phase space measure in the Carroll model. It should be noted that the construction of the measure is not entirely canonical; it involves the choice of a metric on shape space (which, in turn, can be defined through a scale-invariant metric on absolute phase space). In general, the justification of the typicality measure remains a critical step that would require a more in-depth discussion.[6] For instance, we are rather skeptical of a "principle of indifference" as

---

[5]Mathematically, this is the cotangent bundle of shape space S, just as Hamiltonian phase space is the cotangent bundle of Newtonian configuration space.

[6]Dürr, Goldstein, and Zanghì (2019) provide an insightful discussion of typicality measures on shape space, though focussing on the quantum case.

a motivation for a uniform measure.[7] For the time being, we just take the pragmatic attitude that the typicality measure proposed by Barbour, Koslowski, and Mercati will be justified by its empirical and explanatory success. Deviating from their notation, we denote this measure on the reduced phase space by μ_ε.

10.6.1. *Shape complexity and the gravitational arrow*

To describe the macro-evolution of a gravitating system on shape space, Barbour and collaborators introduce a dimensionless (scale-invariant) macrovariable C_S, which they call *shape complexity*:

$$C_S = -V \cdot \sqrt{I}. \tag{10.13}$$

Comparison with (10.6) (setting $E = 0$ and noting that $U = -V$) shows a remarkable relationship between this shape complexity and the gravitational entropy that we computed on absolute phase space. Recalling our previous discussion (or noting that $C_S \approx R/r$, where R is the largest and r the smallest distance between particles), we also see that low shape complexity corresponds to dense homogeneous states in absolute space, while high shape-complexity indicates "structure" — dilute configurations of multiple clusters.

On shape space, considering the simplest case of $N = 3$ particles, the configuration of minimal shape complexity is the equilateral triangle, while the configuration of maximal shape complexity corresponds to "binary coincidences" in which the distance between two particles — relative to their distance to the third — is zero. This is to say that 3-particle configurations with high shape complexity will, in general, contain a Kepler pair (a gravitational bound state of two particles) with the third particle escaping to infinity.

In Section 10.5.1, we have discussed the typical evolution of $-V \cdot I$ and found it to be roughly parabolic or U-shaped. Analogously, one can conclude that the evolution of $C_S = -V \cdot \sqrt{I}$ (in Newtonian time) will typically exhibit a V-shaped profile: it has a global minimum at central time ($D = 0$), from which it grows roughly linearly (modulo

---

[7]However, see the contribution of Bricmont to this volume, which defends a modern version of the principle.

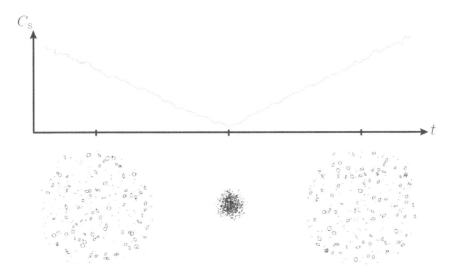

Fig. 10.6.   Top: evolution of the shape complexity C_S found by numerical simulation for $N = 1000$ and Gaussian initial data. Bottom: schematic conception (not found by numerical simulation) of three corresponding configurations on Newtonian space-time. (Source: Barbour *et al.* (2015))

fluctuations) in both time directions (see Fig. 10.6). In the terminology of Barbour, Koslowski, and Mercati, this defines two opposite *gravitational arrows of time* with the Janus point as their common past. Note that these are not *entropic* arrows, though our previous discussion strongly suggests that the evolution of the shape complexity on shape space will align with the evolution of the gravitational entropy (10.7) on absolute space.

A remarkable feature of the relational theory, however, is that it reveals the origin of the gravitational arrow to be *dynamical* rather than *statistical*: the negative of the shape complexity corresponds to the potential that generates the gravitational dynamics on shape space. There is thus a dynamical tendency towards higher values of C_S (lower values of the shape potential). In contrast, Boltzmannian statistical mechanics suggests that entropy increase is typical because there are a great many more high-entropy than low-entropy configurations that a system could evolve into. It does not suggest that physical forces are somehow driving the system towards higher entropy states.

Turning to the statistical analysis of the shape space theory, we are interested in the measure assigned to mid-point data (Janus point configurations) with low shape complexity

$$C_S \in [C_{min}, \, \alpha \cdot C_{min}] := I_1, \tag{10.14}$$

and, respectively, high shape complexity

$$C_S \in (\alpha \cdot C_{min}, \infty) := I_\infty. \tag{10.15}$$

Here, $1 < \alpha \ll \infty$ is some positive constant and C_{min} is the smallest possible value of C_S. The key result of the relational statistical analysis (not yet rigorously proven but strongly substantiated by the 3-particle case and numerical experiments for large N) is now that already for small values of α ($\alpha < 2$ for large N),

$$\frac{\mu_\varepsilon(I_\infty)}{\mu_\varepsilon(PT^*S)} \approx 0, \tag{10.16}$$

and consequently,

$$\frac{\mu_\varepsilon(I_1)}{\mu_\varepsilon(PT^*S)} \approx 1. \tag{10.17}$$

This means that it is typical that a universe at the Janus point (the beginning of our macro-history) is in a very homogeneous state!

Regardless of the philosophical merits of relationalism, the shape space theory of Barbour, Koslowski, and Mercati thus comes with two great virtues: First, it provides a sensible *normalizable* measure on the set of possible micro-evolutions that still establishes an arrow of time as typical. Even more spectacularly, typical evolutions with respect to this measure go through very homogeneous configurations at their Janus point ($\sim$ the "big bang"). In other words, initial states that have very low entropy from the absolutist point of view come out as typical in the shape space description — provided that one accepts the proposed measure on PT^*S as a natural typicality measure. This would resolve the two potential problems that we have identified for the Carroll model: the mysteriously low entropy of our universe and the justification for locating our present state reasonably far away from the entropy minimum.

10.6.2. *Entaxy and Entropy*

On the other hand, Barbour *et al.* introduce another concept called (instantaneous) *entaxy* that we find much less compelling. The instantaneous entaxy (the authors also use the term *solution entaxy* for the measure μ_ε on PT^*S) is supposed to be the measure of a set of shape configurations corresponding to a given value of shape complexity. It thus seems *prima facie* analogous to the Boltzmann entropy defined in terms of the macrovariable C_S, with the notable exception that it *decreases* in time as the shape complexity increases. Recall, however, that the measure μ_ε was only defined on mid-point data, by cutting through the set of solution trajectories at their Janus points, so to speak. Barbour *et al.* now extend it to arbitrary (internal) times by stipulating that the entaxy associated with a particular value of shape complexity at *any* point in history is the measure of *mid-point* configurations with that value of shape complexity.

This definition seems somewhat *ad hoc* and corresponds to comparing macrostates at different times in terms of a measure that is *not stationary* under the dynamics: A set of mid-point data will have a bigger size than the set of time-evolved configurations (phase space volume gets lost, so to speak). Indeed, on the 3-particle shape space, it is not hard to see that the points of maximal shape complexity are dynamical attractors; hence, a stationary continuous measure on shape phase space does not exist. In general, it is not even clear if a stationary measure is a meaningful concept in relational mechanics since there is no absolute (metaphysical) sense in which configurations on different solution trajectories with the same internal time are *simultaneous*. They merely happen to agree on whatever part or feature of the configuration plays the role of an internal "clock". For all these reasons, the entaxy should not be regarded as a shape analogon of the Boltzmann entropy (which is always defined in terms of a stationary measure). In particular, the fact that the gravitational arrows point in the direction of decreasing, rather than increasing, entaxy is not in contradiction with Boltzmannian arguments.

Finally, one may wonder whether we could compute on absolute phase space the Boltzmann entropy associated to the shape complexity or other scale-invariant macrovariables. Note that for $E = 0$, our gravitational entropy (10.7) is a function of $-VI$. Couldn't we have just computed an entropy for the macrovariable $C_S = -V\sqrt{I}$, instead? Interestingly, the answer is negative, and the reason is the following simple result, showing that macroregions defined by scale-invariant macrovariables would have measure zero or infinity.

Lemma 10.1. *Let μ be a measure on Ω that is homogeneous of degree d, i.e., $\mu(\lambda A) = \lambda^d \mu(A)$ for any measurable $A \subset \Omega$ and $\lambda > 0$. Let $F : \Omega \to \mathbb{R}^n$ be a measurable function, homogeneous of degree k, i.e., $F(\lambda x) = \lambda^k F(x) \; \forall x \in \Omega$. Then, we have for any interval $I \subset \mathbb{R}^n$ (possibly consisting of a single point):*

$$\mu\left(\{x \mid F(x) \in \lambda^k I\}\right) = \lambda^d \mu(\{x \mid F(x) \in I\}). \qquad (10.18)$$

Proof.

$$\mu\left(\{x \mid F(x) \in I\}\right) = \mu\left(\{x \mid F(\lambda x) \in \lambda^k I\}\right)$$

$$= \mu\left(\{\lambda^{-1} y \mid F(y) \in \lambda^k I\}\right)$$

$$= \lambda^{-d} \mu\left(\{x \mid F(x) \in \lambda^k I\}\right). \qquad \square$$

From this, we can immediately conclude:

Corollary 10.2. *If the measure μ is homogeneous of degree $d \neq 0$ and F is homogeneous of degree 0 (i.e., scale-invariant), then*

$$\mu\left(F^{-1}(I)\right) \in \{0, +\infty\}. \qquad (10.19)$$

Proof. Applying equation (10.18) with $k = 0$ and $d \neq 0$ yields $\mu\left(F^{-1}(I)\right) = \lambda^d \mu\left(F^{-1}(I)\right)$ for any $\lambda > 0$. $\square$

Hence, using a homogeneous phase space measure (such as the Liouville measure or the microcanonical measure for a homogeneous potential and $E = 0$), macroregions defined in terms of scale-invariant macrovariables must have measure zero or infinity, so that the corresponding Boltzmann entropy would be ill-defined.

This suggests that the concept of entropy is intimately linked to absolute scales, and thus not manifestly relational. Note, in particular, that *expansion* and *heating* — processes that are paradigmatic for entropy increase (especially, but not exclusively in our analysis of gravitating systems) — require absolutes scales of distance and velocity, respectively.

This emphasizes once again that the gravitational arrow of Barbour *et al.* is not an entropic arrow, although it matches — maybe accidentally, maybe for reasons we don't yet understand — the entropic arrow that we identified on absolute phase space. The result also leaves the relationalist with the following interesting dilemma: Either the notion of entropy is meaningful only for subsystems — for which the environment provides extrinsic scales — or we have to explain why the entropy of the universe is a useful and important concept *despite* the fact that it is related to degrees of freedom that are strictly speaking unphysical, corresponding to mere gauge in the shape space theory.

10.7. Conclusion

The works of Carroll and Chen as well as Barbour, Koslowski, and Mercati showed that it is possible to establish an arrow of time as typical, without the need to postulate special boundary conditions or any other form of Past Hypothesis. By proposing the definition of a Boltzmann entropy for a classical gravitating universe, we argued that Newtonian gravity provides a relevant realization of Carroll's entropy model that can be compared to the shape space formulation of Barbour *et al.* We found, in particular, that the gravitational arrows identified by Barbour and collaborators in terms of shape complexity will match the entropic arrows in the theory on absolute space. The extension to other microscopic theories (and/or macroscopic state functions) will require further research. The relationalist and the absolutist approach both provide the resources to avoid the reversibility paradox and ground sensible inferences about the past and future of the universe. However, while certain ambiguities remain in the Carroll model, resulting from the non-normalizability of the phase space measure, those issues are resolved by the shape space theory of Barbour *et al.* — provided one accepts their choice of typicality measure. In any case, for a Newtonian gravitating universe, their analysis suggests that homogeneous configurations at the "big bang" (Janus point) are *typical*, explaining why the universe started in what looks like a very low-entropy state from an absolutist perspective. However, if the shape space theory is actually fundamental, the "entropy of the universe" turns out to be a somewhat spurious concept whose status remains to be discussed in more detail.

Appendix: Computation of the gravitational entropy

We compute the phase space volume of the macroregion $\Gamma(E, I \pm \epsilon I, U \pm \epsilon U)$, that is:

$$\frac{1}{N! h^{3N}} \int d^{3N} p \int d^{3N} q \; \delta\big(H(\mathbf{q}, \mathbf{p}) - E\big)$$

$$\times \mathbb{1}\Big\{(1 - \epsilon)U \le \sum_{\substack{i<j \\ i,j=1}} \frac{Gm^2}{|\mathbf{q}_i - \mathbf{q}_j|} \le (1 + \epsilon)U\Big\}$$

$$\times \mathbb{1}\Big\{(1 - \epsilon)I \le \sum_{i=1}^{N} m\mathbf{q}_i^2 \le (1 + \epsilon)I\Big\}, \tag{10.20}$$

with the Hamiltonian

$$H(q, p) = \sum_{i=1}^{N} \frac{\mathbf{p}_i^2}{2m} - \sum_{1 \le i < j \le N} \frac{Gm^2}{|\mathbf{q}_i - \mathbf{q}_j|}.$$

We shall prove the following:

Proposition. *If the scale-invariant quantity $\sqrt{IU}$ is large enough that*

$$\sqrt{IU} \ge 4Gm^{5/2} \log(N) N^{5/2}, \tag{10.21}$$

we obtain the bounds:

$$|\Gamma(E, I \pm \epsilon I, U \pm \epsilon U)| \geq CN^2 \, e^{-\frac{7}{2}N} \left(\frac{Gm^2\epsilon}{U}\right)^3$$

$$\times (E + (1-\epsilon)U)^{\frac{3N-2}{2}} \left(\frac{I}{m}\right)^{\frac{3N-3}{2}},$$

$$|\Gamma(E, I \pm \epsilon I, U \pm \epsilon U)| \leq \frac{2C}{3N} \left[e^{\frac{3N\epsilon}{2}} - 1\right] (E + (1+\epsilon)U)^{\frac{3N-2}{2}} \left(\frac{I}{m}\right)^{\frac{3N}{2}},$$

for any $\epsilon > \frac{2}{N}$ *and* $N \geq 4$, *where* $C = \frac{(2m)^{\frac{3N-2}{2}}}{2 \cdot N! h^{3N}} [\Omega^{3N-1}]^2$, *with* Ω^{3N-1} *the surface element of the* $(3N-1)$-*dimensional unit sphere.*

For non-negative E, *this can be simplified further by using* $(E + (1+\epsilon)U)^n \leq (1+\epsilon)^n (E+U)^n \leq e^{\epsilon n}(E+U)^n$, *respectively* $(E+(1-\epsilon)U)^n \geq (1-\epsilon)^n (E+U)^n \geq e^{-2\epsilon n}(E+U)^n$, *for* $\epsilon < \frac{1}{2}$.

Proof. We first perform the integral over the momentum variables and are left with

$$\frac{(2m)^{\frac{3N-2}{2}}}{2N! h^{3N}} \Omega^{3N-1} \int d^{3N}q \left(E + \sum_{i<j} \frac{Gm^2}{|\mathbf{q}_i - \mathbf{q}_j|}\right)^{\frac{3N-2}{2}}$$

$$\times \mathbb{1}\left\{(1-\epsilon)U \leq \sum_{\substack{i<j \\ i,j=1}} \frac{Gm^2}{|\mathbf{q}_i - \mathbf{q}_j|} \leq (1+\epsilon)U\right\}$$

$$\times \mathbb{1}\left\{(1-\epsilon)I \leq \sum_{i=1}^{N} m\mathbf{q}_i^2 \leq (1+\epsilon)I\right\}.$$

From this, it is straightforward to obtain the upper bound:

$$(10.20) \leq \frac{(2m)^{\frac{3N-2}{2}}}{2N! h^{3N}} \Omega^{3N-1} \frac{2\Omega^{3N-1}}{3N} (E + (1+\epsilon)U)^{\frac{3N-2}{2}}$$

$$\times \left[\left((1+\epsilon)\frac{I}{m}\right)^{\frac{3N}{2}} - \left((1-\epsilon)\frac{I}{m}\right)^{\frac{3N}{2}}\right]$$

$$\leq \frac{C}{3N}[e^{\frac{3N\epsilon}{2}} - 1](E + (1+\epsilon)U)^{\frac{3N-2}{2}} \left(\frac{I}{m}\right)^{\frac{3N}{2}},$$

where we used $(1+\epsilon)^N = e^{\log(1+\epsilon)N} \leq e^{\epsilon N}$.

For the lower bound, we consider the set $\mathcal{B} := B_1 \times \ldots \times B_N \subset \mathbb{R}^{3N}$ defined by

$$B_j := \{|\mathbf{q}_j| \in [(2j-2)\xi, (2j-1)\xi]\},$$

with $\xi = \xi(I, N)$ to be determined soon. That is, we consider a series of concentric spheres around the origin, their radii being an increasing multiple of ξ, and configurations for which the volume between two spheres is alternately empty or occupied by a single particle. For $\mathcal{B}$, we have $m q^2 \in [I_+ - \Delta I, I_+]$ with

$$I_+ = \sum_{i=2}^{N} m\mathbf{q}_i^2 \leq m \sum_{i=2}^{N} (2i-1)^2 \xi^2 = \frac{m}{3} N(4N^2 - 1)\xi^2 \qquad (10.22)$$

and

$$\Delta I = m \sum_{i=1}^{N} [(2i-1)^2 - (2i-2)^2]\xi^2$$

$$= m \sum_{i=1}^{N} [4i - 3]\xi^2 \leq 2mN^2\xi^2 \leq \frac{2}{N} I_+.$$

We thus set

$$\xi := \sqrt{\frac{3I}{mN(4N^2 - 1)}}, \qquad (10.23)$$

so that $I_+ = I$.

Furthermore, for $q \in \mathcal{B}$, the distance between two particles is bounded from below by $|\mathbf{q}_i - \mathbf{q}_j| \geq (2(j-i) - 1)\xi$, and for each $1 \leq k \leq N-1$, there exists less than N particle pairs with $j - i = k$. Hence, the potential energy is bounded by

$$\sum_{i<j} \frac{Gm^2}{|\mathbf{q}_i - \mathbf{q}_j|} \leq N \sum_{k=1}^{N-1} \frac{Gm^2}{(2k-1)\xi} \leq 2N \log(N) \frac{Gm^2}{\xi} < U \qquad (10.24)$$

by (10.21), where we used

$$\sum_{k=1}^{N-1} \frac{1}{(2k-1)} \leq 1 + \sum_{k=2}^{N-1} \frac{1}{k} \leq 1 + \int_1^N \frac{1}{x} \mathrm{d}x$$
$$= 1 + \log(N) \leq 2\log(N), \quad \text{for } N \geq 4.$$

In particular, we know that for $q \in \mathcal{B}$ and, e.g., $\mathbf{q}_1 = \mathbf{0}$, we have $|V(q)| < U$, while $\lim_{\mathbf{q}_1 \to \mathbf{q}_2} |V(q)| = +\infty$. Hence, by the mean value theorem, there exists for any $(\mathbf{q}_2, \ldots, \mathbf{q}_N) \in B_2 \times \ldots \times B_N$ a $\lambda \in (0,1)$ such that $-V(\lambda \mathbf{q}_2, \mathbf{q}_2, \ldots, \mathbf{q}_N) = U$. Moreover, we have

$$\nabla_{\mathbf{q}_1} V(\mathbf{q}_1, \mathbf{q}_2, \ldots, \mathbf{q}_N) = Gm^2 \sum_{i=2}^N \frac{\mathbf{q}_1 - \mathbf{q}_i}{|\mathbf{q}_1 - \mathbf{q}_i|^3},$$

and thus $|\nabla_{\mathbf{q}_1} V| \leq \frac{|V|^2}{Gm^2}$. Hence, if $\mathbf{q}_1$ is inside a ball of radius $r = \frac{\epsilon Gm^2}{2U}$ around $\lambda \mathbf{q}_2$, we certainly have $|\nabla_{\mathbf{q}_1} V| < (1+\epsilon)^2 U^2 < 2U^2$ and $|V - U| \leq \epsilon U$. Moreover, $m\sum_{i=1}^N \mathbf{q}_i^2$ increases by less than $m(3\xi)^2 < \epsilon I$, and thus remains within the interval $[I \pm \epsilon I]$. We denote this ball by $K_{r(q)}$. Its volume is $\frac{4\pi}{3}\left(\frac{Gm^2\epsilon}{2U}\right)^3$. The volume of a set B_j is

$$|B_j| = \frac{4\pi}{3}[(2j-1)^3 - (2j-2)^3]\xi^3 = \frac{4\pi}{3}[j(12j-18)+7] \geq \frac{16\pi}{3}j^2.$$

Hence:

$$|B_2 \times \cdots \times B_N| = \prod_{j=2}^N |B_j| = \left(\frac{16\pi}{3}\right)^{N-1} \xi^{3(N-1)} \prod_{j=2}^N j^2$$
$$= \left(\frac{16\pi}{3}\right)^{N-1} \xi^{3(N-1)}(N!)^2.$$

Now, we have on the one hand $\xi^{3(N-1)} \geq \left(\frac{3}{4}\right)^{\frac{3(N-1)}{2}} \left(\frac{I}{m}\right)^{\frac{3(N-1)}{2}}$ $N^{-\frac{9(N-1)}{2}}$. On the other hand, comparing with the area of the unit sphere, $\Omega^{3N-1} = \frac{2\pi^{3N/2}}{\Gamma\left(\frac{3N}{2}\right)}$, one checks that $\left(\frac{16\pi}{3}\right)^{N-1}\left(\frac{3}{4}\right)^{\frac{3(N-1)}{2}} > \Omega^{3N-1}\Gamma\left(\frac{3N}{2}\right)e^{\frac{N}{2}}$. Thus:

$$(10.20) \geq \frac{(2m)^{\frac{3N-2}{2}}}{2N!h^{3N}}\,\Omega^{3N-1}\int_{K_{r(q)}\times B_2\times\ldots\times B_N} d^{3N}q$$

$$\times \left(E + \sum_{i<j}\frac{Gm^2}{|\mathbf{q}_i - \mathbf{q}_j|}\right)^{\frac{3N-2}{2}} \mathbb{1}\{\ldots\}\mathbb{1}\{\ldots\}$$

$$\geq C\,e^{\frac{N}{2}}N^{-\frac{9(N-1)}{2}}(N!)^2\,\Gamma\left(\frac{3N}{2}\right)\left(\frac{Gm^2\epsilon}{2U}\right)^3$$

$$\times (E + (1-\epsilon)U)^{\frac{3N-2}{2}}\left(\frac{I}{m}\right)^{\frac{3N-3}{2}}.$$

Summing over all possible permutations of the particles over the rings in the definition of $\mathcal{B}$, we get an additional factor of $N!$. With the Sterling approximation $\Gamma(n+1) = n! > \sqrt{2\pi n}\left(\frac{n}{e}\right)^n$, we finally obtain a lower bound of the form

$$(10.20) \geq C\,N^2 e^{-\frac{7}{2}N}\left(\frac{Gm^2\epsilon}{U}\right)^3 (E + (1-\epsilon)U)^{\frac{3N-2}{2}}\left(\frac{I}{m}\right)^{\frac{3N-3}{2}}. \qquad \Box$$

Acknowledgements

We thank Julian Barbour, Detlef Dürr, Valia Allori, and two anonymous referees for valuable comments.

References

Albert, D. (2001) *Time and Chance*, Harvard University Press, Cambridge, Mass.

Albert, D. (2015) *After Physics*, Harvard University Press, Cambridge, Mass.

Barbour, J. (2003) Scale-invariant gravity: Particle dynamics, *Classical and Quantum Gravity*, **20**, 1543.

Barbour, J., Koslowski, T. Mercati, F. (2013) A gravitational origin of the arrows of time. ArXiv: 1310.5167 [gr-qc].

Barbour, J., Koslowski, T. Mercati, F. (2014) Identification of a gravitational arrow of time, *Physical Review Letters*, **113**, 181101.

Barbour, J., Koslowski, T. Mercati, F. (2015) Entropy and the typicality of universes. ArXiv: 1507.06498 [gr-qc].

Boltzmann, L. (1896) *Vorlesungen über Gastheorie*, Verlag v. J. A. Barth, Leipzig (Nabu Public Domain Reprints).

Boltzmann, L. (1897) Zu Hrn. Zermelos Abhandlung 'Über die mechanische Erklärung irreversibler Vorgänge, *Annalen der Physik*, **60**, 392–398. Reprinted and translated as Chapter 10 in Brush (1966).

Bricmont, J. (1995) Science of chaos or chaos in science, *Physicalia Magazine*, **17**, 159–208.

Brush, S. G. (1966) *Kinetic Theory*, Pergamon, Oxford.

Callender, C. (2004) There is no puzzle about the low-entropy past, In: ed. by C. Hitchcock, *Contemporary Debates in Philosophy of Science*, Blackwell, pp. 240–255.

Callender, C. (2009) The past hypothesis meets gravity, In: eds. by G. Ernst and A. Hutteman, *Time, Chance and Reduction: Philosophical Aspects of Statistical Mechanics*, Cambridge University Press, Cambridge.

Carroll, S., Chen, J. (2004) Spontaneous Inflation and the Origin of the Arrow of Time. ArXiv: 0410270 [hep-th].

Carroll, S. (2010) *From Eternity to Here: The Quest for the Ultimate Theory of Time*, Dutton, Penguin, USA.

Chalmers, D. (1995) Facing Up to the problem of consciousness, *Journal of Consciousness Studies*, **2**(3), 200–19.

Dürr, D., Goldstein, S., Zanghì, N. (2019) Quantum Motion on Shape Space and the Gauge Dependent Emergence of Dynamics and Probability in Absolute Space and Time. *Journal of Statistical Physics*. https://doi.org/10.1007/s10955-019-02362-9.

Feynman, R. (1967) *The Character of Physical Law*, The MIT Press, Cambridge, Mass.

Goldstein, S. (2001) Boltzmann's approach to statistical mechanics, In: eds. by J. Bricmont, D. Dürr, *et al.*, *Chance in Physics. Foundations and Perspectives*, Springer, Berlin, pp. 39–54.

Goldstein, S., Tumulka, R., Zanghì, N. (2016) Is the hypothesis about a low entropy initial state of the universe necessary for explaining the arrow of time? *Physical Review D* **94**(2).

Heggie, D., Hut, P. (2003) *The Gravitational Million-Body Problem*, Cambridge University Press, Cambridge, UK.

Kiessling, M. (2001) How to implement Boltzmann's probabilistic ideas in a relativistic world? In: eds. by J. Bricmont, D. Dürr, *et al.*, *Chance in Physics. Foundations and Perspectives*, Springer, Berlin, pp. 39–54.

Lazarovici, D., Reichert, P. (2015) Typicality, irreversibility and the status of macroscopic laws, *Erkenntnis*, **80**(4), 689–716.

Lebowitz, J. (1993) Macroscopic laws, microscopic dynamics, time's arrow and Boltzmann's entropy, *Physica A* **194**, 1–27.

Loewer, B. (2007) Counterfactuals and the second law, In: eds. by H. Price and R. Corry, *Causation, Physics, and the Constitution of Reality. Russell's Republic Revisited*, Oxford University Press, Oxford.

Loewer, B. (2012) Two accounts of laws and time, *Philosophical Studies*, **160**, 115–37.

Marchal, C., Saari, D. (1974) On the final evolution of the N-body problem, *Journal of Differential Equations*, **20**, 150–186.

Maudlin, T. (2012) *Philosophy of Physics: Space and Time*, Princeton University Press, Princeton.

Padmanabhan, T. (1990) Statistical mechanics of gravitating systems, *Physics Reports*, **188**(5), 285–362.

Penrose, R. (1999) *The Emperor's New Mind: Concerning Computers, Minds, and the Laws of Physics*, Oxford University Press.

Pollard, H. (1967) Gravitational systems, *Journal of Mathematics and Mechanics*, **17**, 601–612.

Reichert, P. (2012) Can a Parabolic-like Evolution of the Entropy of the Universe Provide the Foundation for the Second Law of Thermodynamics?" Master thesis (Math. Institut LMU München).

Saari, D. (1971) Expanding gravitational systems, *Transactions of the American Mathematical Society*, **156**, 219–240.

Wallace, D. (2010) Gravity, entropy, and cosmology: In Search of clarity, *British Journal for Philosophy of Science*, **61**, 513–540.

Winsberg, E. (2012) Bumps on the road to here (from eternity), *Entropy*, **14**(3), 390–406.

Chapter 11

The Influence of Gravity on the Boltzmann Entropy of a Closed Universe

Michael K.-H. Kiessling

Department of Mathematics, Rutgers University
Piscataway NJ 08854, USA
miki@math.rutgres.edu

This contribution inquires into Clausius' proposal that "the entropy of the world tends to a maximum." The question is raised whether the entropy of 'the world' actually does have a maximum; and if the answer is "Yes!," what such states of maximum entropy look like, and if the answer is "No!," what this could entail for the fate of the universe. Following R. Penrose, 'the world' is modelled by a closed Friedman–Lemaître type universe, in which a three-dimensional spherical 'space' is filled with 'matter' consisting of N point particles, their large-scale distribution being influenced by their own gravity. As 'entropy of matter', the Boltzmann entropy for a (semi-)classical macrostate, and Boltzmann's ergodic ensemble formulation of it for an isolated thermal equilibrium state, are studied. Since the notion of a Boltzmann entropy is not restricted to classical non-relativistic physics, the inquiry will take into account quantum theory as well as relativity theory; we also consider black hole entropy. Model universes having a maximum entropy state and those which don't will be encountered. It is found that the answer to our maximum entropy question is not at all straightforward at the general-relativistic level. In particular, it is shown that the increase in Bekenstein–Hawking entropy of general-relativistic black holes does not always compensate for the Boltzmann entropy of a piece of matter swallowed by a black hole.

11.1. Introduction

The notion of *entropy*, roughly meaning "amount of transformation",
was introduced into science in 1865 by Rudolf Clausius in his path-
breaking paper [Clausius (1865)] on the mechanical theory of heat,
where he recast what was known as *the first and second laws of the
theory of heat* in a format still featuring prominently in today's thermodynamics. Although Clausius' discovery of the entropy concept
was based on his sound mathematical reasoning, and on his careful
analysis of empirical evidence obtained in laboratory experiments,
he was thinking in much bigger terms. Indeed, at the very end of his
paper, he argued that the fate of the whole universe is ruled by the
two laws as follows:

> Law 1: *The energy of the world is constant.*
> Law 2: *The entropy of the world tends to a maximum.*

In Law 2), it is tacitly understood that this tendency is monotonic.

Since the fate of the whole universe is at stake, it may be worthwhile to examine Clausius' bold proposal more closely. In this contribution, we ask:

Q1: *Does Clausius' world (or its kin) have an entropy maximum?*
Q2: *If the answer to Q1 is "Yes!", what is its maximum entropy
state?*
Q3: *If the answer to Q1 is "No!", then what is the fate of the universe?*

Since Clausius' notion of the world (universe) has long been
superseded — thanks to the revolutionary developments in observational, experimental, and theoretical physics —, to avoid a pointless
academic exercise, we will examine Clausius' proposal not merely
from the perspective of classical Newtonian-type theory which ruled
supreme at Clausius' times, but we also take quantum theory and relativity theory into account. Thus, we use a sequence of increasingly
more realistic model worlds with finite matter and energy content.
No pretense is made that our findings correctly anticipate the fate of
our real universe; yet, we hope that they offer some insights into its
inner workings.

We follow R. Penrose [Penrose (1989)] and carry out our inquiry in the setting of a closed universe, defined in Section 11.2 and Appendix A. In Section 11.3, we recall Boltzmann's entropy. In Section 11.4, we study the three-dimensional classical model with Newtonian gravity and find the Boltzmann entropy to be without upper bound. In an appendix to Section 11.4 (Appendix B), we demonstrate that the Boltzmann entropy would actually have a maximum if the Newtonian gravitional pair energy would only diverge logarithmically with the separation of the particles rather than reciprocally; if space were two-dimensional, then a spherical Newtonian universe would be just like that. Section 11.5 deals with a non-relativistic quantum-mechanical improvement over the model of Section 11.4. It is found that quantum mechanics stabilizes the divergent Newtonian gravitational attraction at short distance, as a consequence of which the quantum analog of Boltzmann's entropy features a maximum. In Section 11.6, we take special relativity into account, while gravity is still Newtonian. Remarkably, special relativity destabilizes, and in a world containing more than about one solar mass, the quantum Boltzmann entropy does not have a maximum. At long last, in Section 11.7, we inquire into general-relativistic issues and the role of the Bekenstein–Hawking black hole entropy. Roger Penrose has argued that the Bekenstein–Hawking entropy of a black hole which has swallowed all the matter is the maximum entropy state of a closed universe. We dispute this proposal by showing that the sum of the Bekenstein–Hawking black hole entropy and the quantum Boltzmann entropy of the cosmic microwave background radiation ("matter") outside the black hole's event horizon does not always increase, if the black hole swallows up some of the radiation. To rescue the second law, we suggest that the (quantum) Boltzmann entropy of matter inside the horizon has to be taken into account.

11.2. The closed universe framework

We will, for the most part, ignore that our universe seems to have had a beginning in the "Big Bang," and that it is expanding, and that it may well not be closed but open. Here, by "closed", we mean "of finite spatial extent," while "open" means "unbounded," which is the usual terminology in cosmology. Since an unbounded universe with an infinite, or finite, amount of matter in it would almost inevitably have no upper limit to its entropy, and since this was almost surely clear to Clausius, it stands to reason that Clausius had in mind a closed universe of finite matter content. We contemplate, following James Hopwood Jeans [Jeans (1902)] (and subsequently Albert Einstein [Einstein (1917)] and Roger Penrose [Penrose (1989)]), that space is spherical; that is, a three-dimensional sphere of radius R, viz. $\mathbb{S}_R^3$. The easiest way to think about $\mathbb{S}_R^3$ is as the subset of Euclidean space vectors $\mathbf{s} \in \mathbb{R}^4$ restricted to have Euclidean length $|\mathbf{s}| = R$; writing $\mathbf{s} \in \mathbb{S}_R^3$ means just this. We also assume that cosmological time $t \in \mathbb{R}$.

We will, for simplicity, assume that matter in this universe consist of a single species of 'fundamental' particles, which we call 'atoms,' represented by N point particles with gravitational interactions between them. For most of the discussion, we will use Newtonian gravity (see Appendix A), with general relativity taken into account at a later stage.

To characterize a microstate of N Newtonian point particles, it is not enough to give all their positions. One also needs their velocities, respectively, their momenta. The notion of Newton's mechanics extended to particle motions in non-Euclidean spaces was worked out by Killing, see [Killing (1885)], but in our simple geometrical setup, we do not need to invoke the abstract formalism of differential geometry. True, since physical space in our spherical universe model is not flat but a Riemann manifold with constant curvature, we cannot simply add or subtract two points in $\mathbb{S}_R^3$ to get a new point in $\mathbb{S}_R^3$. However, since we can think of points $\mathbf{s} \in \mathbb{S}_R^3$ as vectors $\mathbf{s} \in \mathbb{R}^4$ of length R, we can add or subtract such four-dimensional vectors to get a new four-dimensional vector in $\mathbb{R}^4$. In particular, we can define the particle velocities $\mathbf{v}(t) := \frac{d}{dt}\mathbf{q}(t)$ at $\mathbf{q}(t) \in \mathbb{R}^4$ in the usual vector calculus way as a limit when $\Delta t \to 0$ of the vector differential quotient $\frac{\mathbf{q}(t+\Delta t)-\mathbf{q}(t)}{\Delta t}$ in $\mathbb{R}^4$. And for motions which take place in the subset $\mathbb{S}_R^3 \subset \mathbb{R}^4$, find that $\mathbf{v}(t)$ is always tangential to $\mathbb{S}_R^3$ at $\mathbf{q}(t) \in \mathbb{S}_R^3$. Thus, given $\mathbf{q}(t) \in \mathbb{S}_R^3$ at time t, one only needs three numbers to characterize its velocity $\mathbf{v}(t)$, a three-dimensional vector in the tangent space $T_{\mathbf{q}}\mathbb{S}_R^3$ of the instantaneous particle position $\mathbf{q} \in \mathbb{S}_R^3$. Of course, for each $\mathbf{q}$, the attached tangent space $T_{\mathbf{q}}\mathbb{S}_R^3 \sim \mathbb{R}^3$, but to compare vectors in tangent spaces for different $\mathbf{q}$ and $\mathbf{q}'$, one needs a *connection*. We don't need to worry about this notion in its generality here, because we can think of each $T_{\mathbf{q}}\mathbb{S}_R^3$ naturally as a three-dimensional Euclidean subspace of $\mathbb{R}^4$ in which $\mathbb{S}_R^3$ is embedded.

As for the particle momenta, in Newtonian physics, they are simply the product of the particle velocities with their inert mass m; yet, because of the subsequent developments (Lagrange and Hamilton formulations), momenta are nowadays considered to not live in the tangent space but in the co-tangent space $T_{\mathbf{q}}^*(\mathbb{S}^3) \sim \mathbb{R}^3$ of the particle position $\mathbf{q} \in \mathbb{S}_R^3$. Elements of the co-tangent space act on elements of the tangent space as bounded linear maps into the real numbers. In the Newtonian setting, there is of course the obvious identification $\mathbf{p}(\mathbf{u}) = m\mathbf{v} \cdot \mathbf{u}$ (Euclidean inner product in $\mathbb{R}^3$), with both $\mathbf{v}$ and $\mathbf{u}$ velocities in the same $\mathbb{R}^3$. This is a fine point, which is made for the sake of mathematical accuracy; it is not important for the conceptual developments in this paper.

11.3. Boltzmann's classical entropy formulas

We need a formula which allows us to study the entropy of such a closed universe. While Clausius was an atomist, he did not propose a formula for computing entropy based on a mechanical atomistic theory of matter. This step had to wait until 1872 when Ludwig Boltzmann introduced his "H function" (essentially the negative of the entropy) into the kinetic theory of gases; cf. [Boltzmann (1872)]. We here adapt Boltzmann's H function formula to our setting of a dilute, monatomic, ideal classical gas in $\mathbb{S}_R^3$. We also take into account the subsequent adjustments physicists have made to Boltzmann's entropy formula: We add Gibbs' [Gibbs (1902)] combinatorial term $-\ln N!$, which accounts for the permutations of the N identical particles, and we quantify the entropy in units of the Boltzmann constant k_{B} and the phase space measure of a particle in units of the cube of Planck's constant[1] h. The Boltzmann entropy of such a classical gas reads

$$S_{\mathrm{B}}(f; N) = -k_{\mathrm{B}} N \ln N - k_{\mathrm{B}} N \int_{\mathbb{S}_R^3} \left(\int_{T_{\mathbf{s}}^*(\mathbb{S}_R^3)} f \ln(h^3 f / e) \mathrm{d}^3 p \right) \mathrm{d}^3 s.$$

$$(11.1)$$

Here, $f(\mathbf{s}, \mathbf{p}, t)$ is a continuum approximation to the *normalized empirical* (i.e. actual) density of the N-atom gas in the six-dimensional co-tangent bundle[2] $T^*(\mathbb{S}_R^3) := \bigcup_{\mathbf{s} \in \mathbb{S}_R^3} T_{\mathbf{s}}^*(\mathbb{S}_R^3)$ of the physical space $\mathbb{S}_R^3$, at "cosmic time" t.

We pause for a moment to comment on (11.1).

In the physics literature, the co-tangent bundle $T^*(\mathbb{S}_R^3)$ of physical space $\mathbb{S}_R^3$ is frequently called the "one-particle phase space," but this terminology is misplaced in the context of (11.1). Namely, in the interpretation of $T^*(\mathbb{S}_R^3)$ as one-particle phase space we would have $T^*(\mathbb{S}_R^3) = \bigcup_{\mathbf{q} \in \mathbb{S}_R^3} T_{\mathbf{q}}^*(\mathbb{S}_R^3)$ (N.B., $\mathbf{q}$ denotes the position of a particle, whereas $\mathbf{s}$ denotes a point in space irrespectively of whether that point $\mathbf{s}$ is occupied by a point particle or not). And then the non-negative function $f(\mathbf{q}, \mathbf{p}, t)$ would not be a continuum approximation to an actual normalized density, but would instead have the meaning of a probability density for finding that single particle having position $\mathbf{q} \in \mathbb{S}_R^3$ and momentum $\mathbf{p} \in T_{\mathbf{q}}^*(\mathbb{S}_R^3)$. Except for the factor N and the additive $-N \ln N$, this would turn the integral in (11.1) into an expected value of $\ln f$ with respect to f, essentially. In other words, the integral in (11.1) would be an ensemble entropy — a Gibb's entropy for a single-particle ensemble with ensemble probability density f. This would make its factor N quite incomprehensible, not to speak of the $N \ln N$ term.

---

[1] The Boltzmann constant k_{B} and the Planck constant h were introduced by Max Planck.

[2] If, instead of spherical space $\mathbb{S}_R^3$, we would work with flat space $\mathbb{R}^3$, the co-tangent spaces at different points $\mathbf{q}_1$ and $\mathbf{q}_2$ would just be Euclidean translates of each other, and the co-tangent bundle of $\mathbb{R}^3$ would become just the Cartesian product $\mathbb{R}_{(q)}^3 \times \mathbb{R}_{(p)}^3$, where the suffix (q), (p) indicates position space and momentum space, respectively. But $\mathbb{S}_R^3$ is not a linear space, so its co-tangent bundle $\bigcup_{\mathbf{s} \in \mathbb{S}_R^3} T_{\mathbf{s}}^*(\mathbb{S}_R^3)$ is a more complicated manifold. Yet thanks to the embedding $\mathbb{S}_R^3 \subset \mathbb{R}^4$, the bundle $\bigcup_{\mathbf{s} \in \mathbb{S}_R^3} T_{\mathbf{s}}^*(\mathbb{S}_R^3)$ is a six-dimensional subset of $\mathbb{R}_{(q)}^4 \times \mathbb{R}_{(p)}^4$.

Clearly, thinking of the integral in (11.1) as an ensemble entropy means to completely miss Boltzmann's point that (11.1) is the physical entropy of an individual N body system with $N f(\mathbf{s}, \mathbf{p}, t)$ a continuum approximation to the particle density in the co-tangent bundle $\bigcup_{\mathbf{s} \in \mathbb{S}_R^3} T_\mathbf{s}^*(\mathbb{S}_R^3)$ of *physical space* (here: $\mathbb{S}_R^3$). To obtain $f(\mathbf{s}, \mathbf{p}, t)$ at time t from a system with $N \gg 1$ particles, for each $\mathbf{s} \in \mathbb{S}_R^3$, consider a small sphere centered at $\mathbf{s}$ containing itself $n \gg 1$ particles, but with $n \ll N$. The momenta of the particles in this little sphere can be distributed into bins, forming a histogram over $T_\mathbf{s}^*(\mathbb{S}_R^3)$ that is given a continuum approximation. In the binning process, the $\mathbf{p}$ vectors at different points $\mathbf{q}$ need to be compared with $\mathbf{p}$ vectors at $\mathbf{s}$ for which the earlier-mentioned connection is needed. By contrast, no such connection would be needed if one would merely consider a theoretical probability of a particle at a given $\mathbf{q}$ having a momentum $\mathbf{p}$.

Boltzmann's formula (11.1) results from Boltzmann's more general definition in [Boltzmann (1896)] of the entropy of a macrostate of an individual system, famously summarized by Max Planck as

$$S = k_\mathrm{B} \log W, \qquad (11.2)$$

where W (for the German "Wahrscheinlichkeit") is "the probability of the macrostate;" cf. [Goldstein and Lebowitz (2004)]. The reference to "probability," if understood in terms of "relative frequency" of occurrence in independent and identically distributed trials, is somewhat problematic in a setting where the N body system is all the matter in the one and only universe. It is more appropriate to refer to W as a *typicality measure* for the macrostate; that is the size of the region in N particle phase space consisting of microstates which all give the same macrostate under consideration. If one were to consider, say, a simple fluid in local thermal equilibrium [Garrido, Goldstein and Lebowitz (2004)], a macrostate at time t would be the collection $(\nu(\mathbf{s}, t), \epsilon(\mathbf{s}, t), \mathbf{u}(\mathbf{s}, t))$ consisting of the particle density (ν), energy density (ϵ), and velocity field $(\mathbf{u})$ of the fluid. If the system is a dilute gas, possibly not in local thermal equilibrium, then its (kinetic) macrostate at time t is given by $N f(\mathbf{s}, \mathbf{p}, t)$ (sometimes referred to as a "mesoscopic state").

Boltzmann also had the important insight that, for a macroscopic system of $N \gg 1$ particles not at a phase transition, the measure of the N particle phase space of microstates which correspond to the thermal equilibrium state is essentially the full size of the available region in N body phase space. Ignoring angular momentum conservation for simplicity, in our setting this phase space region is the hypersurface $\{H^{(N)} = E\}$ in $T^*(\mathbb{S}_R^3)^N$, where E is the energy of the universe and where

$$H^{(N)} = \sum_{1 \leq k \leq N} \frac{|\mathbf{p}_k|^2}{2m} + \sum_{1 \leq k < l \leq N} U(|\mathbf{q}_k - \mathbf{q}_l|) \qquad (11.3)$$

is the Hamilton function. Right now $H^{(N)}$ is the sum of Newtonian kinetic energies and bounded continuous pair interaction energies $U(|\mathbf{q}_k - \mathbf{q}_l|)$, expressed in terms of the phase space variables $\mathbf{q}_k$ and $\mathbf{p}_k$, $k \in \{1, ..., N\}$. Recall that particle positions in $\mathbb{S}_R^3$ have been identified with vectors $\mathbf{q}_k \in \mathbb{R}^4$ of Euclidean length $|\mathbf{q}_k| = R$. The Euclidean distance $|\mathbf{q}_k - \mathbf{q}_l|$ is called their *chordal distance* between points $\mathbf{q}_k$, $\mathbf{q}_l \in \mathbb{S}_R^3$; cf. Appendix A.

To compute (in principle at least) the measure of a hypersurface in a high dimensional space, we recall that, in school, we learn that the volume of a three-dimensional ball of radius r in Euclidean space is $V(r) = \frac{4}{3}\pi r^3$, and its surface area is $A(r) = 4\pi r^2$. Next, recalling our college calculus courses, we notice that $A(r) = V'(r)$, where the $'$ means derivative. In a similar vein, if $\mathbf{1}_X$ denotes the so-called *indicator function* of the set $X \subset T^*(\mathbb{S}_R^3)^N$, which takes the value 1 on X and the value 0 outside of X, then

$$\Phi(E) := \int_{T^*(\mathbb{S}_R^3)^N} \mathbf{1}_{\{H^{(N)} \leq E\}} \frac{\mathrm{d}^3 p_1 ... \mathrm{d}^3 q_N}{h^{3N}} \qquad (11.4)$$

denotes the N body phase space measure of the region $\{H^{(N)} \leq E\}$ in $T^*(\mathbb{S}_R^3)^N$, normalized by h^{3N}. Therefore, $\Phi'(E)$ now yields the hypersurface measure of the hypersurface $\{H^{(N)} = E\}$.

To obtain Boltzmann's thermal equilibrium entropy (11.2) of such an N body system having energy E, we now have to take the logarithm of $\Phi'(E)$ — essentially, though not quite! Since $\Phi'(E)$ is not a dimensionless quantity, we multiply it by a reference energy unit, say mc^2. Also, (11.4) overcounts the physically relevant phase space size by a factor $N!$ (physical particles do not carry labels), which we divide out. And so, for the thermal equilibrium state of a classical system of N point particles in a spherical universe, Boltzmann's definition of entropy (11.2) (updated with Planck's h, k_B, and Gibbs' $N!$) essentially becomes

$$S_\mathrm{B}(E, N) = k_\mathrm{B} \ln \left[\frac{1}{N!} mc^2 \Phi'(E) \right]. \tag{11.5}$$

We next connect (11.5) with (11.1). For non-singular pair interactions U, it was rigorously shown in [Kiessling (2009)] that, if N is large enough, with energy scaling $E = N^2 \varepsilon$, where ε is a fixed parameter, and momenta rescaled as $\mathbf{p} \mapsto \sqrt{N} \mathbf{p}$, then (with $o(N)$ meaning: $o(N)/N \to 0$ as $N \to \infty$) we have

$$S_\mathrm{B}(N^2 \varepsilon, N) = \max_{f \in \mathfrak{A}_\varepsilon} S_\mathrm{B}(f; N) + o(N), \tag{11.6}$$

where $\mathfrak{A}_\varepsilon$ is the admissible set of normalized density functions $f(\mathbf{s}, \mathbf{p})$ for which $f \ln f$ is integrable and for which the energy of f, given by

$$\mathcal{E}(f) = \iint \frac{1}{2m} |\mathbf{p}|^2 f(\mathbf{s}, \mathbf{p}) \mathrm{d}^3 p \, \mathrm{d}^3 s$$
$$+ \iiiint \frac{1}{2} U(|\mathbf{s} - \tilde{\mathbf{s}}|) f(\mathbf{s}, \mathbf{p}) f(\tilde{\mathbf{p}}, \tilde{\mathbf{s}}) \mathrm{d}^3 p \, \mathrm{d}^3 s \, \mathrm{d}^3 \tilde{p} \, \mathrm{d}^3 \tilde{s}, \tag{11.7}$$

satisfies $\mathcal{E}(f) = \varepsilon$. Here, each $\iint$ means an integral over $T^* \mathbb{S}_R^3$, cf. (11.1). Note that formula (11.7) is the kinetic theory analog of (11.3).

Formula (11.6), which links Boltzmann's entropy (11.5) of the statistical (thermal) equilibrium state of the spherical N body universe with the entropy functional (11.1) (i.e. Boltzmann's H function), explicates the celebrated

MAXIMUM ENTROPY PRINCIPLE:
The thermal equilibrium state of an isolated system
is a macrostate of highest possible Boltzmann entropy.

We are now ready to inquire into Clausius' proposal that the fate of the universe is to end up in its highest possible entropy state. We will ignore all but the purely Newtonian gravitational interactions, treated as singular limit of bounded continuous pair interactions. On $\mathbb{S}_R^3$, they read $U(|\mathbf{q}_k - \mathbf{q}_l|) = -\frac{Gm^2}{|\mathbf{q}_k - \mathbf{q}_l|}$, which appears to be the same as in $\mathbb{R}^3$ except that the distance is the four-dimensional Euclidean distance of points in $\mathbb{S}_R^3 \subset \mathbb{R}^4$ (Appendix A).

11.4. The non-relativistic classical universe

Consider a finite number N of identical particles with Newtonian gravitational interactions on $\mathbb{S}_R^3$, having total energy $H = E$ at a fixed time t, where

$$
H^{(N)} = \sum_{1 \leq k \leq N} \frac{|\mathbf{p}_k|^2}{2m} - \sum_{1 \leq k < l \leq N} \frac{Gm^2}{|\mathbf{q}_k - \mathbf{q}_l|} \tag{11.8}
$$

is the Hamilton function. Recall that points on $\mathbb{S}_R^3$ have been identified with vectors $\mathbf{q}_k \in \mathbb{R}^4$ of Euclidean length $|\mathbf{q}_k| = R$, and that $|\mathbf{q}_k - \mathbf{q}_l|$ is the chordal distance function on $\mathbb{S}_R^3$.

In the previous section, we mentioned that when $-\frac{Gm^2}{|\mathbf{q}_k - \mathbf{q}_l|}$ is replaced by a continuous $U(|\mathbf{q}_k - \mathbf{q}_l|)$, the asymptotic large N expansion of the Boltzmann entropy (11.6)–(11.7), with $S_B(f; N)$ given by (11.1), has been established in [Kiessling (2009)]; note that it does not seem solvable in closed form. In the following, when we simply write the Newtonian gravitational pair interaction, it has to be understood as singular limit of a family of bounded continuous pair interactions.

The expression (11.5) for Boltzmann's entropy of the statistical equilibrium state can be integrated in the $\mathbf{p}$ variables over N copies of $\mathbb{R}^3$, which gives us

$$
S_B(E, N) = k_B \ln \left[C \int_{(\mathbb{S}_R^3)^N} \left(E + \sum_{k<l} \frac{Gm^2}{|\mathbf{q}_k - \mathbf{q}_l|} \right)_+^{\frac{3N-2}{2}} d^3 q_1 ... d^3 q_N \right] \tag{11.9}
$$

Here, $C = \frac{mc^2\sqrt{2\pi m}^{3N}}{N!\Gamma\left(\frac{3N}{2}\right)h^{3N}}$ and the subscript $_+$ means the positive part (i.e. negative values are replaced by 0). By direct inspection, one sees that integral (11.9) exists only for $N \leq 2$, while it diverges for $N \geq 3$.

Thus, in particular:

The Boltzmann entropy of a finite classical universe of $N \gg 1$ Newtonian point particles in $\mathbb{S}^3_R$ is unbounded above for any finite E.

Conclusions: Clausius' first law of the universe 1) is still mathematically meaningful for this 3-dimensional classical toy universe, but his law 2) is not, because there is no maximum entropy state. Yet we can accomodate the spirit of Clausius' law 2) — the increase of entropy — by replacing it with

> Law 2′: *The entropy of the world increases beyond any bound.*

In a three-dimensional spherical space $\mathbb{S}^3_R$, according to laws 1) and 2′), the fate of the classical toy universe is not to reach a maximum entropy state after which all macroscopic evolution ceases forever (in the sense that the dynamics becomes static: the famous "heat death"), but a never-ending entropy-raising evolution — unless the entropy blows up to infinity in a finite amount of time, at which point the macroscopic evolution may cease in a different sense, having reached a "singular state," which may be considered to be a different type of "heat death."

While entropic considerations do not yield the time scales involved, they do offer insights into the qualitative type of evolution. The first insight in this direction came in 1962 in the celebrated paper [Antonov (1962)]. Recall that (11.9) is the singular limit of a family of similar integrals in which $\frac{1}{|q-q'|}$ is replaced by a regularized interaction. As we already know, for the regularized interactions, the asymptotic (as $N \to \infty$) expansion of (11.9) is given by (11.6)–(11.7), with $\mathcal{S}_B(f; N)$ given by (11.1). And so, it follows that the singular limit of this maximum Boltzmann entropy variational principle, namely to maximize (11.1) constrained with $\mathcal{E}(f) = \varepsilon$ where

$$
\begin{aligned}
\mathcal{E}(f) = &\frac{1}{2m} \iint |\mathbf{p}|^2 f(\mathbf{s}, \mathbf{p}) \mathrm{d}^3 p \, \mathrm{d}^3 s \\
&- \frac{Gm^2}{2} \iiiint \frac{f(\mathbf{s}, \mathbf{p}) f(\tilde{\mathbf{p}}, \tilde{\mathbf{s}})}{|\mathbf{s} - \tilde{\mathbf{s}}|} \mathrm{d}^3 p \, \mathrm{d}^3 s \, \mathrm{d}^3 \tilde{p} \, \mathrm{d}^3 \tilde{s},
\end{aligned} \quad (11.10)
$$

has no maximizing solution f_ε. Antonov did not argue in this manner, but proved directly that the Boltzmann entropy functional (11.1) constrained with $\mathcal{E}(f) = \varepsilon$, with $\mathcal{E}(f)$ given in (11.10), has no upper bound; in fact, he proved this for a gravitating ideal gas in a spherical container [Antonov (1962)], but his proof can be adapted to our model on $\mathbb{S}_R^3$.

We outline his strategy of proof. It is to break up the system into a small localized core, which collapses and whose gravitational energy becomes more and more negative, and a uniform halo which picks up that energy, thus heating up and in the course boosting its entropy beyond any bound. Also the core gets hotter, but not as hot as the halo. The curious thing about Antonov's proof is that the core loses mass while it shrinks, which is picked up by the halo. In this continuum approximation, the mass of the core formally converges to zero when the entropy tends to ∞.

Of course, in an N body system in which each particle has mass m, there is a smallest possible mass of a "collapsing core" from which an infinite amount of gravitational energy could be extracted, namely a single pair of particles whose separation distance converges to zero. A very detailed dynamical study of this scenario has been carried out in a monumental work by Heggie [Heggie (1975)]. Heggie's work indicates that an N-body system develops a tightly bound binary plus an expanded halo containing all other particles. The halo is heated at the expense of the binary, which gets bound together ever tighter through close encounters with an occasional third particle. In the course of infinitely many such encounters, the binary system shrinks to a single point, thereby liberating an unlimited amount of gravitational binding energy. The halo of the $N - 2$ remaining point particles picks up the liberated energy in form of kinetic energy, and this carries the entropy to infinity.

Final remark: Even if the process which shrinks the binary to a point will take forever, in such a Newtonian universe, matter becomes arbitrarily hot in the process and eventually distribute uniformly on $\mathbb{S}_R^3$.

11.5. The non-relativistic quantum universe

The finding in the previous section of an unbounded entropy relies on a single pair of gravitating point particles being able to orbit arbitrarily close to one another, which is allowed in classical physics but not in quantum physics; just recall the hydrogen atom. And so, one would expect that a system of N gravitating point particles in quantum mechanics should have an upper bound to their entropy, given by the quantum analog of Boltzmann's entropy (11.9),

$$S_{\mathrm{B}}^{Q}(E, N) = k_{\mathrm{B}} \ln \mathrm{Tr} P_{E, \triangle E}. \qquad (11.11)$$

Here, $P_{E, \triangle E}$ is the projector onto the subspace of Hilbert space spanned by energy eigenstates, with energy in a small interval $\triangle E$ centered on E. (The size of $\triangle E$ does not matter as long as it contains very many eigenvalues. It can be chosen as small as one pleases (not zero) provided one makes N large enough, correspondingly.) In fact, there should be a maximum entropy state.

Now consider non-relativistic quantum mechanics for fermions interacting via Newtonian gravity. One may think of the particles as neutrinos, which have a tiny rest mass and spin 1/2 but no charge. The Hamiltonian is (11.8), except that now $\mathbf{p}_k = (\hbar/i)\boldsymbol{\nabla}_k$, where $\hbar$ is Planck's constant h divided by 2π, and $\boldsymbol{\nabla}_k$ the gradient operator in the k-th position variable. H acts on antisymmetric N particle wave functions (the Pauli exclusion principle for fermions). By the same proof as given in [Lévy-Leblond (1969)], for N such particles in $\mathbb{R}^3$, one finds that our Hamiltonian for N particles in $\mathbb{S}_R^3$ is also bounded below with *inf spec H* $\propto -N^{7/3}$; see also [Lieb (1990)]. Moreover, H has a self-adjoint Friedrichs extension, and since $\mathbb{S}_R^3$ has finite volume, H has purely discrete spectrum for which Weyl's asymptotic law for the counting of eigenvalues holds. As a consequence, for finite N and $E > E_g$, where E_g is the ground state energy, the non-relativistic quantum mechanical analog (11.11) of Boltzmann's entropy is finite.

The fermionic quantum analog of Boltzmann's entropy $\mathcal{S}_{\mathrm{B}}(f; N)$ is

$$S_{\mathrm{B}}^{Q}(f; N) = 2\mathcal{S}_{\mathrm{B}}(f; N)$$
$$- 2Nk_{\mathrm{B}}h^{-3} \int_{\mathbb{S}_R^3} \left(\int_{T_{\mathbf{s}}^*(\mathbb{S}_R^3)} (1 - h^3 f) \ln(1 - h^3 f) \mathrm{d}^3 p \right) \mathrm{d}^3 s;$$

$$(11.12)$$

the factor of 2 is due to the two spin states of each fermion. Note that the density functions $f(\mathbf{s}, \mathbf{p})$ are now restricted by the stabilizing bound $h^3 f < 1$.

The quantum analog of Boltzmann's maximum entropy principle (11.6)–(11.7) for gravitating fermions confined to a box $\subset \mathbb{R}^3$ has been rigorously derived in the 1970s by Walter Thirring and his school; see [Thirring (2002)], [Messer (1981)]. Their analysis, adapted to our setting, yields the asymptotic expansion for the entropy (11.11) in terms of the quantum entropy of a monatomic, self-gravitating ideal Fermi gas on $\mathbb{S}_R^3$, viz.

$$S_{\mathrm{B}}^{Q}(N^{7/3}\varepsilon, N) = \max_{f \in \mathfrak{A}_\varepsilon} S_{\mathrm{B}}^{Q}(f; N) + o(N). \qquad (11.13)$$

Here, $\mathfrak{A}_\varepsilon$ is the set of normalized $f(\mathbf{s}, \mathbf{p})$ for which $f \ln f$ and $(1 - h^3 f) \ln(1 - h^3 f)$ are integrable and for which $\mathcal{E}(f) = \varepsilon$, with $\mathcal{E}(f)$ given in (11.10).

The Euler–Lagrange equations for this maximum quantum entropy principle yield for each $\mathbf{s} \in \mathbb{S}_R^3$, that $f(\mathbf{s}, \mathbf{p})$ is a well-known Fermi–Dirac density on $T_{\mathbf{s}}^*(\mathbb{S}_R^3)$ of an ideal Fermi gas. The normalized particle density

$$\rho(\mathbf{s}) := \int_{\mathbb{R}^3} f(\mathbf{s}, \mathbf{p}) \mathrm{d}^3 p \qquad (11.14)$$

in turn satisfies a nonlinear system of integral equations which, to the best of my knowledge, have not been discussed in the literature. It is easy to see that they always admit the spatially uniform solution of the non-gravitating ideal Fermi gas, with the difference that the gravitational interactions may shift the chemical potential by a contant amount. For sufficiently large energy ε, their solution is unique — hence, the uniformly distributed ideal Fermi gas is the maximum entropy state when ε is large enough. It is also straightforward to show that when R is large enough, then there is a special value $\varepsilon = \varepsilon_J$ at which the uniformly distributed ideal Fermi gas becomes linearly unstable to spatially non-uniform disturbances — this is precisely the analog of the Jeans criterion, see [Jeans (1902); Kiessling (2003)]. At this Jeans energy ε_J, the entropy maximizers will exhibit a second-order phase transition at which an $\mathbb{S}^3$-parametrized family of $SO(3)$ invariant states bifurcates off of the spatially uniform perfect gas, breaking its $SO(4)[= SO(3) \times SO(3)]$ symmetry. One can also anticipate what happens at even lower energies, for sufficiently large R, by taking guidance from the detailed numerical studies in [Stahl *et al.* (1995); Chavanis (2002)] of similar gravitating systems in a spherical container $\subset \mathbb{R}^3$. Namely, with decreasing ε, the maximum entropy states become more and more concentrated in a continuous manner, until a special value of ε is reached at which two different $\mathbb{S}^3$-parametrized families of maximum entropy states exist. This is the point of a first-order phase transition in the merely $SO(3)$ invariant states, where the *state* of maximum entropy changes discontinuously as function of ε: Fixing the parameter in $\mathbb{S}^3$, the maximum entropy state in the family connected to the Jeans bifurcation is moderately condensed and has the lower temperature; the other maximum entropy state has a strongly condensed core and a dilute halo, and is much hotter. The first-order transition will be associated with local entropy maximizers in its neighborhood, which can be interpreted as metastable states. These will terminate at their respective spinodal points which, for the low temperature states, is determined by the analog of the Emden-Jeans criterion [Emden (1907)]. Continuing to lower energies yet the core-halo state carries the highest entropy and eventually condenses onto a white dwarf type ground state of "monumental size," containing all matter of the universe.

Figure 2 displays a qualitative sketch of the entropy s as function of the energy ε, of various states of the self-gravitating ideal Fermi gas on $\mathbb{S}^3_R$ with R much larger than the particles' Compton length. The maximum entropy graph is shown in green. Also shown (in blue) is the entropy for the meta-stable states which are merely local entropy

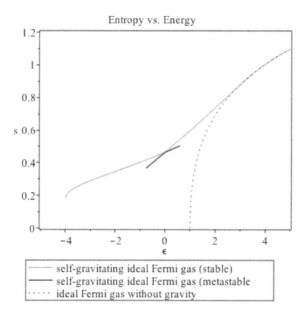

Entropy vs. Energy

self-gravitating ideal Fermi gas (stable)
self-gravitating ideal Fermi gas (metastable
ideal Fermi gas without gravity

maximizers, and (in red, dotted) the maximum entropy of the ideal Fermi gas without gravity (i.e., $G = 0$).

Conclusions: Clausius' two laws of the universe, 1) and 2), are mathematically meaningful for this three-dimensional non-relativistic quantum toy universe. They predict its fate to be a "heat death" where all macroscopic evolution ceases. Depending on the energy content of the world, this "heat death" can be the uniform distribution of matter in $\mathbb{S}_R^3$ (having $SO(4)$ invariance) with $\approx$ Maxwellian velocity distribution and a (hot) temperature — this would happen at high energies.[3] Or it can be a non-uniform, but small gradients distribution of matter in $\mathbb{S}_R^3$ (having $SO(3)$ invariance), with less Maxwellian and more Fermi–Dirac like velocity distribution, and moderate temperatures — this would happen at intermediate energies. Or it can be a very strongly condensed core distribution of

---

[3] Alternatively, the uniform matter state also maximizes entropy when R is very small. This raises the question whether the most typical (i.e. representative) macrostate of the early universe was perhaps uniform — eliminating any need for an inflationary phase to explain the essential uniformity of the cosmic microwave background radiation!

matter in $\mathbb{S}_R^3$ surrounded by a very diluted halo (both parts still having $SO(3)$ invariance); the core will have a recognizable Fermi-Dirac velocity distribution, and low temperature — this would happen at very low energies.

At the first order phase transition point, the description is more complicated. In particular, the state with a highly condensed core and very dilute atmosphere is hotter than the less condensed, small gradients state.

11.6. The "special-relativistic" quantum universe

In this final section, before coming to general relativity, inspired by Chandrasekhar's theory of white dwarf stars, we take special relativity into account while still working with Newtonian gravity. Yet, instead of using a many-body Dirac equation, we work with the "pseudo-relativistic" Hamiltonian

$$H^{(N)} = \sum_{1 \leq k \leq N} \sqrt{m^2 c^4 - \hbar^2 c^2 \Delta_k} - G m^2 \sum_{1 \leq k < l \leq N} \frac{1}{|\mathbf{q}_k - \mathbf{q}_l|},$$
(11.15)

acting on anti-symmetric wave functions for N fermions in $\mathbb{R}^3$. A rigorous analysis for the ground state of (11.15) has been carried out by Lieb and Thirring [Lieb and Thirring (1984)]. The Hamiltonian (11.15) does have a ground state only if $N < N_{\text{Ch}}$, with

$$N_{\text{Ch}} = C \frac{9}{16\sqrt{\pi}} \left(\frac{hc}{Gm^2} \right)^{3/2},$$
(11.16)

where C is determined by solving a nonlinear PDE numerically; $C = 0.87$, if we neglect spin state counting. If m is the neutron mass, then $N_{\text{Ch}} \approx 10^{58}$, which essentially yields Chandrasekhar's maximum mass of a white dwarf star.

What does this imply for the entropy (11.11)?

For $N < N_{\text{Ch}}$, the analogous reasoning as in the non-relativistic case, leads again to a finite quantum Boltzmann entropy (11.11) of the statistical equilibrium state. However, this argument fails in the supercritical case $N > N_{\text{Ch}}$ relevant to our universe. Indeed, as argued in [Kiessling (2001)], for $N > 1.2N_{\text{Ch}}$ and any given E, the quantum Boltzmann entropy (11.11) diverges; the factor 1.2 is not optimal. We do not repeat the full argument here but note some important points.

The argument that the quantum Boltzmann entropy (11.11) diverges for $N > 1.2N_{\text{Ch}}$ in the Hamiltonian (11.15) is based on a rigorous proof that the quantum analog of Boltzmann's maximum entropy principle for the semi-classical continuum approximation to this N fermion system has no upper bound. The strategy of proof is adapted from Antonov's [Antonov (1962)], see Section 11.4. Note that the existence of Chandrasekhar's limit mass, viz. N_{Ch}, makes the proof more tricky than for the classical non-relativistic continuum approximation, where no such limit mass exists. One now takes $N > 1.2N_{\text{Ch}}$ fermions and split them into two subsystems, one with $N_1 > N_{\text{Ch}}$ particles and one with N_2 particles, such that $N_1 + N_2 = N$. While N_2 need not exceed N_{Ch}, it cannot be arbitrarily small for a semi-classical approximation to hold, but $N_2 > 10^{-20}N_{\text{Ch}}$ will do. Both subsystems are placed far apart so that their mutual gravitational energy can be ignored in the argument. Then, one can find semi-classical continuum approximations, f_1 and f_2, to the fermion densities in the two subsystems which obey the exclusion principle $h^3 f_n < 1$ for $n = 1, 2$, such that the total quantum Boltzmann entropy of f_1 and f_2, given by $S_{\text{B}}^Q(f_1, N_1) + S_{\text{B}}^Q(f_2, N_2)$, surpasses any prescribed value S while obeying the energy constraint $\mathcal{E}^{SR}(f_1, N_1) + \mathcal{E}^{SR}(f_2, N_2) = E$, where now

$$\mathcal{E}^{SR}(f; N) = N^{\frac{4}{3}} \iint \sqrt{m^2 c^4 + p^2 c^2}\, f(\mathbf{s}, \mathbf{p}) \mathrm{d}^3 p \mathrm{d}^3 s$$

$$- N^2 \frac{Gm^2}{2} \iiiint \frac{f(\mathbf{s}, \mathbf{p})f(\tilde{\mathbf{p}}, \tilde{\mathbf{s}})}{|\mathbf{s} - \tilde{\mathbf{s}}|} \mathrm{d}^3 p \mathrm{d}^3 s \mathrm{d}^3 \tilde{p} \mathrm{d}^3 \tilde{s} \quad (11.17)$$

is the (pseudo-) special-relativistic energy functional of f. Note that here we have retained N — the existence of N_{Ch} makes it clear that a simple rescaling of space and energy scales cannot be used to absorb N into a scaling factor as done in the non-relativistic models. The extra $N^{1/3}$ factor at the kinetic energy is all that remains from Pauli's exclusion principle for N fermions which, at the microscopic level, rules out that two or more fermions are in the same single-particle state. In this sequence, subsystem 1 is a massive core which collapses, not losing any mass in the process, but liberating gravitational energy, which is picked up by a halo that does not collapse, heating up instead and in the course boosting its entropy beyond any bound.

Conclusions: Clausius' first law of the universe 1) is again mathematically meaningful for this 3-dimensional quantum toy universe of $N \gg 1$ particles, but his law 2) holds only if $N < N_{\mathrm{Ch}}$. Since there is no maximum entropy state for $N > 1.2 N_{\mathrm{Ch}}$ fermions, once again, one has to accomodate the entropy increase in the spirit of Clausius by replacing law 2) with the law 2') stated in Section 11.4.

This (pseudo-) special-relativistic three-dimensional quantum toy universe resembles its non-relativistic version only if the universe contains not many more particles than our sun (if m is the neutron mass). But our galaxy alone contains about 10^{11} stars, and there are hundreds of billions of galaxies in the universe and, for such huge N, our (pseudo-) special-relativistic three-dimensional quantum toy universe resembles the non-relativistic three-dimensional classical toy universe of Section 11.4 more closely. The main difference is that, in the classical setting, a single pair of particles converging onto the same location suffices to boost the entropy beyond any bound, while in the special-relativistic quantum setting one needs more than N_{Ch} fermions.

Additional remarks: For the regime $N < N_{\mathrm{Ch}}$, the semi-classical continuum approximation to the ground state has been rigorously justified by Lieb and Yau [Lieb and Yau (1987)], who derived Chandrasekhar's structure equations of a white dwarf star from an analysis of the ground state of (11.15) in a suitable large-N continuum limit; see also [Lieb (1990)].

11.7. On the significance of the black hole entropy

So far we have exclusively considered Newtonian gravity of N point particles in a static spacetime $\mathbb{S}_R^3 \times \mathbb{R}$, and only the dynamics of the particles became more and more realistic from section to section: we began with non-relativistic Newtonian mechanics; next we used, first non- and then (pseudo-), special-relativistic quantum mechanics. However, Newton's theory misses the important contribution of black holes to classical gravity theory, discovered in general relativity.

The first nontrivial exact solution to Einstein's field equations discovered, the Schwarzschild solution, is a spacetime containing a black hole (and a white hole, and which has two spacelike separated asymptotically flat (Minkowskian) regions). In Schwarzschild coordinates (t, r, θ, ϕ), the asymptotically flat region $\{r > 2GM/c^2\}$ has metric [Pauli (1958); Misner Thorne and Wheeler (1973)]

$$ds^2 = -\left(1 - \frac{2GM}{c^2 r}\right) c^2 dt^2 + \left(1 - \frac{2GM}{c^2 r}\right)^{-1} dr^2$$
$$+ r^2 \left(d\theta^2 + \sin^2\theta d\phi^2\right), \tag{11.18}$$

where M is the mass "as seen from far away" from the hole, i.e. when $r \gg 2GM/c^2$. The metric (11.18) is singular at $r = 2GM/c^2$, but this is an artifact of the coordinates used. The spacetime itself is regular at the three-dimensional hypersurface $\{r = 2GM/c^2\}$, which defines the event horizon of the black hole. What makes the black hole "black" is the feature that no lightlike geodesic, which starts at a spacetime event inside the event horizon, will be able to reach what relativists call "future null infinity" of the spacetime; while for every event outside the event horizon, there exists a lightlike geodesic which does. In more colloquial terms, once inside a black hole, you can't communicate with the world outside of it using electromagnetic radiation.

This raises an important question: "Even though it cannot send electromagnetic signals to the outside world, can matter which falls into a black hole transfer energy to, and thereby raise the entropy of, matter outside of it?"

Roger Penrose has shown how a certain amount of energy could be extracted, in principle, from a rotating black hole. So at least in the most general sense of the first important question, the answer is "Yes."

A curious aspect of the Schwarzschild metric is the following.

A $t = const.$ section of its event horizon is a two-dimensional sphere with area $4\pi(2GM/c^2)^2$, and the famous *Schwarzschild radius*

$$R_S := \frac{2GM}{c^2} \qquad (11.19)$$

is the Euclidean radius of a sphere $\mathbb{S}^2_{R_S} \subset \mathbb{R}^3$ with the same area. Any sphere $\{(t, r, \theta, \phi) : t = const. \& r = R_S\}$ is a so-called *marginally trapped surface*. It is common in the astrophysics community to think of black holes as evolutionary three-dimensional "quasi-objects" hidden inside a marginally trapped two-dimensional surface, and the folklore says that two such black holes can merge, but no black hole can split into two. In this vein, if one compares a spacetime which contains two distant Schwarzschild-type black holes of masses M_1 and M_2, with a spacetime containing a single such hole whose mass is $M = M_1 + M_2$, then the surface area of a $t = const.$ section of its event horizon is

$$A = 4\pi R_S^2 = 16\pi \frac{G^2}{c^4}(M_1 + M_2)^2. \qquad (11.20)$$

And by the simple inequality $(M_1 + M_2)^2 > M_1^2 + M_2^2$ (both $M_k > 0$), we have $A > A_1 + A_2$. Accepting the folklore then, we have in front of us, the black hole analogues of the first and second laws of thermodynamics, the so-called first and second laws of black hole thermodynamics: black hole mass plays the role of energy, the area of the marginally trapped surface plays the role of entropy.[4] This is the gist of it; for rotating black holes, there are some modifications [Heusler (1996)].

This purely formal analogy to conventional thermodynamics has led Bekenstein [Bekenstein (1973)] to suggest that a black hole of mass M actually *has* an entropy $\propto M^2$, which Hawking

---

[4]Einstein's $E = mc^2$ makes it plain that mass must play the role energy, but entropy as a surface quantity is a novelty.

[Hawking (1975)] computed to be[5]

$$S_{\mathrm{BH}} = k_{\mathrm{B}} \frac{4\pi G}{\hbar c} M^2. \tag{11.21}$$

To arrive at (11.21), Hawking went beyond classical general relativity, but it is fair to say that his heuristic derivation still awaits a rigorous foundation.

Accepting the Bekenstein–Hawking entropy as a physical entropy, the second important question one will have to address is: "Which role does black hole entropy play in the assessment of Clausius' law 2)?"

In his book "The emperor's new mind" [Penrose (1989)], Roger Penrose argues (p.338) that when atomistic matter starts uniformly distributed over a spherical space $\mathbb{S}_R^3$, it will develop clumps under its gravity, and those clumps will coalesce, and the universe will evolve in the spirit of Clausius' law 2) into a state of highest entropy — which according to Penrose will be a black hole that has swallowed up all the matter of that universe, having a Bekenstein–Hawking entropy (11.21) with M the mass of the universe. Taking $M = 10^{80} m$, with m the mass of a proton, Penrose computes (pp.342/3) the maximum possible entropy of our universe to be $10^{123} k_{\mathrm{B}}$. In Penrose's scenario, the two laws of Clausius, 1) and 2), are assumed valid, resulting in the "heat death" of the universe, although this "heat death" has little in common with what Clausius and his contemporaries had envisioned.

Incidentally, like Boltzmann, ignoring Poincaré's recurrence time as irrelevant to the explanation of the fate of our material universe, Penrose points out that the evaporation of a black hole by Hawking radiation is irrelevant to his argument because of the stupendously large time scales involved.

Astoundingly, while the entropy of a universe filled with N gravitating fermions which have a special-relativistic kinetic energy but interact with Newtonian gravity is unbounded above (for $N > 1.2 N_{\mathrm{Ch}}$), its more realistic general relativistic version seems to have a bounded entropy! However, if we would assume that the fate of the universe in the semi-classical (pseudo-)special-relativistic theory is a

---

[5]In the pertinent formula (1) of [Kiessling (2001)], a factor $8\pi^2$ is missing.

collapse of all the matter onto itself (the closest analog in that model to a universe whose matter has formed a single black hole), we would get an upper bound on its entropy [Kiessling (2001)], viz.

$$S_B^Q(f; N) \leq k_B 2N \ln N + O(N). \tag{11.22}$$

This raises the question whether in general relativity, one might also get higher entropies by splitting the system into a collapsing core and a halo which receives the liberated energy from the core and carries the entropy to infinity. We need to inspect the Bekenstein–Hawking entropy more closely.

Bekenstein's reasoning for his black-hole entropy formula is based on the proposition that physics is concerned only with the world outside the event horizons of all the black holes in the universe, as most physicists seem to have argued it would — back then. But then, whenever a black hole swallows a piece of matter, it also swallows with it its entropy, in the process of which the entropy of the matter outside the event horizons decreases — violating the second law even in its weakest form:

> 2″) *The world evolves such that its entropy does not decrease.*

To rescue 2″), Bekenstein [Bekenstein (1973)] proposed that the "entropy of the universe" (at any instant of some cosmic time) is the sum, of the entropy of the "matter" outside the event horizons of the black holes, and of the entropy of those black holes. If this sum obeys the second law at least in its form 2″), insisting that this is a law of the universe, then it is logically conceivable that the hole will gobble up all "matter," leaving only the black hole entropy; hence, Penrose's estimate for the maximum entropy of this $\mathbb{S}_R^3$ based universe model.

However, the Bekenstein-Hawking entropy of a black hole which has just swallowed a piece of "matter" is not always larger than, or even equal to, the entropy of the orginal black hole plus the entropy of the piece of "matter" before it was swallowed. This is readily demonstrated by considering a variation of the reasoning where one compares the areas of the marginally trapped surfaces of two Schwarzschild-type black holes with the surface area of the marginally trapped surface resulting after merger. Namely, we now compare the sum of the Bekenstein–Hawking entropy of a Schwarzschild-type black hole of mass M and the entropy of a black body radiation occupying a stove of volume V, with the black hole entropy after this chunk of black body radiation has been swallowed up (we idealize the walls of the stove to have neglible mass and entropy; in fact, one doesn't need a "stove" because the universe is already filled with a black body radiation.) The entropy S_{bb} of such a photon gas expressed in terms of its energy $U_{bb} =: M_{bb}c^2$ is

$$ S_{bb} = k_{\mathrm{B}} \frac{4}{3} \left(\frac{\pi^2 V c^3}{15\hbar^3} \right)^{\frac{1}{4}} M_{bb}^{\frac{3}{4}}. \tag{11.23}$$

And so, writing $S_{bb}(M_{bb}) := C_{bb} M_{bb}^{\frac{3}{4}}$ and also $S_{\mathrm{BH}}(M) := C_{\mathrm{BH}} M^2$, we find

$$ C_{\mathrm{BH}}(M + M_{bb})^2 - C_{\mathrm{BH}} M^2 - C_{bb} M_{bb}^{\frac{3}{4}} $$

$$ = C_{\mathrm{BH}}(2M M_{bb} + M_{bb}^2) - C_{bb} M_{bb}^{\frac{3}{4}}, \tag{11.24}$$

which makes it plain that

$S_{\mathrm{BH}}(M + M_{bb})$

$\quad < S_{\mathrm{BH}}(M) + S_{bb}(M_{bb})$ *if M_{bb} is sufficiently small!* (11.25)

Using the current temperature T_{CMB} of the cosmic microwave background radiation, the criterion (11.25) can be rephrased thus: *if the black hole mass $M < \frac{\hbar c^3}{6\pi G k_{\mathrm{B}} T_{CMB}} \approx 6 \times 10^{22}\mathrm{kg}$, then the BH entropy decreases by swallowing some CMB radiation.* The borderline mass is about that of our moon.

Curiously, Bekenstein in [Bekenstein (1973)] already came to an equivalent conclusion, but then argued that statistical quantum fluctuations would invalidate the conclusion. This may very well be so, but given the preliminary state of any investigation into the realm of quantum physics in black hole spacetimes it seems fair to say that the jury seems still out on this case.

Conclusions: There are several important conclusions to be drawn from our discussion, about the evolution of such a universe model on the time scale of its existence, from the "big bang" until the "big crunch," which certainly are much shorter than any "Poincaré recurrence time" or such.

First of all, assuming law 2″) holds in a general-relativistic universe model which describes the evolution of matter that was uniformly distributed over $\mathbb{S}_R^3$ initially, with entropy understood as the sum of the entropy of the matter outside of the event horizons of all the black holes, plus the entropy of the black holes, it is not yet clear whether the stronger Clausius' law 2) or its weakened version 2′) hold, too. What is clear, though, by letting $M + M_{bb}$ in (11.25) be the total mass in this universe model, is the following:

> *A black hole which contains all "matter" of the universe is*
> *not the maximum entropy state of such a universe!*

Second, it is also conceivable that the fate of any "matter" initially outside of the event horizon of a black hole in this closed universe model is eventually to end up inside a single black hole. It may well turn out that the law 2″) does not hold if one defines physics to be concerned only with what's outside of an event horizon, but 2″), and possibly 2′) or even 2), may well hold (on the stipulated time scales) if we do not ignore the entropy of "matter" inside the event horizons of the black holes! Why should physics end at the event horizon? General relativity allows us to inquire into the fate of matter which has crossed an event horizon. True, according to general relativity, we will not be able to have a space probe explore this fate in situ and have its findings sent to us who reside outside of the event horizon. But to insist that each and every logically coherent consequence of a physical theory, for it to be acceptable as physics has to be "directly measurable" and "communicable to wherever we are," seems to me too narrow a definition of what physics is about.

Appendices

A. The Laplace operator on d-dimensional spheres

In $\mathbb{R}^3$, the Newtonian pair interaction energy $-Gm^2 \frac{1}{|\hat{s}-\tilde{s}|}$ is, up to the factor Gm^2, the Green function for the Laplace operator $\Delta_{\mathbb{R}^3}$ on

$\mathbb{R}^3$; i.e.

$$-\Delta_{\mathbb{R}^3}\frac{1}{|\mathbf{s}-\check{\mathbf{s}}|} = 4\pi\delta_{\{\check{\mathbf{s}}\}}^{(3)}(\mathbf{s}) \tag{A.1}$$

in the sense of measures. Here, $|\mathbf{s}-\check{\mathbf{s}}|$ denotes Euclidean distance in $\mathbb{R}^3$, and $\delta_{\{\check{\mathbf{s}}\}}^{(3)}(\mathbf{s})$ denotes the Dirac point measure at $\check{\mathbf{s}} \in \mathbb{R}^3$ which means that, for any open Lebesgue set $\Lambda \subset \mathbb{R}^3$, we have

$$\int_{\Lambda}\delta_{\{\check{\mathbf{s}}\}}(\mathbf{s})\mathrm{d}^3s = \begin{cases} 1 & \text{if } \check{\mathbf{s}} \in \Lambda \\ 0 & \text{if } \check{\mathbf{s}} \notin \Lambda \end{cases} \tag{A.2}$$

Similarly, $-\frac{Gm^2}{R}\ln\frac{R}{|\check{\mathbf{s}}-\check{\mathbf{s}}|}$, with $|\mathbf{s}-\check{\mathbf{s}}|$ again Euclidean distance in $\mathbb{R}^3$, is the Green function for the Laplace-Beltrami operator $\Delta_{\mathbb{S}_R^2}$ on the sphere $\mathbb{S}_R^2$, i.e.

$$-\Delta_{\mathbb{S}_R^2}\ln\frac{R}{|\mathbf{s}-\check{\mathbf{s}}|} = 2\pi\left(\delta_{\{\check{\mathbf{s}}\}}^{(2)}(\mathbf{s}) - \frac{1}{4\pi R^2}\right) \tag{A.3}$$

in the sense of measures. The additive constant at right hand side (A.3) is inevitable, due to the topology of $\mathbb{S}^2$ — note that $\int_{\mathbb{S}^2}\Delta_{\mathbb{S}^2}\ln\frac{1}{|\mathbf{s}-\check{\mathbf{s}}|}\mathrm{d}^2s = 0$, and that $\int_{\mathbb{S}_R^2}\mathrm{d}^2s = 4\pi R^2$.

Physicists would be tempted to interpret the term $-\frac{1}{4\pi R^2}$ at right hand side (A.3) as a "negative background mass density" per particle which permeates "space" $\mathbb{S}_R^2$, but mathematically it's just encoding the constant positive Gauss curvature of standard $\mathbb{S}_R^2$.

Analogously, $-\frac{1}{|\check{\mathbf{s}}-\check{\mathbf{s}}|}$, with $|\mathbf{s}-\check{\mathbf{s}}|$ now Euclidean distance in $\mathbb{R}^4$, is the Green function for the Laplace-Beltrami operator $\Delta_{\mathbb{S}_R^3}$ on the sphere $\mathbb{S}_R^3$; i.e.,

$$-\Delta_{\mathbb{S}_R^3}\frac{1}{|\mathbf{s}-\check{\mathbf{s}}|} = 4\pi\left(\delta_{\{\check{\mathbf{s}}\}}^{(3)}(\mathbf{s}) - \frac{1}{2\pi^2 R^3}\right) \tag{A.4}$$

in the sense of measures; note that $\int_{\mathbb{S}_R^3}\mathrm{d}^3s = 2\pi^2 R^3$.

To verify the Green function formulas for the spheres, note that for unit spheres, $|\mathbf{s}-\check{\mathbf{s}}|^2 = 2 - 2\cos\psi = 4\sin^2\frac{\psi}{2}$, where ψ is the angle between $\mathbf{s}$ and $\check{\mathbf{s}}$; now use the spherical angle representations for $\Delta_{\mathbb{S}^d}$.

B. The classical entropy of N Newtonian particles in $\mathbb{S}_R^2$

The understanding of complicated issues is often aided by exactly solvable toy models — caricatures of the real problem, yet clearly recognizable as such! In this vein, it may be useful to record that the classical maximum entropy state can be computed exactly for a hypothetical world in which the dimension of physical space is two instead of three; for the flat space analog, see [Aly and Perez (1999)]. The Newtonian gravitational interaction between a pair of particles with positions $\mathbf{q}_k$ and $\mathbf{q}_l$ in $\mathbb{S}_R^2$ reads (Appendix A)

$$U(|\mathbf{q}_k - \mathbf{q}_l|) = U_0 - \frac{Gm^2}{R} \ln \frac{R}{|\mathbf{q}_k - \mathbf{q}_l|}, \qquad (B.1)$$

where $|\mathbf{q}_k - \mathbf{q}_l|$ is now the chordal distance on $\mathbb{S}_R^2$. The constant U_0 is chosen for convenience so that $\int_{\mathbb{S}_R^3} U(|\mathbf{q} - \mathbf{q}'|)\mathrm{d}^2 q = 0$ for each $\mathbf{q}' \in \mathbb{S}_R^2$.

Following verbatim [Kiessling (2011)], the analog of the maximum entropy principle (11.6)–(11.7), with $\mathcal{S}_\mathrm{B}(f; N)$ given by (11.1) with $\mathbb{S}_R^3$ replaced by $\mathbb{S}_R^2$ and h^3 by h^2, can (a) be rigorously derived from (11.5) and (11.3), with U given by (B.1), without any regularization, and (b) be evaluated completely, as follows:

Any maximizer f_ε of $\mathcal{S}_\mathrm{B}(f; N)$ over the admissible set $\mathfrak{A}_\varepsilon$ is of the form

$$f_\varepsilon(\mathbf{s}, \mathbf{p}) = \sigma_\varepsilon(\mathbf{p}|\mathbf{s})\rho_\varepsilon(\mathbf{s}), \qquad (B.2)$$

with $\rho_\varepsilon(\mathbf{s})$ solving the Euler–Lagrange equation

$$\rho(\mathbf{s}) = \frac{\exp\left(- \int U(|\mathbf{s} - \tilde{\mathbf{s}}|)\rho(\tilde{\mathbf{s}})\mathrm{d}^2\tilde{s}/k_\mathrm{B}\vartheta_{\varepsilon,\lambda}(\rho)\right)}{\int \exp\left(- \int U(|\hat{\mathbf{s}} - \tilde{\mathbf{s}}|)\rho(\tilde{\mathbf{s}})\mathrm{d}^2\tilde{s}/k_\mathrm{B}\vartheta_{\varepsilon,\lambda}(\rho)\right)\mathrm{d}^2\hat{s}}, \qquad (B.3)$$

where

$$k_\mathrm{B}\vartheta_\varepsilon(\rho) = \varepsilon - \frac{1}{2}\iint U(|\mathbf{s} - \tilde{\mathbf{s}}|)\rho(\mathbf{s})\rho(\tilde{\mathbf{s}})\mathrm{d}^2 s \mathrm{d}^2\tilde{s} \qquad (B.4)$$

is ($k_\mathrm{B}\times$) the strictly positive "temperature of ρ;" in (B.3) and (B.4), all integrals are over $\mathbb{S}_R^2$. The function $\sigma_\varepsilon(\mathbf{p}|\mathbf{s})$ is a scalar on $T_\mathbf{s}^*\mathbb{S}_R^2$ and given by

$$\sigma_\varepsilon(\mathbf{p}|\mathbf{s}) = [2\pi m k_\mathrm{B}\vartheta_\varepsilon(\rho_\varepsilon)]^{-1} \exp\big(-|\mathbf{p}|^2/2mk_\mathrm{B}\vartheta_\varepsilon(\rho_\varepsilon)\big). \qquad (B.5)$$

The Euler–Lagrange equation itself has many solutions, but two families of solutions can be explicitly stated in terms of elementary functions. Happily, these two families contain all the maximizers of $S_\mathrm{B}(f; N)$ over $\mathfrak{A}_\varepsilon$:

$$f_\varepsilon(\mathbf{s}, \mathbf{p}) = \begin{cases} \dfrac{1}{8\pi^2 m\varepsilon} \exp\big(-\dfrac{1}{2m\varepsilon}|\mathbf{p}|^2\big) R^{-2}; & \varepsilon \geq \dfrac{1}{4}\gamma \\[2ex] \dfrac{1}{2\pi^2 m\gamma} \exp\big(-\dfrac{2}{m\gamma}|\mathbf{p}|^2\big) & \\[2ex] \big(R\cosh\zeta(\varepsilon) - \mathbf{a}\cdot\mathbf{s}\sinh\zeta(\varepsilon)\big)^{-2}; & \varepsilon < \dfrac{1}{4}\gamma \end{cases} \qquad (B.6)$$

where we have set $\gamma := Gm^2/R$, and where $\mathbf{a} \in \mathbb{S}^2 \subset \mathbb{R}^3$ spans the arbitrary axis of rotational symmetry of $\rho_\varepsilon(\mathbf{s})$, while $\zeta(\varepsilon) > 0$ is the unique positive solution of the fixed point equation

$$\zeta = \left(\frac{3}{2} - \frac{2\varepsilon}{\gamma}\right) \tanh\zeta. \qquad (B.7)$$

Thus, for $\varepsilon \geq \gamma/4$, the entropy maximizer is unique and identical with the spatially (on $\mathbb{S}_R^2$) uniform thermal equilibrium of the non-gravitational ideal gas, having a Maxwellian momentum distribution at each $\mathbf{s} \in \mathbb{S}_R^2$. For $\varepsilon < \gamma/4$, the entropy maximizer is not unique and given by any one of the rotation-invariant (about $\mathbf{a} \in \mathbb{S}^2$) members of this two-parameter family. All these states have the same temperature, $k_\mathrm{B}\vartheta = \frac{1}{4}\gamma$, so their heat capacity is infinite.

Introducing the dimensionless energy $\epsilon := 4\varepsilon/\gamma$, and writing for the entropy $k_\mathrm{B}^{-1}S_\mathrm{B}(f_\varepsilon, N) = -N\ln N - N\mathfrak{s}(\epsilon) + o(N)$, we find that $\mathfrak{s}(\epsilon)$ is given by

$$\mathfrak{s}(\epsilon) = \begin{cases} \ln\epsilon & \text{for } \epsilon \geq 1 \\ \epsilon - 1 & \text{for } \epsilon < 1 \end{cases}. \qquad (B.8)$$

At $\epsilon \ (:= 4\varepsilon/\gamma) = 1$, the function $\epsilon \mapsto \mathfrak{s}(\epsilon)$ is continuous together with its first derivative, but its second derivative is not, giving the signature of a second-order phase transition in the sense of Ehrenfest, associated with a symmetry-breaking bifurcation. We remark that by evaluating the dispersion-relation obtained from the Jeans (a.k.a. Vlasov-Poisson) equations for perturbations about the uniform perfect gas state one directly finds that the second-order phase transition point is determined by the Jeans criterion [Jeans (1902); Kiessling (2003)] adapted to our two-dimensional spherical toy universe.

Figure 1 (adapted from [Kiessling (2011)]) displays $s = \mathfrak{s}(\epsilon)$ versus ϵ:

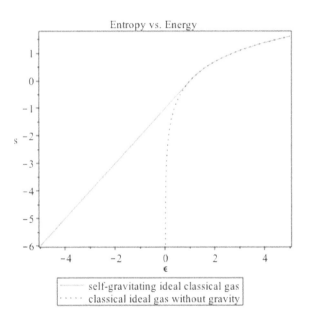

Conclusions: Clausius' two laws of the universe 1) and 2) are mathematically meaningful for this two-dimensional classical toy universe. They predict its fate to be a "heat death" where all macroscopic evolution ceases. In this Hamiltonian system, it might do so forever in the limit $N \to \infty$, though only for a finite (but unpronounceably large) time span of the order of the Poincaré recurrence time when $N < \infty$. When $\epsilon \geq 1$, the "state of heat death" looks precisely as envisioned in the earliest papers on this subject: a spatially uniform distribution of matter with Maxwellian momentum distribution. However, when $\epsilon < 1$, the "state of heat death" is spatially non-uniform, showing gravitational condensation in one hemisphere and rarification in the other, yet rotational symmetry about some arbitrary axis $\mathbf{a}$ (which presumably is determined by the initial data). Since, for $0 < \epsilon < 1$, the spatially uniform Maxwellian is also a solution to the Euler–Lagrange equations of the maximum entropy principle, to anyone who has learned that an increase in entropy means a decrease in structure it may seem paradoxical that the spatially uniform Maxwellian doesn't have the largest entropy. Yet note that the gravitationally condensed states have a higher temperature than the uniform Maxwellian state of the ideal gas with same energy ϵ (or ε). In fact, the decrease of entropy due to the increase in spatial structure inflicted by gravity is overcompensated by an increase of entropy due to the accompanying decrease of structure in momentum space — paradox resolved.

First comment: The reason for why an unbounded entropy does not feature in our model world on $\mathbb{S}_R^2$ has nothing to do with the dimensionality of the space itself, but with the strength of the singularity of the gravitational pair interactions in these different-dimensional worlds. Thus, replacing $\frac{1}{|\mathbf{q}-\mathbf{q}'|}$ by $\ln\frac{1}{|\mathbf{q}-\mathbf{q}'|}$ in the $(\mathbb{S}_R^3)^N$ integral in (11.9) renders (11.9) finite for all N [Kiessling (2000)].

Final comment: In [Kiessling (2011)] also, the asymptotic large N expansion of the entropy of the maximum entropy state with prescribed energy (scaling like $E = N^2\varepsilon \in \mathbb{R}$) and prescribed angular momentum (scaling like $\mathbf{L} = N^{3/2}\boldsymbol{\lambda} \in \mathbb{R}^3$) is studied. In this case, the maximum entropy state is spatially non-uniform whenever $\boldsymbol{\lambda} \neq 0$; moreover, when $\boldsymbol{\lambda} \neq 0$ this state is also not static but either stationary (exhibiting stationary flow) or (quasi-) periodic in time — see [Kiessling (2008)]. For large ε, the entropy maximizer is rotation-symmetric about $\boldsymbol{\lambda}$, but for sufficiently low ε it's not. However, in contrast to the ergodic ensemble (i.e., ignoring angular momentum as constant of motion), the maximum entropy principle of this ergodic subensemble does not seem to be explicitly solvable in terms of known functions, and the complete classification of all maximum entropy states is open.

Acknowledgment

I thank Valia Allori for the invitation to write this article. It is based partly on [Kiessling (2001)], supported by NSF grant DMS-9623220, and partly on [Kiessling (2011)], supported by grant DMS-0807705, updated by findings since. The opinions of the author expressed in this paper are not necessarily those of the NSF. I am indebted to Shelly Goldstein and Joel Lebowitz for teaching me Boltzmann's insights into the inner workings of the universe. I also thank Friedrich Hehl for reference [Heusler (1996)]. Lastly, I thank two anonymous referees for their helpful comments.

References

Aly, J.-J., Perez, J. (1999) Thermodynamics of a two-dimensional unbounded self-gravitating system, *Phys. Rev. E.*, **60**, 5185–5190.

Antonov, V. A. (1962) Vest. Leningrad Gas. Univ. **7**, 135 (English transl.: 'Most probable phase distribution in spherical star systems and conditions for its existence.' In: *Dynamics of Star Clusters,* ed. by J. Goodman and P. Hut, IAU 1985), pp. 525–540)

Bekenstein, J. D. (1973) Black holes and entropy, *Phys. Rev.*, **D7**, 2333–2346.

Bekenstein, J. D. (1974) Generalized second law of thermodynamics in black hole physics, *Phys. Rev.*, **D9**, 3292–3300.

Boltzmann, L. (1872) Weitere Studien über das Wärmegleichgewicht unter Gasmolekülen, *Wiener Berichte*, **66**, 275–370.

Boltzmann, L. (1896) *Vorlesungen über Gastheorie*, J.A. Barth, Leipzig; English translation: Lectures on Gas Theory (S.G. Brush, transl.), Univ. California Press, Berkeley (1964).

Chavanis, P. H. (2002) Phase transitions in self-gravitating systems. Self-gravitating fermions and hard spheres models, *Phys. Rev. E*, **65**, 056123.

Clausius, R. (1865) Über verschiedene für die Anwendung bequeme Formen der Hauptgleichungen der mechanischen Wärmetheorie, *Annalen Phys.*, **125**, 353–400.

Einstein, A. (1917) Kosmologische Betrachtungen zur allgemeinen Relativittstheorie, Sitzungb. König. Preuss. Akad. 142–152.

Emden, R. (1907) *Gaskugeln*, Teubner-Verlag, Leipzig.

Garrido, P., Goldstein, S., Lebowitz, J. L. (2004) Boltzmann Entropy for Dense Fluids Not in Local Equilibrium, *Phys. Rev. Lett.*, **92**, art.050602-1-4.

Gibbs, J. W. (1902) *Elementary Principles in Statistical Mechanics*, Yale Univ. Press, New Haven; reprinted by Dover, New York (1960).

Goldstein, S. (2001) *Boltzmann's approach to statistical mechanics*, Proceedings of the Conference Chance in Physics: Foundations and Perspectives, Ischia,

Italy, 29.Nov.-3.Dec. 1999, (J. Bricmont, D. Dürr, M. C. Galavotti, G. C. Ghirardi, F. Petruccione, and N. Zanghì, orgs. and eds.), *Lect. Notes Phys.* **574**, pp.39–54, Springer Verlag, Heidelberg.

Goldstein, S., Lebowitz, J. L. (2004) On the (Boltzmann) entropy of nonequilibrium systems, *Physica*, **D193**, 53–66.

Hawking, S. (1975) Particle creation by black holes, *Comm. Math. Phys.*, **43**, 199–220.

Heggie, D. C. (1975) Binary evolution in stellar dynamics, *Mon. Not. R. astr. Soc.*, **173**, 729–787.

Heusler, M. (1996) *Black Hole Uniqueness Theorems*, Cambridge Lect. Notes Phys., **6**, Cambridge Univ. Press.

Jeans, J. H. (1902) The stability of a spherical nebula, *Philos. Trans. Roy. Soc. London*, **199**, 1–53.

Kiessling, M. K.-H. (2000) Statistical mechanics approach to some problems in conformal geometry *Physica*, **A 297**, 353–368.

Kiessling, M. K.-H. (2001) *How to Implement Boltzmann's Probabilistic Ideas in a Relativistic World*, Proceedings of the Conference Chance in Physics: Foundations and Perspectives, Ischia, Italy, 29.Nov.-3.Dec. 1999, (J. Bricmont, D. Dürr, M. C. Galavotti, G. C. Ghirardi, F. Petruccione, and N. Zanghì, orgs. and eds.), *Lect. Notes Phys.*, **574**, pp. 83–100, Springer Verlag, Heidelberg.

Kiessling, M. K.-H. (2003) The 'Jeans swindle': a true story — mathematically speaking, *Adv. in Appl. Math.*, **31**, 132–149.

Kiessling, M. K.-H. (2008) Statistical equilibrium dynamics, pp. 91–108 in *AIP Conf. Proc.*, **97**, A. Campa, A. Giansanti, G. Morigi, and F. Sylos Labini (eds.), *American Inst. Phys.*

Kiessling, M. K.-H. (2009) The Vlasov continuum limit for the classical microcanonical ensemble, *Rev. Math. Phys.*, **21**, 1145–1195.

Kiessling, M. K.-H. (2011) Typicality analysis for the Newtonian N-body problem on $\mathbb{S}^2$ in the $N \to \infty$ limit, *J. Stat. Mech.: Theor. Exp.*, **P01028** 70 pps.

Killing, W. (1885) Die Mechanik in den nicht-*Euklid*ischen Raumformen, *J. Reine Angew. Math.*, **98**, 1–48.

Lévy-Leblond, J. M. (1969) Non-saturation of gravitational forces, *J. Math. Phys.*, **10**, 806–812.

Lieb, E. H. (1990) The stability of matter: From atoms to stars, *Bull. Am. Math. Soc.*, **22**, 1–49.

Lieb, E. H., Thirring, W. (1984) Gravitational collapse in quantum mechanics with relativistic kinetic energy, *Annals Phys.*, **155**, 494–512.

Lieb, E. H., Yau, H. T. (1987) The Chandrasekhar theory of stellar collapse as a limit of quantum mechanics, *Commun. Math. Phys.*, **112**, 147–213.

Messer, J. (1981) *Temperature-dependent Thomas–Fermi theory*, *Lect. Notes Phys.*, **147**.

Messer, J., Spohn, H. (1982) Statistical mechanics of the isothermal Lane-Emden equation, *J. Stat. Phys.*, **29**, 561–578.

Misner, C. W., Thorne, K. S., Wheeler, J. A. (1973) *Gravitation*, (W.H. Freeman Co., New York).

Pauli, W. (1958) *Theory of Relativity*, Dover.

Penrose, R. (1989) *The Emperors New Mind*, (Oxford Univ. Press, Oxford)

Stahl B., Kiessling M. K. H., Schindler K. (1995) Phase transitions in gravitating systems and the formation of condensed objects, *Planet. Space Sci.*, **43**, 271–282.

Thirring, W. (2002) *Quantum Mathematical Physics*, Springer, Berlin Heidelberg.

Part IV
Some Consideration from Quantum Mechanics

Chapter 12

Foundations of Statistical Mechanics and the Status of the Born Rule in de Broglie-Bohm Pilot-Wave Theory

Antony Valentini

Augustus College,
14 Augustus Road, London SW19 6LN, UK
Department of Physics and Astronomy,
Clemson University, Kinard Laboratory,
Clemson, SC 29634-0978, USA

We compare and contrast two distinct approaches to understanding the Born rule in de Broglie-Bohm pilot-wave theory, one based on dynamical relaxation over time (advocated by this author and collaborators) and the other based on typicality of initial conditions (advocated by the 'Bohmian mechanics' school). It is argued that the latter approach is inherently circular and physically misguided. The typicality approach has engendered a deep-seated confusion between contingent and law-like features, leading to misleading claims not only about the Born rule but also about the nature of the wave function. By artificially restricting the theory to equilibrium, the typicality approach has led to further misunderstandings concerning the status of the uncertainty principle, the role of quantum measurement theory, and the kinematics of the theory (including the status of Galilean and Lorentz invariance). The restriction to equilibrium has also made an erroneously-constructed stochastic model of particle creation appear more plausible than it actually is. To avoid needless controversy, we advocate a modest 'empirical approach' to the foundations of statistical mechanics. We argue that the existence or otherwise of quantum nonequilibrium in our world is an empirical question to be settled by experiment.

12.1. Introduction

The pilot-wave theory of de Broglie (1928) and Bohm (1952a,b) is a deterministic theory of motion for individual systems. In the version first given by de Broglie, a system with configuration q and wave function $\psi(q,t)$ has an actual trajectory $q(t)$ determined by de Broglie's equation of motion

$$\frac{dq}{dt} = \frac{j}{|\psi|^2}, \tag{12.1}$$

where ψ obeys the usual Schrödinger equation (units $\hbar = 1$)

$$i\frac{\partial\psi}{\partial t} = \hat{H}\psi \tag{12.2}$$

(with a Hamiltonian operator $\hat{H}$) and j is a current satisfying the continuity equation

$$\frac{\partial|\psi|^2}{\partial t} + \partial_q \cdot j = 0 \tag{12.3}$$

(with ∂_q the gradient operator in configuration space).[1] Equation (12.3) is a straightforward consequence of (12.2), and using (12.1) it may be rewritten as

$$\frac{\partial|\psi|^2}{\partial t} + \partial_q \cdot (|\psi|^2 \dot{q}) = 0 \tag{12.4}$$

(where $\dot{q} = j/|\psi|^2$ is the configuration-space velocity field).

Thus, for example, for a single low-energy spinless particle of mass m we find a current

$$\mathbf{j} = |\psi|^2 \frac{\nabla S}{m} \tag{12.5}$$

(where S is the phase of $\psi = |\psi|\exp(iS)$) and (12.1) reads

$$\frac{d\mathbf{x}}{dt} = \frac{\nabla S}{m}. \tag{12.6}$$

---

[1]This construction applies to any system with a Hamiltonian $\hat{H}$ given by a differential operator on configuration space (Struyve and Valentini, 2009).

Given an initial wave function $\psi(q,0)$, (12.2) determines $\psi(q,t)$ at all times and so the right-hand side of (12.1) is also determined at all times. Given an initial configuration $q(0)$, (12.1) then determines the trajectory $q(t)$. Thus, for example, in a two-slit experiment with a single particle, if the incident wave function is known then (12.6) determines the trajectory $\mathbf{x}(t)$ for any initial position $\mathbf{x}(0)$.

Mathematically, for a given wave function, the law of motion (12.1) defines a trajectory $q(t)$ for each initial configuration $q(0)$. In practice, however, we do not know the value of $q(0)$ within the initial packet. For an ensemble of systems with the same $\psi(q,0)$, the value of $q(0)$ will generally vary from one system to another. We may then consider an initial distribution $\rho(q,0)$ of values of $q(0)$ over the ensemble. As the trajectories $q(t)$ evolve, so will the distribution $\rho(q,t)$. By construction $\rho(q,t)$ will obey the continuity equation

$$\frac{\partial \rho}{\partial t} + \partial_q \cdot (\rho \dot{q}) = 0. \tag{12.7}$$

In principle there is no reason why we could not consider an arbitrary initial distribution $\rho(q,0)$. De Broglie's equation (12.1) determines the time evolution of a trajectory $q(t)$ for any initial $q(0)$, and over the ensemble the continuity equation (12.7) determines the time evolution of a density $\rho(q,t)$ for any initial $\rho(q,0)$. There is certainly no reason of principle why $\rho(q,0)$ should be equal to $|\psi(q,0)|^2$.

As an extreme example, an ensemble of one-particle systems could have the initial distribution $\rho(\mathbf{x},0) = \delta^3(\mathbf{x} - \mathbf{x}_0)$, with every particle beginning at the same point $\mathbf{x}_0$. As the distribution evolves, it will remain a delta-function concentrated on the evolved point $\mathbf{x}(t)$. Every particle in the ensemble would follow the same trajectory. If such an ensemble were fired at a screen with two slits, every particle would land at the same final point $\mathbf{x}_f$ on the backstop and there would be no interference pattern (indeed no pattern at all), in gross violation of quantum mechanics.

If instead it so happens that $\rho(q,0) = |\psi(q,0)|^2$, then since ρ and $|\psi|^2$ obey identical continuity equations ((12.7) and (12.4) respectively) it follows that

$$\rho(q,t) = |\psi(q,t)|^2 \tag{12.8}$$

for all t. This is the usual Born rule. In conventional quantum mechanics (12.8) is taken to be an axiom or law of nature. Whereas in pilot-wave theory (12.8) is a special state of 'quantum equilibrium': if it happens to hold at one time it will hold at all times (for an ensemble of isolated systems). Thus, for example, if such an ensemble of particles is fired at a screen with two slits, the incoming equilibrium ensemble will evolve into an equilibrium ensemble at the backstop, and hence the usual interference pattern $\rho = |\psi|^2$ will be trivially obtained. More generally, as first shown in detail by Bohm (1952b), for systems and apparatus initially in quantum equilibrium, the distribution of outcomes of quantum measurements will agree with the conventional quantum formalism.

But in principle the theory allows for 'quantum nonequilibrium' ($\rho \neq |\psi|^2$). How then can pilot-wave theory explain the success of the Born rule (12.8), which has been confirmed to high accuracy in every laboratory experiment? Most workers in the field (past and present) simply take it as a postulate. Thus, for example, according to Bell (1987, p. 112) '[i]t is *assumed* that the particles are so delivered initially by the source', while according to Holland (1993, p. 67) the Born rule is one of the 'basic postulates'. This is unsatisfactory. There is after all a basic conceptual distinction between equations of motion and initial conditions. The former are regarded as immutable laws (they could not be otherwise), whereas the latter are contingencies (there is no reason of principle why they could not be otherwise). Once the laws are known they are the same for all systems, whereas for a given system the initial conditions must be determined empirically. Thus Newton, for example, wrote down laws that explain the motion of the moon, but he made no attempt to explain the current position and velocity of the moon: the latter are arbitrary or contingent initial conditions to be determined empirically, which may then be inserted into the laws of motion to determine the position and velocity at other times. In pilot-wave theory, if we consider only ensembles restricted by the additional postulate (12.8), then this is closely analogous to considering Newtonian mechanics only for ensembles restricted to thermal equilibrium (with a uniform distribution on the energy surface in phase space). In both theories there is a much wider nonequilibrium physics, which is lost if we simply adopt initial equilibrium as a postulate.

Most workers in the field seem unperturbed by this and continue to treat (12.8) as a postulate. Others are convinced that some further explanation is required and that the question — if pilot-wave theory is true, why do we always observe the Born rule? — requires a more satisfying answer.

There are currently two main approaches to understanding the Born rule in pilot-wave theory, which we briefly summarise here.

The first approach, associated primarily with this author and collaborators, proposes that the Born rule we observe today should be explained by a process of 'quantum relaxation' (analogous to thermal relaxation), whereby initial nonequilibrium distributions $\rho \neq |\psi|^2$ evolve towards equilibrium on a coarse-grained level, $\bar{\rho} \to \overline{|\psi|^2}$ (in terms of coarse-grained densities $\bar{\rho}$ and $\overline{|\psi|^2}$, much as in Gibbs' classical account of thermal relaxation for the coarse-grained density on phase space). This process may be understood in terms of a 'subquantum' coarse-graining H-theorem on configuration space, analogous to the classical coarse-graining H-theorem on phase space (Valentini, 1991a,b). Extensive numerical simulations, carried out with wave functions that are superpositions of different energy states, have confirmed the general expectation that initial densities $\rho(q, 0)$ lacking in fine-grained microstructure rapidly become highly filamentary on small scales and indeed approach the equilibrium density $|\psi|^2$ on a coarse-grained level (Valentini, 1992, 2001; Valentini and Westman, 2005; Towler, Russell and Valentini, 2012; Colin, 2012; Abraham, Colin and Valentini, 2014). This may be quantified by a decrease of the coarse-grained H-function

$$\bar{H}(t) = \int dq \; \bar{\rho} \ln (\bar{\rho}/\overline{|\psi|^2}), \qquad (12.9)$$

which reaches its minimum $\bar{H} = 0$ if and only if $\bar{\rho} = \overline{|\psi|^2}$, and which is found to decay approximately exponentially with time (Valentini and Westman, 2005; Towler, Russell and Valentini, 2012; Abraham, Colin and Valentini, 2014). Similar studies and simulations have been carried out for field theory in an expanding universe, for

which relaxation is found to be suppressed at very long cosmological wavelengths (Valentini, 2007, 2008a, 2010a; Colin and Valentini,
2013, 2015, 2016). This opens the door to possible empirical evidence for quantum nonequilibrium in the cosmic microwave background (Valentini, 2010a; Colin and Valentini, 2015; Vitenti, Peter
and Valentini, 2019) — as well as in relic particles left over today
from the very early universe (Valentini, 2001, 2007; Underwood and
Valentini, 2015, 2016). It has further been shown that if nonequilibrium systems were discovered today, their physics would be radically different from the physics we currently know, involving practical
superluminal signalling, violations of the uncertainty principle, and a
general breakdown of standard quantum constraints (such as expectation additivity and the indistinguishability of non-orthogonal quantum states) (Valentini, 1991a,b, 1992, 2002a, 2004, 2009; Pearle and
Valentini, 2006). On this view, quantum theory is merely a special
'equilibrium' case of a much wider nonequilibrium physics, which
may have existed in the early universe and which could still exist in
some exotic systems today.

The second approach, associated primarily with Dürr, Goldstein,
and Zanghì, as well as with Tumulka and other collaborators, proposes that the Born rule we observe today should be explained in
terms of the 'typicality' of configurations $q_{\text{univ}}(0)$ for the whole universe at the initial time $t = 0$ (Dürr, Goldstein and Zanghì, 1992;
Dürr and Teufel, 2009; Goldstein, 2017; Tumulka, 2018). In this
approach, if $\Psi_{\text{univ}}(q_{\text{univ}}, 0)$ is the initial universal wave function then
$|\Psi_{\text{univ}}(q_{\text{univ}}, 0)|^2$ is assumed to be the natural measure on the set
of possible initial universal configurations $q_{\text{univ}}(0)$. It may then be
shown that the Born rule (12.8) is almost always obtained for ensembles of sub-systems prepared with wave function ψ — where 'almost
always' is defined with respect to the measure $|\Psi_{\text{univ}}(q_{\text{univ}}, 0)|^2$. This
is regarded as an explanation for the empirical success of the Born
rule (12.8). On this view there is no realistic chance of ever observing
quantum nonequilibrium, which is intrinsically unlikely (as defined
with respect to $|\Psi_{\text{univ}}(q_{\text{univ}}, 0)|^2$). The Born rule is in effect regarded
as an intrinsic part of the theory, though instead of postulating
the probability distribution (12.8) for sub-systems this approach

postulates the typicality measure $|\Psi_{\text{univ}}(q_{\text{univ}}, 0)|^2$ for the whole universe at $t = 0$. If this is correct, quantum nonequilibrium will never be observed and de Broglie-Bohm theory will never be experimentally distinguishable from conventional quantum theory.

The typicality approach has given rise to a distinctive physical perspective on pilot-wave theory — concerning for example the status of the uncertainty principle and of Lorentz invariance, among other important topics. These views may be classified under the heading of the 'Bohmian mechanics school', where the term 'Bohmian mechanics' was first introduced by Dürr, Goldstein and Zanghì (1992) to denote the dynamical theory defined by equations (12.1) and (12.2).

It should however be noted that, historically speaking, the dynamics defined by (12.1) and (12.2) was first proposed by de Broglie at the 1927 Solvay conference (for a many-body system with a pilot wave in configuration space) (Bacciagaluppi and Valentini, 2009). De Broglie called his new form of dynamics 'pilot-wave theory'. The theory was revived by Bohm in 1952, though rewritten in a second-order form with a law of motion for acceleration that includes a 'quantum potential' Q. Bohm's version of the dynamics is physically distinct from de Broglie's: in principle it allows for non-standard initial momenta $p \neq \partial_q S$ (Bohm, 1952a, pp. 170, 179; Colin and Valentini, 2014).[2] Thus there are important physical differences between de Broglie's dynamics and Bohm's dynamics. The terminology 'Bohmian mechanics', as applied to de Broglie's equations (12.1) and (12.2), is therefore misleading: it does not give due credit to de Broglie and it misrepresents the views of Bohm. Our concern in this paper is with the status of the Born rule in de Broglie's dynamics which, following de Broglie's own usage (as well as Bell's), we refer to as 'pilot-wave theory'.

---

[2]In Bohm's dynamics there arises the additional question of why we observe today an 'extended quantum equilibrium' in phase space, with momenta satisfying $p = \partial_q S$ as well as configurations distributed according to (12.8). Colin and Valentini (2014) show that extended nonequilibrium does not relax and is unstable, and argue that as a result Bohm's dynamics is physically untenable.

Writings by the Bohmian mechanics school generally fail to recognise the significance and priority of de Broglie's work. For example, in rather glib sections entitled 'History', both Goldstein (2017) and Tumulka (2018) portray de Broglie as having simply proposed or considered the guidance equation at the 1927 Solvay conference. But in fact, as early as 1923 de Broglie had postulated the single-particle guidance equation — as a new law of motion, expressing a unification of the principles of Maupertuis and Fermat — and in that same year de Broglie used his theory to predict electron interference (four years before it was observed by Davisson and Germer). Furthermore, it was de Broglie's early research into his new form of dynamics (with particles guided by waves) that led Schrödinger to the wave equation in 1926.[3]

The distinctive approach of the Bohmian mechanics school has been reiterated and developed in a number of papers and reviews. A textbook has also been published (Dürr and Teufel, 2009). For the sake of perspective it is worth remarking that writings by members of this school generally focus on their own interpretation. There are other approaches to de Broglie-Bohm theory, not only that taken by this author and collaborators but also others that lie outside the scope of this paper.[4] The Bohmian mechanics school has been particularly influential among philosophers of physics. The entry 'Bohmian mechanics' in *The Stanford Encyclopedia of Philosophy* is written by a leading member of the school (Goldstein, 2017) (regularly updated by the same author since 2001). It is noteworthy that such an extensive reference encyclopedia does not contain an entry on de Broglie-Bohm theory generally; only this one particular school is represented, suggesting a skewed perception of the field among philosophers. One of the aims of this paper is to redress this imbalance in the philosophy of physics literature.

We shall compare and contrast the two approaches outlined above, in particular regarding the status of the Born rule and related physical questions. As we shall discuss, in our view the typicality

---

[3]For an extensive historical analysis of de Broglie's remarkable work in the period 1923–27, see Bacciagaluppi and Valentini (2009, chapter 2).

[4]See, for example, the books by Holland (1993) and by Bohm and Hiley (1993).

approach is essentially circular (Valentini, 1996, 2001). With respect to a different initial measure (such as $|\Psi_{\text{univ}}(q_{\text{univ}}, 0)|^4$), we will almost always obtain initial violations of the Born rule (such as $\rho \propto |\psi|^4$). While it may be said that nonequilibrium is 'untypical' (has zero measure) with respect to the univeral Born-rule measure, it may equally be said that nonequilibrium is 'typical' (has unit measure) with respect to a non-Born-rule measure. In effect, in the typicality approach the Born rule is taken as an axiom, albeit at the level of the universe as a whole. This is misleading, not least because a postulate about initial conditions can have no fundamental status in a theory of dynamics.

As we shall see, the typicality approach has engendered a basic confusion between contingent and law-like features. This has led to misleading claims not only about the Born rule but also about the nature of the wave function (or pilot wave). The artificial restriction to equilibrium has led to further misunderstandings concerning the status of the uncertainty principle, the role of quantum measurement theory, and the kinematics of the theory (including the status of Galilean and Lorentz invariance). The restriction to equilibrium has also made an erroneously-constructed stochastic model of particle creation seem more plausible than it actually is.

By considering how hidden variables can account for the Born rule (12.8), workers in quantum foundations find themselves confronted by issues in the foundations of statistical mechanics — a subject which is no less fractious and controversial than quantum foundations itself. We begin by outlining our own views on the subject (Section 12.2), summarise some of the key results in quantum relaxation and how these apply to cosmology (Sections 12.3 and 12.4), and then provide a critique of the typicality approach (Section 12.5) and of related viewpoints (Section 12.6 and 12.7), ending with some concluding remarks (Section 12.8).

12.2. Empirical approach to statistical mechanics

Pilot-wave dynamics is a deterministic theory of motion. As in classical physics, there is a clear conceptual distinction between the laws of motion on the one hand and initial conditions on the other.

The initial conditions (for the wave function ψ and for the configuration q) are in principle arbitrary — which is to say, perhaps more properly, that they are contingent. Whatever the actual initial conditions were, there is no known reason of principle why they could not have been different. In order to find out what the initial conditions actually were, we do not appeal to laws or principles but to simple empiricism: we carry out observations today and on that basis we try (using our knowledge of the dynamical laws) to deduce the initial conditions. This is as true for ensembles as it is for single systems. In pilot-wave theory, for an ensemble of systems with the same initial wave function $\psi(q, 0)$, the initial distribution $\rho(q, 0)$ of actual initial configurations $q(0)$ could in principle be anything. To find out what $\rho(q, 0)$ was in an actual case we must resort to empirical observation.

This raises the subtle question of what it might mean for pilot-wave dynamics to 'explain' the Born rule (12.8). If the distribution ρ is purely empirical, there might then seem to be no question of 'explaining' the particular distribution (12.8): it is not something one explains, it is something one finds empirically.

The matter is further complicated by the time-reversal invariance of pilot-wave dynamics. If all initial conditions are in principle possible, then any distribution $\rho(q, t_0)$ today is in principle possible, since it could have evolved from some appropriate $\rho(q, 0)$ (which can in principle be calculated from $\rho(q, t_0)$ by time-reversal of the equations of motion). Again, there might seem to be no question of 'explaining' (12.8): it is simply a brute 'matter of fact' established by observation.

In our view there is in fact considerable scope for explaining the presently-observed Born rule (12.8), in the sense that it can be explained in terms of past conditions (with the aid of dynamical laws) — where the past conditions are, however, ultimately empirical and not fixed by any fundamental laws or principles. On this view, the present is explained in terms of the past, while the past is itself something we establish empirically. This of course leaves open the possibility of further explanation by peering even further into the past, and in our universe this chain of causal explanations eventually leads us back to the big bang. To understand the origins of the

Born rule, then, we are led to consider conditions in the very early universe. Specifically: what initial conditions (at or close to the big bang) could have given rise to the all-pervasive distribution (12.8) which we see today?

In the context of statistical mechanics, it might be objected that to explain the state (12.8) seen today we must not merely deduce which past conditions (for example, which particular initial states $\rho(q,0) \neq |\psi(q,0)|^2$) could have evolved into (12.8) today, we must instead show that 'all' or 'most' past conditions evolve into (12.8) today. For otherwise, it might be said, we have simply replaced one unexplained empirical fact (conditions today) with another unexplained empirical fact (conditions in the past), so that in a sense we are not really making progress. In our view this objection is misguided and has roots in some unfortunate misunderstandings in the early history of statistical mechanics.

First of all, it is perfectly reasonable to explain the present in terms of the past. This is standard practice across the physical sciences — from astrophysics to geology. As a simple example, suppose that today at time t_0 the moon is observed to have a certain position and momentum, so that it now occupies a particular location (q_0, p_0) in phase space. With the aid of Newton's laws, this fact today may be explained by the fact that the moon was at a location $(q(t), p(t))$ in phase space at some earlier time $t < t_0$. If t is very far in the past, pre-dating direct human observation, then in practice we would deduce that the moon must have been at $(q(t), p(t))$ at time t. That we have had to deduce the past from the present would not undermine our physical intuition that the moon may be said to be where it is now *because* it was in the deduced earlier state at an earlier time. This is normal scientific practice. On the other hand, one can imagine the philosophical objection being raised, that the past state is a mere deduction (or retrodiction) and not a *bona fide* 'explanation' for the observed state today. It might also be suggested that we would have a satisfactory explanation only if we could show that *all* — or in some sense 'most' — possible earlier states $(q(t), p(t))$ of the moon evolve into the moon being in the state (q_0, p_0) today. Needless to say, most physicists would disagree with this objection (not least because, from what we understand of lunar dynamics, such a suggestion has no chance of being correct). The objection seems unfounded. And yet similar objections are frequently heard in the context of statistical mechanics. Why?

In our view the trouble stems from a mistake in the early history of the subject. Boltzmann originally hoped to deduce the second law of thermodynamics from mechanics alone. As is well known, this ambitious project was fundamentally misguided. It is impossible to deduce any kind of necessary uni-directional evolution in time in a time-reversal invariant theory. For any initial molecular state that evolves towards thermal equilibrium, one can always construct a time-reversed initial state that evolves away from thermal equilibrium. Boltzmann's program was dogged by such 'reversibility objections', resulting in heated debate about the foundations of the subject. The debate rages even today.[5] It is now widely accepted that the laws of mechanics alone do not suffice: one must also assume something about the initial conditions (such as an absence of fine-grained microstructure, or an absence of correlations among molecular velocities). Debates then continue about the status of the assumption about the initial conditions, with many authors attempting to justify the assumption on the basis of some fundamental principle or other. Running like a thread through these debates is the expectation that, in order for the program to succeed, it must be shown either that the required initial conditions are 'almost always' satisfied or that they are required by some principle (where such attempts invariably lead to further controversy). In our view, this expectation is misguided and reflects the historical error in Boltzmann's original program. There was never any reason to expect all allowed initial conditions to give rise to thermal relaxation, and subsequent attempts to show that 'most' initial conditions will do so, or that the required initial conditions are consequences of some fundamental principle or other, in effect propagate the original error (albeit in a reduced or weaker form).

---

[5]For an even-handed and scholarly review see, for example, Uffink (2007).

We advocate a more modest — and in our view more reasonable — 'empirical' approach to statistical mechanics (Valentini, 1996, 2001). On this view the observed thermodynamical behaviour is an empirical fact which must be explained (with the aid of dynamical laws) in terms of past conditions, where the latter are themselves also empirical. The past conditions do not need to be 'almost always' or 'typically' true, nor do they need to be true by virtue of some deep principle or other: they simply need to explain or be consistent with what is observed. Just as in the case of the moon, where we try to deduce — or if necessary guess — its state in the past given its state today, in the case of a box of gas that is evolving towards thermal equilibrium we try to deduce or guess the required character of the initial (microscopic) state. For a gas, of course, there will be a set or class of microstates yielding the observed behaviour. Unlike for the moon, we make no attempt to deduce the exact initial micro-state. And with so many variables involved, it is convenient to apply statistical methods. The essential aim and method of statistical mechanics is then this. First, to find a class of initial conditions that yields the observed behaviour. And second, to understand the evolution of those initial conditions towards equilibrium in terms of a general mechanism — without having to solve the exact equations of motion for the huge number of variables involved.

In the case of pilot-wave theory we wish to explain the observed validity of the Born rule (12.8) for laboratory systems — to within a certain experimental accuracy. For example, if we prepare a large number N of hydrogen atoms in the ground state with wave function $\psi_{100}(\mathbf{x})$, and if we measure the electron position $\mathbf{x}$ (relative to the nucleus) for each atom, then for large N we find an empirical distribution $\rho(\mathbf{x})$ of the schematic form

$$\rho = |\psi_{100}|^2 \pm \epsilon, \qquad (12.10)$$

where ϵ characterises the accuracy to which the Born rule has been confirmed (where this will depend on the accuracy of the position measurements as well as on the value of N). How can this be explained in terms of past conditions?

The first thing to note is that, when we encounter a hydrogen atom in the laboratory, the atom has not been floating freely in a vacuum for billions of years prior to us experimenting with it. The atom has a past history during which it has interacted with other things. That past history traces back ultimately to the formation of the earth, the solar system, our galaxy, and ultimately merges with the history of the universe as a whole, which as we know began with a hot and violent phase called the big bang. In fact, every system we have access to in the laboratory has a long and violent astrophysical history. Therefore, when attempting to explain the Born rule today by conditions in the past, we should make use of our knowledge of that history. In other words, when we attempt to deduce what earlier conditions are required to explain what we observe in the laboratory today, we should take into account what we already know about the history of the systems in question.

The second thing to note is that the Born rule for a sub-system such as an atom is a simple consequence of the Born rule applied to a larger system from which the atom may have been extracted. If we consider an ensemble of many-body systems, all with the same wave function $\Psi(q, t_0)$ and with an ensemble distribution $P(q, t_0) = |\Psi(q, t_0)|^2$ of configurations q at some time t_0, then it is readily shown that if an ensemble of sub-systems with configurations x are extracted from the parent ensemble and prepared with an effective (or reduced) wave function $\psi(x, t)$ at $t > t_0$, then the distribution of extracted configurations x will be $\rho(x, t) = |\psi(x, t)|^2$ (Valentini, 1991a).[6] In other words, equilibrium for a many-body system implies equilibrium for extracted sub-systems (a property which is sometimes called 'nesting'). This means that we can explain the Born rule for extracted sub-systems such as atoms if we are able to explain the Born rule for larger parent systems.

Wherever we look, in fact, we find the Born rule — not only in the laboratory today but also further afield. For example, the relative intensities of atomic spectral lines emitted from the outskirts of

---

[6] A similar result was obtained by Dürr, Goldstein and Zanghì (1992).

distant quasars agree with the Born rule (as applied to atomic transitions). The observed cosmological helium abundance agrees with calculations based on the Born rule (as applied to nuclear reactions in the early hot universe). Perhaps the ultimate test of the Born rule is currently taking place in satellite observations of the small temperature and polarization anisotropies in the cosmic microwave background, which were caused by classical inhomogeneities which existed at the time of photon decoupling (around 400,000 years after the big bang), which in turn grew from classical inhomogeneities which existed in the very early universe, and which according to inflationary cosmology were in turn formed from quantum vacuum fluctuations in an 'inflaton' scalar field. Ultimately, on our current understanding, the 'primordial power spectrum' (the spectrum of very early classical inhomogeneities) was generated by a Born-rule spectrum of primordial quantum field fluctuations.

To explain the success of the Born rule, then, we can consider the earliest possible conditions in the history of our universe. Clearly, the initial conditions must be such as to evolve into or imply the Born rule at relevant later times. What initial conditions should we assume?

One possibility, of course, is to simply assume that the universe began in a state of quantum equilibrium. Below we argue that this is, in effect, the assumption made by Dürr, Goldstein and Zanghì (1992). Unlike in the thermal case, we have not observed relaxation to the Born rule actually taking place over time. All we see is the equilibrium Born rule. Since the equations of motion preserve the Born rule over time, a simple way to explain our observation of the Born rule now is to assume that the Born rule was true at the beginning.

But initial equilibrium is only one possibility among uncountably many. Given the known violent history of our universe and of everything in it, and given the results for quantum relaxation summarised in the next section, there clearly exists a large class of initial nonequilibrium states that will evolve to yield the Born rule today to an excellent approximation (in particular at the short wavelengths relevant to local physics). As we shall see, the said initial nonequilibrium distributions may be broadly characterised as having no fine-grained microstructure (with respect to some coarse-graining length) while the initial wave functions are superpositions of at least a few energy eigenstates (in order to guarantee a sufficiently complex de Broglie velocity field). A coarse-graining H-theorem provides a general mechanism in terms of which we can understand how equilibrium is approached, without having to solve the exact equations of motion for the system. By itself, of course, the H-theorem does not prove that equilibrium is actually reached. The rate and extent of relaxation depend on the system and on its initial wave function, as shown by extensive numerical simulations. The Born rule today may then be understood to have arisen dynamically, by a process of relaxation from an earlier nonequilibrium state for which the Born rule was not valid.

Once again, in a time-reversal invariant dynamics it cannot be true that all initial nonequilibrium states relax to equilibrium. But it does not need to be true: non-relaxing initial conditions (with fine-grained microstructure, or with very simple wave functions with trivial velocity fields) are ruled out empirically, not by theoretical fiat.

As we shall see in Section 12.4, modern developments in theoretical and observational cosmology make it possible to test the Born rule for quantum fields at very early times. Thus the existence of initial equilibrium or nonequilibrium is an empirical question — not only in principle but also in practice.

12.3. Overview of quantum relaxation

In this section we provide a brief overview of quantum relaxation. Further details may be found in the cited papers.

12.3.1. *Coarse-graining H-theorem*

For a nonequilibrium ensemble of isolated systems with configurations q, it follows from (12.4) and (12.7) that the ratio $f = \rho/|\psi|^2$ is preserved along trajectories: $df/dt = 0$. (This is the analogue of Liouville's theorem, $d\rho_{\mathrm{cl}}/dt = 0$, for a classical phase-space density ρ_{cl}.) The exact H-function $H(t) = \int dq\, \rho \ln(\rho/|\psi|^2)$ is then constant, $dH/dt = 0$, and there is no fine-grained relaxation. However, if we average the densities ρ and $|\psi|^2$ over small coarse-graining cells of volume δV,[7] we may assume the absence of fine-grained microstructure at $t = 0$,[8]

$$\bar{\rho}(q,0) = \rho(q,0), \quad \overline{|\psi(q,0)|^2} = |\psi(q,0)|^2, \tag{12.11}$$

and consider the time evolution of the coarse-grained H-function (12.9). Defining $\tilde{f} \equiv \bar{\rho}/\overline{|\psi|^2}$, straightforward manipulations show that

$$\bar{H}_0 - \bar{H} = \int dq\, |\psi|^2 (f \ln(f/\tilde{f}) - f + \tilde{f}) \tag{12.12}$$

(where the subscript 0 denotes a quantity at $t = 0$ and the absence of a subscript denotes a quantity at a general time t). Use of the inequality $x \ln(x/y) - x + y \geq 0$ — for all real and non-negative x, y, with equality if and only if $x = y$ — then implies the coarse-graining H-theorem (Valentini, 1991a, 1992)[9]

$$\bar{H}(t) \leq \bar{H}(0). \tag{12.13}$$

From (12.12) it also follows that (12.13) becomes a strict inequality, $\bar{H}(t) < \bar{H}(0)$, when $f \neq \tilde{f}$. Since $|\psi|^2$ remains smooth this occurs when $\rho \neq \bar{\rho}$, that is, when ρ develops fine-grained structure — which it generally will for non-trivial velocity fields that vary over the coarse-graining cells.

The quantity $\bar{H}$ is equal to minus the relative entropy of $\bar{\rho}$ with respect to $\overline{|\psi|^2}$. As already noted, $\bar{H}$ is bounded from below by zero and the minimum $\bar{H} = 0$ is attained if and only if $\bar{\rho} = \overline{|\psi|^2}$. Thus a decrease of $\bar{H}$ quantifies relaxation to the Born rule.[10]

The result (12.13) formalises an intuitive understanding of relaxation in terms of a mixing of two 'fluids' with densities ρ and $|\psi|^2$

---

[7] For systems with N degrees of freedom, $q = (q_1, q_2, \ldots, q_N)$ and $\delta V = (\delta q)^N$.

[8] This will hold to arbitrary accuracy as $\delta V \to 0$ if ρ_0 and $|\psi_0|^2$ are smooth functions.

[9] As is well known for the classical case, the result (12.13) is time-symmetric, with $t = 0$ a local maximum of $\bar{H}(t)$.

[10] The quantity $\bar{H}$ is also equal to the well-known (in mathematical statistics) Kullback-Leibler divergence $D_{\mathrm{KL}}(\bar{\rho} \| \overline{|\psi|^2})$, which measures how $\bar{\rho}$ differs from $\overline{|\psi|^2}$.

in configuration space. These obey the same continuity equation and are therefore 'stirred' by the same velocity field $\dot{q}$. For a sufficiently complicated flow, ρ and $|\psi|^2$ tend to become indistinguishable on a coarse-grained level (Valentini, 1991a). This is similar to the classical stirring of two fluids that was famously discussed by Gibbs (1902), where his fluids were analogous to the classical phase-space densities ρ_{cl} (for a general ensemble) and $\rho_{eq} = $ const. (for an equilibrium ensemble) on the energy surface, and where the mixing of ρ_{cl} and ρ_{eq} may be quantified by a decrease of the classical H-function $\bar{H}_{cl} = \int dq\, \bar{\rho}_{cl} \ln(\bar{\rho}_{cl}/\overline{\rho_{eq}})$. In both cases the nonequilibrium density develops fine-grained structure while the equilibrium density remains smooth. The increase of the 'subquantum entropy' $\bar{S} = -\bar{H}$ may be associated with the mixing of ρ and $|\psi|^2$ in configuration space, just as the increase of the Gibbs entropy $\bar{S}_{Gibbs} = -\bar{H}_{cl}$ may be associated with the mixing of ρ_{cl} and ρ_{eq} in phase space.

Note that we use the word 'mixing' informally in the simple sense of 'stirring' (as in Gibbs' original analogy). Mathematical mixing is defined by an infinite-time limit.[11] Its physical relevance is therefore questionable, and in any case it might not apply to realistic systems even in that limit (Uffink, 2007). In our view the above process — whereby ρ develops fine-grained structure while $|\psi|^2$ does not — is the essential physical mechanism that drives quantum relaxation over realistic timescales. Though as noted, the actual rate and extent of relaxation will depend on the system.

An assumption about initial conditions is of course necessary to explain relaxation in a time-reversal invariant theory. We offer no 'principle' to justify the initial conditions (12.11). In the spirit of our empirical approach, they are justified only by the extent to which they help us explain observations. We take (12.11) to be matters of fact about our world, while acknowledging that in principle they could be false. Their truth or falsity is ultimately a matter for experiment.

---

[11]Formally, a dynamical system (with measure μ on a set Γ subject to measure-preserving transformations T_t) is 'mixing' if and only if $\lim_{t\to\infty} \mu(T_t A \cap B) = \mu(A)\mu(B)$ for all relevant subsets A and B of Γ.

It is also worth remarking that our approach (like its classical Gibbsean counterpart) does not rely on any particular interpretation of probability theory. The density ρ might represent a subjective probability for a single system, a distribution over a theoretical ensemble, or the distribution of a real ensemble of existing systems, according to taste or requirement.

Note also that, for a finite real ensemble of N systems, the actual density ρ will be a sum of delta-functions, which can approach a smooth function only in the large-N limit (Valentini, 1992, pp. 18, 36). To obtain a density with no fine-grained structure on a coarse-graining scale δV, we must of course take the appropriate large-N limit *before* considering small δV.[12]

12.3.2. *Numerical simulations*

Extensive numerical simulations demonstrate that quantum relaxation takes place efficiently for wave functions ψ that are superpositions of multiple energy eigenstates (Valentini and Westman, 2005; Towler, Russell and Valentini, 2012; Colin, 2012; Abraham, Colin and Valentini, 2014). An example is shown in Fig. 12.1, for a two-dimensional oscillator in a superposition of $M = 25$ modes and with an initial Gaussian nonequilibrium density ρ_0. The top row displays the time evolution of the (coarse-grained) equilibrium density $|\psi|^2$, while the bottom row displays relaxation of the actual (coarse-grained) density ρ. Comparable results are obtained for superpositions with as little as $M = 4$ modes, and also for a two-dimensional box.

In all of these simulations $\bar{H}$ is found to decay approximately exponentially with time: $\bar{H}(t) \approx \bar{H}_0 e^{-t/\tau}$ for some constant τ whose value depends on the initial wave function as well as on the coarse-graining length (Towler, Russell and Valentini, 2012). An example is shown in Fig. 12.2.

---

[12]This resolves a concern raised by Norsen (2018, p. 16) that the condition (12.11) on ρ_0 cannot be satisfied for realistic finite ensembles. The same concern arises, of course, in the classical Gibbsean approach and has the same resolution.

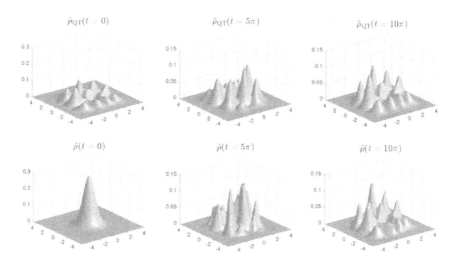

Fig. 12.1. Illustration of numerical quantum relaxation for an oscillator in a superposition of $M = 25$ modes (Abraham, Colin and Valentini, 2014). The wave function has period 2π. The top row shows the evolving coarse-grained density $\tilde{\rho}_{\mathrm{QT}}$ as predicted by quantum theory, while the bottom row shows the coarse-grained relaxation of an initial nonequilibrium density $\rho(x, y, 0) = (1/\pi)e^{-(x^2+y^2)}$ (where tildes denote a smoothed coarse-graining with overlapping cells). After five periods the coarse-grained densities are almost indistinguishable. (Note the different vertical scale at $t = 0$.)

Some simulations show a small but discernible non-zero 'residue' in $\bar{H}$ at large times (unlike the case displayed in Fig. 12.2), indicating that equilibrium is not reached exactly (Abraham, Colin and Valentini, 2014). For these cases, the trajectories tend to show some degree of confinement (not fully exploring the support of $|\psi|^2$). Numerical evidence shows that this is less likely to happen for larger M. Because all laboratory systems have a long and violent astrophysical history, during which the relevant value of M will have been very large, we may expect that in the remote past they will have reached equilibrium on a very small coarse-graining scale.

12.3.3. *The early universe*

This motivates us to consider quantum relaxation in the early universe. This may be discussed for a free massless scalar field ϕ on flat

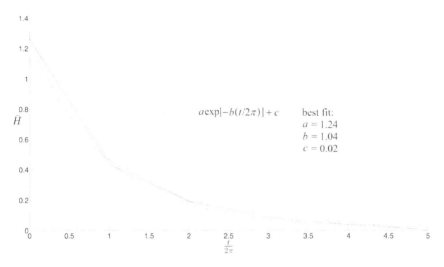

Fig. 12.2.   Approximate exponential decay of $\bar{H}(t)$ for the same simulation displayed in Fig. 12.1 (Abraham, Colin and Valentini, 2014). The error in $\bar{H}$ is estimated by running three separate simulations with different numerical grids (the solid curves). Fitting to an exponential (dashed curve) yields a best-fit residue $c = 0.02$ that is comparable to the late-time error, indicating no discernible late-time residue in $\bar{H}$. We find a decay timescale $\tau = 2\pi/b \simeq 6$ (units $\hbar = m = \omega = 1$).

expanding space with spacetime metric

$$d\tau^2 = dt^2 - a^2 d\mathbf{x}^2 \qquad (12.14)$$

and a scale factor $a(t) \propto t^{1/2}$ (corresponding to a radiation-dominated expansion), where t is standard cosmological time and physical wavelengths $\lambda_{\text{phys}}(t)$ are proportional to $a(t)$. The Fourier components $\phi_{\mathbf{k}}(t)$ may be written as

$$\phi_{\mathbf{k}} = \frac{\sqrt{V}}{(2\pi)^{3/2}}(q_{\mathbf{k}1} + i q_{\mathbf{k}2}), \qquad (12.15)$$

where $q_{\mathbf{k}1}$, $q_{\mathbf{k}2}$ are real and V is a normalisation volume. The field Hamiltonian then takes the form $\hat{H} = \sum_{\mathbf{k}r} \hat{H}_{\mathbf{k}r}$, where $\hat{H}_{\mathbf{k}r}$ $(r = 1, 2)$ coincides with the Hamiltonian of a harmonic oscillator of mass $m = a^3$ and angular frequency $\omega = k/a$ (Valentini, 2007, 2008a, 2010a). Thus a single (unentangled) field mode $\mathbf{k}$ in the early universe is mathematically equivalent to a two-dimensional oscillator with a time-dependent mass. This system is in turn equivalent to an ordinary oscillator (with constant mass and angular frequency) but with t replaced by a 'retarded time' $t_{\mathrm{ret}} = t_{\mathrm{ret}}(t, k)$ that depends on k (Colin and Valentini, 2013). It is found that quantum relaxation depends crucially on how λ_{phys} compares with the Hubble radius $H^{-1} = a/\dot{a}$. For short wavelengths $\lambda_{\mathrm{phys}} \ll H^{-1}$ we find $t_{\mathrm{ret}}(t, k) \to t$ and we recover physics on static flat space, with the same rapid relaxation illustrated in Fig. 12.1 for the oscillator. Whereas for long wavelengths $\lambda_{\mathrm{phys}} \gtrsim H^{-1}$ we find $t_{\mathrm{ret}}(t, k) < t$ and relaxation is retarded or suppressed (Valentini, 2008a; Colin and Valentini, 2013).

The suppression of quantum relaxation at long (super-Hubble) wavelengths is illustrated in Fig. 12.3, where we show the time evolution of a nonequilibrium field mode with $\lambda_{\mathrm{phys}} = 10H^{-1}$ (at initial time t_i). Over the given time interval (t_i, t_f) relaxation proceeds but is incomplete: in particular, the final nonequilibrium width (or variance) is smaller than the final equilibrium width. In contrast, in Fig. 12.4 we show relaxation with the same initial conditions and over the same time interval (t_i, t_f) but with no spatial expansion: relaxation now takes place essentially completely and the final widths match closely.

Cosmologically speaking, the reduced width of the final distribution in Fig. 12.3 is of particular interest. For a mode with wavenumber k we may write

$$\langle |\phi_{\mathbf{k}}|^2 \rangle = \langle |\phi_{\mathbf{k}}|^2 \rangle_{\mathrm{QT}} \xi(k), \tag{12.16}$$

where $\langle \ldots \rangle$ and $\langle \ldots \rangle_{\mathrm{QT}}$ denote respective nonequilibrium and equilibrium expectation values. The function $\xi(k)$ quantifies the degree of primordial quantum nonequilibrium as a function of k. For the case shown in Fig. 12.3, at the final time we clearly have $\xi(k) < 1$ (corresponding to a 'power deficit'). As a general trend we expect $\xi(k)$ to be smaller for smaller k, where longer wavelengths imply less relaxation. This has been broadly confirmed by running extensive simulations for varying values of k and plotting the function $\xi = \xi(k)$. The resulting curves show oscillations of magnitude $\lesssim 10\%$. As a first approximation we may ignore these, in which case we find a good fit to the

time evolution of quantum equilibrium on expanding space

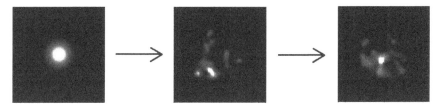

incomplete relaxation on expanding space

Fig. 12.3. Incomplete quantum relaxation for a super-Hubble field mode $\phi_{\mathbf{k}}$ on expanding space over a time interval (t_i, t_f) (Colin and Valentini, 2013). The final nonequilibrium width is noticeably smaller than the final equilibrium width.

time evolution of quantum equilibrium on static space

essentially complete relaxation on static space

Fig. 12.4. Essentially complete quantum relaxation for a field mode $\phi_{\mathbf{k}}$ on static space over the same time interval (t_i, t_f) as in Fig. 12.3 and with the same initial conditions (Colin and Valentini, 2013). The final nonequilibrium width closely matches the final equilibrium width.

function

$$\xi(k) = \tan^{-1}\left(c_1\frac{k}{\pi} + c_2\right) - \frac{\pi}{2} + c_3, \qquad (12.17)$$

where the parameters c_1, c_2 and c_3 depend on the initial state and on the time interval (Colin and Valentini, 2015). This function tends to a maximum $\xi \to c_3$ for $k \to \infty$, and decreases smoothly for smaller k. Simulations for a range of different initial nonequilibria — all assumed to have a narrower-than-quantum initial width[13] — show a good fit to the curve (12.17) (ignoring the oscillations) (Colin and Valentini, 2016). Thus, with some mild assumptions about the initial state, the 'deficit function' (12.17) is a robust approximate prediction of the cosmological quantum relaxation scenario (for a free scalar field at the end of a radiation-dominated era).

12.4. Testing the primordial Born rule with cosmological data

We now outline how the Born rule may be tested at very early times with cosmological data.

The temperature of the cosmic microwave background (CMB) today is slightly anisotropic. It is customary to write a spherical harmonic expansion

$$\frac{\Delta T(\hat{\mathbf{n}})}{\bar{T}} = \sum_{l=2}^{\infty} \sum_{m=-l}^{+l} a_{lm} Y_{lm}(\hat{\mathbf{n}}) \qquad (12.18)$$

for the measured anisotropy $\Delta T(\hat{\mathbf{n}}) \equiv T(\hat{\mathbf{n}}) - \bar{T}$, where the unit vector $\hat{\mathbf{n}}$ labels points on the sky and $\bar{T}$ is the mean temperature. As noted the CMB was formed around 400,000 years after the big bang, and its small anisotropies reflect small inhomogeneities of the universe at that time. Thus the coefficients a_{lm} are generated by the

---

[13]Heuristically, it seems natural to assume initial conditions with a subquantum statistical spread (so the initial state contains less noise than a conventional quantum state), but in principle this assumption could of course be incorrect.

Fourier-space 'primordial curvature perturbation' $\mathcal{R}_{\mathbf{k}}$ according to the formula (Lyth and Riotto, 1999)

$$a_{lm} = \frac{i^l}{2\pi^2} \int d^3\mathbf{k} \; \mathcal{T}(k,l) \mathcal{R}_{\mathbf{k}} Y_{lm}(\hat{\mathbf{k}}), \qquad (12.19)$$

where the 'transfer function' $\mathcal{T}(k,l)$ encodes the relevant astrophysics.

Cosmologists generally assume that the measured function $\Delta T(\hat{\mathbf{n}})$ is a single realisation of a random variable with a probability distribution $P[\Delta T(\hat{\mathbf{n}})]$ associated with a 'theoretical ensemble' (which may be interpreted according to taste). It is usual to assume 'statistical isotropy', which means that P is invariant under a rotation $\hat{\mathbf{n}} \to \hat{\mathbf{n}}'$ (that is, $P[\Delta T(\hat{\mathbf{n}}')] = P[\Delta T(\hat{\mathbf{n}})]$). This implies that the ensemble average $\langle \Delta T(\hat{\mathbf{n}}_1) \Delta T(\hat{\mathbf{n}}_2) \rangle$ depends only on the angle between $\hat{\mathbf{n}}_1$ and $\hat{\mathbf{n}}_2$, which in turn implies (Hajian and Souradeep, 2005, appendix B)

$$\langle a^*_{l'm'} a_{lm} \rangle = \delta_{ll'} \delta_{mm'} C_l, \qquad (12.20)$$

where

$$C_l \equiv \langle |a_{lm}|^2 \rangle \qquad (12.21)$$

(the 'angular power spectrum') is independent of m. Thus, while for fixed l there are $2l+1$ different quantities $|a_{lm}|^2$, statistical isotropy implies that they each have the same ensemble mean C_l.

For our one observed sky we may define a measured mean statistic

$$C_l^{\text{sky}} \equiv \frac{1}{2l+1} \sum_{m=-l}^{+l} |a_{lm}|^2.$$

This obviously satisfies $\langle C_l^{\text{sky}} \rangle = C_l$. Thus C_l^{sky} (measured for one sky) is an unbiased estimate of C_l for the ensemble. Furthermore, assuming a Gaussian distribution it may be shown that C_l^{sky} has a 'cosmic variance'

$$\frac{\Delta C_l^{\text{sky}}}{C_l} = \sqrt{\frac{2}{2l+1}}. \qquad (12.22)$$

Thus for large l we expect to find $C_l^{\mathrm{sky}} \approx C_l$, whereas for small l the accuracy is limited.

It is also usually assumed that the theoretical ensemble for $\mathcal{R}$ is statistically homogeneous (that is, the probability distribution $P[\mathcal{R}(\mathbf{x})]$ is invariant under spatial translations). This implies that $\langle \mathcal{R}(\mathbf{x})\mathcal{R}(\mathbf{x}')\rangle$ depends only on $\mathbf{x} - \mathbf{x}'$, which implies

$$\langle \mathcal{R}^*_{\mathbf{k}'}\mathcal{R}_{\mathbf{k}}\rangle = \delta_{\mathbf{k}\mathbf{k}'}\langle|\mathcal{R}_{\mathbf{k}}|^2\rangle. \tag{12.23}$$

From (12.19) and (12.23) it follows that

$$C_l = \frac{1}{2\pi^2}\int_0^\infty \frac{dk}{k}\, \mathcal{T}^2(k,l)\mathcal{P}_{\mathcal{R}}(k), \tag{12.24}$$

where

$$\mathcal{P}_{\mathcal{R}}(k) \equiv \frac{4\pi k^3}{V}\langle|\mathcal{R}_{\mathbf{k}}|^2\rangle \tag{12.25}$$

is the 'primordial power spectrum' for $\mathcal{R}_{\mathbf{k}}$. Thus measurements for a single sky constrain the spectrum $\mathcal{P}_{\mathcal{R}}(k)$ — which is a property of the theoretical ensemble.

Note that in the discussion so far the quantities a_{lm}, C_l and $\mathcal{R}_{\mathbf{k}}$ are treated classically.[14]

What is the origin of the spectrum (12.25)? According to inflationary cosmology, during a very early period of approximately exponential expansion a perturbation $\phi_{\mathbf{k}}$ of the 'inflaton field' generates a curvature perturbation $\mathcal{R}_{\mathbf{k}} \propto \phi_{\mathbf{k}}$ (once the physical wavelength of the mode exceeds the Hubble radius) (Liddle and Lyth, 2000). The quantum-theoretical variance $\langle|\phi_{\mathbf{k}}|^2\rangle_{\mathrm{QT}}$ is calculated from quantum field theory assuming the Born rule for an appropriate vacuum state. From this we readily obtain the corresponding variance $\langle|\mathcal{R}_{\mathbf{k}}|^2\rangle_{\mathrm{QT}}$ and hence a quantum-theoretical prediction $\mathcal{P}_{\mathcal{R}}^{\mathrm{QT}}(k)$ for the spectrum of $\mathcal{R}_{\mathbf{k}}$ (approximately flat with a slight tilt). If instead we have

---

[14]Goldstein, Struyve and Tumulka (2015) give a rather confused account of the relation between primordial perturbations and CMB anisotropies, missing in particular the crucial role played by statistical isotropy.

a nonequilibrium variance (12.16) for $\phi_\mathbf{k}$, the predicted spectrum for $\mathcal{R}_\mathbf{k}$ is corrected by the factor $\xi(k)$:

$$\mathcal{P}_\mathcal{R}(k) = \mathcal{P}_\mathcal{R}^{\mathrm{QT}}(k)\xi(k). \tag{12.26}$$

Inserting this into (12.24) yields a corrected angular power spectrum C_l. From measurements of the CMB we may then set observational bounds on $\xi(k)$ — that is, on corrections to the Born rule in the very early universe (Valentini, 2010a).

To obtain a prediction for $\xi(k)$, we may for example assume that quantum relaxation took place during a pre-inflationary era. It is not uncommon for cosmologists to assume that such an era was radiation-dominated.[15] In this case, at the end of pre-inflation we expect the nonequilibrium variance to be corrected by a deficit function of the approximate form (12.17). It may be shown that relaxation does not take place during inflation itself (Valentini, 2010a). If we make the simplifying assumption that the spectrum is unaffected by the transition from pre-inflation to inflation, we obtain a prediction for a corrected primordial power spectrum (12.26) with $\xi(k)$ of the form (12.17) (with three unknown parameters). Note, however, that if the lengthscale c_1 is too large the dip in the spectrum will be essentially unobservable (even if it exists).

CMB data from the *Planck* satellite show hints of a power deficit at large scales (small k and low l) (Aghanim *et al.*, 2016). Because of the large cosmic variance in this region, it is however difficult to draw firm conclusions. Extensive data analysis shows that the predicted deficit (12.17) fits the data more or less as well as the standard 'power-law' model (where the evaluated significance takes into acount the larger number of parameters) (Vitenti, Peter and Valentini, 2019). This is a modest success, in the sense that with three extra parameters the significance could have been worse. But evidence from this fitting alone neither supports nor rules out the prediction (12.17). If one is inclined to invoke Ockham's razor in favour of the simplest cosmological model, then the observed low-power anomaly may reasonably be regarded as a statistical fluctuation. To obtain evidence

---

[15]See, for example, Wang and Ng (2008).

for or against our model we must include more detailed predictions, such as oscillations around the curve (12.17) (Colin and Valentini, 2015; Kandhadai and Valentini, 2020) or possible violations of statistical isotropy (Valentini, 2015).

12.5. Critique of 'typicality' as an explanation for the Born rule

In this section we provide a critical assessment of the typicality approach to understanding the Born rule.

12.5.1. *Typicality, probability, and intrinsic likelihood*

As we noted in the Introduction, the Bohmian mechanics school attempts to explain the success of the Born rule today by appealing to a notion of 'typicality' for the initial configuration $q_{\mathrm{univ}}(0)$ of the universe. In this approach, if $\Psi_{\mathrm{univ}}(q_{\mathrm{univ}}, 0)$ is the initial universal wave function then $|\Psi_{\mathrm{univ}}(q_{\mathrm{univ}}, 0)|^2$ is assumed to be the 'natural measure' on the set of initial universal configurations $q_{\mathrm{univ}}(0)$.

In our view the typicality approach amounts to assuming, without justification, that the universe as a whole began in quantum equilibrium (Valentini, 1996, 2001). The approach then seems circular. Defenders of the approach might attempt to avoid the charge of circularity by claiming that typicality and probability are conceptually distinct. In our view, however, typicality is synonomous with probability. The Bohmian mechanics school employs the word 'typicality' when referring to probability for the whole universe and employs the word 'probability' when referring to probability for sub-systems. But in our view the two words mean the same thing.

Typicality and the Born rule

Dürr, Goldstein and Zanghì (1992) showed that, if we consider sub-systems within the universe, we will obtain the Born rule for 'almost all' initial configurations $q_{\mathrm{univ}}(0)$ — where 'almost all' is defined with respect to $|\Psi_{\mathrm{univ}}(q_{\mathrm{univ}}, 0)|^2$. Thus it may be said that for sub-systems the quantum equilibrium distribution $\rho = |\psi|^2$ is 'typical', where the notion of 'typicality' is defined with respect to the measure $|\Psi_{\text{univ}}(q_{\text{univ}}, 0)|^2$.

This result may be illustrated by a simple example. We consider a model universe at $t = 0$ containing a large number n of unentangled sub-systems all with the same initial wave function $\psi(q, 0)$ and with initial configurations $q_1(0), q_2(0), \ldots, q_n(0)$ that generally vary from one sub-system to another. We may write

$$q_{\text{univ}}(0) = (q_1(0), q_2(0), \ldots, q_n(0)) \tag{12.27}$$

and

$$\Psi_{\text{univ}}(q_{\text{univ}}, 0) = \psi(q_1, 0)\psi(q_2, 0) \cdots \psi(q_n, 0). \tag{12.28}$$

For large n, a given initial configuration $q_{\text{univ}}(0)$ determines an initial distribution $\rho(q, 0)$ over the ensemble of sub-systems. Thus we have a schematic correspondence

$$q_{\text{univ}}(0) \longleftrightarrow \rho(q, 0) \quad (n \to \infty).$$

Note that the induced distribution $\rho(q, 0)$ need not be equal to (or even close to) $|\psi(q, 0)|^2$. On the other hand, if we adopt the measure $|\Psi_{\text{univ}}(q_{\text{univ}}, 0)|^2$ on the set of possible initial configurations $q_{\text{univ}}(0)$, then it is easy to see that, with respect to this measure, in the limit $n \to \infty$ almost all points $q_{\text{univ}}(0)$ correspond to the Born-rule distribution $\rho(q, 0) = |\psi(q, 0)|^2$. This is because, with respect to the universal measure

$$|\Psi_{\text{univ}}(q_{\text{univ}}, 0)|^2 = |\psi(q_1, 0)|^2 |\psi(q_2, 0)|^2 \cdots |\psi(q_n, 0)|^2,$$

in effect we have n independent and identically-distributed random variables $q_1, q_2, q_3, \ldots, q_n$, each with the same probability distribution $|\psi(q, 0)|^2$. In the limit $n \to \infty$ we then necessarily find $\rho(q, 0) = |\psi(q, 0)|^2$.

It might then appear that nonequilibrium configurations — that is, configurations $q_{\text{univ}}(0)$ corresponding to distributions $\rho(q, 0) \neq |\psi(q, 0)|^2$ — comprise a vanishingly small set (in the limit $n \to \infty$) and may then be regarded as intrinsically unlikely or 'untypical'. But this conclusion rests crucially on the choice of measure $|\Psi_{\text{univ}}(q_{\text{univ}}, 0)|^2$. For example, if instead we choose a measure

$|\Psi_{\mathrm{univ}}(q_{\mathrm{univ}}, 0)|^4$ (up to overall normalisation), then by the same argument almost all configurations $q_{\mathrm{univ}}(0)$ now correspond to a nonequilibrium distribution $\rho(q, 0) \propto |\psi(q, 0)|^4$ for sub-systems. More generally, if we choose a measure $|\Psi_{\mathrm{univ}}(q_{\mathrm{univ}}, 0)|^p$ (for constant $p > 0$), we will almost always obtain a nonequilibrium distribution $\rho(q, 0) \propto |\psi(q, 0)|^p$ for sub-systems. The typicality approach then seems circular: by assuming a universal Born-rule measure $|\Psi_{\mathrm{univ}}(q_{\mathrm{univ}}, 0)|^2$, one is simply assuming the Born rule at the initial time $t = 0$ (Valentini, 1996, 2001).

In our illustrative example $\Psi_{\mathrm{univ}}(q_{\mathrm{univ}}, 0)$ takes the simple form (12.28) and we only consider measurements at $t = 0$. The original argument by Dürr, Goldstein and Zanghì (1992) is more general than this and includes a discussion of time ensembles of measurements. But the key objection remains: the Born rule is guaranteed to hold for sub-systems in the early universe only because the Born rule is assumed to hold for the whole universe at $t = 0$.

It is important to emphasise that there is a qualitative difference between cases with large-but-finite n and the literal limit $n \to \infty$. For any finite n, however large, a set of points $q_{\mathrm{univ}} = (q_1, q_2, \ldots, q_n)$ that has zero measure with respect to $|\Psi_{\mathrm{univ}}(q_{\mathrm{univ}}, 0)|^2$ will arguably have zero measure with respect to any reasonable density function on configuration space. Thus, for example, if a particle moving in two spatial dimensions has an initial wave function $\psi(x, y, 0)$ then points and lines in the two-dimensional configuration space will have zero measure with respect to $|\psi(x, y, 0)|^2$. Those same points and lines will also have zero Lebesgue measure (or zero area), and furthermore they will have zero measure with respect to any density proportional to $|\psi(x, y, 0)|^p$ (with $p > 0$). Much the same may be said for general configuration spaces. This means that, for finite n, if a set of points $q_{\mathrm{univ}}(0)$ has zero measure with respect to $|\Psi_{\mathrm{univ}}(q_{\mathrm{univ}}, 0)|^2$ then that same set of points may reasonably be regarded as objectively small and hence physically negligible. But this objectivity vanishes when we take the limit $n \to \infty$ (the limit where the typicality argument is applied). For example, as we noted above, for $n \to \infty$ the nonequilibrium set

$$S_{\mathrm{noneq}} = \{q_{\mathrm{univ}}(0) \mid \rho(q, 0) \propto |\psi(q, 0)|^4\} \qquad (12.29)$$

(the set of initial points $q_{\text{univ}}(0)$ yielding a nonequilibrium sub-system density $\rho(q,0) \propto |\psi(q,0)|^4$ has zero measure with respect to $|\Psi_{\text{univ}}(q_{\text{univ}},0)|^2$ (that is, $\mu_{\text{eq}}[S_{\text{noneq}}] = 0$ where $d\mu_{\text{eq}} = |\Psi_{\text{univ}}(q_{\text{univ}},0)|^2 dq_{\text{univ}}$). On the other hand, the same set S_{noneq} has unit measure with respect to $|\Psi_{\text{univ}}(q_{\text{univ}},0)|^4$ (suitably normalised) (that is, $\mu_{\text{noneq}}[S_{\text{noneq}}] = 1$ where $d\mu_{\text{noneq}} \propto |\Psi_{\text{univ}}(q_{\text{univ}},0)|^4 dq_{\text{univ}}$). Thus, in the limit $n \to \infty$, there is *no* objective sense in which the nonequilibrium set S_{noneq} is 'small'.

Technically, the qualitative difference between cases with finite n and cases where the literal limit $n \to \infty$ is taken may be highlighted in terms of the notion of 'absolute continuity'. For finite n, the alternative measure $|\Psi_{\text{univ}}(q_{\text{univ}},0)|^4$ is absolutely continuous with respect to the equilibrium measure $|\Psi_{\text{univ}}(q_{\text{univ}},0)|^2$. This simply means, by definition, that the alternative measure of a set S is equal to zero whenever the equilibrium measure of S is equal to zero ($\mu_{\text{eq}}[S] = 0$ implies $\mu_{\text{noneq}}[S] = 0$). But this will *not* hold for $n \to \infty$, where we can have both $\mu_{\text{eq}}[S_{\text{noneq}}] = 0$ and $\mu_{\text{noneq}}[S_{\text{noneq}}] = 1$.

In this context one should beware of statements appealing to absolute continuity. For example, Dürr and Teufel (2009, p. 222) write:

> "... any measure ... which is absolutely continuous with respect to the equivariant measure $[|\Psi_{\text{univ}}|^2]$ ... defines the same sense of typicality."

Taken by itself this statement is correct and simply amounts to a statement of the definition of absolute continuity. But to avoid misunderstandings in this context, it is important to add (as Dürr and Teufel omit to) that absolute continuity fails in the relevant limit $n \to \infty$. In that limit, running the argument with an alternative typicality measure will *not* yield the Born rule for sub-systems, because the alternative measure will not be absolutely continuous with respect to the Born-rule measure.

Intrinsic likelihood and unlikelihood

We have seen that at $t = 0$ nonequilibrium is 'untypical' (has zero measure) with respect to the equilibrium measure, and that equally nonequilibrium is 'typical' (has unit measure) with respect to a corresponding nonequilibrium measure. It might then be said that initial nonequilibrium is unlikely with respect to the equilibrium measure, and that it is *likely* with respect to a nonequilibrium measure.

With these clarifications in mind, there is clearly no sense in which quantum nonequilibrium is instrinsically unlikely — contrary to claims made by the Bohmian mechanics school, according to which the theory will always yield the Born rule. For example, in their founding paper Dürr *et al.* claim that

> "...in a universe governed by Bohmian mechanics it is in principle impossible to know more about the configuration of any subsystem than what is expressed by [the Born rule]." (Dürr, Goldstein and Zanghì, 1992)

Similarly, Tumulka writes that

> "...in a universe governed by Bohmian mechanics, observers will see outcomes with exactly the probabilities specified by the usual rules of quantum mechanics..." (Tumulka, 2018)

But as we have noted this is true only in quantum equilibrium: for general nonequilibrium ensembles the Born rule is violated (Valentini, 1991a,b) and it is perfectly possible to know more about a sub-system than is allowed by the Born rule or by the associated uncertainty principle (Valentini, 2002a).

The claims made by the Bohmian mechanics school amount to an artificial and unjustified restriction of the theory to equilibrium only. This is most explicit in the presentation by Tumulka, who writes that it is

> "...a fundamental law of Bohmian mechanics, to demand that [q] be *typical* with respect to $|\Psi|^2$." (Tumulka, 2018)

But as we have emphasised, only the dynamical equations have the status of fundamental laws. In a theory of dynamics it is physically nonsensical to regard a restriction on the initial state as a 'fundamental law'.

Dürr and Teufel (2009, p. 224) go so far as to compare searching for nonequilibrium to waiting for a stone to jump up spontaneously in the air:

> "...one could sit in front of a stone and wait for the stone to jump into the air, because in a very atypical world, that could happen, now, tomorrow, maybe the day after tomorrow." (Dürr and Teufel, 2009, p. 224)

But there is no scientific basis for the claim that quantum nonequilibrium is intrinsically unlikely. This claim stems, as we have seen, from a circular argument in which the Born-rule measure is taken to *define* 'typicality' for the initial conditions of the universe. In our view, in contrast, initial conditions are ultimately empirical.

Typicality and probability

If we replace the word 'typicality' by its synonym 'probability', it becomes apparent that the argument given by Dürr *et al.* simply assumes a universal Born-rule probability density $|\Psi_{\text{univ}}|^2$, which implies the Born-rule probability density $|\psi|^2$ for sub-systems.[16]

Goldstein (2001) has, however, defended the view that typicality and probability are *not* synonymous. Goldstein argues that for a set S the precise value of the typicality measure $\mu(S)$ (say $1/2$ or $3/4$) is immaterial: the only thing that matters is whether $\mu(S)$ is very small or not. If $\mu(S)$ is very small, this is regarded as a sufficient explanation for why events in S do not occur. According to Goldstein, this concept of typicality plays an important role in scientific explanation. In our view such attempts to elevate the notion of typicality from a synonym for probability to a fundamentally new kind of explanatory principle raise more questions than they answer. It is claimed that what counts is only whether $\mu(S)$ is very small or not. How small is small enough? Questions remain as to precisely how typicality differs from probability. One may also ask why the notion of probability alone does not suffice. The nature of probability is

---

[16]For measurements at a single time this result is a restatement of the 'nesting' property discussed in Section 12.2 (Valentini, 1991a).

already controversial: it seems misguided to introduce a conceptual variant which proves to be even more controversial.

Other responses to circularity

Dürr and Teufel (2009, pp. 220–222) make some attempt to respond to our charge of circularity.

The first response appeals to the equivariance of the Born-rule measure over time. This measure has the special property that its form as a function of Ψ is preserved by the dynamics (that is, it is 'equivariant'). Dürr and Teufel argue that this property singles out the Born-rule measure as the preferred measure of typicality:

> "Would another measure...say the one with density $|\Psi|^4$...not yield typicality for the empirical distribution $|\varphi|^4$ by the same argument? In fact, it would, but only at *exactly* that moment of time where the measure has the $|\Psi|^4$ density. Since the measure is not equivariant, its density will soon change to something completely different.... The equivariant measure of typicality on the other hand is special... Typicality defined by this measure does not depend on time." (Dürr and Teufel, 2009, pp. 220–221)

This defence is physically unconvincing. As we have seen, in their argument for the typicality of the Born rule for sub-systems the universal Born-rule measure is applied to the initial universal configuration $q_{\mathrm{univ}}(0)$ at only one time (the initial time $t = 0$). From this we may immediately derive the Born rule for sub-systems at that same initial time $t = 0$. The time evolution of the measure at $t > 0$ is irrelevant to the statistics of sub-systems at $t = 0$. Furthermore, as Dürr and Teufel themselves admit, if an alternative universal measure were applied at $t = 0$ then by the same derivation we would obtain an alternative (non-Born-rule) distribution for sub-systems at $t = 0$ — regardless of how the measure evolves at later times. Dürr and Teufel wish to argue in favour of Born-rule typicality for initial configurations at $t = 0$, but their argument appeals to properties of the time evolution of the measure at $t > 0$. Physically speaking, it is hard to understand how initial conditions at $t = 0$ can be dictated by, or influenced by, a convenient mathematical property of the evolution at $t > 0$.

In a similar vein, while commenting on the quantum relaxation approach advocated by this author, and in particular on the idea of primordial quantum nonequilibrium, Tumulka claims outright that an initial nonequilibrium measure would be intrinsically unnatural and puzzling because it would not be equivariant:

> "...the $|\Psi|^2$ distribution is special since it is equivariant...if we found empirically (which we have not) that it was necessary to assume that $[q(0)]$ was $|\Psi|^4$-distributed, then it would be a big puzzle needing explanation why it was $|\Psi|^4$ of all distributions instead of the natural, equivariant $|\Psi|^2$." (Tumulka, 2018)

The argument seems to be that initial equilibrium is natural and needs no explanation because it is equivariant. Whereas, because initial nonequilibrium is not equivariant, it is unnatural and so — if it were observed — would present a major puzzle. Again, physically it is difficult to see how initial conditions at $t = 0$ can be dictated by a convenient mathematical property of the evolution at later times.

A second response to the charge of circularity appeals to ease of proof:

> "...the equivariant measure is a highly valuable technical tool, because this is the measure which allows us to prove the theorem! At time t, let us say today when we do the experiment, any other measure would look so odd (it would depend on Ψ_t in such an intricate way) that we would have no chance of proving anything! And if it did not look odd today, then it would look terribly odd tomorrow! The equivariant measure always looks the same and what we prove today about the empirical distribution will hold forever." (Dürr and Teufel, 2009, p. 222)

But we are concerned with objective physical facts about the initial configuration of our universe. In a scientific theory, the initial state of the universe is not determined by mathematical convenience or ease of proof of a certain theorem, but by empirical observation and measurement.

Typicality in classical and quantum statistical mechanics

In defence of the typicality approach it might be asserted that, in statistical mechanics generally, certain undesirable initial conditions are often ruled out on the grounds that they are exceptional (or untypical) with respect to a particular measure. Some authors do indeed justify the required restrictions on initial conditions along these lines. But it should be admitted that all initial conditions are allowed in principle. The actual realised initial conditions are ultimately an empirical matter to be constrained by experiment. Furthermore, the mere use of a different word does not entail the use of a genuinely different concept: in our view ruling out undesirable initial conditions as 'untypical' simply amounts to ruling them out as 'improbable'.

The use of typicality in classical statistical mechanics has also been defended by Goldstein (2001), who argues that to explain thermal relaxation it suffices to note that phase-space points corresponding to thermal equilibrium occupy an overwhelmingly larger volume than phase-space points corresponding to thermal nonequilibrium. According to this argument, if a system begins at an initial point corresponding to nonequilibrium, then it is overwhelmingly likely to evolve (and quickly) to a final point corresponding to equilibrium, merely by virtue of the much larger volume occupied by the latter points. It may then be said that thermal relaxation is 'typical' with respect to the phase-space volume measure. However, as noted by Uffink (2007, pp. 979–980), a specific system trajectory $(q(t), p(t))$ traces out a set of points S_0 of measure zero (with respect to phase-space volume), whereas the set of points S_1 never visited by the trajectory is of measure one. By definition the system remains in the zero-measure set S_0 for all time, and does *not* move into the set S_1 even though the latter has an overwhelmingly larger phase-space measure. It is then untenable to claim that a system is more likely to move into a set of points simply because that set has a larger phase-space volume. Uffink concludes — in our view correctly — that a *bona fide* explanation for relaxation must appeal to properties of the dynamics and not merely to a measure-theoretic counting of states.

In our view measure-theoretic arguments are misleading and by themselves give no indication of likelihood. We emphasise once again that initial conditions are ultimately a matter for experiment. As scientists we need to understand which initial conditions are consistent with present observations. To this end we must consider the dynamics, and also take into account our knowledge of past history (whether locally in the laboratory or at cosmological scales).

12.5.2. *Probability and ensembles in cosmology*

According to the Bohmian mechanics school it is meaningless to consider probabilities or ensembles for the whole universe:

> What physical significance can be assigned to a probability distribution on the initial configurations for the entire universe? What can be the relevance to physics of such an ensemble of universes? (Dürr, Goldstein and Zanghì, 1992)
> ... since we only have access to one universe... an ensemble of universes is meaningless for physics. (Dürr and Teufel, 2009, p. 224)

This claim conflicts with common practice in both theoretical and observational cosmology. In recent decades hundreds of millions of dollars have been spent on satellite observations of the CMB (as well as on galaxy surveys) with the express aim of putting empirical constraints on the probability distribution for primordial cosmological perturbations. These observations do have a well-defined meaning, in spite of the above claim.

First of all, to suggest that there is only one universe is correct only as a trivial tautology. In practice cosmologists employ the word 'universe' to denote the totality of what we are currently able to observe. Given our current knowledge it is perfectly plausible that what we see is only one element of a very large and perhaps even infinite ensemble. In current observational cosmology, the data are well described by a 'standard' cosmological model according to which what we see is only a tiny patch within an infinite flat (expanding) space — a conclusion that is arguably the most conservative option at the present time. Further afield, contemporary theoretical cosmology includes 'eternal' inflationary models with an infinity of pocket universes, while string theories are widely believed to imply the existence of a 'multiverse'. Thus in theory there is no difficulty in imagining that the universe we see is merely one of a large ensemble, and this may well be the case as a matter of physical fact.

Secondly, the meaning of a probability distribution 'for the universe' is by no means as problematic as the Bohmian mechanics school portrays it. As we outlined in Section 12.4, practising cosmologists routinely test the predictions of such distributions via measurements of the CMB. As we explained, such measurements probe the primordial power spectrum (12.25) for a theoretical ensemble. By this means, whole classes of cosmological models have been ruled out by observation because they predict an incorrect spectrum.

This is not to say that one cannot or should not question the meaning of a theoretical ensemble. But we would argue that such foundational questions have no special connection with pilot-wave theory or cosmology; they arise in any practical application of probability theory or statistical inference, whether one is considering genetic populations on earth or the distribution of galaxies in space. Furthermore, the relevant mathematical properties of the assumed 'probability distribution over a theoretical ensemble' do not depend on any particular interpretation of probability theory. One could, for example, regard the distribution as expressing a subjective degree of belief, or as representing a really existing ensemble; it would make no difference to how the distribution is employed in mathematical practice. Indeed, if one prefers one may avoid the notion of a theoretical ensemble of 'universes' and instead consider a real ensemble of approximately independent sub-regions of a single universe. Given that the 'universe' we see may in any case be just such a sub-region, which approach one takes is immaterial.

In this context one should also beware of the claim (sometimes made by the Bohmian mechanics school[17]) that pilot-wave dynamics is in some special sense fundamentally a dynamics of the whole universe. If this were true, we would need a complete theory of cosmology

---

[17]See, for example, Goldstein and Zanghì (2013).

to work with and apply pilot-wave theory. In a trivial sense, of course, there will always be small interactions between even very distant systems, and in all known theories of physics it could be said that fundamentally the theory is a theory of the whole universe. But this is no more true in pilot-wave theory than it is in classical mechanics, classical field theory, or general relativity. In principle one needs to consider the whole universe in all these theories; but in practice, the universe divides into approximately independent pieces, at least in the real situations occurring in our actual world. The situation in pilot-wave theory shows no essential difference from that of other physical theories.

In our view, the question of the meaning of probability and of ensembles in cosmology has its place as a valid and interesting philosophical question, but the emphasis placed on this question by the Bohmian mechanics school has proved to be misleading and (we would argue) a distraction. The question of the existence or otherwise of primordial quantum nonequilbrium is empirical. It will be answered by detailed work in theoretical and observational cosmology, not by foundational debates about the meaning of probability and related topics.

On a related note we comment on a recent paper by Norsen (2018), which attempts to combine quantum relaxation ideas with typicality arguments. In our view considerations of typicality add nothing substantial to a quantum relaxation scenario, and merely introduce a new word to denote the probability measure for a universal theoretical ensemble.[18] Furthermore if, as Norsen (p. 24) advocates, we also consider 'reasonably smooth' non-Born-rule typicality measures on the initial universal configuration, then while (as Norsen notes) our relaxation results suggest that the Born rule will still be obtained on a coarse-grained level at later times, the fact remains that such initial measures will imply nonequilibrium for sub-systems in the early universe (with all the novel physical implications we have described).

---

[18]Norsen (p. 15) follows the Bohmian mechanics school in making the misleading claim that "...there is simply no such thing as the probability distribution P for particle configurations of the universe as a whole, because there is just one universe".

This then contradicts Norsen's claim (p. 22) that early nonequilibrium is intrinsically unlikely (a claim made, confusingly, by appealing to the initial Born-rule measure). To avoid such needless controversy, we should bear in mind that there simply is no intrinsic typicality (or probability) measure for initial conditions, and that in the end the existence or non-existence of early nonequilibrium can only be established by observation.

12.6. Contingency and the nature of the universal wave function

The typicality approach has led to misunderstandings not only of the Born rule in pilot-wave theory but also of the nature of the wave function (or pilot wave).

As we have emphasised, in a theory of dynamics the initial conditions are contingent and only the laws of motion are law-like. And yet, in the typicality approach the initial configuration $q_{univ}(0)$ of the universe is restricted by the requirement that it be typical with respect to the measure $|\Psi_{univ}(q_{univ}, 0)|^2$. The latter measure is treated as if it had a law-like status (as explicitly claimed by Tumulka (2018)). In our view, in contrast, the initial probability distribution for the universe is a contingency which can be constrained only by cosmological observation.

This confusion between contingent and law-like entities has been taken to an extreme in claims made by the Bohmian mechanics school regarding the nature of the universal wave function Ψ_{univ}, specifically: (1) that Ψ_{univ} cannot be regarded as contingent, and (2) that Ψ_{univ} is not a physical object but a law-like entity ('nomological' rather than 'ontological').

The argument that Ψ_{univ} cannot be regarded as contingent is essentially this: the wave function for the whole universe

"is not controllable: it is what it is." (Goldstein, 2010)

There are several problems with this argument. Firstly, the same could be said of the universal configuration q_{univ}, resulting in the remarkable conclusion that no properties of our universe can be regarded as contingent (not even the position of the moon). Secondly, and in a similar vein, one could just as well say that our universe has only one spacetime geometry, and indeed only one intergalactic magnetic field. Each of these objects 'is not controllable' and 'is what it is'. And yet, according to standard thinking in physics and cosmology, the detailed form of either object is not completely determined by physical laws: each has a strong element of contingency. Thirdly and finally, as noted in Section 12.5.2, in this context we ought to beware of statements that there is 'only one universe': in principle such statements are trivially and tautologically true, but in practice the universe studied by cosmologists may well be one element of a large (and possibly infinite) ensemble, where the object which we call Ψ_{univ} can vary contingently across the ensemble.

The argument that Ψ_{univ} is not a physical object but a law-like entity is based on three assertions: (a) that Ψ_{univ} cannot be regarded as contingent (claim (1) above, which we have argued to be unfounded), (b) that Ψ_{univ} is static, and (c) that Ψ_{univ} is uniquely determined by the laws of quantum gravity.[19] On these grounds it has been claimed that Ψ_{univ} is a law-like entity roughly analogous to a classical Hamiltonian (Dürr, Goldstein and Zanghì, 1997; Goldstein and Zanghì, 2013). Thus:

> "...the wave function is a component of physical law rather than of the reality described by the law." (Dürr, Goldstein and Zanghì, 1997, p. 33)

But the arguments (a)–(c) do not bear scrutiny. We have already seen that argument (a) is spurious. Argument (b) is also questionable, based as it is on the time-independence of the Wheeler-DeWitt equation (the analogue of the Schrödinger equation) in canonical quantum gravity.[20] However, the physical meaning and consistency

---

[19]Other authors express concern about regarding a field on configuration space as a physical object. In our view it is not unreasonable for configuration space to be the fundamental arena of a realistic physics, with physical objects propagating on it (cf. footnote 23).

[20]The Wheeler-DeWitt equation takes the schematic atemporal form $\hat{\mathcal{H}}\Psi = 0$, where $\hat{\mathcal{H}}$ is an appropriate operator for the Hamiltonian density (Rovelli, 2004).

of the quantum-gravitational formalism remains in doubt, in particular because of the notorious 'problem of time' (the problem of explaining the emergence of apparent temporal evolution in some appropriate limit). Many workers have suggested that a physical time parameter is in effect hidden within the formalism and that, when correctly written as a function of physical degrees of freedom, the wave function is in fact time dependent.[21] As for argument (c), in canonical quantum gravity the solutions for Ψ_{univ} (satisfying the Wheeler-DeWitt equation as well as the other required constraints) are in fact highly *non*-unique (Rovelli 2004). In quantum cosmological models, for example, the solutions for Ψ_{univ} have the same kind of contingency that we are used to for quantum states in other areas of physics (Bojowald 2015).

It is worth emphasising that even if a consistent theory of quantum gravity did require Ψ_{univ} to be static, this still would not by any means establish that Ψ_{univ} is law-like. The key aspect of Ψ_{univ} that makes it count as a physical object is its contingency, in other words its under-determination by known physical laws. This implies that Ψ_{univ} contains a lot of independent and contingent structure — just like the electromagnetic field or the universal spacetime geometry — and so should be regarded as part of the physical state of the world (Valentini, 1992, p. 17; Brown and Wallace, 2005, p. 532; Valentini, 2010b).

12.7. Further criticisms of 'Bohmian mechanics'

Pilot-wave theory is, in general, a nonequilibrium physics that violates the statistical predictions of quantum theory (Valentini, 1991a,b, 1992). It can only be properly understood from this general perspective. The Bohmian mechanics school has instead promoted the belief that pilot-wave theory is instrinsically a theory of equilibrium. We now consider the principal physical misunderstandings that have arisen from this mistaken belief.

---

[21]See, for example, Roser and Valentini (2014) and the exhaustive review of the problem of time by Anderson (2017).

12.7.1.   *'Absolute uncertainty'*

The Bohmian mechanics school has asserted that typicality with respect to the Born-rule measure is 'the origin of absolute uncertainty'. On this view the uncertainty principle is an 'absolute' and 'irreducible' limitation on our knowledge:

> "In a universe governed by Bohmian mechanics there are sharp, precise, and irreducible limitations on the possibility of obtaining knowledge... absolute uncertainty arises as a necessity, emerging as a remarkably clean and simple consequence of the existence of trajectories." (Dürr, Goldstein and Zanghì, 1992)
>
> "...in a Bohmian universe we have an *absolute uncertainty*...the [Born rule] is a sharp expression of the inaccessibility in a Bohmian universe of micro-reality, of the unattainability of knowledge of the configuration of a system that transcends the limits set by its wave function ψ." (Goldstein, 2010)
>
> "In Bohmian mechanics... there are sharp limitations to knowledge and control: inhabitants of a Bohmian universe cannot know the position of a particle more precisely than allowed by the $|\psi|^2$ distribution .... Furthermore, they cannot measure the position at time t without disturbing the particle..." (Tumulka, 2018)

But as we pointed out in Section 12.5.1, the typicality argument in effect inserts the Born rule by hand at the initial time. Furthermore, as we noted in the Introduction, the uncertainty principle is not absolute or irreducible but merely a peculiarity of the state of quantum equilibrium. In general, the uncertainty principle would be violated if we had access to quantum nonequilibrium systems (Valentini, 1991b, 2002a).

12.7.2.   *The status of quantum measurement theory*

As also briefly noted in the Introduction, in the presence of quantum nonequilibrium key quantum constraints are violated: these include statistical locality, expectation additivity for quantum observables, and the indistinguishability of non-orthogonal quantum states (Valentini, 1991a,b, 1992, 2002a, 2004; Pearle and Valentini, 2006). The physics of nonequilibrium is radically different from the physics of equilibrium, the latter being merely a highly restricted special case of the former. It should then come as no surprise that the nonequilibrium theory of measurement differs radically from its equilibrium counterpart. As philosophers of physics are well aware, measurement is 'theory laden': we need some body of theory in order to know how to perform measurements correctly. As Einstein put it, in an often-quoted conversation with Heisenberg:

> "It is the theory which decides what we can observe."
> (Heisenberg, 1971, p. 63)

In the presence of quantum nonequilibrium systems, pilot-wave theory itself tells us how to perform correct measurements. It is found, for example, that if we possessed an ensemble of 'apparatus pointers' with an arbitrarily narrow nonequilibrium distribution (much narrower than the standard quantum width as defined by the initial wave function of the pointer), then it would be possible to use the apparatus to perform 'subquantum measurements': in particular, we would be able to measure the position and trajectory of a particle without disturbing its wave function (to arbitrary accuracy) (Valentini, 2002a; Pearle and Valentini, 2006).

From this perspective, the physics of quantum equilibrium is highly misleading — and so is the associated equilibrium theory of measurement (also known as 'quantum measurement theory'). In fact, the detailed dynamics of pilot-wave theory shows that the procedures known as 'quantum measurements' are generally not correct measurements. Instead, those procedures are merely special kinds of experiments which have been designed to respect a formal analogy with classical measurements (where the analogy is implemented by a mathematical correspondence between classical and quantum Hamiltonians) (Valentini, 1992, 1996, 2010b).

To put the so-called quantum theory of 'measurement' in a proper perspective, we must consider the more general physics of quantum nonequilibrium and its associated theory of subquantum measurement. But because the Bohmian mechanics school believes that the theory is fundamentally grounded in equilibrium, they are led to believe that the equilibrium theory *is* the theory — and that the associated quantum theory of measurement has a fundamental status. Thus Dürr, Goldstein and Zanghì (1996, 2004) argue that the quantum theory of measurement arises as an account of what they call 'reproducible experiments' and reproducible 'measurement-like' experiments. Measurements that lie outside of the domain of the quantum formalism are not considered. Thus both quantum equilibrium and its associated theory of measurement are in effect regarded as fundamental features of pilot-wave theory. In our view this is deeply mistaken. The physics of equilibrium is a special case of a much wider physics in which new kinds of measurements are possible. If instead we artificially restrict ourselves to the equilibrium domain, the result is a distorted understanding of measurement and an overstatement of the significance of the conventional quantum formalism.

12.7.3. *The misleading kinematics of quantum equilibrium*

It is worth noting how the artificial restriction to quantum equilibrium makes the idea of fundamental Lorentz invariance (at the level of the underlying equations of motion) seem much more plausible than it really is. As we have remarked, in general nonequilibrium gives rise to instantaneous signaling between remote entangled systems (Valentini, 1991b).[22] The reality in principle of superluminal communication between widely-separated experimenters strongly suggests the existence of an absolute simultaneity associated with a preferred slicing of spacetime (Valentini, 2008b). And indeed most versions of pilot-wave dynamics (and in particular of quantum field theory) are defined with respect to a preferred frame with a preferred time parameter t — where effective Lorentz invariance emerges only at the statistical level of quantum equilibrium (Bohm, Hiley and Kaloyerou, 1987; Valentini, 1992; Bohm and Hiley, 1993; Holland,

---

[22]Similar conclusions hold in all nonlocal and deterministic hidden-variables theories (Valentini, 2002b).

1993). If instead the theory is always and everywhere artificially restricted to equilibrium, locality will always hold at the statistical level and practical nonlocal signalling will be impossible. It may then seem plausible to search for a version of pilot-wave theory in which the dynamics is fundamentally Lorentz invariant, since one will never be faced directly with the awkward question of what happens when practical superluminal signals are viewed from a Lorentz-boosted frame and appear to travel backwards in time (potentially generating causal paradoxes). Even so, despite several attempts, a fundamentally Lorentz-invariant pilot-wave theory remains elusive and problematic (Dürr *et al.*, 1999; Tumulka, 2007; Dürr *et al.*, 2014).

The attachment to fundamental Lorentz invariance has in turn encouraged a misunderstanding of the role of Galilean invariance, which the Bohmian mechanics school mistakenly regards as a fundamental symmetry of the low-energy theory (Dürr, Goldstein and Zanghì, 1992; Allori *et al.*, 2008; Dürr and Teufel, 2009; Goldstein, 2017; Tumulka, 2018). Pilot-wave theory is a first-order or 'Aristotelian' dynamics with a law of motion (12.1) for velocity (as first envisaged by de Broglie in 1923), in contrast with Newtonian theory which is a second-order dynamics with a law of motion for acceleration. Because of this fundamental difference, the natural kinematics of pilot-wave theory is also Aristotelian with a preferred state of rest (Valentini, 1997). Galilean invariance may be shown to be a fictitious symmetry of the low-energy pilot-wave theory of particles — just as invariance under uniform acceleration is well known to be a fictitious symmetry of Newtonian mechanics. If instead one tries to insist on Galilean invariance being a physical symmetry of the low-energy theory, the result is a conceptually incoherent combination of an Aristotelian dynamics with a Galilean kinematics.

In response it might be claimed that Galilean invariance plays an important role in selecting the form of the low-energy guidance equation (Dürr, Goldstein and Zanghì, 1992; Dürr and Teufel, 2009; Goldstein, 2017). But in fact the de Broglie velocity $v = j/|\psi|^2$ is generally determined by the quantum current j, which may be derived as a Noether current associated with a global phase symmetry $\psi \to \psi e^{i\theta}$ on configuration space (Struyve and Valentini, 2009).

The derivation takes place in one frame of reference, with no need to consider boosts. The relevant symmetry is in configuration space, not in space or spacetime.[23]

12.7.4. *Particle creation and indeterminism*

It is also worth noting how the artificial restriction to quantum equilibrium makes a fundamentally stochastic model of particle creation — developed by the Bohmian mechanics school — appear more plausible than it really is. For if one denies the general contingency of the Born rule for initial conditions, it may seem no great loss to introduce a fixed and non-contingent probability into the dynamics as well.

The stochastic model promoted by the Bohmian mechanics school was constructed as follows. Bell (1986; 1987, chapter 19) had already proposed a discrete model of fermion numbers evolving stochastically on a lattice and had suggested that taking the continuum limit might yield a deterministic theory. The Bohmian mechanics school studied the continuum limit of Bell's model and arrived at a theory of particle trajectories with stochastic jumps at events where the particle numbers change (Dürr *et al.*, 2004, 2005). They named their approach 'Bell-type quantum field theory', and have attempted to apply it to bosons as well as to fermions. The fundamental probability rule for the jumps is chosen so as to preserve the Born rule.

Any interacting quantum field theory will contain a plethora of events where the particle numbers change (photon emission, electron-positron pair creation, and so on), and the Bohmian mechanics school has suggested that determinism must be abandoned to describe them:

> "In Bell-type [quantum field theories], God does play dice. There are no hidden variables which would fully predetermine the time and destination of a jump." (Dürr *et al.*, 2004, p. 3)
>
> "The quantum equilibrium distribution, playing a central role in Bohmian mechanics, then more or less dictates

---

[23]This reinforces our view that configuration space is the fundamental physical arena of pilot-wave dynamics (cf. footnote 19).

that creation of a particle occurs in a stochastic manner..." (Dürr *et al.*, 2005, p. 2)

It would, however, be remarkable indeed if indeterminism were
required to describe particle creation — when determinism suffices to
describe all other quantum-mechanical processes. But in fact, indeterminism is not required. The Bohmian mechanics school obtained
a stochastic continuum limit of Bell's model because they adopted an
erroneous definition of fermion number F. In quantum field theory, F
is conventionally defined as the number of particles minus the number of anti-particles.[24] As a particle physicist, this is what Bell would
have meant by fermion number. Unfortunately, Dürr *et al.* mistakenly took Bell's 'fermion number' to mean the number of particles
plus the number of anti-particles.

The correct continuum limit of Bell's model was taken by Colin
(2003), who employed the standard definition of F. As a result Colin
obtained a 'Dirac sea' theory of fermions — anticipated by Bohm and
Hiley (1993, p. 276) — in which particle trajectories are determined
by a pilot wave that obeys the many-body Dirac equation.[25] The
resulting model is fully deterministic, as Bell suggested it would be.
There are no fixed or fundamental stochastic elements, and the usual
contingency of probabilities applies to all processes.

For completeness we note that, for bosons, it is straightforward
to develop a deterministic pilot-wave field theory, in which the time
evolution of a (for example scalar) field ϕ is determined by the
Schrödinger wave functional $\Psi[\phi, t]$ (Holland, 1993). In such a the-
ory, again, the Born rule $P = |\Psi|^2$ is contingent and may be under-
stood as arising from a process of dynamical relaxation (Valentini,
2007). In contrast, the Bohmian mechanics school encounters difficulties defining particle trajectories and a Born-rule position-space
density for single bosons, as briefly noted by Dürr *et al.* (2005, p. 13).

---

[24]In particle physics, for historical reasons F is defined as the sum $F = L + B$
of lepton and baryon numbers — where L is the number of leptons minus the
number of antileptons and similarly for B.

[25]The Dirac-sea model requires regularisation (for example a cutoff) (Colin, 2003;
Colin and Struyve, 2007). The same is true of the models developed by Dürr *et al.*
in the presence of interactions.

Such problems recall the long history (in standard quantum theory) of controversial attempts to define a position-space 'wave function' for single photons and other bosons, attempts which invariably lead to negative probabilities and superluminal wave packet propagation. Without a solution to this — probably insoluble — problem, so-called 'Bell-type quantum field theory' remains undefined for bosons.[26]

12.7.5. *The problem of falsifiability*

Finally, the artificial restriction to quantum equilibrium has compromised the status of pilot-wave theory as a falsifiable scientific theory. For it is then impossible to measure the trajectory of a system without disturbing its wave function; hence it is impossible to test the de Broglie equation of motion, which associates a specific set of trajectories with each given wave function. There are alternative pilot-wave theories, with alternative velocity fields, which nevertheless preserve the Born distribution and which therefore imply the same empirical predictions for the equilibrium state (Deotto and Ghirardi, 1998). The physics of equilibrium is insensitive to the details of the trajectories. Thus, if we have access to equilibrium only, the alternative theories can never be tested against de Broglie's original theory. Indeed, in equilibrium, pilot-wave theories are forever experimentally indistinguishable from conventional quantum theory. As Dürr *et al.* put it:

> "For every conceivable experiment, whenever quantum mechanics makes an unambiguous prediction, Bohmian mechanics makes exactly the same prediction. Thus, the two cannot be tested against each other." (Dürr, Goldstein, Tumulka and Zanghì, 2009)

It is of course logically and mathematically possible for the world to be governed by pilot-wave theory and to be always and everywhere in quantum equilibrium. But from a scientific point of view such a

---

[26]Dürr *et al.* (2005, p. 13) cite two papers 'in preparation' (their refs. [18] and [28]) purporting to address this problem. To the author's knowledge, and unsurprisingly, neither paper was completed.

theory is unfalsifiable and therefore unacceptable. This shortcoming
is, however, not a feature of pilot-wave theory itself — which abounds
in new and potentially-observable physics — but rather stems from
a misunderstanding of the status of the Born rule in this theory.

12.8. Conclusion

The foundations of statistical mechanics are notoriously controversial, and overlap with difficult questions concerning the nature of
probability, the justification for standard methods of statistical inference, and even with philosophical questions concerning the foundations of the scientific method. However, important as these questions
are, in our view they are not especially relevant to either pilot-wave
theory or cosmology but instead arise generally across the sciences.
We claim that attempts to forge a special link between such questions and pilot-wave theory are at best a distraction and at worst
deeply misleading.

Something comparable took place during the early development of
atomic theory in the late nineteenth century. At that time theoretical
physics was divided between what we might now call 'operationalists' (who saw the macroscopic laws of thermodynamics as paradigmatic for physics generally) and 'realists' (who thought those laws
required a deeper explanation in terms of atoms and kinetic theory).
Boltzmann in particular was especially passionate about the philosophical importance of atomism as a basis for explanation in physics,
vis à vis the competing operationalist views of Mach and Ostwald.[27]
In retrospect it seems unfortunate that Boltzmann became embroiled
in foundational controversies concerning probability, time reversal,
and so on — important and interesting questions which, in hindsight,
proved to be a distraction from the main goal of demonstrating the
reality of atoms. The eventual atomistic explanation for Brownian
motion by Einstein in 1905 owed little to such foundational debates
and more to technical developments in kinetic theory — and the

---

[27]See, for example, Boltzmann's selected writings in the collection *Theoretical
Physics and Philosophical Problems* (Boltzmann, 1974).

foundational debates persist to this day, more than a century after the existence of atoms was firmly established.

Similarly, theoretical physics today is again divided between operationalists (who see quantum mechanics as an operational theory of macroscopic observations) and realists (who think there must be a reality behind the formalism). Among the various realist approaches, pilot-wave theory is the most closely analogous to kinetic theory. Once again, it seems unfortunate that the subject has become embroiled in foundational controversies in statistical mechanics and probability theory, when surely the main goal is to find out whether the trajectories posited by pilot-wave theory really exist or not. In our view, while those foundational controversies are important and interesting, in the context of pilot-wave theory they have proved to be a distraction from the even more important question of whether pilot-wave theory itself is true or not. Furthermore, the viewpoint championed by the Bohmian mechanics school (and widely followed by philosophers of physics) has played a major role in obscuring the physics of the theory — which is fundamentally a nonequilibrium physics that violates quantum mechanics. We emphasise, once again, that the existence or non-existence of quantum nonequilibrium in our universe (past and present) is an empirical question that will be settled only by detailed theoretical and observational work.

Acknowledgement

I am grateful to Valia Allori for the invitation to contribute to this volume.

References

Abraham, E., Colin, S., Valentini, A. (2014) Long-time relaxation in pilot-wave theory, *Journal of Physics A*, **47**, 395306. [arXiv:1310.1899]

Aghanim, N. *et al.* (Planck Collaboration) (2016) *Planck* 2015 results. XI. CMB power spectra, likelihoods, and robustness of parameters, *Astronomy and Astrophysics*, **594**, A11. [arXiv:1507.02704]

Allori, V., Goldstein, S., Tumulka, R., Zanghì, N. (2008) On the common structure of Bohmian mechanics and the Ghirardi–Rimini–Weber theory, *British Journal for the Philosophy of Science*, **59**, 353–389. [arXiv:quant-ph/0603027]

Anderson, E. (2017) *The Problem of Time*, Springer-Verlag.

Bacciagaluppi, G., Valentini, A. (2009) *Quantum Theory at the Crossroads: Reconsidering the 1927 Solvay Conference*, Cambridge University Press, Cambridge. [arXiv:quant-ph/0609184]

Bell, J. S. (1986) Quantum field theory without observers, *Physics Reports*, **137**, 49–54.

Bell, J. S. (1987) *Speakable and Unspeakable in Quantum Mechanics*, Cambridge University Press, Cambridge.

Bohm, D. (1952a) A suggested interpretation of the quantum theory in terms of 'hidden' variables. I, *Physical Review*, **85**, 166–179.

Bohm, D. (1952b) A suggested interpretation of the quantum theory in terms of 'hidden' variables. II, *Physical Review*, **85**, 180–193.

Bohm, D., Hiley, B. J. (1993) *The Undivided Universe: An Ontological Interpretation of Quantum Theory*, Routledge.

Bohm, D., Hiley, B. J., Kaloyerou, P. N. (1987). An ontological basis for the quantum theory, *Physics Reports*, **144**, 321–375.

Bojowald, M. (2015) Quantum cosmology: A review, *Reports on Progress in Physics*, **78**, 023901. [arXiv:1501.04899]

Boltzmann, L. (1974) *Theoretical Physics and Philosophical Problems*, ed. by B. McGuinness, Reidel, Dordrecht.

Colin, S. (2003) A deterministic Bell model, *Physics Letters A*, **317**, 349–358. [arXiv:quant-ph/0310055]

Colin, S. (2012) Relaxation to quantum equilibrium for Dirac fermions in the de Broglie-Bohm pilot-wave theory, *Proceedings of the Royal Society A*, **468**, 1116–1135. [arXiv:1108.5496]

Colin, S., Struyve, W. (2007) A Dirac sea pilot-wave model for quantum field theory, *Journal of Physics A*, **40**, 7309–7341. [arXiv:quant-ph/0701085]

Colin, S., Valentini, A. (2013) Mechanism for the suppression of quantum noise at large scales on expanding space, *Physical Review D*, **88**, 103515. [arXiv:1306.1579]

Colin, S., Valentini, A. (2014) Instability of quantum equilibrium in Bohm's dynamics, *Proceedings of the Royal Society A*, **470**, 20140288.

Colin, S., Valentini, A. (2015) Primordial quantum nonequilibrium and large-scale cosmic anomalies, *Physical Review D*, **92**, 043520. [arXiv:1407.8262]

Colin, S., Valentini, A. (2016) Robust predictions for the large-scale cosmological power deficit from primordial quantum nonequilibrium, *International Journal of Modern Physics D*, **25**, 1650068. [arXiv:1510.03508]

de Broglie, L. (1928) La nouvelle dynamique des quanta, In: *Électrons et Photons: Rapports et Discussions du Cinquième Conseil de Physique*. Paris: Gauthier-Villars, pp. 105–132. [English translation: Bacciagaluppi, G. and Valentini, A. (2009).]

Deotto, E., Ghirardi, G. C. (1998) Bohmian mechanics revisited, *Foundations of Physics*, **28**, 1–30. [arXiv:quant-ph/9704021]

Dürr, D., Goldstein, S., Zanghì, N. (1992) Quantum equilibrium and the origin of absolute uncertainty, *Journal of Statistical Physics*, **67**, 843–907. [arXiv:quant-ph/0308039]

Dürr, D., Goldstein, S., Zanghì, N. (1996) Bohmian mechanics as the foundation of quantum mechanics, In: eds. J. T. Cushing, A. Fine and S. Goldstein, *Bohmian Mechanics and Quantum Theory: An Appraisal*, Kluwer, Dordrecht, pp. 21–44. [arXiv:quant-ph/9511016]

Dürr, D., Goldstein, S., Zanghì, N. (1997) Bohmian mechanics and the meaning of the wave function, In: eds. R. S. Cohen, M. Horne and J. Stachel, *Experimental Metaphysics: Quantum Mechanical Studies for Abner Shimony*, Kluwer, Dordrecht, pp. 25–38. [arXiv:quant-ph/9512031]

Dürr, D., Goldstein, S., Münch-Berndl, K., Zanghi, N. (1999) Hypersurface Bohm-Dirac models, *Physical Review A*, **60**, 2729–2736.

Dürr, D., Goldstein, S., Zanghì, N. (2004) Quantum equilibrium and the role of operators as observables in quantum theory, *Journal of Statistical Physics*, **116**, 959–1055. [arXiv:quant-ph/0308038]

Dürr, D., Goldstein, S., Tumulka, R., Zanghì, N. (2004) Bohmian mechanics and quantum field theory, *Physical Review Letters*, **93**, 090402. [arXiv:quant-ph/0303156]

Dürr, D., Goldstein, S., Tumulka, R., Zanghì, N. (2005) Bell-type quantum field theories, *Journal of Physics A*, **38**, R1–R43. [arXiv:quant-ph/0407116]

Dürr, D., Goldstein, S., Tumulka, R., Zanghì, N. (2009) Bohmian mechanics, In: eds. D. Greenberger, K. Hentschel and F. Weinert, *Compendium of Quantum Physics: Concepts, Experiments, History and Philosophy*, Springer-Verlag, pp. 47–55. [arXiv:0903.2601]

Dürr, D., Goldstein, S., Norsen, T., Struyve, W., Zanghì, N. (2014) Can Bohmian mechanics be made relativistic? *Proceedings of the Royal Society A*, **470**, 20130699. [arXiv:1307.1714]

Dürr, D., Teufel, S. (2009) *Bohmian Mechanics: The Physics and Mathematics of Quantum Theory*, Springer-Verlag.

Gibbs, J. W. (1902) *Elementary Principles in Statistical Mechanics*, Charles Scribner's Sons, New York.

Goldstein, S. (2001) Boltzmann's approach to statistical mechanics, In: eds. J. Bricmont *et al.*, *Chance in Physics: Foundations and Perspectives*, Springer-Verlag. [arXiv:cond-mat/0105242]

Goldstein, S. (2010) Bohmian mechanics and quantum information, *Foundations of Physics*, **40**, 335–355. [arXiv:0907.2427]

Goldstein, S. (2017) Bohmian mechanics, In: ed. E. N. Zalta, *The Stanford Encyclopedia of Philosophy* (Summer 2017 Edition). [https://plato.stanford.edu/archives/sum2017/entries/qm-bohm]

Goldstein, S., Struyve, W., Tumulka, R. (2015) The Bohmian approach to the problems of cosmological quantum fluctuations. ArXiv:1508.01017.

Goldstein, S., Zanghì, N. (2013) Reality and the role of the wavefunction in quantum theory, In: eds. D. Albert and A. Ney, *The Wave Function: Essays in the Metaphysics of Quantum Mechanics*, Oxford University Press, Oxford, pp. 91–109. [arXiv:1101.4575]

Hajian, A., Souradeep, T. (2005) The cosmic microwave background bipolar power spectrum: Basic formalism and applications. ArXiv:astro-ph/0501001.

Heisenberg, W. (1971) *Physics and Beyond*, Harper & Row, New York.

Holland, P. R. (1993) *The Quantum Theory of Motion: An Account of the de Broglie-Bohm Causal Interpretation of Quantum Mechanics*, Cambridge University Press, Cambridge.

Kandhadai, A., Valentini, A. (2020) In preparation.

Liddle, A. R., Lyth, D. H. (2000) *Cosmological Inflation and Large-Scale Structure*, Cambridge University Press, Cambridge.

Lyth, D. H., Riotto, A. (1999) Particle physics models of inflation and the cosmological density perturbation, *Physics Reports*, **314**, 1–146.

Norsen, T. (2018) On the explanation of Born-rule statistics in the de Broglie-Bohm pilot-wave theory, *Entropy*, **20**, 422.

Pearle, P., Valentini, A. (2006) Quantum mechanics: Generalizations. In: eds. J.-P. Françoise *et al.*, *Encyclopaedia of Mathematical Physics*, Elsevier, North-Holland. [arXiv:quant-ph/0506115]

Roser, P., Valentini, A. (2014) Classical and quantum cosmology with York time, *Classical and Quantum Gravity*, **31**, 245001. [arXiv:1406.2036]

Rovelli, C. (2004) *Quantum Gravity*, Cambridge University Press, Cambridge.

Struyve, W., Valentini, A. (2009) De Broglie-Bohm guidance equations for arbitrary Hamiltonians, *Journal of Physics A*, **42**, 035301. [arXiv:0808.0290]

Towler, M. D., Russell, N. J., Valentini, A. (2012) Time scales for dynamical relaxation to the Born rule, *Proceedings of the Royal Society A*, **468**, 990–1013. [arXiv:1103.1589]

Tumulka, R. (2007) The 'unromantic pictures' of quantum theory, *Journal of Physics A*, **40**, 3245–3273. [arXiv:quant-ph/0607124]

Tumulka, R. (2018) Bohmian mechanics, In: eds. E. Knox and A. Wilson, *The Routledge Companion to the Philosophy of Physics*, Routledge, New York. [arXiv:1704.08017]

Uffink, J. (2007) Compendium of the foundations of classical statistical physics, In: eds. J. Butterfield and J. Earman, *Philosophy of Physics (Handbook of the Philosophy of Science)*, Elsevier, North-Holland, Amsterdam.

Underwood, N. G., Valentini, A. (2015) Quantum field theory of relic nonequilibrium systems, *Physical Review D*, **92**, 063531. [arXiv:1409.6817]

Underwood, N. G., Valentini, A. (2016) Anomalous spectral lines and relic quantum nonequilibrium. ArXiv:1609.04576 (*Physical Review D*, in press).

Valentini, A. (1991a) Signal-locality, uncertainty, and the subquantum H-theorem. I, *Physics Letters A*, **156**, 5–11.

Valentini, A. (1991b) Signal-locality, uncertainty, and the subquantum H-theorem, II, *Physics Letters A*, **158**, 1–8.

Valentini, A. (1992) On the pilot-wave theory of classical, quantum and subquantum physics. Ph.D. thesis, International School for Advanced Studies, Trieste, Italy. [http://hdl.handle.net/20.500.11767/4334]

Valentini, A. (1996) Pilot-wave theory of fields, gravitation and cosmology, In: eds. J. T. Cushing, A. Fine and S. Goldstein, *Bohmian Mechanics and Quantum Theory: An Appraisal*, Kluwer, Dordrecht.

Valentini, A. (1997) On Galilean and Lorentz invariance in pilot-wave dynamics, *Physics Letters A*, **228**, 215. [arXiv:0812.4941]

Valentini, A. (2001) Hidden variables, statistical mechanics and the early universe, In: eds. J. Bricmont *et al.*, *Chance in Physics: Foundations and Perspectives*, Springer-Verlag. [arXiv:quant-ph/0104067]

Valentini, A. (2002a) Subquantum information and computation, *Pramana— Journal of Physics*, **59**, 269–277. [arXiv:quant-ph/0203049].

Valentini, A. (2002b) Signal-locality in hidden-variables theories, *Physics Letters A*, **297**, 273–278. [arXiv:quant-ph/0106098]

Valentini, A. (2004) Universal signature of non-quantum systems, *Physics Letters A*, **332**, 187–193. [arXiv:quant-ph/0309107]

Valentini, A. (2007) Astrophysical and cosmological tests of quantum theory, *Journal of Physics A*, **40**, 3285–3303. [arXiv:hep-th/0610032]

Valentini, A. (2008a) De Broglie-Bohm prediction of quantum violations for cosmological super-Hubble modes. ArXiv:0804.4656.

Valentini, A. (2008b) Hidden variables and the large-scale structure of spacetime, In: eds. W. L. Craig and Q. Smith, *Einstein, Relativity and Absolute Simultaneity*, Routledge, London, pp. 125–155. [arXiv:quant-ph/0504011]

Valentini, A. (2009) Beyond the quantum, *Physics World*, **22N11**, 32–37. [arXiv:1001.2758]

Valentini, A. (2010a) Inflationary cosmology as a probe of primordial quantum mechanics, *Physical Review D*, **82**, 063513. [arXiv:0805.0163]

Valentini, A. (2010b) De Broglie-Bohm pilot-wave theory: Many-worlds in denial? In: eds. S. Saunders *et al.*, *Many Worlds? Everett, Quantum Theory, and Reality*, Oxford University Press, Oxford, pp. 476–509. [arXiv:0811.0810]

Valentini, A. (2015) Statistical anisotropy and cosmological quantum relaxation. ArXiv:1510.02523.

Valentini, A., Westman, H. (2005) Dynamical origin of quantum probabilities, *Proceedings of the Royal Society A*, **461**, 253. [arXiv:quant-ph/0403034]

Vitenti, S., Peter, P., Valentini, A. (2019) Modeling the large-scale power deficit with smooth and discontinuous primordial spectra, *Physical Review D*, **100**, 043506. [ArXiv:1901.08885]

Wang, I.-C., Ng, K.-W. (2008) Effects of a preinflation radiation-dominated epoch to CMB anisotropy, *Physical Review D*, **77**, 083501.

Chapter 13

Time's Arrow in a Quantum Universe: On the Status of Statistical Mechanical Probabilities

Eddy Keming Chen

Department of Philosophy, University of California, San Diego, 9500 Gilman Drive, La Jolla, CA 92093-0119, USA
eddykemingchen@ucsd.edu

In a quantum universe with a strong arrow of time, it is standard to postulate that the initial wave function started in a particular macrostate — the special low-entropy macrostate selected by the Past Hypothesis. Moreover, there is an additional postulate about statistical mechanical probabilities according to which the initial wave function is a "typical" choice in the macrostate (the Statistical Postulate). Together, they support a probabilistic version of the Second Law of Thermodynamics: typical initial wave functions will increase in entropy. Hence, there are two sources of randomness in such a universe: the quantum-mechanical probabilities of the Born rule and the statistical mechanical probabilities of the Statistical Postulate.

I propose a new way to understand time's arrow in a quantum universe. It is based on what I call the Thermodynamic Theories of Quantum Mechanics. According to this perspective, there is a *natural* choice for the initial quantum state of the universe, which is given not by a wave function but by a density matrix. Importantly, that density matrix plays two roles in my theories. First, it plays a microscopic role by appearing in the fundamental dynamical equations. Second, it plays a macroscopic/thermodynamical role by being exactly the normalized projection onto the Past Hypothesis subspace (inside the Hilbert space of the universe). This also entails that, given an initial subspace, we obtain a unique choice of the

479

initial density matrix. I call this property *the conditional unique-ness* of the initial quantum state. The conditional uniqueness provides a new and general strategy to eliminate statistical mechanical probabilities in the fundamental physical theories, by which we can reduce the two sources of randomness to only the quantum mechanical one. The dual role of the initial density matrix also enables us to explore the possibility of an *absolutely unique* initial quantum state, in a way that might realize Penrose's idea (1989) of a strongly deterministic universe.

13.1. Introduction

In a quantum universe with a strong arrow of time (large entropy gradient), it is standard to attribute the temporal asymmetry to a special boundary condition. This boundary condition is a macrostate of extremely low entropy. David Albert (2000) calls it the Past Hypothesis. The quantum version of the Past Hypothesis dictates that the initial wave function of the universe lies within the low-entropy macrostate. Mathematically, the macrostate is represented by a low-dimensional subspace in the total Hilbert space, which corresponds to low Boltzmann entropy.

The Past Hypothesis is accompanied by another postulate about statistical mechanical probabilities, which specifies a probability distribution over wave functions compatible with the low entropy macrostate (the Statistical Postulate). Together, the Past Hypothesis and the Statistical Postulate support a probabilistic version of the Second Law of Thermodynamics: typical initial wave functions will increase in entropy.

The standard theory, which has its origin in Boltzmann's statistical mechanics and has been substantially developed in the last century, is a simple and elegant way of understanding time's arrow in a quantum universe. It has two features:

(1) Although the theory restricts the choices of initial quantum states of the universe, it does not select a unique one.
(2) The theory postulates statistical mechanical probabilities (or a typicality measure) on the fundamental level. They are *prima*

facie on a par with and in addition to the quantum mechanical probabilities.

Hence, there are two sources of randomness in such a universe: the quantum-mechanical probabilities in the Born rule and the statistical mechanical probabilities in the Statistical Postulate. It is an interesting conceptual question how we should understand the two kinds of probabilities in the standard theory.

In this paper, I propose a new way to understand time's arrow in a quantum universe. It is based on what I call the Thermodynamic Theories of Quantum Mechanics (TQM). According to this perspective, there is a *natural* choice for the initial quantum state of the universe, which is given not by a wave function but by a density matrix. Moreover, the density matrix enters into the fundamental equations of quantum mechanics. Hence, it plays a microscopic role. Furthermore, the density matrix is exactly the (normalized) projection operator onto the Past Hypothesis macrostate, represented by a subspace in the Hilbert space of the universe.

In Chen (2018b), I introduced the Initial Projection Hypothesis (IPH) as a new postulate about a natural initial density matrix of the universe. In that paper, I focused on its relevance to the debate about the nature of the quantum state. In Chen (2019), I discussed its empirical equivalence with the Past Hypothesis. In this paper, I shall explore its relevance to the status of statistical mechanical probabilities. According to the Initial Projection Hypothesis, given a choice of an initial subspace, we have a unique choice of the initial density matrix. This property is called *conditional uniqueness*, which will be sufficient to eliminate statistical mechanical probabilities at the level of fundamental physics. Thus, in contrast to the standard theory, TQM has the following two features:

(1) The theory selects a unique initial quantum state of the universe, given a choice of the Past Hypothesis subspace. [Conditional Uniqueness]

(2) Because of conditional uniqueness, the theory does not postulate statistical mechanical probabilities (or a typicality measure) at the level of fundamental physics.

That is, TQM provides a new and general strategy for eliminating statistical mechanical probabilities at the fundamental level. To be sure, statistical mechanical probabilities can still emerge as useful tools of analysis. TQM satisfies conditional uniqueness, which is sufficient for the purpose at hand. But we might also wonder whether TQM may lead to absolute uniqueness, in the sense that there is a unique initial quantum state of the universe that does not depend on a prior selection of a unique initial subspace. Such absolute uniqueness will go some way towards realizing a strongly deterministic universe in the sense of Penrose (1989).

Here is the roadmap of the paper. In Section 13.2, I will review the standard (Boltzmannian) way of understanding temporal asymmetry in classical and quantum mechanics. In Section 13.3, I introduce the Thermodynamic Theories of Quantum Mechanics. First, I formulate the Initial Projection Hypothesis, which is the key component of TQM and an alternative to the Past Hypothesis. Next, I focus on the property of conditional uniqueness, the elimination of statistical mechanical probabilities at the fundamental level, the emergence of such probabilities as useful tools of analysis, and the possibility of absolute uniqueness and strong determinism. I also discuss the (lack of) analogues in classical statistical mechanics, implications for the status of the quantum state, Lorentz invariance, theoretical unity, generalizations to other cosmological initial conditions, and relation to other proposals in the literature.

13.2. Statistical mechanics and time's arrow

Statistical mechanics concerns macroscopic systems such as a gas in a box. It is an important subject for understanding the arrow of time, since thermodynamic systems are irreversible. A gas in a box can be described as a system of N particles, with $N > 10^{20}$. If the system is governed by classical mechanics, although it is difficult to solve the equations exactly, we can still use classical statistical mechanics (CSM) to describe its statistical behaviors, such as the approach to thermal equilibrium suggested by the Second Law of Thermodynamics. Similarly, if the system is governed by quantum mechanics, we can use quantum statistical mechanics (QSM) to describe its statistical behaviors. Generally speaking, there are two different views on CSM: the individualistic view and the ensemblist view. For concreteness and simplicity, we will adopt the individualistic view in this paper, and we will start with a review of CSM, which should be more familiar to researchers in the foundations of physics.

13.2.1. *Classical statistical mechanics*

Let us review the basic elements of CSM on the individualistic view.[1] For concreteness, let us consider a classical-mechanical system with N particles in a box $\Lambda = [0, L]^3 \subset \mathbb{R}^3$ and a Hamiltonian $H = H(X) = H(\boldsymbol{q}_1, \ldots, \boldsymbol{q}_N; \boldsymbol{p}_1, \ldots, \boldsymbol{p}_n)$.

(1) Microstate: at any time t, the microstate of the system is given by a point in a $6N$-dimensional phase space,

$$X = (\boldsymbol{q}_1, \ldots, \boldsymbol{q}_N; \boldsymbol{p}_1, \ldots, \boldsymbol{p}_n) \in \Gamma_{total} \subseteq \mathbb{R}^{6N}, \qquad (13.1)$$

where Γ_{total} is the total phase space of the system.
(2) Dynamics: the time dependence of $X_t = (\boldsymbol{q}_1(t), \ldots, \boldsymbol{q}_N(t); \boldsymbol{p}_1(t), \ldots, \boldsymbol{p}_n(t))$ is given by the Hamiltonian equations of motion:

$$\frac{d\boldsymbol{q}_i}{dt} = \frac{\partial H}{\partial \boldsymbol{p}_i}, \ \frac{d\boldsymbol{p}_i}{dt} = -\frac{\partial H}{\partial \boldsymbol{q}_i}. \qquad (13.2)$$

(3) Energy shell: the physically relevant part of the total phase space is the energy shell $\Gamma \subseteq \Gamma_{total}$ defined as:

$$\Gamma = \{X \in \Gamma_{total} : E \leq H(x) \leq E + \delta E\}. \qquad (13.3)$$

We only consider microstates in Γ.
(4) Measure: the measure μ_V is the standard Lebesgue measure on phase space, which is the volume measure on $\mathbb{R}^{6N}$.

---

[1] In this subsection and the next one, I follow the discussion in Goldstein and Tumulka (2011). Sections 13.2.1.1 and 13.2.2 are not intended to be rigorous axiomatizations of CSM and QSM. They are only summaries of the key concepts that are important for appreciating the main ideas in the later sections.

(5) Macrostate: with a choice of macro-variables, the energy shell Γ can be partitioned into macrostates Γ_ν:

$$\Gamma = \bigcup_\nu \Gamma_\nu. \tag{13.4}$$

A macrostate is composed of microstates that share similar macroscopic features (similar values of the macro-variables), such as volume, density, and pressure.

(6) Unique correspondence: every phase point X belongs to one and only one Γ_ν. (This point is implied by (5). But we make it explicit to better contrast it with the situation in QSM.)

(7) Thermal equilibrium: typically, there is a dominant macrostate Γ_{eq} that has almost the entire volume with respect to μ_V:

$$\frac{\mu_V(\Gamma_{eq})}{\mu_V(\Gamma)} \approx 1. \tag{13.5}$$

A system is in thermal equilibrium if its phase point $X \in \Gamma_{eq}$.

(8) Boltzmann Entropy: the Boltzmann entropy of a classical-mechanical system in microstate X is given by:

$$S_B(X) = k_B \log(\mu_V(\Gamma(X))), \tag{13.6}$$

where $\Gamma(X)$ denotes the macrostate containing X. The thermal equilibrium state thus has the maximum entropy.

(9) Low-Entropy Initial Condition: when we consider the universe as a classical-mechanical system, we postulate a special low-entropy boundary condition, which David Albert calls *the Past Hypothesis*:

$$X_{t_0} \in \Gamma_{PH} \, , \, \mu_V(\Gamma_{PH}) \ll \mu_V(\Gamma_{eq}) \approx \mu_V(\Gamma), \tag{13.7}$$

where Γ_{PH} is the Past Hypothesis macrostate with volume much smaller than that of the equilibrium macrostate. Hence, $S_B(X_{t_0})$, the Boltzmann entropy of the microstate at the boundary, is very small compared to that of thermal equilibrium.

(10) A central task of CSM is to establish mathematical results that demonstrate (or suggest) that μ_V — most microstates will approach thermal equilibrium.

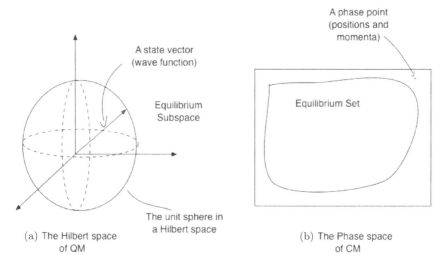

Fig. 13.1. QM and CM employ different state spaces. These pictures are not drawn to scale, and they are drawn with much fewer dimensions. In (a), we only draw a three-dimensional equilibrium subspace, which is embedded in a higher-dimensional energy shell. In (b), we only draw a two-dimensional energy surface, which includes a large equilibrium set.

13.2.2. *Quantum statistical mechanics*

Let us now turn to quantum statistical mechanics (QSM). For concreteness, let us consider a quantum-mechanical system with N fermions (with $N > 10^{20}$) in a box $\Lambda = [0, L]^3 \subset \mathbb{R}^3$ and a Hamiltonian $\hat{H}$. For concreteness, I will focus on the individualistic view of QSM. The main difference with CSM is that they employ different state spaces. The classical state space is the phase space while the quantum state space is the Hilbert space. However, CSM and QSM are conceptually similar, as we can see from below.

(1) Microstate: at any time t, the microstate of the system is given by a normalized (and anti-symmetrized) wave function:

$$\psi(\boldsymbol{q_1}, \ldots, \boldsymbol{q_N}) \in \mathscr{H}_{total} = L^2(\Lambda^N, \mathbb{C}^k), \quad \| \psi \|_{L^2} = 1, \quad (13.8)$$

where $\mathscr{H}_{total} = L^2(\Lambda^N, \mathbb{C}^k)$ is the total Hilbert space of the system.

(2) Dynamics: the time dependence of $\psi(\boldsymbol{q_1}, \ldots, \boldsymbol{q_N}; t)$ is given by the Schrödinger equation:

$$i\hbar \frac{\partial \psi}{\partial t} = \hat{H}\psi. \tag{13.9}$$

(3) Energy shell: the physically relevant part of the total Hilbert space is the subspace ("the energy shell"):

$$\mathscr{H} \subseteq \mathscr{H}_{total} \,, \ \mathscr{H} = \mathrm{span}\{\phi_\alpha : E_\alpha \in [E, E + \delta E]\}, \tag{13.10}$$

This is the subspace (of the total Hilbert space) spanned by energy eigenstates ϕ_α whose eigenvalues E_α belong to the $[E, E + \delta E]$ range. Let $D = \dim \mathscr{H}$, the number of energy levels between E and $E + \delta E$. We only consider wave functions ψ in $\mathscr{H}$.

(4) Measure: given a subspace $\mathscr{H}$, the measure μ_S is the surface area measure on the unit sphere in that subspace $\mathscr{S}(\mathscr{H})$.[2]

(5) Macrostate: with a choice of macro-variables,[3] the energy shell $\mathscr{H}$ can be orthogonally decomposed into macro-spaces:

$$\mathscr{H} = \oplus_\nu \mathscr{H}_\nu \,, \ \sum_\nu \dim \mathscr{H}_\nu = D \tag{13.11}$$

Each $\mathscr{H}_\nu$ corresponds more or less to small ranges of values of macro-variables that we have chosen in advance.

(6) Non-unique correspondence: typically, a wave function is in a superposition of macrostates and is not entirely in any one of the macrospaces. However, we can make sense of situations where

---

[2]For simplicity, here we assume that the subspaces we deal with are finite-dimensional. In cases where the Hilbert space is infinite-dimensional, it is an open and challenging technical question. For example, we could use Gaussian measures in infinite-dimensional spaces, but we no longer have uniform probability distributions.

[3]For technical reasons, Von Neumann (1955) suggests that we round up these macro-variables (represented by quantum observables) so as to make the observables commute. See Goldstein et al. (2010c) Section 2.2 for a discussion of von Neumann's ideas.

ψ is (in the Hilbert space norm) very close to a macrostate $\mathscr{H}_\nu$:

$$\langle\psi|\, P_\nu\, |\psi\rangle \approx 1, \tag{13.12}$$

where P_ν is the projection operator into $\mathscr{H}_\nu$. This means that $|\psi\rangle$ lies almost entirely in $\mathscr{H}_\nu$. (This is different from CSM, where every phase point lies entirely within some macrostate.)

(7) Thermal equilibrium: typically, there is a dominant macro-space $\mathscr{H}_{eq}$ that has a dimension that is almost equal to D:

$$\frac{\dim\mathscr{H}_{eq}}{\dim\mathscr{H}} \approx 1. \tag{13.13}$$

A system with wave function ψ is in equilibrium if the wave function ψ is very close to $\mathscr{H}_{eq}$ in the sense of (13.12): $\langle\psi|\, P_{eq}\, |\psi\rangle \approx 1$.

Simple Example. Consider a gas consisting of $n = 10^{23}$ atoms in a box $\Lambda \subseteq \mathbb{R}^3$. The system is governed by quantum mechanics. We orthogonally decompose the Hilbert space $\mathscr{H}$ into 51 macro-spaces: $\mathscr{H}_0 \oplus \mathscr{H}_2 \oplus \mathscr{H}_4 \oplus \ldots \oplus \mathscr{H}_{100}$, where $\mathscr{H}_\nu$ is the subspace corresponding to the macrostate such that the number of atoms in the left half of the box is between $(\nu - 1)\%$ and $(\nu + 1)\%$ of n, with the endpoints being the exceptions: $\mathscr{H}_0$ is the interval of $0\% - 1\%$ and $\mathscr{H}_{100}$ is the interval of $99\% - 100\%$. In this example, $\mathscr{H}_{50}$ has the overwhelming majority of dimensions and is thus the equilibrium macro-space. A system whose wave function is very close to $\mathscr{H}_{50}$ is in equilibrium (for this choice of macrostates).

(8) Boltzmann Entropy: the Boltzmann entropy of a quantum-mechanical system with wave function ψ that is very close to a macrostate ν is given by:

$$S_B(\psi) = k_B\log(\dim\mathscr{H}_\nu), \tag{13.14}$$

where $\mathscr{H}_\nu$ denotes the subspace containing almost all of ψ in the sense of (13.12). The thermal equilibrium state thus has the maximum entropy:

$$S_B(eq) = k_B\log(\dim\mathscr{H}_{eq}) \approx k_B\log(D), \tag{13.15}$$

where eq denotes the equilibrium macrostate.

(9) Low-Entropy Initial Condition: when we consider the universe as a quantum-mechanical system, we postulate a special low-entropy boundary condition on the universal wave function — the quantum-mechanical version of *the Past Hypothesis*:

$$\Psi(t_0) \in \mathcal{H}_{PH} \ , \ \dim \mathcal{H}_{PH} \ll \dim \mathcal{H}_{eq} \approx \dim \mathcal{H} \qquad (13.16)$$

where $\mathcal{H}_{PH}$ is the Past Hypothesis macro-space with dimension much smaller than that of the equilibrium macro-space.[4] Hence, the initial state has very low entropy in the sense of (13.23).

(10) A central task of QSM is to establish mathematical results that demonstrate (or suggest) that most (maybe even all) μ_S-wave functions of small subsystems, such as gas in a box, will approach thermal equilibrium.

If the microstate ψ of a system is close to some macro-space $\mathcal{H}_\nu$ in the sense of (13.12), we can say that the macrostate of the system is $\mathcal{H}_\nu$. The macrostate $\mathcal{H}_\nu$ is naturally associated with a density matrix:

$$\hat{W}_\nu = \frac{I_\nu}{\dim \mathcal{H}_\nu}, \qquad (13.17)$$

where I_ν is the projection operator onto $\mathcal{H}_\nu$. The density matrix also obeys a dynamical equation; it evolves according to the von Neumann equation:

$$i\hbar \frac{d\hat{W}(t)}{dt} = [\hat{H}, \hat{W}]. \qquad (13.18)$$

Given the canonical correspondence between a subspace and its normalized projection operator, $\hat{W}_\nu$ is also a representation of the subspace $\mathcal{H}_\nu$. It can be decomposed into wave functions, but the decomposition is not unique. Different measures can give rise to the same density matrix. One such choice is $\mu_S(d\psi)$, the uniform

---

[4] Again, we assume that $\mathcal{H}_{PH}$ is finite-dimensional, in which case we can use the surface area measure on the unit sphere as the typicality measure for # 10. It remains an open question in QSM about how to formulate the low-entropy initial condition when the initial macro-space is infinite-dimensional.

distribution over wave functions:

$$\hat{W}_\nu = \int_{\mathscr{S}(\mathscr{H}_\nu)} \mu_S(d\psi)\,|\psi\rangle\,\langle\psi|\,. \tag{13.19}$$

In (13.19), $\hat{W}_\nu$ is defined with a choice of measure on wave functions in $\mathscr{H}_\nu$. However, we should not be misled into thinking that the density matrix is derivative of wave functions. What is intrinsic to a density matrix is its geometrical meaning in the Hilbert space. In the case of $\hat{W}_\nu$, as shown in the canonical description (13.17), it is just a normalized projection operator.

13.2.3. Ψ_{PH}-*Quantum theories*

If we treat the universe as a quantum system, then we can distill a picture of the fundamental physical reality from the standard Boltzmannian QSM (individualistic perspective). The universe is described by a quantum state represented by a universal wave function. It starts in a Past Hypothesis macrostate (Section 13.2.2, number 9; also see Fig. 1 below) and is selected randomly according to the probability specified by the Statistical Postulate (Section 13.2.2, number 4). It evolves by the quantum dynamics — the Schrödinger equation

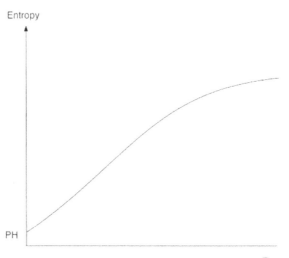

(Section 13.2.2, number 2). That is, it has three fundamental postulates (fundamental lawlike statements[5]):

(1) The Schrödinger equation.
(2) The Past Hypothesis.
(3) The Statistical Postulate.

However, this theory by itself faces the measurement problem. To solve the measurement problem, we can combine it with three well-known strategies:

• Bohmian mechanics (BM): the fundamental ontology in addition to the quantum state also includes point particles with precise locations, and the fundamental dynamics also includes a guidance equation that relates the wave function to the velocity of the particles.
• Everettian mechanics (S0): the fundamental ontology consists in just the quantum state evolving unitarily according to the Schrödinger equation.
• GRW spontaneous collapse theory (GRW0): the fundamental ontology consists in just the quantum state evolving by the Schrödinger equation, but the unitary evolution is interrupted by a spontaneous collapse mechanism.

The universal wave function Ψ is central to standard formulations of the above theories. The Past Hypothesis and the Statistical Postulate need to be added to them to account for the arrow of time. Let

---

[5] Barry Loewer calls the joint system — the package of laws that includes PH and SP in addition to the dynamical laws of physics — the Mentaculus Vision. For developments and defenses of the nomological account of the Past Hypothesis and the Statistical Postulate, see Albert (2000); Loewer (2007); Wallace (2011, 2012) and Loewer (2016). Albert and Loewer are writing mainly in the context of CSM. The Mentaculus Vision is supposed to provide a "probability map of the world." As such, it requires one to take the probability distribution very seriously. On our approach, which we introduce below as the thermodynamic theories of QM, there is no fundamental postulate about a probability distribution over initial quantum states. However, we can still extract objective probabilities from the initial density matrix. In our approach, we may call it the Wentaculus, where W stands for the density matrix, the probability of the Mentaculus is objectively grounded in the microdynamics. See section 13.3.2.3 for more details.

us label them Ψ_{PH}-BM, Ψ_{PH}-S0, and Ψ_{PH}-GRW0. They are all instances of what I call Ψ_{PH}-quantum theories.

There are also Everettian and GRW theories with additional fundamental ontologies in physical space-time, such as Everettian theory with a mass-density ontology (S0), GRW theory with a mass-density ontology (GRWm), and GRW theory with a flash ontology (GRWf). See Allori *et al.* (2008, 2010) for discussions. Thus, we also have Ψ_{PH}-Sm, Ψ_{PH}-GRWm, and Ψ_{PH}-GRWf.

13.3. The thermodynamic theories of quantum mechanics

The Ψ_{PH}-quantum theories attempt to solve the measurement problem and account for the arrow of time. However, each Ψ_{PH}-quantum theory admits many choices for the initial quantum state. This is because the Past Hypothesis subspace $\mathscr{H}_{PH}$, although being small compared to the full energy shell, is still compatible with many wave functions. One might wonder whether there is some hypothesis that could determine a unique initial quantum state of the universe. There is an interesting consequence if that can be done: it may bring us closer to what Penrose (1989) calls *strong determinism*, the idea that the entire history of the universe is fixed by the theory, with laws that specify both a deterministic dynamics and a (nomologically) unique initial state of the universe (see Section 13.3.5). As we shall see, there are other benefits of having a unique initial quantum state of the universe, such as the elimination of statistical mechanical probabilities (Section 13.3.2.1).

Is it possible to narrow down the initial choices to exactly one? I believe that we can, but we need to make some conceptual changes to the standard theory. Our proposal consists in three steps:

(1) Allow the universe to be in a fundamental mixed state represented by a density matrix.
(2) Let the density matrix enter into the fundamental dynamical equations.
(3) Choose a natural density matrix associated with the Past Hypothesis subspace $\mathscr{H}_{PH}$.

The natural density matrix together with its fundamental dynamical equations define a new class of quantum theories, which we will call the *Thermodynamic Theories of Quantum Mechanics*. As we explain below, all of them will have a conditionally unique initial quantum state.

13.3.1. *The initial projection hypothesis*

The standard approach to the foundations of quantum mechanics assumes that the universe is described by a *pure* quantum state $\Psi(t)$, which is represented as a vector in the (energy shell of the) Hilbert space of the system. However, it is also possible to describe the universe by a *mixed* quantum state $W(t)$, which has its own geometric meaning in the Hilbert space.

On the latter perspective, the quantum state of the universe is in a mixed state W, which becomes the central object of quantum theory. W obeys its own unitary dynamics — the von Neumann equation (13.18), which is a generalization of the Schrödinger equation (13.9). Moreover, we can write down density-matrix versions of Bohmian mechanics, Everettian mechanics, and GRW spontaneous collapse theory. (See Appendix for the mathematical details.) Given these equations, the density matrix of the universe plays a microscopic role: it enters into the fundamental dynamical equations of those theories. As a result, it guides Bohmian particles, gives rise to GRW collapses, and realizes the emergent multiverse of Everett. I call this view *Density Matrix Realism*, in contrast with Wave Function Realism, the view that the central object of quantum theory is the universal wave function.[6]

---

[6] Density Matrix Realism may be unfamiliar to some people, as we are used to take the mixed states to represent our *epistemic uncertainties of the actual pure state* (a wave function). The proposal here is that the density matrix directly represents the actual quantum state of the universe; there is no further fact about which is the actual wave function. In this sense, the density matrix is "fundamental." In fact, this idea has come up in the foundations of physics. See, for example, Dürr *et al.* (2005); Maroney (2005), Wallace (2011, 2012), and Wallace (2016). I discuss this idea in more detail in Chen (2018b) Section 13.3.

According Density Matrix Realism, the initial quantum state is also represented by a density matrix. Now, we know that to account for the arrow of time we need to postulate a low-entropy initial condition — the Past Hypothesis. Moreover, under Wave Function Realism, it is also necessary to postulate a Statistical Postulate.

Now, we make a crucial observation. If the initial quantum state is a density matrix, then there is a natural choice for it given the Past Hypothesis subspace. The natural choice is the normalized projection operator onto $\mathscr{H}_{PH}$:

$$\hat{W}_{IPH}(t_0) = \frac{I_{PH}}{dim\mathscr{H}_{PH}}, \tag{13.20}$$

where t_0 represents a temporal boundary of the universe, I_{PH} is the projection operator onto $\mathscr{H}_{PH}$, dim counts the dimension of the Hilbert space, and $dim\mathscr{H}_{PH} \ll dim\mathscr{H}_{eq}$. Since the quantum state at t_0 has the lowest entropy, we call t_0 the initial time. In Chen (2018b), I call (13.20) the *Initial Projection Hypothesis* (IPH). In words: the initial density matrix of the universe is the normalized projection onto the PH-subspace.

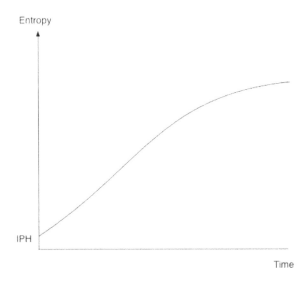

The projection operator onto $\mathscr{H}_{PH}$ contains no more and no less information than what is contained in the subspace itself. There is a

natural correspondence between a subspace and its projection operator. If we specify the subspace, we know what its projection operator
is, and vice versa. Since the projection operator onto a subspace carries no more information than that subspace itself, the projection
operator is no more complex than $\mathscr{H}_{PH}$. This is different from Ψ_{PH},
which is confined by PH to be a vector inside $\mathscr{H}_{PH}$. A vector carries more information than the subspace it belongs to, as specifying
a subspace is not sufficient to determine a vector. (For example, to
determine a vector in an 18-dimensional subspace of a 36-dimensional
vector space, we need 18 coordinates in addition to specifying the
subspace. The higher the dimension of the subspace, the more information is needed to specify the vector.) If PH had fixed Ψ_{PH} (the
QSM microstate), it would have required much more information and
become a much more complex posit. PH as it is determines Ψ_{PH} only
up to an equivalence class (the QSM macrostate).

I propose, then, that we replace the Past Hypothesis with the
Initial Projection Hypothesis, the Schrödinger equation with the von
Neumann equation, and the universal wave function with a universal
density matrix. In contrast to Ψ_{PH} theories, there are only two corresponding fundamental postulates (fundamental laws of physics) in
the new theory:

(1) The von Neumann equation.
(2) The Initial Projection Hypothesis.

Notice that IPH defines a unique initial quantum state, given a
choice of the initial PH subspace. There is only one state that is
possible, not a collection of states, in the PH subspace. We call
this *conditional uniqueness*. The quantum state $\hat{W}_{IPH}(t_0)$ is infor-
mationally equivalent to the constraint that PH imposes on the initial microstates. Given the PH subspace, $\hat{W}_{IPH}(t_0)$ is singled out by
the data in PH (we will come back to this point in Section 13.3.2).
Consequently, on the universal scale, we do not need to impose an
additional probability or typicality measure on the Hilbert space.
$\hat{W}_{IPH}(t_0)$ is mathematically equivalent to an integral over projection
onto each normalized state vectors (wave functions) compatible with
PH *with respect to* $\mu_S(d\psi)$. Of course, we are not defining $\hat{W}_{IPH}(t_0)$

in terms of state vectors. Rather, we are thinking of $\hat{W}_{IPH}(t_0)$ as a geometric object in the Hilbert space: the (normalized) projection operator onto $\mathscr{H}_{PH}$. That is the *intrinsic* understanding of the density matrix. (In Section 13.3.2 and Section 13.3.3, we will discuss different versions of PH and their relevance to conditional uniqueness. In Section 13.3.5, we will discuss why this strategy does not work so well in CSM.)

As before, the above theory by itself faces the measurement problem. To solve the measurement problem, we can combine it with the well-known strategies of BM, EQM, and GRW. We will label them as W_{IPH}-BM, W_{IPH}-EQM, and W_{IPH}-GRW. (See the Appendix for mathematical details.) Together, we call them *Density-Matrix Quantum Theories with the Initial Projection Hypothesis* (W_{IPH}-QM). Moreover, since they are motivated by considerations of the thermodynamic macrostate of the early universe, we also call them the *Thermodynamic Theories of Quantum Mechanics* (TQM).

In W_{IPH}-QM, the density matrix takes on a central role as the quantum microstate. Besides the low-entropy initial condition, it is also necessary to reformulate some definitions in quantum statistical mechanics:

6′ Non-unique correspondence: typically, a density matrix is in a superposition of macrostates and is not entirely in any one of the macrospaces. However, we can make sense of situations where W is very close to a macrostate $\mathscr{H}_\nu$:

$$\mathrm{tr}(W P_\nu) \approx 1, \qquad (13.21)$$

where P_ν is the projection operator onto $\mathscr{H}_\nu$. This means that almost all of W is in $\mathscr{H}_\nu$.

7′ Thermal equilibrium: typically, there is a dominant macro-space $\mathscr{H}_{eq}$ that has a dimension that is almost equal to D:

$$\frac{\dim \mathscr{H}_{eq}}{\dim \mathscr{H}} \approx 1. \qquad (13.22)$$

A system with density matrix W is in equilibrium if W is very close to $\mathscr{H}_{eq}$ in the sense of (13.21): $\mathrm{tr}(W P_{eq}) \approx 1$.

8′ Boltzmann Entropy: the Boltzmann entropy of a quantum-mechanical system with density matrix W that is very close to a macrostate ν is given by:

$$S_B(W) = k_B \log(\dim \mathscr{H}_\nu), \qquad (13.23)$$

where $\mathscr{H}_\nu$ denotes the subspace containing almost all of W in the sense of (13.21).

For the Bohmian version, we have the additional resource of the particle configuration, which can enable us to further define the effective macrostate and the effective Boltzmann entropy of a quantum system in a way that is analogous to the effective wave function of the system.[7]

13.3.2. *Conditional uniqueness*

13.3.2.1. *Eliminating statistical postulate with conditional uniqueness*

The Past Hypothesis, given a choice of an initial subspace, is compatible with a continuous infinity of initial wave functions. However, the Initial Projection Hypothesis determines a unique initial density matrix given an initial subspace; the initial quantum state is the normalized projection onto the subspace we select. I call this feature the *conditional uniqueness* of the initial quantum state in W_{IPH} theories.

In the standard theory, the Statistical Postulate determines a uniform probability distribution (μ_S with a normalization constant) over wave functions once we fix an initial subspace. This is because (it is reasonable to think that) there are some wave functions compatible

---

[7]Here is a flat-footed way to do it. First, we start from the wave-function picture of BM. If Ψ_i is the effective wave function of the universe, and if Ψ_i is almost entirely in $\mathscr{H}_\nu$, then the universe is in an effective macrostate of $\mathscr{H}_\nu$ and it has effective Boltzmann entropy of $k_B \log(\dim \mathscr{H}_\nu)$. Second, we note that we can define an analogous notion of the effective density matrix of the universe. Third, we can transpose the first point to the density-matrix perspective. If W_i is the effective density matrix of the universe, and if W_i is almost entirely in $\mathscr{H}_\nu$, then the universe is in an effective macrostate of $\mathscr{H}_\nu$ and it has effective Boltzmann entropy of $k_B \log(\dim \mathscr{H}_\nu)$. To be sure, further analysis is required.

with PH that are anti-entropic, i.e. they will evolve into lower-entropy states. Having a uniform probability distribution will provide a way to say that they are unlikely, since there are very few of them compared to the entropic wave functions. In the new theory (W_{IPH}), if we fix an initial subspace, there is only one choice for the initial quantum state because of conditional uniqueness. If typical (in the sense of μ_S) wave functions starting in $\mathscr{H}_{PH}$ will increase in entropy, then we know that almost all of the normalized projection onto $\mathscr{H}_{PH}$ will increase in entropy.

Therefore, in W_{IPH} theories, we no longer need the Statistical Postulate; we no longer need a probability/typicality distribution over initial quantum states, since only one state is possible (given an initial subspace). The only kind of probability in those theories will be the quantum mechanical probabilities. We expect the following to be true. In W_{IPH}-BM, by the W-version of the quantum equilibrium postulate (see Appendix), most initial particle configurations will increase in entropy. In W_{IPH}-GRW, by the W-version of the spontaneous collapse postulate, most likely the future collapses will result in higher-entropy states. In W_{IPH}-EQM, most emergent branches/worlds will evolve into higher-entropy ones.[8]

The property "conditional uniqueness" is a conditional property: the uniqueness of the initial quantum state depends on a choice of an initial subspace, representing the PH low-entropy macrostate. There are reasons to believe that the choice of the initial subspace will not be unique in the standard wave-function theory Ψ_{PH}. After all, the PH selects a low-entropy macrostate given macroscopic variables such as volume, density, and volume (Section 13.2.2, number 5). The values of the macroscopic variables do not have precise correspondence with subspaces in the Hilbert space. Two slightly different subspaces $\mathscr{H}_1$ and $\mathscr{H}_2$ that share most wave functions but differ regarding some wave functions will not make a difference to the role that PH plays.

---

[8]Much more work need to be done to show these results rigorously. A step in this direction is to show the empirical equivalence between Ψ_{PH} theories and W_{IPH} theories. For an overview of some existing results and some new analysis, see Chen (2019).

The role of PH is to explain the macroscopic regularity of entropic arrow of time and perhaps other arrows of time, which by themselves do not require a maximal level of precision. (Macroscopic facts are notoriously vague and imprecise. See the sorites problem surveyed in Hyde and Raffman (2018).) The precision in the subspace will not correspond to anything in the physical world. If the actual wave function Ψ is contained in two slightly different subspaces $\mathscr{H}_1$ and $\mathscr{H}_2$, the actual velocity of Bohmian particles, the actual collapse probability distribution, and the structure of emergent Everettian branches are all determined by Ψ. The slight differences between the two subspaces (i.e. slight differences in the boundary of the macrostate) have no influence on the microscopic facts in the world, since any microscopic influence is screened off by the actual wave function. Moreover, the non-unique correspondence in QSM (Section 13.2.2, 6) makes the boundaries between macrostates even more fuzzy. It would seem highly arbitrary to pick a precise boundary of the macrostate. If there are many choices of the initial subspace, all differing from each other by a small amount, then perhaps there will be many choices of the initial projection operators that are hard to distinguish on a macroscopic level.

However, if there are merely multiple choices of the initial subspace, and it is *determinate* which ones are the admissible choices, perhaps we can just collect all the admissible choices of the initial subspace and make PH a disjunctive statement: it is one of the admissible choices. But the problem is that what is admissible and what is not admissible will also be themselves *indeterminate*. This is analogous to the higher-order problem in vagueness see Sorensen (2018). So the most realistic version of PH will be allowing the fuzzy boundary of the initial macrostate. We discuss these different versions in more detail below.

13.3.2.2. *Three versions of PH and IPH*

In what follows, I formulate three versions of PH, with different levels of precision: the Strong PH, the Weak PH, and the Fuzzy PH. For each version, I formulate a version of IPH that matches the level of precision in PH. Although the Fuzzy PH might be the best choice for the standard theory, the Strong IPH might be viable for the new theories of W_{IPH}.

The Strong Past Hypothesis relies on a particular decomposition of the energy shell into macrostates (Section 2.2, 5). It selects a unique initial subspace — $\mathscr{H}_{PH}$. The initial wave function has to start from $\mathscr{H}_{PH}$.

Strong Past Hypothesis

$$\Psi(t_0) \in \mathscr{H}_{PH} \qquad (13.24)$$

Moreover, it is a well-defined probability space on which we can impose the surface area measure μ_S (on the unit sphere of vectors). If the Strong PH makes into the Ψ_{PH} theories, then we can impose a similarly strong IPH in the W_{IPH} theories:

Strong Initial Projection Hypothesis

$$\hat{W}_{IPH}(t_0) = \frac{I_{PH}}{dim\,\mathscr{H}_{PH}} \qquad (13.25)$$

The Strong IPH selects a unique initial quantum state, which does not require a further specification of statistical mechanical probabilities.

As mentioned before, there may be several low-dimensional subspaces that can do the job of low-entropy initial condition equally well. They do not have to be orthogonal to each other. The Past Hypothesis is formulated in macroscopic language which is not maximally precise. The inherent imprecision can result in indeterminacy of the initial subspace. There are two kinds of indeterminacy here: (1) there is a determinate set of admissible subspaces to choose from, and (2) the boundary between what is admissible and what is not is also fuzzy; some subspaces are admissible but some others are borderline cases.

The first kind of indeterminacy corresponds to the Weak Past Hypothesis:

Weak Past Hypothesis

$$\Psi(t_0) \in \mathscr{H}_1 \text{ or } \Psi(t_0) \in \mathscr{H}_2 \text{ or } ... \text{ or } \Psi(t_0) \in \mathscr{H}_k \qquad (13.26)$$

To be sure, the set of admissible subspaces may be infinite. For simplicity, we shall assume it is finite for now. For the Weak PH, there is the corresponding Weak IPH:

Weak Initial Projection Hypothesis

$$\hat{W}_{WPH_i}(t_0) = \frac{I_{\mathscr{H}_i}}{dim\mathscr{H}_i} \qquad (13.27)$$

where $\mathscr{H}_{WPH_i}$ is the normalized projection onto the subspace $\mathscr{H}_i$. For each admissible initial subspace, there is exactly one choice of the initial quantum state. Even though there is only one actual initial density matrix, there are many possible ones that slightly differ from each other $(\hat{W}_{WPH_1}(t_0), \hat{W}_{WPH_2}(t_0), \ldots, \hat{W}_{WPH_k}(t_0))$. Weak IPH does not need to be supplemented with an additional probability distribution over possible initial quantum states in that subspace.

The second kind of indeterminacy corresponds to the Fuzzy Past Hypothesis:

Fuzzy Past Hypothesis

The universal wave function started in some low-entropy state. The boundary of the macrostate is vague. (The Fuzzy PH does not determinately pick out any set of subspaces.)

Given the Fuzzy PH, we can formulate a Fuzzy IPH:

Fuzzy Initial Projection Hypothesis

$$\hat{W}_{FPH}(t_0) = \frac{I_{\mathscr{H}_i}}{dim\mathscr{H}_i} \qquad (13.28)$$

where $\mathscr{H}_i$ is an admissible precisification of the Fuzzy PH. Just like the Weak IPH, the Fuzzy IPH does not select a unique initial quantum state, since there are many admissible ones. (There is an actual initial density matrix, and it is compatible with the Fuzzy IPH.) However, unlike Weak IPH, there is no sharp cut-off between what is admissible and what is not admissible. Hence, for theories that include the Fuzzy IPH, it may be indeterminate whether some normalized projection operator is possible or not possible.

In any case, neither the Weak IPH nor the Fuzzy IPH requires the further specification of statistical mechanical probabilities. To see why, consider what is required for the Statistical Postulate. It requires a probability space to define the measure. The measure can only be defined relative to an admissible precisification. That is, the measure is over possible state vectors in a precise state space. In other words, the statistical mechanical probabilities are only meaningful relative to a precisification of the initial subspace. To be sure, the exact probability may be empirically inaccessible to us — what we have access to may only be certain interval-valued probabilities.[9] However, without the precisification of the boundary, we do not even have an interval-valued probabilities, as where to draw the boundary is unsettled. Given this fact, we can see that the original statistical mechanical probabilities require admissible precisifications — the statement will have to be:

- For a choice of an initial subspace, and given a uniform measure μ_S on it, μ_S−most (maybe even all) wave functions in that subspace will approach thermal equilibrium.

So the corresponding statement for the Weak IPH and the Fuzzy IPH will be:

- Every admissible density matrix will approach thermal equilibrium, where an admissible density matrix is a normalized projection operator onto an admissible initial subspace.

Since every admissible density matrix will approach thermal equilibrium, there is no need for an additional probability measure over initial quantum states in order to neglect the anti-entropic states. What is different on Fuzzy IPH is that what is admissible is a vague matter.

To summarize: all three versions of the IPH get rid of the fundamental statistical mechanical probabilities. The Strong IPH selects a unique initial quantum state. Although the Weak IPH and the Fuzzy

---

[9]See Bradley (2016) for a review of imprecise probabilities.

IPH do not, the initial state is unique given an admissible precisification of the PH subspace. Hence, all three versions — Strong IPH, Weak IPH, and Fuzzy IPH — satisfy conditional uniqueness of the initial quantum state. That is, given a choice of an initial subspace, there is a unique initial quantum state, i.e. its normalized projection. Since conditional uniqueness is sufficient to eliminate statistical mechanical probabilities at the fundamental level, all three versions will be able to reduce the two kinds of probabilities to just the quantum mechanical probabilities. However, for Strong IPH, the choice of the initial subspace is unique while it is not unique for Weak IPH or Fuzzy IPH. Therefore, Strong IPH trivially satisfies conditional uniqueness.

13.3.2.3. *The emergence of statistical mechanical probabilities*

Although the W_{IPH} theories (with any of the three versions of IPH) eliminate statistical mechanical probabilities at the fundamental level, they can still emerge as useful tools for analysis. As we saw in (13.19), the normalized projection onto a subspace can be decomposed as an integral of pure state density matrices with respect to the uniform probability distribution on the unit sphere. So mathematical results proven for a statistical ensemble of wave functions can be directly applied to normalized projections. The statistical perspective could prove useful for analysis of typicality and long-time behaviors of a statistical ensemble of wave functions, which could be translated to the long-time behavior of the mixed-state density matrix.

However, we recall that the decomposition into an integral of pure states is not unique. For the same density matrix in (13.19), it can be decomposed as a sum of pure states that are the orthonormal basis vectors, which gives us a different statistical ensemble of wave functions (a discrete probability distribution). Different statistical ensembles and the different emergent probabilities can be useful for different purposes. The non-uniqueness in decomposition further illustrates the idea that, in W_{IPH} theories, given an objective (real) density matrix W, the statistical mechanical probabilities over initial wave functions are not fundamental but are only emergent at the level of analysis. Moreover, the probability that emerges from the initial density matrix is in some sense more objectively grounded than the Mentaculus probabilities. This is because different initial density matrix will in general give rise to different microhistories of the world, while many different Mentaculus probabilities will not lead to any objective differences in the micro-histories. We can call our framework the Wentaculus, where W stands for the initial density matrix. The Wentaculus recovers all the exact probabilities without the standard worries about arbitrary measures facing the Mentaculus.

13.3.3. *Absolute uniqueness*

All three versions of IPH satisfy the property called *conditional uniqueness*: given a choice of the initial subspace, there is a unique choice of the initial quantum state — the normalized projection onto that subspace. The Strong IPH satisfies that trivially. Indeed, the Strong IPH satisfies a stronger property that I call *absolute uniqueness*: there is only a unique choice of the initial quantum state. This is because Strong IPH builds in a selection of a unique initial subspace.

As we mentioned in Section 13.3.2.1, the Strong PH in the Ψ_{PH} theories, which is the counterpart to the Strong IPH in W_{IPH} theories, is highly implausible. The precision of the Strong PH corresponds to a precise low-entropy initial macrostate, which has no correspondence to any precision in the microscopic histories of particles, collapses, or field values. The PH macrostate plays merely a macroscopic role, and the actual microstate, i.e. a wave function, screens off the microscopic influences of the PH macrostate.

However, the situation is quite different in W_{IPH} theories with a Strong IPH. Here, the initial density matrix $W_{IPH}(t_0)$ plays two roles. First, it plays the usual macroscopic role corresponding to the low-entropy initial condition. Second, and more importantly, it plays the microscopic role since it enters into the fundamental dynamical equations of W_{IPH} quantum mechanics (see Appendix for details). $W_{IPH}(t_0)$ directly guides Bohmian particles, undergoes GRW collapses, and realizes the emergent branches of Everett. Different choices of $W_{IPH}(t_0)$ will leave different microscopic traces in the world, in the form of different velocity fields for Bohmian particles, different probabilities for GRW collapses, and emergent branches with microscopic differences. Hence, the extra precision that sets Strong IPH apart from Weak or Fuzzy IPH has a worldly correspondence: it matches the microscopic history of the universe. Even if we were to think that different choices of the initial subspaces might not make an observable differences (at least to our current observational technologies), there is still a fact of the matter in the world about the precise microscopic history. Hence, it seems to me it is quite justifiable to postulate the Strong IPH in W_{IPH} theories, even though it is not justifiable to postulate the Strong PH in Ψ_{PH} theories.

The Strong IPH has another interesting consequence. It will make the Everettian theory strongly deterministic in the sense of Penrose (1989): Strong determinism "is not just a matter of the future being determined by the past; *the entire history of the universe is fixed, according to some precise mathematical scheme, for all time.*" This is because the Everettian theory is deterministic for all beables (local or non-local). With the help of a unique initial state, there is no ambiguity at all about what the history could be — there is only one way for it to go: starting from the absolutely unique $\hat{W}_{IPH}(t_0)$ (on Strong IPH) and evolving deterministically by the von Neumann equation (13.18).

13.3.4. *Other applications*

Getting rid of statistical mechanical probabilities at the fundamental level is not the only advantage of W_{IPH}-quantum theories over Ψ_{PH}-quantum theories. I discuss them in more detail in Chen (2018b,c). Here, I briefly summarize the other features of W_{IPH}-quantum theories:

(1) The meaning of the quantum state.

It has been a long-standing puzzle regarding how to understand the meaning of the quantum state, especially because it is defined as a function on a high-dimensional space and it is a non-separable object in physical space. A particularly attractive proposal is to understand it as nomological — being on a par with laws of nature. However, it faces a significant problem since typical wave functions are highly complex and not simple enough to be nomological. In contrast, if the quantum state is given by IPH, the initial quantum state will inherit the simplicity of the PH subspace. If PH is simple enough to be a law, then the initial quantum state is simple enough to be nomological. This feature of simplicity supports the nomological interpretation without relying on specific proposals about quantum gravity (cf: Goldstein and Zanghì (2013)). However, I should emphasize that the nomological interpretation is not the only way to understand the density matrix theory, as other proposals are also valid, such as the high-dimensional field and the low-dimensional multi-field interpretations (Chen (2018b) Section 13.3.3). This is because of the dual role the initial density matrix plays.

(2) Lorentz invariance.

David Albert (2015) observes that there is a conflict among three properties: quantum entanglement, Lorentz invariance, and what he calls narratability. A world is narratable just in case its entire history can be narrated in a linear sequence and every other sequence is merely a geometrical transformation from that. Since narratability is highly plausible, denying it carries significant cost. Hence, the real conflict is between quantum entanglement (a purely kinematic notion) and Lorentz invariance. This applies to all quantum theories that take quantum entanglement to be fundamental. However, given IPH, it is possible to take the nomological interpretation of the quantum state and remove entanglement facts among the facts about the distribution of physical matter. This is especially natural for W_{IPH}-Sm, somewhat less naturally to W_{IPH}-GRWm and W_{IPH}-GRWf, and potentially applicable to a fully Lorentz-invariant version of W_{IPH}-BM.

(3) Kinematic and dynamic unity.

In Ψ_{PH}-quantum theories, many subsystems will not have pure-state descriptions by wave functions due to the prevalence of entanglement. Most subsystems can be described only by a mixed-state density matrix, even when the universe as a whole is described by a wave function. In contrast, in W_{IPH}-quantum theories, there is more uniformity across the subsystem level and the universal level: the universe as a whole as well as most subsystems are described by the same kind of object — a (mixed-state) density matrix. Since state descriptions concern the kinematics of a theory, we say that W_{IPH}-quantum theories have more *kinematic unity* than their Ψ-counterparts:

> KINEMATIC UNITY The state description of the universe is of the same kind as the state descriptions of most quasi-isolated subsystems.

Moreover, in a universe described by Ψ_{PH}-BM, subsystems sometimes do not have conditional wave functions due to the presence of spin. In contrast, in a universe described by W_{IPH}-BM, the universe and the subsystems always have quantum states given by density matrices. This is because we can always define conditional density matrix for a Bohmian subsystem (Dürr *et al.* (2005)). That is, in W_{IPH}-BM, the W-guidance equation is always valid for the universe and the subsystems. In Ψ_{PH}-BM, the wave-function version of the guidance equation is not always valid. Thus, the W-BM equations are valid in more circumstances than the BM equations. We say that W_{IPH}-BM has more dynamic unity than Ψ_{PH}-BM:

> DYNAMIC UNITY The dynamical laws of the universe are the same as the effective laws of most quasi-isolated subsystems.

Both kinematic unity and dynamic unity come in degrees, and they are only defeasible reasons to favor one theory over another. But it is nonetheless interesting that merely adopting the density-matrix framework will make the theory more "unified" in the above senses.

13.3.5. *Generalizations*

IPH is not the only principle that leads to the selection of a unique or effectively unique initial quantum state of the universe. It is just one example of a simple principle. It is easy to generalize the strategy discussed here to other simple principles about the cosmological initial condition:

- Start from the full Hilbert space (energy shell) $\mathscr{H}$.
- Use simple principles (if there are any) to determine an initial subspace $\mathscr{H}_0 \subset \mathscr{H}$.
- Choose the natural quantum state in that subspace — the normalized projection $\hat{W}_0(t_0) = \frac{I_0}{dim\mathscr{H}_0}$.
- The natural choice will be simple and unique.

Another related but different example is the quantum version of the Weyl Curvature Hypothesis proposed by Ashtekar and Gupt (2016) based on the proposal of the classical version in Penrose (1989). However, Ashtekar and Gupt's hypothesis results in an infinite-dimensional unit ball of initial wave functions, which may not be normalizable. It is not clear to me whether there will be significant dimension reduction if we intersect it with the energy shell. In any case, the problem of non-normalizability is a general problem in cosmology which may require an independent solution. Perhaps we do not need to normalize the initial projection operator. In that case, the statistical analysis will be less transparent, but the initial state will still be well defined, giving rise to well-defined dynamics.

13.3.6. *Other proposals*

I would like to contrast my proposal with three other proposals in the literature.

Albert (2000) proposes that it is plausible that Ψ_{PH}-GRW theories do not need the Statistical Postulate. This is because the GRW jumps may be large enough (in Hilbert space) to render every initial wave function entropic in a short time. An anti-entropic wave function of a macroscopic system that evolves forward will be quickly hit by a GRW jump. As long as the GRW jump has a certain width that is large compared to the width of the anti-entropic set, the wave function will collapse into an entropic one. This relies on a conjecture about GRW theory: the final and empirically adequate GRW models will have a collapse width that is large enough. If that conjecture can be established, then, for every initial wave function, it is with a GRW probability that it will be entropic. This may correspond to the right form of the probabilistic version of the Second Law of Thermodynamics. I believe this is a very plausible conjecture. But it is an additional conjecture nonetheless, and it only works for collapse theories. It is an empirically open question whether GRW will survive experimental tests in the next 30 years. In contrast, my proposal is fully general — it works for GRW theories, Bohmian theories, and Everettian theories, and it does not rely on additional conjectures beyond those already postulated in QSM.

Wallace (2011) proposes that we can replace the Past Hypothesis and the Statistical Postulate with a Simple Dynamical Conjecture. In essence, it says that every Simple wave function will evolve to higher entropy. Here, "Simple" is a technical notion here meaning that the wave function is simple to specify and not specified by using time-reverse of an anti-entropic wave function. The idea is to replace the statistical postulate, which gives us reasons to neglect certain initial wave functions, with another postulate about simplicity, which also gives us reasons to set aside certain initial wave functions. This is a very interesting conjecture, which I think one should seriously investigate. But it is an additional conjecture nonetheless, although it has applicability to all quantum theories.

Wallace (2016) proposes that we can allow quantum states to be either pure or mixed. Moreover, he suggests that we can reinterpret probability distributions over wave functions as part of the state description and not an additional postulate about probability. There is much in common between Wallace's (2016) proposal and my proposal. However, the way that I get rid of statistical mechanical probabilities is not by way of a reinterpretation strategy but by using the uniqueness (or conditional uniqueness) of the initial quantum state. To be sure, there are many ways to achieve the goal, and the two approaches are quite related. Another difference is that in

Wallace's interpretation the universe or the system has two quantum states — one pure and the other mixed. For me, the universe has only one quantum state — a mixed state.

I would like to briefly mention the possibility of implementing my proposal in CSM. The upshot is that it is much less natural to do so in CSM than in QSM, for several reasons. On the classical mechanical phase space, the object that plays a similar role to the density matrix is the probability function $\rho(x)$. However, it is not clear how to understand its meaning as something ontic. If it is to be understood as a high-dimensional field and the only object in the ontology, then we have a Many-Worlds theory, unless we revise the Hamiltonian dynamics and make it stochastic. If it is to be understood as a low-dimensional multi-field (cf: Chen (2017), and Hubert and Romano (2018), and the references therein), it is not clear what corresponds to the momentum degrees of freedom. Second, the dynamics is not as natural in CSM. If it is to be understood as a high-dimensional field guiding a point particle (the phase point), it is not clear why we increase the complexity of the ontology. If it is to be understood as a nomological object, it is not clear what role it plays in the dynamics — it certainly does not give rise to a velocity field, because that is the job of the Hamiltonian function.[10]

13.4. Conclusion

The Thermodynamic Theories of Quantum Mechanics (TQM or W_{IPH}-QM) provide a new strategy to eliminate statistical mechanical probabilities from the fundamental postulates of the physical theory. They do so in a simple way not relying on any further conjectures about quantum mechanics or statistical mechanics. Moreover, they lead to several other applications and generalizations, which we only touched on briefly in this paper. Most importantly, we have found another deep connection between the foundations of quantum mechanics and the foundations of statistical mechanics.

---

[10]See McCoy (2018) for a recent attempt in this direction. Despite what I say here, I believe it is worth exploring, as McCoy does, alternative interpretations of the statistical state in CSM and QSM.

Appendix

(1) Ψ_{PH}-*Quantum Theories*

Ψ_{PH}-*Bohmian mechanics:* (Q, Ψ_{PH})

The Past Hypothesis:

$$\Psi_{PH}(t_0) \in \mathscr{H}_{PH} \qquad (13.29)$$

The Initial Particle Distribution:

$$\rho_{t_0}(q) = |\psi(q, t_0)|^2 \qquad (13.30)$$

The Schrödinger Equation:

$$i\hbar \frac{\partial \psi}{\partial t} = H\psi \qquad (13.31)$$

The Guidance Equation (Dürr *et al.* 1992):

$$\frac{dQ_i}{dt} = \frac{\hbar}{m_i} \mathrm{Im} \frac{\nabla_i \psi(q)}{\psi(q)} (q = Q) \qquad (13.32)$$

Ψ_{PH}-*Everettian mechanics*

The Past Hypothesis:

$$\Psi_{PH}(t_0) \in \mathscr{H}_{PH} \qquad (13.33)$$

The Schrödinger Equation:

$$i\hbar \frac{\partial \psi}{\partial t} = H\psi \qquad (13.34)$$

The Mass Density Equation:

$$m(x,t) = \langle \psi(t)| M(x) |\psi(t)\rangle, \qquad (13.35)$$

W_{PH}-S0: only Ψ_{PH}.
W_{PH}-Sm: $m(x,t)$ and Ψ_{PH}.

Ψ_{PH}-*GRW theory*

The Past Hypothesis:

$$\Psi_{PH}(t_0) \in \mathscr{H}_{PH} \qquad (13.36)$$

The linear evolution of the wave function is interrupted randomly (with rate $N\lambda$, where λ is of order 10^{-15} s^{-1}) by collapses:

$$\Psi_{T+} = \frac{\Lambda_{I_k}(X)^{1/2}\Psi_{T-}}{||\Lambda_{I_k}(X)^{1/2}\Psi_{T-}||}, \tag{13.37}$$

with the collapse center X being chosen randomly with probability distribution $\rho(x) = ||\Lambda_{i_k}(x)^{1/2}\Psi_{T-}||^2 dx$. The collapse rate operator is defined as:

$$\Lambda_{I_k}(x) = \frac{1}{(2\pi\sigma^2)^{3/2}} e^{-\frac{(Q_k-x)^2}{2\sigma^2}} \tag{13.38}$$

where Q_k is the position operator of "particle" k, and σ is a new constant of nature of order 10^{-7} m postulated in current GRW theories.

Ψ_{PH}-GRWm and Ψ_{PH}-GRWf: defined with local beables $m(x,t)$ and F.

(2) W_{IPH}-Quantum Theories

W_{IPH}-*Bohmian mechanics:* (Q, W_{IPH})

The Initial Projection Hypothesis:

$$\hat{W}_{IPH}(t_0) = \frac{I_{PH}}{dim\mathcal{H}_{PH}} \tag{13.39}$$

The Initial Particle Distribution:

$$P(Q(t_0) \in dq) = W_{IPH}(q, q, t_0)dq \tag{13.40}$$

The Von Neumann Equation:

$$i\hbar\frac{\partial\hat{W}}{\partial t} = [\hat{H}, \hat{W}] \tag{13.41}$$

The W_{PH}-Guidance Equation (Dürr *et al.* 2005):

$$\frac{dQ_i}{dt} = \frac{\hbar}{m_i}\mathrm{Im}\frac{\nabla_{q_i}W_{IPH}(q, q', t)}{W_{IPH}(q, q', t)}(q = q' = Q) \tag{13.42}$$

W_{IPH}-Everettian mechanics

The Initial Projection Hypothesis:

$$\hat{W}_{IPH}(t_0) = \frac{I_{PH}}{dim\mathscr{H}_{PH}} \tag{13.43}$$

The Von Neumann Equation:

$$i\hbar\frac{\partial \hat{W}}{\partial t} = [\hat{H}, \hat{W}] \tag{13.44}$$

The Mass Density Equation:

$$m(x,t) = tr(M(x)W(t)), \tag{13.45}$$

W_{IPH}-S0: only W_{IPH}.
W_{IPH}-Sm: $m(x,t)$ and W_{IPH}.

W_{IPH}-GRW theory

The Initial Projection Hypothesis:

$$\hat{W}_{IPH}(t_0) = \frac{I_{PH}}{dim\mathscr{H}_{PH}} \tag{13.46}$$

The linear evolution of the density matrix is interrupted randomly (with rate $N\lambda$, where λ is of order 10^{-15} s^{-1}) by collapses:

$$W_{T+} = \frac{\Lambda_{I_k}(X)^{1/2}W_{T-}\Lambda_{I_k}(X)^{1/2}}{tr(W_{T-}\Lambda_{I_k}(X))} \tag{13.47}$$

with X distributed by the following probability density:

$$\rho(x) = tr(W_{T-}\Lambda_{I_k}(x)) \tag{13.48}$$

The collapse rate operator is defined as:

$$\Lambda_{I_k}(x) = \frac{1}{(2\pi\sigma^2)^{3/2}}e^{-\frac{(Q_k-x)^2}{2\sigma^2}} \tag{13.49}$$

where Q_k is the position operator of "particle" k, and σ is a new constant of nature of order 10^{-7} m postulated in current GRW theories. The dynamical equations of W-GRW were introduced in Allori *et al.* (2013) Section 4.5.

W_{IPH}-GRWm and W_{IPH}-GRWf: defined with local beables $m(x,t)$ and F.

Acknowledgement

I would like to thank two anonymous reviewers of this volume for their helpful feedback on an earlier draft of this paper. I am grateful for stimulating discussions with Valia Allori, Sean Carroll, Detlef Dürr, Michael Esfeld, Veronica Gomez, Alan Hájek, Dustin Lazarovici, Matthias Lienert, Tim Maudlin, Sebastian Murgueitio, Wayne Myrvold, Jill North, Daniel Rubio, Ted Sider, Michael Tooley, Roderich Tumulka, David Wallace, Isaac Wilhelm, Nino Zanghì, and especially David Albert, Sheldon Goldstein, and Barry Loewer. I also benefited from the discussion at the Philosophy of Time Society at the 2019 Central APA in Denver.

References

Albert, D. (ms). Laws and physical things.

Albert, D. Z. (2000) *Time and Chance.* Harvard University Press.

Albert, D. Z. (2015) *After Physics.* Harvard University Press.

Allori, V. (2013) Primitive ontology and the structure of fundamental physical theories, *The Wave Function: Essays on the Metaphysics of Quantum Mechanics*, 58–75.

Allori, V., Goldstein, S., Tumulka, R., Zanghì, N. (2008) On the common structure of bohmian mechanics and the Ghirardi–Rimini–Weber theory: Dedicated to Giancarlo Ghirardi on the occasion of his 70th birthday, *The British Journal for the Philosophy of Science*, **59**(3), 353–389.

Allori, V., Goldstein, S., Tumulka, R., Zanghì, N. (2010) Many worlds and schrödinger's first quantum theory, *British Journal for the Philosophy of Science*, **62**(1), 1–27.

Allori, V., Goldstein, S., Tumulka, R., Zanghì, N. (2013) Predictions and primitive ontology in quantum foundations: a study of examples, *The British Journal for the Philosophy of Science*, **65**(2), 323–352.

Ashtekar, A. and Gupt, B. (2016) Initial conditions for cosmological perturbations, *Classical and Quantum Gravity*, **34**(3), 035004.

Bell, J. S. (1980) De Broglie-Bohm, delayed-choice, double-slit experiment, and density matrix, *International Journal of Quantum Chemistry*, **18**(S14), 155–159.

Bradley, S. (2016) Imprecise probabilities, In Zalta, E. N., editor, *The Stanford Encyclopedia of Philosophy*. Metaphysics Research Lab, Stanford University, winter 2016 edition.

Callender, C. (2011) Thermodynamic asymmetry in time, In Zalta, E. N., editor, *The Stanford Encyclopedia of Philosophy*. Metaphysics Research Lab, Stanford University, fall 2011 edition.

Chen, E. K. (2017) Our fundamental physical space: An essay on the metaphysics of the wave function, *Journal of Philosophy*, **114**, 7.

Chen, E. K. (2018a) The intrinsic structure of quantum mechanics, *Philsci-archive preprint:15140*.

Chen, E. K. (2018b) Quantum mechanics in a time-asymmetric universe: On the nature of the initial quantum state, *The British Journal for the Philosophy of Science*, forthcoming.

Chen, E. K. (2018c) Time asymmetry and quantum entanglement: A humean unification, *Manuscript*.

Chen, E. K. (2019) Quantum states of a time-asymmetric universe: Wave function, density matrix, and empirical equivalence, *arXiv:1901.08053*.

Coen, E., Coen, J. (2010) *A serious man*, Faber & Faber.

Dürr, D., Goldstein, S., Tumulka, R., Zanghì, N. (2005) On the role of density matrices in bohmian mechanics, *Foundations of Physics*, **35**(3), 449–467.

Dürr, D., Goldstein, S., Zanghì, N. (1992) Quantum equilibrium and the origin of absolute uncertainty, *Journal of Statistical Physics*, **67**(5–6), 843–907.

Dürr, D., Goldstein, S., Zanghì, N. (2012) *Quantum Physics without Quantum Philosophy*, Springer Science & Business Media.

Goldstein, S. (2001) Boltzmann's approach to statistical mechanics, In *Chance in Physics*, pp. 39–54. Springer.

Goldstein, S. (2012) Typicality and notions of probability in Physics, In *Probability in physics*, pp. 59–71. Springer.

Goldstein, S., Lebowitz, J. L., Mastrodonato, C., Tumulka, R., Zanghì, N. (2010a) Approach to thermal equilibrium of macroscopic quantum systems, *Physical Review E*, **81**(1), 011109.

Goldstein, S., Lebowitz, J. L., Mastrodonato, C., Tumulka, R., Zanghì, N. (2010b) Normal typicality and von neumann's quantum ergodic theorem, In *Proceedings of the Royal Society of London A: Mathematical, Physical and Engineering Sciences*, volume 466, pp. 3203–3224. The Royal Society.

Goldstein, S., Lebowitz, J. L., Tumulka, R., Zanghì, N. (2010c) Long-time behavior of macroscopic quantum systems: Commentary accompanying the english translation of John von Neumann's 1929 article on the quantum ergodic theorem, *The European Physical Journal H*, **35**(2), 173–200.

Goldstein, S., Tumulka, R. (2011) Approach to thermal equilibrium of macroscopic quantum systems, In *Non-Equilibrium Statistical Physics Today: Proceedings of the 11th Granada Seminar on Computational and Statistical Physics, AIP Conference Proceedings*, volume 1332, pp. 155–163. American Institute of Physics, New York.

Goldstein, S., Zanghì, N. (2013) Reality and the role of the wave function in quantum theory, *The Wave Function: Essays on the Metaphysics of Quantum Mechanics*, pages 91–109.

Hubert, M., Romano, D. (2018) The wave-function as a multi-field, *European Journal for Philosophy of Science*, **8**(3), 521–537.

Hyde, D., Raffman, D. (2018) Sorites paradox, In Zalta, E. N., editor, *The Stanford Encyclopedia of Philosophy*. Metaphysics Research Lab, Stanford University, summer 2018 edition.

Lebowitz, J. L. (2008) Time's arrow and boltzmann's entropy, *Scholarpedia*, **3**(4), 3448.

Lewis, D. (1986) *Philosophical Papers, Volume 2*, Oxford University Press, Oxford.

Loewer, B. (2004) David lewis's humean theory of objective chance, *Philosophy of Science*, **71**(5), 1115–1125.

Loewer, B. (2007) Counterfactuals and the second law, In Price, H. and Corry, R., editors, *Causation, Physics, and the Constitution of Reality: Russell's Republic Revisited*. Oxford University Press.

Loewer, B. (2016) The Mentaculus Vision. In Barry Loewer, Brad Weslake, and Eric Winsberg (eds.) *Time's Arrow and World's Probability Structure*. Harvard University Press, forthcoming.

Maroney, O. (2005) The density matrix in the de broglie–bohm approach, *Foundations of Physics*, **35**(3), 493–510.

McCoy, C. D. (2018) An alternative interpretation of statistical mechanics, *Erkenntnis*, forthcoming.

Ney, A., Albert, D. Z. (2013) *The Wave Function: Essays on the Metaphysics of Quantum Mechanics*, Oxford University Press.

North, J. (2011) Time in thermodynamics, *The Oxford Handbook of Philosophy of Time*, pp. 312–350.

Penrose, R. (1989) *The Emperor's New Mind: Concerning Computers, Minds, and the Laws of Physics*, Oxford University Press.

Sorensen, R. (2018) Vagueness, In Zalta, E. N., editor, *The Stanford Encyclopedia of Philosophy*. Metaphysics Research Lab, Stanford University, summer 2018 edition.

Von Neumann, J. (1955) *Mathematical Foundations of Quantum Mechanics*, Number 2. Princeton University Press.

Wallace, D. (2011) The logic of the past hypothesis.

Wallace, D. (2012) *The Emergent Multiverse: Quantum Theory According to the Everett Interpretation*, Oxford University Press.

Wallace, D. (2016) Probability and irreversibility in modern statistical mechanics: Classical and quantum, *Quantum Foundations of Statistical Mechanics (Oxford University Press, forthcoming)*.

Part V

Boltzmann and Gibbs

Chapter 14

Gibbs and Boltzmann Entropy in Classical and Quantum Mechanics

Sheldon Goldstein[*], Joel L. Lebowitz[†], Roderich Tumulka[‡]
and Nino Zanghì[§]

[*]*Departments of Mathematics, Physics, and Philosophy,
Rutgers University, Hill Center, 110 Frelinghuysen Road,
Piscataway, NJ 08854-8019, USA
oldstein@math.rutgers.edu,*
[†]*Departments of Mathematics and Physics, Rutgers University,
Hill Center, 110 Frelinghuysen Road, Piscataway,
NJ 08854-8019, USA
lebowitz@math.rutgers.edu*
[‡]*Mathematisches Institut, Eberhard-Karls-Universität,
Auf der Morgenstelle 10, 72076 Tübingen, Germany
roderich.tumulka@uni-tuebingen.de*
[§]*Dipartimento di Fisica, Università di Genova, and Istituto
Nazionale di Fisica Nucleare (Sezione di Genova),
Via Dodecaneso 33, 16146 Genova, Italy
zanghi@ge.infn.it*

The Gibbs entropy of a macroscopic classical system is a function of a probability distribution over phase space, i.e., of an ensemble. In contrast, the Boltzmann entropy is a function on phase space, and is thus defined for an individual system. Our aim is to discuss and compare these two notions of entropy, along with the associated ensemblist and individualist views of thermal equilibrium. Using the Gibbsian ensembles for the computation of the Gibbs entropy, the two notions yield the same (leading order) values for the

entropy of a macroscopic system in thermal equilibrium. The two
approaches do not, however, necessarily agree for non-equilibrium
systems. For those, we argue that the Boltzmann entropy is the
one that corresponds to thermodynamic entropy, in particular, in
connection with the second law of thermodynamics. Moreover, we
describe the quantum analog of the Boltzmann entropy, and we
argue that the individualist (Boltzmannian) concept of equilibrium
is supported by the recent works on thermalization of closed quantum systems.

14.1. Introduction

Disagreement among scientists is often downplayed, and science is
often presented as an accumulation of discoveries, of universally
accepted contributions to our common body of knowledge. But in
fact, there is substantial disagreement among physicists, not only
concerning questions that we have too little information about to
settle them, such as the nature of dark matter, but also concerning
conceptual questions about which all facts have long been in the literature, such as the interpretation of quantum mechanics. Another
question of the latter type concerns the definition of entropy and
some related concepts. In particular, two different formulations are
often given in the literature for how to define the thermodynamic
entropy (in equilibrium and non-equilibrium states) of a macroscopic
physical system in terms of a microscopic, mechanical description
(classical or quantum).

14.1.1. *Two definitions of entropy in classical
statistical mechanics*

In classical mechanics, the *Gibbs entropy* of a physical system with
phase space $\mathscr{X}$, for example $\mathscr{X} = \mathbb{R}^{6N} = \{(\boldsymbol{q}_1, \boldsymbol{v}_1, \ldots, \boldsymbol{q}_N, \boldsymbol{v}_N)\}$
for N point particles in $\mathbb{R}^3$ with positions $\boldsymbol{q}_j$ and velocities $\boldsymbol{v}_j$, is
defined as

$$S_{\mathrm{G}}(\rho) = -k \int_{\mathscr{X}} dx \, \rho(x) \, \log \rho(x), \qquad (14.1)$$

where k is the Boltzmann constant, $dx = N!^{-1}d^3\boldsymbol{q}_1\,d^3\boldsymbol{v}_1\cdots d^3\boldsymbol{q}_N$ $d^3\boldsymbol{v}_N$ the (symmetrized) phase space volume measure, log the natural logarithm,[1] and ρ a probability density on $\mathscr{X}$.[2]

The *Boltzmann entropy* of a macroscopic system is defined as

$$S_{\mathrm{B}}(X) = k\,\log\mathrm{vol}\,\Gamma(X), \tag{14.2}$$

where $X \in \mathscr{X}$ is the actual phase point of the system, vol means the volume in $\mathscr{X}$, and $\Gamma(X)$ is the set of all phase points that "look macroscopically the same" as X. Obviously, there is no unique precise definition for "looking macroscopically the same," so we have a certain freedom to make a reasonable choice. It can be argued that for large numbers N of particles (as appropriate for macroscopic physical systems), the arbitrariness in the choice of $\Gamma(X)$ shrinks and becomes less relevant.[3] A convenient procedure is to partition the phase space into regions Γ_ν we call macro sets (see Fig. 14.1),

$$\mathscr{X} = \bigcup_\nu \Gamma_\nu, \tag{14.3}$$

and to take as $\Gamma(X)$ the Γ_ν containing X. That is,

$$S_{\mathrm{B}}(X) = S_{\mathrm{B}}(\nu) := k\,\log\mathrm{vol}\,\Gamma_\nu. \tag{14.4}$$

We will give more detail in Section 14.5. Boltzmann's definition (14.2) is often abbreviated as "$S_{\mathrm{B}} = k\,\log W$" with $W = \mathrm{vol}\,\Gamma(X)$. In every energy shell, there is usually one macro set $\Gamma_\nu = \Gamma_{\mathrm{eq}}$ that corresponds to thermal equilibrium and takes up by far most (say, more than 99.99%) of the volume (see Section 14.5.1).

---

[1]One actually takes the expression $u\log u$ to mean the continuous extension of the function $u \mapsto u\log u$ from $(0,\infty)$ to the domain $[0,\infty)$; put differently, we follow the convention to set $0\log 0 = 0$.

[2]Changing the unit of phase space volume will change $\rho(x)$ by a constant factor, and thus $S_{\mathrm{G}}(\rho)$ by addition of a constant, an issue that does not matter for the applications and disappears anyway in the quantum case.

[3]For example, for a dilute gas and large N, $-S_{\mathrm{B}}(X)/k + N$ equals approximately the H functional, i.e., the integral in (14.9) below, which does not refer any more to a specific choice of boundaries of $\Gamma(X)$.

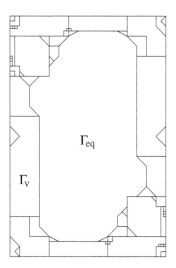

Fig. 14.1. Partition of phase space, or rather an energy shell therein, into macro sets Γ_ν, with the thermal equilibrium set taking up most of the volume (not drawn to scale). Reprinted from (Goldstein et al., 2017b).

14.1.2. X vs. ρ

An immediate problem with the Gibbs entropy is that while every classical system has a definite phase point X (even if we observers do not know it), a system does not "have a ρ"; that is, it is not clear which distribution ρ to use. For a system in thermal equilibrium, ρ presumably means a Gibbsian equilibrium ensemble (micro-canonical, canonical, or grand-canonical). It follows that, for thermal equilibrium states, S_B and S_G agree to leading order, see (14.31) below. In general, several possibilities for ρ come to mind:

(a) ignorance: $\rho(x)$ expresses the strength of an observer's belief that $X = x$.

(b) preparation procedure: A given procedure does not always reproduce the same phase point, but produces a random phase point with distribution ρ.

(c) coarse graining: Associate with every $X \in \mathscr{X}$ a distribution $\rho_X(x)$ on $\mathscr{X}$ that captures how macro-similar X and x are (or perhaps, how strongly an ideal observer seeing a system with phase point X would believe that $X = x$).

Correspondingly, there are several different notions of Gibbs entropy, which we will discuss in Sections 14.4 and 14.6. Here, maybe (c) could be regarded as a special case of (a), and thermal equilibrium ensembles as a special case of (b). In fact, it seems that Gibbs himself had in mind that any system in thermal equilibrium has a random phase point whose distribution ρ should be used, which is consistent with option (b); in his words (Gibbs, 1902, p. 152):

> "[...] we shall find that [the] distinction [between inter-action of a system S_1 with a system S_2 with determined phase point X_2 and one with distribution ρ_2] corresponds to the distinction in thermodynamics between mechanical and thermal action."

In our discussion, we will also address the status of the Gibbsian ensembles (see also Goldstein, 2019). We will argue that S_B qualifies as a definition of thermodynamic entropy whereas version (a) of S_G does not; (b) is not correct in general; and (c) is acceptable to the extent that it is not regarded as a special case of (a).

Different views about the meaning of entropy and the second law have consequences about the explanatory and predictive power of the second law that we will consider in Section 14.4. They also have practical consequences in the formulation of hydrodynamic equations, e.g., Navier-Stokes equations, for macroscopic variables (Goldstein and Lebowitz, 2004): such macroscopic equations can be compatible with a microscopic Hamiltonian evolution only if they make sure that the Boltzmann entropy increases.

The remainder of this article is organized as follows. In Section 14.2, we raise the question of the status of the second law for S_G and S_B. In Section 14.3, we consider the analogs of Gibbs and Boltzmann entropy in quantum mechanics. The following Sections 14.4–14.8 focus again on the classical case for simplicity. In Section 14.4, we discuss and criticize option (a), the idea that entropy is about subjective knowledge. In Section 14.5, we explain why Boltzmann entropy indeed tends to increase with time and discuss doubts and objections to this statement. In Section 14.6, we discuss an individualist understanding of Gibbs entropy as a generalization of Boltzmann entropy. In Section 14.7, we discuss the status of Gibbs's ensembles. In Section 14.8, we comment on a few proposals for how entropy increase should work for Gibbs entropy. In Section 14.9, we add some deeper considerations about entropy in quantum mechanics. In Section 14.10, we conclude.

14.2. Status of the second law

14.2.1. *Gibbs entropy*

Another immediate problem with the Gibbs entropy is that it does not change with time,

$$\frac{dS_G(\rho_t)}{dt} = 0, \tag{14.5}$$

if ρ is taken to evolve in accord with the microscopic Hamiltonian dynamics, that is, according to the Liouville equation

$$\frac{\partial \rho}{\partial t} = -\sum_{i=1}^{6N} \frac{\partial}{\partial x_i} (\rho(x) \, v(x)), \tag{14.6}$$

where $v(x)$ is the vector field on $\mathscr{X}$ that appears in the equation of motion

$$\frac{dX_i}{dt} = v_i(X(t)) \tag{14.7}$$

for the phase point $X(t) \in \mathscr{X}$, such as $v = \omega \nabla H$ with $6N \times 6N$ matrix $\omega = \left(\begin{smallmatrix} 0 & I \\ -I & 0 \end{smallmatrix} \right)$ in position and momentum coordinates and H the Hamiltonian function. Generally, the Gibbs entropy does not change when ρ gets transported by any volume-preserving bijection $\Phi : \mathscr{X} \to \mathscr{X}$,

$$S_G(\rho \circ \Phi^{-1}) = S_G(\rho). \tag{14.8}$$

In particular, by Liouville's theorem of the conservation of phase space volume, this applies when $\Phi = \Phi_t$ is the Hamiltonian time evolution, with $X(t) = \Phi_t(X(0))$. The time independence of $S_G(\rho)$ conflicts with the formulation of the second law given by Clausius, the person who coined the "laws of thermodynamics" as follows

(Clausius, 1865, p. 365):

    1. The energy of the universe is constant.
    2. The entropy of the universe tends to a maximum.

Among the authors who took entropy to be the Gibbs entropy, some (e.g., Khinchin, 1941) have argued that Clausius's wording of the second law is inappropriate or exaggerated, others (e.g., Mackey, 1989) argued that the Liouville evolution (14.6) is not the relevant evolution here. We will come back to this point in Section 14.8. As we will explain in Section 14.5, Clausius's statement is actually correct for the Boltzmann entropy.

14.2.2. *Boltzmann's H and k log W*

We should address right away a certain confusion about Gibbs and Boltzmann entropy that has come from the fact that Boltzmann used, in connection with the Boltzmann equation, the definition

$$S = -k \int_{\mathscr{X}_1} d^3\boldsymbol{q}\, d^3\boldsymbol{v}\, \tilde{f}(\boldsymbol{q}, \boldsymbol{v}) \log \tilde{f}(\boldsymbol{q}, \boldsymbol{v}) \tag{14.9}$$

for entropy. Boltzmann used the notation H (not to be confused with the Hamiltonian function) for the integral in (14.9); this functional gave name to the H-theorem, which asserts that H always decreases — except in thermal equilibrium, when H is constant in time.

    Here, $\mathscr{X}_1$ is the 1-particle phase space (for definiteness, $\mathbb{R}^6$), $\tilde{f} = Nf$, and f is a normalized distribution density in $\mathscr{X}_1$. The formula (14.9) obviously looks much like the Gibbs entropy (and has presumably inspired Gibbs's definition, which was published after (14.9)). In fact, (14.9) is also the Gibbs entropy of $\rho(x_1, \ldots, x_N) := N! f(x_1) \cdots f(x_N)$ (up to addition of the constant $k \log(N^N/N!)$). But here it is relevant that f means the empirical distribution of N points in $\mathscr{X}_1$ (after smoothing, or in the limit $N \to \infty$), and is thus computed from $X \in \mathscr{X}$ (see Section 14.5.4 below for more detail). So, the H functional or the S in (14.9) is indeed a function on $\mathscr{X}$, and in fact a special case of (14.2) (up to addition of the constant kN) corresponding to a particular choice of dividing $\mathscr{X}$ into macro sets Γ_ν (see Section 14.5.1).

14.3. The quantum case

Consider a macroscopic quantum system (e.g., a gas of $N > 10^{23}$ atoms in a box) with Hilbert space $\mathscr{H}$. We write $\mathbb{S}(\mathscr{H})$ for the unit sphere in the Hilbert space $\mathscr{H}$.

14.3.1. *Entropy in quantum mechanics*

The natural analog of the Gibbs entropy is the *quantum Gibbs entropy* or *von Neumann entropy* of a given density matrix $\hat{\rho}$ on $\mathscr{H}$ (von Neumann, 1927),

$$S_{\text{vN}}(\hat{\rho}) = -k \operatorname{tr}(\hat{\rho} \log \hat{\rho}). \qquad (14.10)$$

Von Neumann (1929) himself thought that this formula was not the most fundamental one but only applicable in certain circumstances; we will discuss his proposal for the fundamental definition in Section 14.9 below. In a sense, the density matrix $\hat{\rho}$ plays the role analogous to the classical distribution density ρ, and again, the question arises as to what exactly $\hat{\rho}$ refers to: an observer's ignorance or what? Our discussion of options (a)–(c) above for the Gibbs entropy will apply equally to the von Neumann entropy. In addition, there is also a further possibility:

(d) reduced density matrix: The system $\mathscr{S}_1$ is entangled with another system $\mathscr{S}_2$, and $\hat{\rho}$ is obtained from the (perhaps even pure) state $\Psi \in \mathbb{S}(\mathscr{H}_1 \otimes \mathscr{H}_2)$ of $\mathscr{S}_1 \cup \mathscr{S}_2$ through a partial trace, $\hat{\rho} = \operatorname{tr}_2 |\Psi\rangle\langle\Psi|$.

As we will discuss in Section 14.9.3, S_{vN} with option (d) does not yield a good notion of thermodynamic entropy, either.

In practice, systems are never isolated. But *even if* a macroscopic system were perfectly isolated, heat would flow in it from the hotter to the cooler parts, and, as we will explain in this section, there is a natural sense in which entropy can be defined and increases. The idealization of an isolated system helps us focus on this sense. In Section 14.3.2, we will point out why the results also cover non-isolated systems.

The closest quantum analog of the Boltzmann entropy is the following. A macro state ν should correspond to, instead of a subset Γ_ν of phase space, a subspace $\mathcal{H}_\nu$ of Hilbert space $\mathcal{H}$, called a *macro space* in the following: For different macro states $\nu' \neq \nu$, the macro spaces should be mutually orthogonal; thus, yielding a decomposition of Hilbert space into an orthogonal sum (von Neumann, 1929; Goldstein *et al.*, 2010a; 2010b),

$$\mathcal{H} = \bigoplus_\nu \mathcal{H}_\nu, \tag{14.11}$$

instead of the partition (14.3). Now the dimension of a subspace of $\mathcal{H}$ plays a role analogous to the volume of a subset of $\mathcal{X}$, and correspondingly we define the *quantum Boltzmann entropy* of a macro state ν by (Griffiths, 1994; Lebowitz, 2008; Goldstein *et al.*, 2010b)

$$S_{\mathrm{qB}}(\nu) = k \log \dim \mathcal{H}_\nu. \tag{14.12}$$

In fact, already Einstein (1914, Eq. (4a)) argued that the entropy of a macro state should be proportional to the log of the "number of elementary quantum states" compatible with that macro state; it seems that (14.12) fits very well with this description. To a system with wave function $\psi \in \mathcal{H}_\nu$ (or, for that matter, a density matrix concentrated in $\mathcal{H}_\nu$) we attribute the entropy value $S_{\mathrm{qB}}(\psi) := S_{\mathrm{qB}}(\nu)$.

It seems convincing that $S_{\mathrm{qB}}(\nu)$ yields the correct value of thermodynamic entropy. On one hand, it is an extensive, or additive quantity: If we consider two systems $\mathscr{S}_1, \mathscr{S}_2$ with negligible interaction, then the Hilbert space of both together is the tensor product, $\mathcal{H} = \mathcal{H}_{\mathscr{S}_1 \cup \mathscr{S}_2} = \mathcal{H}_{\mathscr{S}_1} \otimes \mathcal{H}_{\mathscr{S}_2}$, and it seems plausible that the macro states of $\mathscr{S}_1 \cup \mathscr{S}_2$ correspond to specifying the macro state of $\mathscr{S}_1$ and $\mathscr{S}_2$, respectively; that is, $\nu = (\nu_1, \nu_2)$ with $\mathcal{H}_\nu = \mathcal{H}_{\nu_1} \otimes \mathcal{H}_{\nu_2}$. As a consequence, the dimensions multiply, so

$$S_{\mathrm{qB}}(\nu) = S_{\mathrm{qB}}(\nu_1) + S_{\mathrm{qB}}(\nu_2). \tag{14.13}$$

On the other hand, it is plausible that, analogously to the classical case, in every energy shell (i.e., the subspace $\mathcal{H}_{(E-\Delta E, E]}$ corresponding to an energy interval $(E - \Delta E, E]$ with ΔE the resolution of macroscopic energy measurements), there is a dominant macro space $\mathcal{H}_{\tilde{\nu}}$ whose dimension is near (say, greater than 99.99% of) the dimension of the energy shell, and this macro state $\tilde{\nu}$ corresponds to thermal equilibrium (e.g., Tasaki, 2016), $\mathcal{H}_{\tilde{\nu}} = \mathcal{H}_{\mathrm{eq}}$. As a consequence,

$$S_{\mathrm{qB}}(\mathrm{eq}) \approx k \log \dim \mathcal{H}_{(E-\Delta E, E]}, \qquad (14.14)$$

and the right-hand side is well known to yield appropriate values of thermodynamic entropy in thermal equilibrium. In fact, the right-hand side agrees with $S_{\mathrm{vN}}(\hat{\rho}_{\mathrm{mc}})$, the von Neumann entropy associated with the micro-canonical density matrix $\hat{\rho}_{\mathrm{mc}}$ (i.e., the normalized projection to $\mathcal{H}_{(E-\Delta E, E]}$) and with thermal equilibrium at energy E.

Of course, a general pure quantum state $\psi \in \mathbb{S}(\mathcal{H})$ will be a non-trivial superposition of contributions from different $\mathcal{H}_{\nu}$'s, $\psi = \sum_{\nu} \hat{P}_{\nu}\psi$, where $\hat{P}_{\nu}$ is the projection to $\mathcal{H}_{\nu}$. One can say that ψ is a superposition of different entropy values, and in analogy to other observables one can define the self-adjoint operator

$$\hat{S} = \sum_{\nu} S_{\mathrm{qB}}(\nu)\, \hat{P}_{\nu}, \qquad (14.15)$$

whose eigenvalues are the $S_{\mathrm{qB}}(\nu)$ and eigenspaces the $\mathcal{H}_{\nu}$.

Here, the question arises as to which entropy value we should attribute to a system in state ψ. At this point, the foundations of statistical mechanics touch the foundations of quantum mechanics, as the problem of Schrödinger's cat (or the measurement problem of quantum mechanics) concerns precisely the status of wave functions ψ that are superpositions of macroscopically different contributions, given that our intuition leads us to expect a definite macroscopic situation. The standard "Copenhagen" formulation of quantum mechanics does not have much to say here that would be useful, but several proposed "quantum theories without observers" have long solved the issue in clean and clear ways (Goldstein, 1998):

• Bohmian mechanics (in our view, the most convincing proposal) admits further variables besides ψ by assuming that quantum particles have definite positions, too. Since Schrödinger's cat then has an actual configuration, there is a fact about whether it is dead or alive, even though the wave function is still a superposition. In the same way, the configuration usually selects one of the macroscopically different contributions $\hat{P}_\nu \psi$; in fact, this happens when

> the significantly nonzero $\hat{P}_\nu \psi$ do not overlap much in configuration space. $\qquad\qquad$ (14.16)

Artificial examples can be constructed for which this condition is violated, but it seems that (14.16) is fulfilled in practice. In that case, it seems natural to regard the entropy of that one contribution selected by the Bohmian configuration as the actual entropy value.

- Collapse theories modify the Schrödinger equation in a non-linear and stochastic way so that, for macroscopic systems, the evolution of ψ avoids macroscopic superpositions and drives ψ towards (a random) one of the $\mathcal{H}_\nu$. Then the question about the entropy of a macroscopic superposition does not arise.

- Maybe a many-worlds view (Everett, 1957; Wallace, 2012), in which each of the contributions corresponds to a part of reality, can be viable (see Allori *et al.* (2011) for critical discussion). Then the system's entropy has different values in different worlds.

We have thus defined the quantum Boltzmann entropy of a macroscopic system,

$$S_{\mathrm{qB}}(\mathscr{S}), \qquad\qquad (14.17)$$

in each of these theories.

Some authors (e.g., von Neumann (1929); Safranek *et al.* (2019a, 2019b)) have proposed averaging $S_{\mathrm{qB}}(\nu)$ with weights $\|\hat{P}_\nu \psi\|^2$, but that would yield an *average* entropy, not *the* entropy (see also Section 14.4.1 and our discussion of (14.66) in Section 14.9.2).

Let us turn to the question of what the second law asserts in quantum mechanics. Since S_{vN} remains constant under the unitary time evolution in $\mathcal{H}$, the discussion given in Section 14.8 for S_{G} applies equally to S_{vN}. About S_{qB}, we expect the following.

Conjecture 1. *In Bohmian mechanics and collapse theories, for macroscopic systems $\mathscr{S}$ with reasonable Hamiltonian $\hat{H}$ and decomposition $\mathscr{H} = \oplus_\nu \mathscr{H}_\nu$, every ν^* and most $\psi \in \mathbb{S}(\mathscr{H}_{\nu^*})$ are such that with probability close to 1, $S_{\mathrm{qB}}(\mathscr{S})$ is non-decreasing with time, except for infrequent, shallow, and short-lived valleys, until it reaches the thermal equilibrium value. Moreover, after reaching that value, $S_{\mathrm{qB}}(\mathscr{S})$ stays at that value for a very long time.*

Careful studies of this conjecture would be of interest but are presently missing. We provide a bit more discussion in Section 14.9.

14.3.2. *Open systems*

For open (non-isolated) systems, the quantum Boltzmann entropy can still be considered and should still tend to increase — not for the reasons that make the von Neumann entropy of the reduced density matrix increase (evolution from pure to mixed states), but for the reasons that make the quantum Boltzmann entropy of an isolated system in a pure state increase (evolution from small to large macro spaces). Let us elaborate.

In fact, the first question about an open system $\mathscr{S}_1$ (interacting with its environment $\mathscr{S}_2$) would be how to define its state. A common answer is the *reduced density matrix*

$$\hat{\rho}_1 := \mathrm{tr}_2 |\Psi\rangle\langle\Psi| \qquad (14.18)$$

(if $\mathscr{S}_1 \cup \mathscr{S}_2$ is in the pure state Ψ); another possible answer is the *conditional density matrix* (Dürr *et al.*, 2005)

$$\hat{\rho}_{\mathrm{cond}} := \mathrm{tr}_{s_2} \langle x_2|\Psi\rangle\langle\Psi|x_2\rangle\big|_{x_2 = X_2}. \qquad (14.19)$$

Here, tr_{s_2} denotes the partial trace over the non-positional degrees of freedom of $\mathscr{S}_2$ (such as spin), the scalar products are partial scalar products involving only the position degrees of freedom of $\mathscr{S}_2$, and X_2 is the Bohmian configuration of $\mathscr{S}_2$. (To illustrate the difference between $\hat{\rho}_1$ and $\hat{\rho}_{\mathrm{cond}}$, if $\mathscr{S}_1$ is Schrödinger's cat and $\mathscr{S}_2$ its environment, then $\hat{\rho}_1$ is a mixture of a live and a dead state, whereas $\hat{\rho}_{\mathrm{cond}}$ is either a live or a dead state, each with the appropriate probabilities.)

Second, it seems that the following analog of (14.16) is usually fulfilled in practice for both the reduced density matrix $\hat{\rho} = \hat{\rho}_1$ and the conditional density matrix $\hat{\rho} = \hat{\rho}_{\text{cond}}$:

> For those ν for which $\hat{P}_\nu \hat{\rho} \hat{P}_\nu$ is significantly nonzero,
> the functions g_ν on configuration space given by $\qquad$ (14.20)
> $g_\nu(x) := \langle x | \hat{P}_\nu \hat{\rho} \hat{P}_\nu | x \rangle$ do not overlap much.

As with (14.16), if (14.20) holds, then the Bohmian configuration X_1 of $\mathscr{S}_1$ selects the actual macro state $\nu(X_1)$ of $\mathscr{S}_1$. Moreover, $\nu(X_1)$ should usually be the same for $\hat{\rho} = \hat{\rho}_1$ as for $\hat{\rho} = \hat{\rho}_{\text{cond}}$. Then this macro state determines the entropy,

$$S_{\text{qB}}(\mathscr{S}_1) := S_{\text{qB}}(\nu(X_1)). \qquad (14.21)$$

In short, the concept of quantum Boltzmann entropy carries over from isolated systems in pure states to open systems.

14.3.3. *Quantum thermalization*

In the 21st century, there has been a wave of works on the thermalization of closed quantum systems, often connected with the key words "eigenstate thermalization hypothesis" (ETH) and "canonical typicality" (see, e.g., Gemmer *et al.*, 2004; Goldstein *et al.*, 2006; Popescu *et al.*, 2006; Goldstein *et al.*, 2010a; Gogolin and Eisert, 2016; Goldstein *et al.*, 2015; Kaufman *et al.*, 2016 and the references therein). The common theme of these works is that an individual, closed, macroscopic quantum system in a pure state ψ_t that evolves unitarily will, under conditions usually satisfied and after a sufficient waiting time, behave very much as one would expect a system in thermal equilibrium to behave. More precisely, on such a system $\mathscr{S}$ with ψ_0 in an energy shell, relevant observables yield their thermal equilibrium values up to small deviations with probabilities close to 1. For example, this happens simultaneously for all observables referring to a small subsystem $\mathscr{S}_1$ of $\mathscr{S}$ that interacts weakly with the remainder $\mathscr{S}_2 := \mathscr{S} \setminus \mathscr{S}_1$, with the further consequence ("canonical typicality") that the reduced density matrix of $\mathscr{S}_1$,

$$\hat{\rho}_1 := \mathrm{tr}_2 \, |\psi_t\rangle\langle\psi_t|, \qquad (14.22)$$

is close to a canonical one (Goldstein *et al.*, 2006),

$$\hat{\rho}_1 \approx \frac{1}{Z} e^{-\beta \hat{H}_1}, \qquad (14.23)$$

for suitable β and normalizing constant Z. Here, $\hat{H}_1$ is the Hamiltonian of $\mathscr{S}_1$. For another example, every initial wave function of $\mathscr{S}$ in an energy shell will, after a sufficient waiting time and for most of the time in the long run, be close to the thermal equilibrium macro space $\mathscr{H}_{\mathrm{eq}}$ (Goldstein *et al.*, 2010a), provided that the Hamiltonian $\hat{H}$ is non-degenerate and (ETH) all eigenvectors of $\hat{H}$ are very close to $\mathscr{H}_{\mathrm{eq}}$.

These works support the idea that the approach to thermal equilibrium need not have anything to do with an observer's ignorance. In fact, the system $\mathscr{S}$ always remains in a pure state, and thus has von Neumann entropy $S_{\mathrm{vN}} = 0$ at all times. This fact illustrates that the kind of thermalization relevant here involves neither an increase in von Neumann entropy nor a stationary density matrix of $\mathscr{S}$. Rather, ψ_t reaches, after a sufficient waiting time, the ε-neighborhood of the macro space $\mathscr{H}_{\tilde{\nu}} = \mathscr{H}_{\mathrm{eq}}$ corresponding to thermal equilibrium in the energy shell (Goldstein *et al.*, 2010a). That is the "individualist" or "Boltzmannian" version of approach to thermal equilibrium in quantum mechanics.

In fact, there are two individualist notions of thermal equilibrium in quantum mechanics, which have been called "macroscopic" and "microscopic thermal equilibrium" (Goldstein *et al.*, 2017b). Boltzmann's approach requires only that macro observables assume their equilibrium values (Goldstein *et al.*, 2010a), whereas a stronger statement is actually true after a long waiting time: that all micro observables assume their equilibrium values (Goldstein *et al.*, 2015). This is true not only for macroscopic systems, but also for small systems (Goldstein *et al.*, 2017a), and has in fact been verified experimentally for a system with as few as 6 degrees of freedom (Kaufman *et al.*, 2016).

We have emphasized earlier in this subsection that thermalization does not require mixed quantum states. We should add that this does not mean that pure quantum states are fully analogous to phase points in classical mechanics. In Bohmian mechanics, for example, the analog of a phase point would be the pair (ψ, Q) comprising the system's wave function ψ and its configuration Q.

14.4. Subjective entropy is not enough

By "subjective entropy" we mean Gibbs entropy under option (a): the view that the Gibbs entropy is the thermodynamic entropy, and that the distribution ρ in the Gibbs entropy represents an observer's subjective perspective and limited knowledge (e.g., Jaynes, 1965; Krylov, 1979; Mackey, 1989; Garibyan and Tegmark, 2014). We would like to point to three major problems with this view.

14.4.1. *Cases of wrong values*

The first problem is that in some situations, the subjective entropy does not appropriately reproduce the thermodynamic entropy. For example, suppose an isolated room contains a battery-powered heater, and we do not know whether it is on or off. If it is on, then after ten minutes the air will be hot, the battery empty, and the entropy of the room has a high value S_3. Not so if the heater is off, then the entropy has the low initial value $S_1 < S_3$. In view of our ignorance, we may attribute a subjective probability of 50 percent to each of "on" and "off." After ten minutes, our subjective distribution ρ over phase space will be spread over two regions with macroscopically different phase points, and its Gibbs entropy $S_G(\rho)$ will have a value S_2 between S_1 and S_3 (in fact, slightly above the average of S_1 and S_3).[4] But the correct thermodynamic value is not S_2, it is either

---

[4]Generally, if several distribution functions ρ_i have mutually disjoint supports, and we choose one of them randomly with probability p_i, then the resulting distribution $\rho = \sum_i p_i \rho_i$ has Gibbs entropy $S_G(\sum_i p_i \rho_i) = -k \int dx (\sum_i p_i \rho_i(x)) \log(\sum_i p_i \rho_i(x)) = -k \sum_i \int dx\, p_i\, \rho_i \log(p_i\, \rho_i(x)) = \sum_i p_i S_G(\rho_i) - k \sum_i p_i \log p_i > \sum_i p_i S_G(\rho_i)$. In our example, $S_2 = (S_1 + S_3)/2 + k \log 2$.

S_1 (if the heater was off) or S_3 (if the heater was on). So subjective entropy yields the wrong value.

The same problem arises with option (b), which concerns a system prepared by some procedure that, when repeated over and over, will lead to a distribution ρ of the system's phase point X. Suppose that the isolated room also contains a mechanism that tosses a coin or generates a random bit $Y \in \{0, 1\}$ in some other mechanical way; after that, the mechanism turns on the heater or does not, depending on Y. We would normally say that the entropy S after ten minutes is random, that $S = S_1$ with probability $1/2$ and $S = S_3$ with probability $1/2$. But the distribution ρ created by the procedure (and the canonical distribution for each given value of Y) is the same as in the previous paragraph, and has Gibbs entropy $S_G(\rho) = S_2$, the wrong value.

14.4.2. *Explanatory and predictive power*

The second major problem with subjective entropy concerns the inadequacy of its explanatory and predictive power. Consider, for example, the phenomenon that by thermal contact, heat always flows from the hotter to the cooler body, not the other way around. The usual explanation of this phenomenon is that entropy decreases when heat flows to the hotter body, and the second law excludes that. Now, that explanation would not get off the ground if entropy meant subjective entropy: In the absence of observers, does heat flow from the cooler to the hotter? In distant stars, does heat flow from the cooler to the hotter? In the days before humans existed, did heat flow from the cooler to the hotter? After the human race becomes extinct, will heat flow from the cooler to the hotter? If not, why would observers be relevant at all to the explanation of the phenomenon?

And as with explanation, so with prediction: Can we predict that heat will flow from the hotter to the cooler also in the absence of observers? If so, why would observers be relevant to the prediction?

So, subjective entropy does not seem relevant to either explaining or predicting heat flow. That leaves us with the question, what is subjective entropy good for? The study of subjective entropy is a subfield of psychology, not of physics. It is all about beliefs.

Some ensemblists may be inclined to say that the explanation of heat flow is that it occurs the same way *as if* an observer observed it. But the fact remains that observers actually have nothing to do with it.

Once the problem of explanatory power is appreciated, it seems obvious that subjective entropy is inappropriate: How could an objective physical phenomenon such as heat flow from the hotter to the cooler depend on subjective belief? In fact, since different observers may have different, incompatible subjective beliefs, how could coherent consequences such as physical laws be drawn from them? And what if the subjects made mistakes, what if they computed the time-evolved distribution $\rho \circ \Phi_t$ incorrectly; what if their beliefs were irrational — would that end the validity of subjective entropy? Somebody may be inclined to say that subjective entropy is valid only if it is rational (e.g., Bricmont, 2019), but that means basically to back off from the thought that entropy is subjective. It means that it does not play much of a role whether anybody's actual beliefs follow that particular ρ, but rather that there is a correct ρ that should be used; we will come back to this view at the end of the next subsection.

Another drawback of the subjective entropy, not unrelated to the problem of explanatory power, is that it draws the attention away from the fact that the universe must have very special initial conditions in order to yield a history with a thermodynamic arrow of time. While the Boltzmann entropy draws attention to the special properties of the initial state of the universe, the subjective entropy hides any such objective properties under talk about knowledge.

14.4.3. *Phase points play no role*

The third problem with subjective entropy is that $S_G(\rho)$ has nothing to do with the properties of the phase points x at which ρ is significantly non-zero. $S_G(\rho)$ measures essentially the width of the distribution ρ, much like the standard deviation of a probability distribution, except that the standard deviation yields the radius of the set over which ρ is effectively distributed, whereas the Gibbs entropy yields the log of its volume (see (14.26) below). The problem is reflected in the fact, mentioned around (14.8), that any volume-preserving transformation $\Phi : \mathscr{X} \to \mathscr{X}$ will leave the Gibbs entropy unchanged. It does not matter to the Gibbs entropy how the x's on which ρ is concentrated behave physically, although this behavior is crucial to thermodynamic entropy and the second law.

Some ensemblists may be inclined to say that the kind of ρ that occurs in practice is not any old density function, but is approximately concentrated on phase points that look macroscopically similar. This idea is essentially option (c) of Section 14.1.2, which was to take ρ as a kind of coarse graining of the actual phase point X. Specifically, if $\Gamma(X)$ denotes again the set of phase points that look macroscopically similar to X, then we may want to take $\rho = \rho_X$ to be the flat distribution over $\Gamma(X)$,

$$\rho_X(x) = \frac{1}{\operatorname{vol}\Gamma(X)}1_{\Gamma(X)}(x) = \begin{cases} (\operatorname{vol}\Gamma(X))^{-1} & \text{if } x \in \Gamma(X) \\ 0 & \text{if } x \notin \Gamma(X). \end{cases}$$
(14.24)

With this choice we obtain exact agreement between the Gibbs and Boltzmann entropies,

$$S_{\mathrm{G}}(\rho_X) = S_{\mathrm{B}}(X).$$
(14.25)

Indeed, whenever ρ is the flat distribution over any subset A of phase space $\mathscr{X}$, $\rho(x) = (\operatorname{vol} A)^{-1} 1_A(x)$, then

$$S_{\mathrm{G}}(\rho) = -k \int_A dx\, \frac{(-1)}{\operatorname{vol} A} \log \operatorname{vol} A = k \log \operatorname{vol} A.$$
(14.26)

(This fact also illustrates the mathematical meaning of the Gibbs entropy of any distribution ρ as the log of the volume over which ρ is effectively distributed.)

Of course, if we associate an entropy value $S_{\mathrm{G}}(\rho_X)$ with every $X \in \mathscr{X}$ in this way, then the use of Gibbs's definition (14.1) seems like an unnecessary detour. In fact, we have associated with every $X \in \mathscr{X}$ an entropy value $S(X)$, and talk about the knowledge of observers is not crucial to the definition of the function S, as is obvious from the fact that the function S is nothing but the Boltzmann entropy, which was introduced without mentioning observers.

This brings us once more to the idea that the ρ in $S_{\mathrm{G}}(\rho)$ is the subjective belief of a *rational* observer as advocated by Bricmont (2019). One could always use the Boltzmann entropy and add a narrative about observers and their beliefs, such as: Whenever $X \in \Gamma_\nu$, a rational observer should use the flat distribution over Γ_ν, and the Gibbs entropy of that observer's belief is what entropy really means. One could say such words. But they are also irrelevant, as observers' knowledge is irrelevant to which way heat flows, and the resulting entropy value agrees with $S_{\mathrm{B}}(X)$ anyway.

14.4.4. *What is attractive about subjective entropy*

Let us turn to factors that may seem attractive about the subjective entropy view: First, it seems like the obvious interpretation of the density ρ that comes up in all ensembles. But the Boltzmannian individualist view offers an alternative interpretation, as we will explain in Section 14.5.

Second, it is simple and elegant. That may be true but does not do away with the problems mentioned.

Third, the subjective view mixes well with certain interpretations of quantum mechanics such as Copenhagen and quantum Bayesianism, which claim that quantum mechanics is all about information or that one should not talk about reality. These interpretations are problematical as well, mainly because all information must ultimately be information about things that exist, and it does not help to leave vague and unspecified which things actually exist (Goldstein, 1998).

Fourth, the subjective view may seem to mix well with the work of Shannon (1948), as the Shannon entropy is a discrete version of Gibbs entropy and often regarded as quantifying the information content of a probability distribution. But actually, there is not a strong link, as Shannon regarded the probabilities in his study of optimal coding of data for transmission across a noisy channel as objective and did not claim any connection with thermodynamics. (By the way, it is dangerous to loosely speak of the "amount of information" in the same way as one speaks of, e.g., the amount of sand; after all, the sand grains are equal to each other, and one does not care about whether one gets this or that grain, whereas different pieces of information are not equivalent to each other.)

Fifth and finally, a strong pull towards subjective entropy comes from the belief that "objective entropy" either does not work or is ugly — a wrong belief, as we will explain in Section 14.5.

14.4.5. *Remarks*

Further critiques of subjective entropy can be found in Callender, 1999; Lebowitz and Maes, 2003; Goldstein *et al.*, 2017a; Goldstein, 2019.

We would like to comment on another quote. Jaynes (1965), a defender of subjective entropy, reported a dictum of Wigner's:

> "Entropy is an anthropomorphic concept."

Of course, this phrase can be interpreted in very different ways. Jaynes took it to express that entropy refers to the knowledge of human observers — the subjective view that we have criticized. But we do admit that there is a trait in entropy that depends partly on human nature, and that is linked to a certain (though limited and usually unproblematical) degree of arbitrariness in the definition of "looking macroscopically the same." This point will come up again in the next section.

14.5. Boltzmann's vision works

Many authors expressed disbelief that Boltzmann's understanding of entropy and the second law could possibly work. Von Neumann (1929, Sec. 0.6) wrote:

> "As in classical mechanics, also here [in the quantum case] there is no way that entropy could always increase, or even have a predominantly positive sign of its [time] derivative (or difference quotient): the time reversal objection as well as the recurrence objection are valid in quantum mechanics as well as in classical mechanics."

Khinchin (1941, Section 33, p. 139):

> "[One often] states that because of thermal interaction
> of material bodies the entropy of the universe is con-
> stantly increasing. It is also stated that the entropy of
> a system "which is left to itself" must always increase;
> taking into account the probabilistic foundation of ther-
> modynamics, one often ascribes to this statement a sta-
> tistical rather than absolute character. This formulation is
> wrong if only because the entropy of an isolated system is a
> thermodynamic function — not a phase-function — which
> means that it cannot be considered as a random quan-
> tity; if E and all [external parameters] λ_s remain constant
> the entropy cannot change its value whereas by changing
> these parameteres in an appropriate way we can make the
> entropy increase or decrease at will. Some authors (foot-
> note: Comp. Borel, Mécanique statistique classique, Paris
> 1925.) try to generalize the notion of entropy by consider-
> ing it as being a phase function which, depending on the
> phase, can assume different values for the same set of ther-
> modynamical parameters, and try to prove that entropy
> so defined must increase, with overwhelming probability.
> However, such a proof has not yet been given, and it is
> not at all clear how such an artificial generalization of
> the notion of entropy could be useful to the science of
> thermodynamics."

Jaynes (1965):

> "[T]he Boltzmann H theorem does not constitute a demon-
> stration of the second law for dilute gases[.]"

Even Boltzmann himself was at times unassured. In a letter to Felix
Klein in 1889, he wrote:

> "Just when I received your dear letter I had another
> neurasthenic attack, as I often do in Vienna, although I
> was spared them altogether in Munich. With it came the
> fear that the whole H-curve was nonsense."

But actually, the H-curve (i.e., the time evolution of entropy)
makes complete sense, Boltzmann's vision does work, and von
Neumann, Khinchin, and Jaynes were all mistaken, so it is perhaps

worth elucidating this point. Many other, deeper discussions can be found in the literature (e.g., qualitative, popular accounts in Penrose, 1989; Lebowitz and Maes, 2003; Carroll, 2010, overviews in Goldstein, 2001; Lebowitz, 2008; Goldstein, 2019, more technical and detailed discussions in Boltzmann, 1898; Ehrenfest and Ehrenfest, 1911; Lanford, 1976; Garrido *et al.*, 2004; Falcioni *et al.*, 2007; Cerino *et al.*, 2016; Goldstein *et al.*, 2017a; Lazarovici, 2018). So we now give a summary of Boltzmann's explanation of the second law.

14.5.1. *Macro states*

We start with a partition $\mathscr{X} = \cup_\nu \Gamma_\nu$ of phase space into macro sets as in Fig. 14.1. A natural way of obtaining such a partition would be to consider several functions $M_j : \mathscr{X} \to \mathbb{R}$ ($j = 1, \ldots, K$) that we would regard as "macro variables." Since macro measurements have limited resolution (say, $\Delta M_j > 0$), we want to think of the M_j as suitably coarse-grained with a discrete set of values, say, $\{n \Delta M_j : n \in \mathbb{Z}\}$. Then two phase points $x_1, x_2 \in \mathscr{X}$ will look macroscopically the same if and only if $M_j(x_1) = M_j(x_2)$ for all $j = 1, \ldots, K$, corresponding to

$$\Gamma_\nu = \Big\{ x \in \mathscr{X} : M_j(x) = \nu_j \; \forall j \Big\}, \qquad (14.27)$$

one for every macro state $\nu = (\nu_1, \ldots, \nu_K)$ described by the list of values of all M_j. We will discuss a concrete example due to Boltzmann in Section 14.5.4. Since coarse-grained energy should be one of the macro variables, say,

$$M_1(x) = [H(x)/\Delta E]\Delta E \qquad (14.28)$$

with H the Hamiltonian function and $[x]$ the nearest integer to $x \in \mathbb{R}$, every Γ_ν is contained in one micro-canonical energy shell

$$\mathscr{X}_{\text{mc}} := \mathscr{X}_{(E-\Delta E, E]} := \Big\{ x \in \mathscr{X} : E - \Delta E < H(x) \le E \Big\}. \qquad (14.29)$$

Of course, this description still leaves quite some freedom of choice and thus arbitrariness in the partition, as different physicists may make different choices of macro variables, and of the way and scale to coarse-grain them; this realization makes an "anthropomorphic" element in S_B explicit. Wallace (2019) complained that this element makes the Boltzmann entropy "subjective" as well, but that complaint does not seem valid: rather, S_B and its increase provide an objective answer to a question that is of interest from the human perspective. Moreover, as mentioned already, this anthropomorphic element becomes less relevant for larger N. It is usually not problematical and not subject to the same problems as the subjective entropy.

Usually in macroscopic systems, there is, for every energy shell $\mathscr{X}_{\mathrm{mc}}$ (or, if there are further macroscopic conserved quantities besides energy, in the set where their values have been fixed as well), one macro set $\Gamma_{\tilde{\nu}}$ that contains most (say, more than 99.99%) of the phase space volume of $\mathscr{X}_{\mathrm{mc}}$ (see, e.g., Boltzmann, 1898; Lanford, 1973; Lazarovici, 2018);[5] in fact, see Goldstein *et al.*, 2017b, Eq. (6),

$$\frac{\operatorname{vol}\Gamma_{\tilde{\nu}}}{\operatorname{vol}\mathscr{X}_{\mathrm{mc}}} \approx 1 - 10^{-cN} \tag{14.30}$$

with positive constant c. The existence of this dominant macro state means that all macro observables are nearly-constant functions on $\mathscr{X}_{\mathrm{mc}}$, in the sense that the set where they deviate (beyond tolerances) from their dominant values has tiny volume. This macro state $\tilde{\nu}$ is the thermal equilibrium state, $\Gamma_{\tilde{\nu}} = \Gamma_{\mathrm{eq}}$, see Fig. 14.1, and the dominant values of the macro observables are their thermal equilibrium values. That fits with thermal equilibrium having maximal entropy, and it has the consequence that

$$S_B(\mathrm{eq}) \approx k \log \operatorname{vol} \mathscr{X}_{\mathrm{mc}} = S_G(\rho_{\mathrm{mc}}), \tag{14.31}$$

where $\rho_{\mathrm{mc}}(x) = (\operatorname{vol}\mathscr{X}_{\mathrm{mc}})^{-1} 1_{\mathscr{X}_{\mathrm{mc}}}(x)$ is the micro-canonical distribution, and the (relative or absolute) error in the approximation tends to zero as $N \to \infty$.

---

[5]There are exceptions, in which none of the macro sets dominates; for example, in the ferromagnetic Ising model with vanishing external magnetic field and not-too-high temperature, there are two macro states (the first having a majority of spins up, the second having a majority of spins down) that together dominate but have equal volume; see also (Lazarovici, 2018). But that does not change much about the discussion.

Moreover, different macro sets Γ_ν have vastly different volumes. In fact, usually the small macro sets of an energy shell taken together are still much smaller than the next bigger one,

$$\mathrm{vol}(\Gamma_{<\nu}^{(E-\Delta E, E]}) \ll \mathrm{vol}(\Gamma_\nu) \tag{14.32}$$

with

$$\Gamma_{<\nu}^{(E-\Delta E, E]} := \bigcup_{\nu': S_\mathrm{B}(\nu') < S_\mathrm{B}(\nu)} \Gamma_{\nu'} \cap \mathscr{X}_{(E-\Delta E, E]}. \tag{14.33}$$

(There are exceptions to this rule of thumb; in particular, symmetries sometimes imply that two or a few macro sets must have approximately the same volume.)

14.5.2. *Entropy increase*

Now increase of Boltzmann entropy means that the phase point $X(t)$ moves to bigger and bigger macro sets Γ_ν. In this framework, the second law can be stated as follows.

> **Mathematical second law.** Given $\nu \neq$ eq, for most phase points $X(0)$ in Γ_ν, $X(t)$ moves to bigger and bigger macro sets as t increases until it reaches Γ_eq, except possibly for entropy valleys that are infrequent, $\qquad$ (14.34) shallow, and short-lived; once $X(t)$ reaches Γ_eq, it stays in there for an extraordinarily long time, except possibly for infrequent, shallow, and short-lived entropy valleys.

The described behavior is depicted in Fig 14.2. Entropy valleys (i.e., periods of entropy decrease and return to the previous level) are also called fluctuations. The *physical second law* then asserts that the actual phase point of a real-world closed system behaves the way described in (14.34) for most phase points.

As an illustration of (14.34) and as a step towards making it plausible, let us consider two times, 0 and $t > 0$. Let $A_t := \Phi_t(\Gamma_\nu)$.

Fig. 14.2. A typical entropy curve $S(X(t))$ according to Boltzmann: It should go up except for infrequent, shallow, short-lived valleys; frequency, depth, and duration of the valleys are exaggerated for better visibility. After very long times, the entropy should go down considerably.

By Liouville's theorem, $\mathrm{vol}(A_t) = \mathrm{vol}(\Gamma_\nu)$, and thus, by (14.32),

$$\frac{\mathrm{vol}(A_t \cap \Gamma_{<\nu}^{(E-\Delta E,E]})}{\mathrm{vol}(A_t)} \ll 1. \tag{14.35}$$

That is, only a small minority of points in A_t will have entropy smaller than $S_\mathrm{B}(\nu)$. That is, for most points $X(0) \in \Gamma_\nu$,

$$S_\mathrm{B}(X(0)) \leq S_\mathrm{B}(X(t)). \tag{14.36}$$

Another simple special case is the one in which the macro evolution is deterministic (Garrido *et al.*, 2004; Goldstein and Lebowitz, 2004; De Roeck *et al.*, 2006). For the sake of concreteness, assume that in a time step of a certain size τ, Γ_{ν_1} gets mapped into Γ_{ν_2}, which in turn gets mapped into Γ_{ν_3}, and so on up to ν_m:

$$\Phi_\tau(\Gamma_{\nu_i}) \subseteq \Gamma_{\nu_{i+1}}. \tag{14.37}$$

Then, by Liouville's theorem, $\mathrm{vol}\,\Gamma_{\nu_i} \leq \mathrm{vol}\,\Gamma_{\nu_{i+1}}$, so

$$S_\mathrm{B}(\nu_i) \leq S_\mathrm{B}(\nu_{i+1}) \tag{14.38}$$

for all $i = 1, \ldots, m - 1$, so entropy does not decrease. Of course, in realistic cases, the macro evolution becomes deterministic only in the limit $N \to \infty$, and as long as N is finite, there are a minority of points in Γ_{ν_i} that do not evolve to $\Gamma_{\nu_{i+1}}$.

Generally, if the Hamiltonian motion is not specially desgined for the given partition of $\mathscr{X}$, then it is quite intuitive that the motion of the phase point should tend to lead to larger macro sets, and not to smaller ones. Numerical simulations exhibiting this behavior are presented in (Falcioni *et al.*, 2007). It is also quite intuitive that the phase point would stay in Γ_{eq} for a very, very long time: If the non-equilibrium set $\Gamma_{\mathrm{noneq}} := \Gamma_{<\mathrm{eq}}^{(E-\Delta E,E]} = \mathscr{X}_{\mathrm{mc}} \setminus \Gamma_{\mathrm{eq}}$ has only the fraction 10^{-cN} of the volume of the energy shell, cf. (14.30), then only a tiny fraction of Γ_{eq} should be able to evolve into the non-equilibrium set Γ_{noneq} in a short time; and if most points in Γ_{noneq} spend a substantial amount of time there, then it will take very, very long until a substantial fraction of Γ_{eq} has visited Γ_{noneq}. The statement that points in Γ_{eq} stay there for a long time fits well with the observed stationarity of thermal equilibrium — which is why it is called "equilibrium."

Let us briefly address two classic objections to the idea that entropy increases:

- Time reversal (Loschmidt's objection) shows that entropy increase cannot hold for *all* phase points in Γ_ν. Concretely, for relevant Hamiltonians the time reversal mapping $R : \mathscr{X} \to \mathscr{X}$, defined by

$$R(\boldsymbol{q}_1, \ldots, \boldsymbol{q}_N, \boldsymbol{v}_1, \ldots, \boldsymbol{v}_N) := (\boldsymbol{q}_1, \ldots, \boldsymbol{q}_N, -\boldsymbol{v}_1, \ldots, -\boldsymbol{v}_N)$$

(14.39)

with $\boldsymbol{q}_i$ the position and $\boldsymbol{v}_i$ the velocity of particle i, has the property

$$R \circ \Phi_t \circ R = \Phi_{-t}. \qquad (14.40)$$

Usually, R maps Γ_ν onto some $\Gamma_{\nu'}$ (where ν' may or may not equal ν), so $S_{\mathrm{B}}(R(x)) = S_{\mathrm{B}}(x)$. So if some $X(0) \in \Gamma_{\nu_1}$ evolves to $X(t) \in \Gamma_{\nu'}$ with $S_{\mathrm{B}}(\nu') > S_{\mathrm{B}}(\nu_1)$, then $R(X(t)) \in \Gamma_\nu$ evolves to $R(X(0))$, and its entropy decreases.
- Recurrence (Zermelo's objection) shows that $S_{\mathrm{B}}(X(t))$ cannot *forever* be non-decreasing; thus, $X(t)$ cannot stay forever in Γ_{eq} once it reaches Γ_{eq}. (The Poincaré recurrence theorem states that under conditions usually satisfied in $\mathscr{X}_{\mathrm{mc}}$, every trajectory $X(t)$, except for a set of measure zero of $X(0)$s, returns arbitrarily close to $X(0)$ at some arbitrarily late time.)

Contrary to von Neumann's statement quoted in the beginning of Section 14.5, the second law as formulated in (14.34) is not refuted by either objection: after all, the second law applies to *most*, not *all*, phase points $X(0)$, and it does not claim that $X(t)$ will stay in thermal equilibrium *forever*, but only for a very, very long time.

14.5.3. *Non-equilibrium*

The term "non-equilibrium" is sometimes understood (Gallavotti, 2003) as referring to so called non-equilibrium steady states (NESS), which concern, for example, a system $\mathscr{S}$ coupled to two infinite reservoirs of different temperature; so $\mathscr{S}$ is an open system heated on one side and cooled on another, and it will tend to assume a macroscopically stationary ("steady") state with a temperature gradient, a nonzero heat current, and a positive rate of entropy production (Onsager, 1931; Bergmann and Lebowitz, 1955; Derrida, 2007; Goldstein *et al.*, 2017a). In contrast, in this Section 14.5, we are considering a closed system (i.e., not interacting with the outside), and "non-equilibrium" refers to any phase point in $\mathscr{X}_{mc} \setminus \Gamma_{eq}$. Examples of non-equilibrium macro states include, but are not limited to, states in local thermal equilibrium but not in (global) thermal equilibrium (such as systems hotter in one place than in another). Other examples arise from removing a constraint or wall; such macro states may have been in thermal equilibrium before the constraint was removed but are not longer so afterwards; for example, think of a macro state in which all particles are in the left half of a box; for another example, suppose we could turn on or off the interaction between two kinds of particles (say, "red ones" and "blue ones"), and think of a macro state that is a thermal equilibrium state when the interaction is off (so that the red energy and the blue energy are separately conserved) but not when it is on, such as when both gases are in the same volume but at different temperatures.

14.5.4. *Concrete example: The Boltzmann equation*

Here is a concrete example of a partition of phase space due to Boltzmann. Divide the 1-particle phase space $\mathscr{X}_1$ into cells C_i (say, small cubes of equal volume) and count (with a given tolerance) the particles in each cell. The macro state is described by the list $\nu = (\nu_1, \ldots, \nu_L)$ of all these occupation numbers; for convenience, we will normalize them:

$$\nu_i(x) := f_i := \left[\frac{\#\{j \in \{1, \ldots, N\} : (\boldsymbol{q}_j, \boldsymbol{v}_j) \in C_i\}}{N \Delta f \, \text{vol} \, C_i} \right] \Delta f \qquad (14.41)$$

with $N \Delta f$ the tolerance in counting and $[\cdot]$ again the nearest integer. This example of a partition $\Gamma_\nu = \{x \in \mathscr{X} : \nu(x) = \nu\}$ is good for dilute, weakly interacting gases but not in general (Garrido *et al.*, 2004; Goldstein and Lebowitz, 2004) (see also Section 14.5.8).

Boltzmann considered N billiard balls of radius a in a container $\Lambda \subset \mathbb{R}^3$, so $\mathscr{X}_1 = \Lambda \times \mathbb{R}^3$. In a suitable limit in which $N \to \infty$, $a \to 0$, and the cells C_i become small, the normalized occupation numbers f_i become a continuous density $f(\boldsymbol{q}, \boldsymbol{v})$. He argued (convincingly) that for most $x \in \Gamma_\nu$ this density, essentially the empirical distribution of the N particles in $\mathscr{X}_1$, will change in time according to the *Boltzmann equation*, an integro-differential equation (Boltzmann, 1872, 1898; Ehrenfest and Ehrenfest, 1911; Lanford, 1976). It reads, in the version appropriate for the hard sphere gas without external forces,

$$\left(\frac{\partial}{\partial t} + \boldsymbol{v} \cdot \nabla_{\boldsymbol{q}} \right) f(\boldsymbol{q}, \boldsymbol{v}, t) = Q(\boldsymbol{q}, \boldsymbol{v}, t) \qquad (14.42)$$

with the "collision term"

$$Q(\boldsymbol{q}, \boldsymbol{v}, t) = \lambda \int_{\mathbb{R}^3} d^3 \boldsymbol{v}_* \int_{\mathbb{S}^2} d^2 \boldsymbol{\omega} \, 1_{\boldsymbol{\omega} \cdot (\boldsymbol{v} - \boldsymbol{v}_*) > 0} \, \boldsymbol{\omega} \cdot (\boldsymbol{v} - \boldsymbol{v}_*)$$
$$\times \left[f(\boldsymbol{q}, \boldsymbol{v}', t) \, f(\boldsymbol{q}, \boldsymbol{v}'_*, t) - f(\boldsymbol{q}, \boldsymbol{v}, t) \, f(\boldsymbol{q}, \boldsymbol{v}_*, t) \right], \quad (14.43)$$

involving a constant $\lambda > 0$ and the abbreviations

$$\boldsymbol{v}' = \boldsymbol{v} - [(\boldsymbol{v} - \boldsymbol{v}_*) \cdot \boldsymbol{\omega}] \boldsymbol{\omega} \qquad (14.44)$$

$$\boldsymbol{v}'_* = \boldsymbol{v}_* + [(\boldsymbol{v} - \boldsymbol{v}_*) \cdot \boldsymbol{\omega}] \boldsymbol{\omega} \qquad (14.45)$$

for the outgoing velocities of a collision between two balls with incoming velocities $\boldsymbol{v}$ and $\boldsymbol{v}_*$ and direction $\boldsymbol{\omega}$ between the centers of the two balls. The Boltzmann equation is considered for $\boldsymbol{v} \in \mathbb{R}^3$ and $\boldsymbol{q} \in \Lambda$ along with a boundary condition representing that balls hitting the boundary of Λ get reflected there. A function $f(\boldsymbol{q}, \boldsymbol{v})$ is a stationary solution if and only if it is independent of $\boldsymbol{q}$ and a

Maxwellian (i.e., Gaussian) in v — that is, if and only if it represents thermal equilibrium. Correspondingly, non-equilibrium macro states correspond to any density function f that is not a global (i.e., q-independent) Maxwellian.

The entropy turns out to be (up to addition of the constant kN and terms of lower order)

$$S(x) = -k \sum_i \mathrm{vol}(C_i)\, N f_i \log\big[N f_i\big]. \qquad (14.46)$$

In the limit of small C_i, this becomes (14.9), i.e., $-kH$ in terms of the H functional. Boltzmann further proved the H-theorem, which asserts that for any solution f_t of the Boltzmann equation,

$$\frac{dH}{dt} \le 0, \qquad (14.47)$$

with equality only if f is a local Maxwellian. The H-theorem amounts to a derivation of the second law relative to the partition $\{\Gamma_\nu\}$ under consideration.

14.5.5. *Rigorous result*

Some authors suspected that Boltzmann's vision, and the Boltzmann equation in particular, was not valid. For example, Khinchin (1941, Section 33, p. 142) complained about individualist accounts of entropy:

> "All existing attempts to give a general proof of this postulate [i.e., $S = k \log W$,] must be considered as an aggregate of logical and mathematical errors superimposed on a general confusion in the definition of the basic quantities."

But actually, the Boltzmann equation (and with it the increase of entropy) is rigorously valid for most phase points in Γ_ν, at least for a short time, as proved by Lanford (1975; 1976). Here is a summary statement of Lanford's theorem, leaving out some technical details:

Theorem 1. *Let $\bar t > 0$ and $\lambda > 0$ (the constant in the Boltzmann equation) be constants. For a very large number N of billiard balls of (very small) radius a with $4Na^2 = \lambda$, for every $0 \le t < \frac{1}{5}\bar t$, for any* *nice density f_0 in $\mathscr{X}_1 = \Lambda \times \mathbb{R}^3$ with mean free time $\geq \bar{t}$, and for a coarse graining of $\mathscr{X}_1$ into cells C_i that are small but not too small, most phase points X with empirical distribution f_0 (relative to $\{C_i\}$ and within small tolerances) evolve in such a way that the empirical distribution of $X(t)$ (relative to $\{C_i\}$) is close to f_t, where $t' \mapsto f_{t'}$ is the solution of the Boltzmann equation with initial datum f_0.*

It is believed but not proven that the Boltzmann equation is valid for a much longer duration, maybe of the order of recurrence times. The method of proof fails after $\frac{1}{5}\bar{t}$, but it does not give reason to think that the actual behavior changes at $\frac{1}{5}\bar{t}$.

Where is the famous *Stosszahlansatz*, or hypothesis of molecular chaos, in this discussion? This hypothesis was stated by Boltzmann as specifying the approximate number of collisions with parameter $\boldsymbol{\omega}$ between particles from cells C_i and C_j within a small time interval. In our discussion, it is hidden in the assumption that the initial phase point $X(0)$ be *typical* in Γ_ν: Both Theorem 1 and the wording (14.34) of the second law talk merely about *most* phase points in Γ_ν, and for most phase points in Γ_ν ($\nu = f$), it is presumably true that the number of upcoming collisions is, within reasonable tolerances, given by the hypothesis of molecular chaos, not just at the initial time but at all relevant times. We will discuss molecular chaos further in Section 14.5.7.

14.5.6. *Empirical vs. marginal distribution*

Many mathematicians (e.g., Kac, 1956; Cercignani *et al.*, 1994 but also Tolman, 1938) considered the Boltzmann equation in a somewhat different context with a different meaning, viz., with f not the *empirical* distribution but the *marginal* distribution. This means the following.

- The empirical distribution, for a given phase point $X = (\boldsymbol{q}_1, \boldsymbol{v}_1, \ldots, \boldsymbol{q}_N, \boldsymbol{v}_N) \in \mathscr{X}$, is the distribution on $\mathscr{X}_1$ with density

$$f_{\text{emp}}^X(\boldsymbol{q}, \boldsymbol{v}) = \frac{1}{N} \sum_{j=1}^N \delta^3(\boldsymbol{q}_j - \boldsymbol{q})\, \delta^3(\boldsymbol{v}_j - \boldsymbol{v}). \tag{14.48}$$

As such, it is not a continuous distribution but becomes roughly continuous-looking only after coarse graining with cells C_i in $\mathscr{X}_1$ that are not too small (so that the occupation numbers are large enough), and it becomes a really continous distribution only after taking a limit in which the cells shrink to size 0 while $N \to \infty$ fast enough for the occupation numbers to become very large.

- The marginal distribution starts from a distribution ρ on phase space $\mathscr{X}$ and is obtained by integrating over the positions and velocities of $N - 1$ particles (and perhaps averaging over the number i of the particle not integrated out, if ρ was not permutation invariant to begin with). The marginal distribution can also be thought of as the average of the (exact) empirical distribution: the empirical distribution f_{emp}^X associated with $X \in \mathscr{X}$ becomes a continuous function when X is averaged over using a continuous ρ.

For example, Kac (1956, p. 1) wrote:

"$f(r, v)dr\, dv$ is the average number of molecules in $dr\, dv$"

whereas Boltzmann (1898, Section 3 p. 36) wrote:

"let $f(\xi, \eta, \zeta, t)d\xi\, d\eta\, d\zeta$ [...] be the number of m-molecules whose velocity components in the three coordinate directions lie between the limits ξ and $\xi + d\xi$, η and $\eta + d\eta$, ζ and $\zeta + d\zeta$[.]"

Note that Kac wrote "average number" and Boltzmann wrote "number": For Kac, f was the marginal and for Boltzmann, the empirical distribution.[6]

Of course, the (coarse-grained) empirical distribution is a function of the phase point X, and so is any functional of it, such as H; thus, the empirical distribution can serve the role of the macro state ν, and H that of Boltzmann entropy. This is not possible for the marginal distribution.

---

[6]That is also why Boltzmann normalized f so that $\int_{\mathscr{X}_1} f = N$, not $\int_{\mathscr{X}_1} f = 1$: a marginal of a probability distribution would automatically be normalized to 1, not N, but if f means the empirical density, then it is natural to take it to mean the density of particles in $\mathscr{X}_1$, which is normalized to N, not 1.

So why would anybody want the marginal distribution? Kac aimed at a rigorous derivation of the Boltzmann equation in whatever context long before Lanford's theorem, and saw better chances for a rigorous proof if he assumed collisions to occur at random times at exactly the rate given by Boltzmann's hypothesis of molecular chaos. This setup replaces the time evolution in phase space (or rather, since Kac dropped the positions, in $3N$-dimensional velocity space) by a stochastic process, in fact Markov jump process. (By the way, as a consequence, any density ρ on $3N$-space tends to get wider over time, and its Gibbs entropy increases, contrary to the Hamiltonian evolution.) So the mathematician's aim of finding statements that are easier to prove leads in a different direction than the aim of discussing the mechanism of entropy increase in nature.

Another thought that may lead authors to the marginal distribution is that $f(\boldsymbol{q}, \boldsymbol{v}) \, d^3\boldsymbol{q} \, d^3\boldsymbol{v}$ certainly cannot be an integer but must be an infinitesimal, so it cannot be the number of particles in $d^3\boldsymbol{q} \, d^3\boldsymbol{v}$ but must be the *average* number of particles. Of course, this thought neglects the idea that as long as N is finite, also the cells C_i should be kept of finite size and not too small, and the correct statement is that $f_i \, \mathrm{vol} \, C_i$ is the number of particles in C_i (or, depending on the normalization of f, N^{-1} times the number of particles); when followers of Boltzmann express the volume of C_i as $d^3\boldsymbol{q} \, d^3\boldsymbol{v}$, they merely express themselves loosely.

14.5.7. *The past hypothesis*

Lanford's theorem has implications also for negative t: For most phase points in Γ_f, the Boltzmann equation also applies in the other time direction, so that entropy increases in both time directions! (See Fig. 14.3.) That is, before time 0 the Boltzmann equation applies with the sign of t reversed.

It is generally the case, not just for a dilute gas of hard spheres, that entropy increases in both time directions for most phase points in Γ_ν. This fact leads to the following worry: If Lanford's theorem, or the statement (14.34), persuaded us to expect that entropy increases after the time we chose to call $t = 0$, should it not persuade us just as much to expect that entropy decreases before $t = 0$? But this

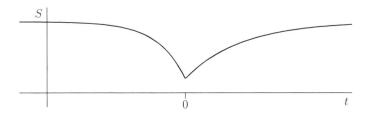

Fig. 14.3.   For most phase points in Γ_ν, $\nu \neq$ eq, entropy increases in both time directions, albeit not necessarily at the same rate.

decrease does not happen. Does that mean we were unjustified in expecting from Lanford's theorem or (14.34) that entropy increases after $t = 0$?

Here is a variant of this worry in terms of explanation. Statement (14.34) may suggest that the explanation for the increase of entropy after $t = 0$ is that this happens for typical phase points $X(0)$. But if we know that entropy was lower before $t = 0$ than at $t = 0$, then $X(0)$ was not typical. This undercuts the explanation considered for the behavior after $t = 0$.

Here is the resolution of this worry. The assumption really made by Boltzmann's followers is not that the phase point $X(0)$ at the beginning of an experiment is typical in Γ_ν but the following:

> **Past hypothesis.** The phase point of the universe at the initial time T_0 of the universe (presumably the big bang) is typical in its macro set Γ_{ν_0}, where ν_0 has very low entropy. $\qquad$ (14.49)

Here, "typical" means that, in relevant ways, it behaves like most points in Γ_{ν_0}. The ways relevant here are features of the macro history of the universe shared by most points in Γ_{ν_0}.

Given that entropy keeps increasing, the initial macro state must be one of extremely low entropy; one estimate (Penrose, 1989) yields 10^{123} Joule/Kelvin less than the thermal equilibrium entropy at the same energy; thus, Γ_{ν_0} must have extremely small volume compared to the relevant energy shell. All the same, we do not know very clearly what ν_0 actually is; one proposal is known as the Weyl curvature hypothesis (Penrose, 1979).

A related worry at this point may arise from the observation (for any macroscopic system or specifically a hard sphere gas as considered for the Boltzmann equation) that if the time evolution Φ_t of Γ_ν lies (say) in a macro set $\Gamma_{\nu'}$ of much greater volume, then phase points $X(t)$ coming from $X(0) \in \Gamma_\nu$ would be *atypical* in $\Gamma_{\nu'}$. So if the prediction of entropy increase after t was based on the assumption that $X(t)$ be *typical* in $\Gamma_{\nu'}$, then it could not be applied to $X(0) \in \Gamma_\nu$. So why should entropy still increase at $t > 0$? Because Lanford's theorem says so — at least until $\bar{t}/5$. But even after that time, it is plausible that the Boltzmann equation continues to be valid (and therefore entropy continues to increase) because it is plausible that the number of upcoming collisions of each type agrees with the value specified by the hypothesis of molecular chaos (see, e.g., Lebowitz, 2008). That is because it is plausible that, for typical $X(0) \in \Gamma_\nu$, $X(t)$ contains very special correlations in the exact positions and velocities of all particles concerning the collisions before t, but not concerning those after t. Likewise, we would expect for a general macroscopic system, unless the dynamics (as given by the Hamiltonian) is specially contrived for the partition $\{\Gamma_\nu\}$, that $X(t) \in \Gamma_{\nu'}$ coming from a typical $X(0) \in \Gamma_\nu$ behaves towards the future, but of course not towards the past, like a typical point in $\Gamma_{\nu'}$, as stated by the second law (14.34). The whole reasoning does not change significantly if $\Phi_t(\Gamma_\nu)$ is distributed over *several* $\Gamma_{\nu'_1}, \ldots, \Gamma_{\nu'_\ell}$ of much greater volume, instead of being contained in one $\Gamma_{\nu'}$.

Putting this consideration together with the past hypothesis, we are led to expect that the Boltzmann entropy of the universe keeps increasing (except for occasional insignificant entropy valleys) to this day, and further until the universe reaches thermal equilibrium. As a consequence for our present-day experiments:

> *Development conjecture.* Given the past hypothesis, an isolated system that, at a time t_0 before thermal equilibrium of the universe, has macro state ν appears macroscopically in the future, but not the past, of t_0 like a system that at time t_0 is in a typical micro state compatible with ν.   (14.50)

This statement follows from Lanford's theorem for times up to $\bar{t}/5$, but otherwise has not been proven mathematically; it summarizes the presumed implications of the past hypothesis (i.e., of the low entropy initial state of the universe) to applications of statistical mechanics. For a dilute gas, it predicts that its macro state will evolve according to the Boltzmann equation in the future, but not the past, of t_0, as long as it is isolated. It also predicts that heat will not flow from the cooler body to the hotter, and that a given macroscopic object will not spontaneously fly into the air although the laws of mechanics would allow that all the momenta of the thermal motion at some time all point upwards.

By the way, the development conjecture allows us to make sense of Tolman's (1938, Section 23, p. 60) "hypothesis of equal a priori probabilities," which asserts

> "that the phase point for a given system is just as likely to be in one region of the phase space as in any other region of the same extent which corresponds equally well with what knowledge we do have as to the condition of the system."

That sounds like X is always uniformly distributed over the Γ_ν containing X, but that statement is blatantly inconsistent, as it cannot be true at two times, given that $\Phi_t(\Gamma_\nu) \neq \Gamma_{\nu'}$ for any ν'. But the subtly different statement (14.50) is consistent, and (14.50) is what Tolman should have written.

The past hypothesis brings out clearly that the origin of the thermodynamic arrow of time lies in a special property (of lying in Γ_{ν_0}) of the physical state of the matter at the time T_0 of the big bang, and not in the observers' knowledge or their way of considering the world. The past hypothesis is the one crucial assumption we make in addition to the dynamical laws of classical mechanics. The past hypothesis may well have the status of a law of physics — not a dynamical law but a law selecting a set of admissible histories among the solutions of the dynamical laws. As Feynman (1965, p. 116) wrote:

> "Therefore I think it is necessary to add to the physical laws the hypothesis that in the past the universe was more

> ordered, in the technical sense, than it is today — I think
> this is the additional statement that is needed to make
> sense, and to make an understanding of the irreversibility."

Making the past hypothesis explicit in the form of (14.49) or a similar one also enables us to understand the question whether the past hypothesis could be *explained*. Boltzmann (1898, Section 90) suggested tentatively that the explanation might be a giant fluctuation out of thermal equilibrium, an explanation later refuted by Eddington (1931) and Feynman (1965; 1995). (Another criticism of this explanation put forward by Popper (1976, Section 35) is without merit.) Some explanations of the past hypothesis have actually been proposed in highly idealized models (Carroll, 2010; Barbour *et al.*, 2013, 2014, 2015; Goldstein *et al.*, 2016); it remains to be seen whether they extend to more adequate cosmological models.

14.5.8. *Boltzmann entropy is not always H*

Jaynes (1965) wrote that

> "the Boltzmann H yields an "entropy" that is in error by
> a nonnegligible amount whenever interparticle forces affect
> thermodynamic properties."

It is correct that the H functional represents (up to a factor $-k$) the Boltzmann entropy (14.2) only for a gas of non-interacting or weakly interacting molecules. Let us briefly explain why; a more extensive discussion is given in (Goldstein and Lebowitz, 2004).

As pointed out in Section 14.5.1, the coarse grained energy (14.28) should be one of the macro variables, so that the partition into macro sets Γ_ν provides a partition of the energy shell. If interaction cannot be ignored, then the H functional does not correspond to the Boltzmann entropy, since restriction to the energy shell is not taken into account by H. When interaction can be ignored there is only kinetic energy, so the Boltzmann macro states based on the empirical distribution alone determine the energy and hence the H functional corresponds to the Boltzmann entropy.

14.6. Gibbs entropy as fuzzy Boltzmann entropy

This section is about another connection between Gibbs and Boltzmann entropy that is not usually made explicit in the literature; it involves interpreting the Gibbs entropy in an individualist spirit as a variant of the Boltzmann entropy for "fuzzy" macro sets. That is, we now describe a kind of Gibbs entropy that is not an ensemblist entropy (as the Gibbs entropy usually is) but an individualist entropy, which is better. This possibility is not captured by any of the options (a)–(c) of Section 14.1.2.

By a fuzzy macro set, we mean using functions $\gamma_\nu(x) \geq 0$ instead of sets Γ_ν as expressions of a macro state ν: some phase points x look a lot like ν, others less so, and $\gamma_\nu(x)$ quantifies how much. The point here is to get rid of the sharp boundaries between the sets Γ_ν shown in Fig. 14.1, as the boundaries are artificial and somewhat arbitrary anyway. A partition into sets Γ_ν is still contained in the new framework as a special case by taking γ_ν to be the indicator function of Γ_ν, $\gamma_\nu = 1_{\Gamma_\nu}$, but we now also allow continuous functions γ_ν. One advantage may be to obtain simpler expressions for the characterization of ν since we avoid drawing boundaries. The condition that the Γ_ν form a partition of $\mathscr{X}$ can be replaced by the condition that

$$\sum_\nu \gamma_\nu(x) = 1 \quad \forall x \in \mathscr{X}. \tag{14.51}$$

Another advantage of this framework is that we can allow without further ado that ν is a continuous variable, by replacing (14.51) with its continuum version

$$\int d\nu \, \gamma_\nu(x) = 1 \quad \forall x \in \mathscr{X}. \tag{14.52}$$

It will sometimes be desirable to normalize the function γ_ν so its integral becomes 1; we write γ_ν^1 for the normalized function,

$$\gamma_\nu^1(x) = \frac{\gamma_\nu(x)}{\|\gamma_\nu\|_1} \quad \text{with} \quad \|\gamma_\nu\|_1 := \int_{\mathscr{X}} dx' \, \gamma_\nu(x'). \tag{14.53}$$

So what would be the appropriate generalization of the Boltzmann entropy to a fuzzy macro state? It should be k times the log of the volume over which γ_ν^1 is effectively distributed — in other words, the Gibbs entropy,

$$S_{\mathrm{B}}(\nu) := S_{\mathrm{G}}(\gamma_\nu^1). \qquad (14.54)$$

Now fix a phase point x. Since x is now not uniquely associated with a macro state, it is not clear what $S_{\mathrm{B}}(x)$ should be. In view of (14.52), one might define $S_{\mathrm{B}}(x)$ to be the average $\int d\nu\, S_{\mathrm{B}}(\nu)\, \gamma_\nu(x)$. Be that as it may, if the choice of macro states γ_ν is reasonable, one would expect that different νs for which $\gamma_\nu(x)$ is significantly non-zero have similar values of $S_{\mathrm{B}}(\nu)$, except perhaps for a small set of exceptional xs.

Another advantage of fuzzy macro states is that they sometimes factorize in a more convenient way. Here is an example. To begin with, sometimes, when we are only interested in thermal equilibrium macro states, we may want to drop non-equilibrium macro states and replace the equilibrium set Γ_{eq} in an energy shell $\mathscr{X}_{(E-\Delta E,E]}$ by the full energy shell, thereby accepting that we attribute wildly inappropriate νs (and S_{B}s) to a few xs. As a second step, the canonical distribution

$$\rho_{\mathrm{can}}(\beta, x) = \frac{1}{Z} e^{-\beta H(x)} \qquad (14.55)$$

is strongly concentrated in a very narrow range of energies, a fact well known as "equivalence of ensembles" (between the micro-canonical and the canonical ensemble). Let us take $\nu = \beta$ and $\gamma_\nu^1(x) = \rho_{\mathrm{can}}(\beta, x)$ as a continuous family of fuzzy macro states. As a third step, consider a system consisting of two non-interacting subsystems, $\mathscr{S} = \mathscr{S}_1 \cup \mathscr{S}_2$, so $\mathscr{X} = \mathscr{X}_1 \times \mathscr{X}_2$, $x = (x_1, x_2)$ and $H(x) = H_1(x_1) + H_2(x_2)$. Then γ_β^1 factorizes,

$$\gamma_\beta^1(x) = \gamma_{1,\beta}^1(x_1)\, \gamma_{2,\beta}^1(x_2), \qquad (14.56)$$

whereas the energy shells do not,

$$\mathscr{X}_{(E-\Delta E,E]} \neq \mathscr{X}_{1,(E_1-\Delta E_1,E_1]} \times \mathscr{X}_{2,(E_2-\Delta E_2,E_2]}, \qquad (14.57)$$

because a prescribed total energy E can be obtained as a sum $E_1 + E_2$ for very different values of E_1 through suitable choice of E_2, corresponding to different macro states for $\mathscr{S}_1$ and $\mathscr{S}_2$. One particular splitting $E = E_1 + E_2$ will have the overwhelming majority of phase space volume in $\mathscr{X}_{(E-\Delta E, E]}$; it is the splitting that maximizes $S_1(E_1) + S_2(E_2)$ under the constraint $E_1 + E_2 = E$, and at the same time the one corresponding to the same β value, i.e., with E_i the expectation of $H_i(x_i)$ under the distribution $\gamma^1_{i,\beta}$ $(i = 1, 2)$. So equality in (14.57) fails only by a small amount, in the sense that the symmetric difference set between the left and right-hand sides has small volume compared to the sets themselves. In short, in this situation, the use of sets Γ_ν forces us to consider certain non-equilibrium macro states, and if we prefer to introduce only equilibrium macro states, then the use of fuzzy macro states γ_ν is convenient.

The quantum analog of fuzzy macro states consists of, instead of the orthogonal decomposition (14.11) of Hilbert space, a POVM (positive-operator-valued measure) $\hat{G}$ on the set M of macro states. That is, we replace the projection $\hat{P}_\nu$ to the macro space $\mathscr{H}_\nu$ by a positive operator $\hat{G}_\nu$ with spectrum in $[0, 1]$ such that $\sum_\nu \hat{G}_\nu = \hat{I}$, where $\hat{I}$ is the identity operator on $\mathscr{H}$. The eigenvalues of $\hat{G}_\nu$ would then express how much the corresponding eigendirection looks like ν. Set $\hat{G}^1_\nu := (\operatorname{tr} \hat{G}_\nu)^{-1} \hat{G}_\nu$; then $\hat{G}^1_\nu$ is a density matrix, and the corresponding entropy value would be

$$S_{\mathrm{qB}}(\nu) := S_{\mathrm{vN}}(\hat{G}^1_\nu). \tag{14.58}$$

As before when using the $\mathscr{H}_\nu$, a quantum state ψ may be associated with several very different νs. As discussed around (14.16), this problem presumably disappears in Bohmian mechanics and collapse theories.

14.7. The status of ensembles

If we use the Boltzmann entropy, then the question of what the ρ in the Gibbs entropy means does not come up. But a closely related question remains: What is the meaning of the Gibbs ensembles (the micro-canonical, the canonical, the grand-canonical) in Boltzmann's "individualist" approach? This is the topic of the present section (see also Goldstein, 2019).

By way of introduction to this section, we can mention that Wallace, an ensemblist, feels the force of arguments against subjective entropy but thinks that there is no alternative. He wrote (Wallace, 2019, Section 10):

> "It will be objected by (e.g.) Albert and Callender that when we say "my coffee will almost certainly cool to room temperature if I leave it" we are saying something objective about the world, not something about my beliefs. I agree, as it happens; that just tells us that the probabilities of statistical mechanics cannot be interpreted epistemically. And then, of course, it is a mystery how they can be interpreted, given that the underlying dynamics is deterministic [. . .]. But (on pain of rejecting a huge amount of solid empirical science [. . .]) some such interpretation must be available."

The considerations of this section may be helpful here.

It is natural to call any measure (or density function ρ) that is normalized to 1 a "probability distribution." But now we need to distinguish more carefully between different roles that such a measure may play:

(i) frequency in repeated preparation: The outcome of an experiment is unpredictable, but will follow the distribution ρ if we repeat the experiment over and over. (This kind of probability could also be called *genuine probability* or *probability in the narrow sense*.)

(ii) degree of belonging: As in Section 14.6, ρ could represent a fuzzy set; then $\rho(x)$ indicates how strongly x belongs to this fuzzy set.

(iii) typicality: This is how the flat distribution over a certain Γ_ν enters the past hypothesis (14.49) and the development conjecture (14.50).

14.7.1. *Typicality*

A bit more elucidation of the concept of typicality may be useful here, as the difference between typicality and genuine probability is subtle. A feature or behavior is said to be *typical* in a set S if it occurs for most (i.e., for the overwhelming majority of) elements of S. Let us elaborate on this by means of an example. The digits of π, 314159265358979..., look very much like a random sequence although there is nothing random about the number π, as it is uniquely defined and thus fully determined. Also, there is no way of "repeating the experiment" that would yield a different sequence of digits; we can only do statistics about the digits in this one sequence. We can set up reasonable criteria for whether a given sequence (finite or infinite) "looks random," such as whether the relative frequency of each digits is, within suitable tolerances, $1/10$, and that of each k-digit subsequence $1/10^k$ (for all k much smaller than the length of the sequence). Then the digits of π will presumably pass the criteria, thereby illustrating the fine but relevant distinction between "looking random" and "being random."

Looking random is an instance of typical behavior, as most sequences of digits of given length look random and most numbers between 3 and 4 have random-looking decimal expansions — "most" relative to the uniform distribution, also known as Lebesgue measure. So with respect to a certain behavior (say, the frequencies of k-digit subsequences), π presumably is like most numbers.

Similarly, with respect to another sort of behavior (say, the future macroscopic history of the universe), the initial phase point of the universe is presumably like most points in Γ_{ν_0}, as demanded by the past hypothesis (14.49); and according to the development conjecture (14.50), the phase point of a closed system now is like most points in its macro set.

In contrast, when considering a random experiment in probability theory, we usually imagine that we can repeat the experiment, with relative frequencies in agreement with the distribution ρ; that is clearly impossible for the universe as a whole (or, if it were possible, irrelevant, because we are not concerned with what happens in other universes). But even if an experiment, such as the preparation of a macroscopic body (say, a gas), can be repeated a hundred or a thousand times, the frequency distribution of the sample points (i.e., the empirical distribution (14.48)) consists of a mere 10^3 delta peaks and thus is far from resembling a continuous distribution in a phase space of 10^{23} dimensions, so we do not have much basis for claiming that it is essentially the same as the micro-canonical or the canonical or any other common ensemble ρ. On the other hand, even a single sample point could meaningfully be said to be typical with respect to a certain high-dimensional distribution ρ (and a certain concept of which kinds of "behavior" are being considered). What we are getting at is that Gibbs's ensembles are best understood as measures *of typicality, not of genuine probability.*

For example, there is a subtle difference between the Maxwell–Boltzmann distribution and the canonical distribution. The former is the empirical distribution of (say) 10^{23} points in 6-dimensional 1-particle phase space $\mathscr{X}_1$, while the latter is a distribution of a single or a few points in high-dimensional phase space. The former is genuine probability, the latter typicality. This contrast is also related to the contrast between empirical and marginal distribution: On the one hand, the canonical distribution of a system $\mathscr{S}_1$ is often derived as a marginal distribution of the micro-canonical distribution of an even bigger system $\mathscr{S}_1 \cup \mathscr{S}_2$, and a marginal of a typicality distribution is still a typicality distribution. On the other hand, the Maxwell–Boltzmann distribution is the empirical distribution that arises from most phase points relative to the micro-canonical distribution. If ρ-most phase points lead to the empirical distribution f, then that explains why we can regard f as a *genuine probability* distribution, in the sense of being objective.

14.7.2. *Equivalence of ensembles and typicality*

It is not surprising that typicality and genuine probability can easily be conflated, and that various conundrums can arise from that. For example, the "mentaculus" of Albert and Loewer (Albert, 2015; Loewer, 2019) is a view based on understanding the past hypothesis (14.49) with genuine probability instead of typicality. That is, in this view, one considers the uniform distribution over the initial macro set Γ_{ν_0} and regards it as defining "the" probability distribution of the micro history $t \mapsto X(t)$ of the universe, in particular as assigning a precisely defined value in $[0, 1]$ as the probability to every conceivable event; for example, the event that intelligent life forms evolve on Earth and travel to the moon.

In contrast, the typicality approach used in (14.49) does not require or claim that the initial phase point $X(T_0)$ of the universe is random, merely that it looks random (just as π merely looks random). Furthermore, the typicality view does not require a uniquely selected measure of typicality, just as π is presumably typical relative to several measures: the uniform distribution on the interval $[3, 4]$ or on $[0, 10]$, or some Gaussian distribution. We elaborate on this aspect in the subsequent paragraphs, again through contrast with Albert and Loewer.

Loewer once expressed (personal communication, 2018) that different measures $\rho_1 \neq \rho_2$ for the initial phase point of the universe would necessarily lead to different observable consequences. That is actually not right; let us explain. To begin with, one may think that ρ_1 and ρ_2 must lead to different observable consequences if one *defines* (following the mentaculus) the probability of any event E at time t as $P(E) = \int_{\Phi_t^{-1}(E)} \rho_i$. However, a definition alone does not ensure that the quantity $P(E)$ has empirical significance; the question arises whether observers inside the universe can determine $P(E)$ from their observations. More basically, we need to ask whether observers can determine from their observations which of ρ_1 and ρ_2 was used as the initial distribution. To this end, let us imagine that the initial phase point $X(T_0)$ of the universe gets chosen truly randomly with distribution either ρ_1 or ρ_2. Now the same point $X(T_0)$ may be compatible with both distributions ρ_1, ρ_2 if they overlap appropriately. Since all observations must be made in the one history arising from $X(T_0)$, and since the universe cannot be re-run with an independent initial phase point, it is impossible to decide empirically whether ρ_1 or ρ_2 was the distribution according to which $X(T_0)$ was chosen. Further thought in this direction shows that observers can only reliably distinguish between ρ_1 and ρ_2 if they can observe an event that has probability near 1 for ρ_1 and near 0 for ρ_2 or vice versa; that is, they can only determine the typicality class of the initial distribution. (As a consequence for the mentaculus, its main work, which is to efficiently convey hard core facts about the world corresponding to actual patterns of events over space and time, is done via typicality, i.e., concerns only events with measure near 0 or near 1.)

In the typicality view, there does not have to be a fact about which distribution ρ is "the right" distribution of the initial phase point of the universe; that is because points typical relative to ρ_1 are also typical relative to ρ_2 and vice versa, if ρ_1 and ρ_2 are not too different. Put differently, in the typicality view, there is a type of equivalence of ensembles, parallel to the well known fact that one can often replace a micro-canonical ensemble by a canonical one without observable consequences. There is room for different choices of ρ, and this fits well with the fact that Γ_{ν_0} has boundaries with some degree of arbitrariness.

14.7.3. Ensemblist vs. individualist notions of thermal equilibrium

The two notions of Gibbs and Boltzmann entropy are parallel to two notions of thermal equilibrium. In the view that we call the ensemblist view, a system is in thermal equilibrium if and only if its phase point X is random with the appropriate distribution (such as the micro-canonical or the canonical distribution). In the individualist view, in contrast, a system is in thermal equilibrium (at a given energy) if and only if its phase point X lies in a certain subset Γ_{eq} of phase space.

The definition of this set may be quite complex; for example, according to the partition due to Boltzmann described in (14.41), the set Γ_{eq} contains those phase points for which the relative number of particles in each cell C_i of $\mathscr{X}_1$ agrees, up to tolerance Δf, with the content of C_i under the Maxwell–Boltzmann distribution. In general, the task of selecting the set Γ_{eq} has similarities with that of selecting the set S of those sequences of (large) length L of decimal digits that "look random." While a reasonable test of "looks random" can be designed (such as outlined in the first paragraph of Section 14.7.1 above), different scientists may well come up with criteria that do not exactly agree, and again there is some arbitrariness in where exactly to "draw the boundaries." Moreover, such a test or definition S is often unnecessary for practical purposes, even for designing and testing pseudo-random number generators. For practical purposes, it usually suffices to specify the distribution (here, the uniform distribution over all sequences of length L) with the understanding that S should contain, in a reasonable sense, the typical points relative to that distribution. Likewise, we usually never go to the trouble of actually selecting Γ_{eq} and the other Γ_ν: it often suffices to imagine that they *could* be selected. Specifically, for thermal equilibrium, it often suffices to specify the distribution (here, the uniform distribution over the energy shell, that is, the micro-canonical distribution) with the understanding that Γ_{eq} should contain, in a reasonable sense, the typical points relative to that distribution (i.e., points that look macroscopically the way most points in $\mathscr{X}_{mc}$ do).

That is why, also for the individualist, specifying an ensemble ρ can be a convenient way of describing a thermal equilibrium state: It would basically specify the set Γ_{eq} as the set of those points x that look macroscopically like ρ-most points. For the same reasons as in Section 14.7.2, this does not depend sensitively on ρ: points typical relative to the micro-canonical ensemble near energy E are also typical relative to the canonical ensemble with mean energy E.

Let us turn to the ensemblist view. The basic problem with the ensemblist definition of thermal equilibrium is the same as with the Gibbs entropy: a system has an X but not a ρ. To say the least, it remains open what should be meant by ρ. Is it subjective? But whether or not a system is in thermal equilibrium is not subjective. Does that mean that in distant places with no observers around, no system can be in thermal equilibrium? That does not sound right.

What if ρ corresponded to a repeated preparation procedure? That is problematical as well. Suppose such a procedure produced a random phase point $X(0)$ that is uniformly distributed in some set M. Now let time evolve up to t; the resulting phase point $X(t)$ is uniformly distributed in $\Phi_t(M)$, the associated Gibbs entropy is constant, and so the distribution of $X(t)$ is far from thermal equilibrium. In other words, the system never approaches thermal equilibrium. That does not sound right. We will come back to this point in Section 14.8.

Moreover, thermal equilibrium should have something to do with physical behavior, which depends on X. Thermal equilibrium should mean that the temperature, let us say the energy per degree of freedom, is constant (within tolerances) throughout the volume of the system, and that is a property of X. Thermal equilibrium should mean that the values of macro variables, which are functions on phase space, are constant over time, and that is true of some X and not of others.

14.7.4. *Ergodicity and mixing*

Ergodic and mixing are two kinds of chaoticity properties of a dynamical system $\mathscr{Y}$ such as an energy surface in phase space. A measure-preserving dynamical system in $\mathscr{Y}$ is called ergodic if almost every trajectory spends time, in the long run, in all regions of $\mathscr{Y}$ according to their volumes. Equivalently, a system is ergodic if time averages almost surely equal phase averages. A measure-preserving dynamical system in $\mathscr{Y}$ is called mixing if for all subsets A, B of $\mathscr{Y}$, $\mathrm{vol}(A \cap \Phi_t(B))$ tends to $\mathrm{vol}(A)\,\mathrm{vol}(B)/\mathrm{vol}(\mathscr{Y})$ as $t \to \infty$. In other words, mixing means that for large t, $\Phi_t(B)$ will be deformed into all regions of phase space according to their volumes.

It is sometimes claimed (e.g., Mackey, 1989) that ergodicity or mixing are crucial for statistical mechanics. We would like to explain now why that is not so — with a few caveats; see also Cerino *et al.*, 2016.

- Mixing would seem relevant to the ensemblist approach to thermal equilibrium because mixing implies that the uniform distribution over $\Phi_t(B)$ converges setwise to the uniform distribution (Liouville measure) over the energy surface $\mathscr{Y}$ (and every density converges weakly to a constant) (Krylov, 1979; Ruelle, 2007). However, for every finite t, the Gibbs entropy has not increased, so some ensemblists are pushed to say that the entropy does not increase within finite time, but jumps at $t = \infty$ to the thermal equilibrium value. Thus, some ensemblists are pushed to regard thermal equilibrium as an idealization that never occurs in the real world. That does not seem like an attractive option.

- The fact that the observed thermal equilibrium value a_{eq} of a macro variable A coincides with its micro-canonical average $\langle A \rangle$ has sometimes been explained through the following reasoning (e.g., Khinchin, 1941, Section 9): *Any macro measurement takes a time that is long compared to the time that collisions take or the time of free flight between collisions. Thus, it can be taken to be infinite on the micro time scale. Thus, the measured value is actually not the value of $A(X(t))$ at a particular t but rather its time average. By ergodicity, the time average is equal to the phase average, QED.*

  This reasoning is incorrect. In fact, ergodicity is neither necessary nor sufficient for $a_{\text{eq}} = \langle A \rangle$. Not necessary because A is nearly constant over the energy surface, so most phase points will yield a value close to $\langle A \rangle$ even if the motion is not ergodic. And not sufficient because the time needed for the phase point of an ergodic system to explore an energy surface is of the order 10^N years and thus much longer than the duration of the measurement. This point is also illustrated by the fact that in a macro system with non-uniform temperature, you can clearly measure unequal temperatures in different places with a thermometer faster than the temperature equilibrates.

- Still, ergodicity and similar properties are not completely unrelated to thermodynamic behavior. Small interactions between different parts of the system are needed to drive the system towards a dominant macro set (i.e., towards thermal equilibrium), and the same small interactions often also have the consequence of making the dynamics ergodic. That is, ergodicity is neither cause nor consequence of thermodynamic behavior, but the two have a common cause.

- In addition, and aside from thermodynamic behavior or entropy increase, mixing and similar properties play a role for the emergence of macroscopic randomness. For example, suppose a coin is tossed or a die is rolled; chaotic properties of the dynamics ensure that the outcome as a function of the initial condition varies over a very small scale on phase space, with the consequence that a probability density over the initial conditions that varies over a larger

scale will lead to a uniform probability distribution of the outcome. (For more extensive discussion see (Dürr and Teufel, 2009,
Section 4.1).)

- Boltzmann himself considered ergodicity, but not for the purposes
  above. He used it as a convenient assumption for estimating the
  mean free path of a gas molecule (Boltzmann, 1898, Section 10).

14.7.5. Remarks

Of course, Boltzmannian individualists use probability, too. There is
no problem with considering procedures that prepare random phase
points, or with using Gibbs ensembles as measures of typicality.

When arguing against the individualist view, Wallace (2019,
Section 4) mentioned as an example that transport coefficients (such
as thermal conductivity) can be computed using the two-time correlation function

$$C_{ij}(t) = \langle X_i(t) X_j(0) \rangle, \tag{14.59}$$

where $\langle \cdot \rangle$ means averaging while taking $X(0)$ as random with micro-
canonical distribution. Wallace wrote:

> "since $C(t)$ is an explicitly probabilistic quantity, it is not
> even defined on the Boltzmannian approach."

Actually, that is not correct. The individualist will be happy as soon
as it is shown that for most phase points in $\mathscr{X}_{\text{mc}}$, the rate of heat
conduction is practically constant and can be computed from $C(t)$
in the way considered.

Another objection of Wallace's (2019, Section 5) concerns a sys-
tem $\mathscr{S}$ for which (say) two macro sets, Γ_{ν_1} and Γ_{ν_2}, each comprise
nearly half of the volume of the energy shell, so that the thermal
equilibrium state is replaced by two macro states ν_1 and ν_2, as can
occur with a ferromagnet (see Footnote 5 in Section 14.5.1). Wallace
asked how an individualist can explain the empirical observation that
the system is, in the long run, equally likely to reach ν_1 or ν_2. To
begin with, if a preparation procedure leads to a random phase point
with distribution ρ, then $\int_{\Phi_t^{-1}(\Gamma_{\nu_1})} \rho$ is the probability of $\mathscr{S}$ being in

ν_1 at time t. As mentioned in Section 14.7.4 (penultimate bullet), a broad ρ will lead to approximately equal probabilities for ν_1 and ν_2. This leads to the question why practical procedures lead to broad ρs, and that comes from typicality as expressed in the development conjecture. Put differently, for a large number of identical ferromagnets it is typical that about half of them are in ν_1 and about half of them in ν_2.

14.8. Second law for the Gibbs entropy?

To the observation that the Gibbs entropy is constant under the Liouville evolution (14.6), different authors have reacted in different ways. We describe some reactions along with some comments of ours. (Of course, our basic reservations about the Gibbs entropy are those we expressed in Section 14.4.)

- Khinchin (1941) thought it was simply wrong to say that entropy increases in a closed system. He thought that the second law should merely assert that when a system is in thermal equilibrium and a constraint is lifted, then the new thermal equilibrium state has higher entropy than the previous one.

  With this attitude, one precludes the explanation of many phenomena.
- Ruelle (2007) expressed the view that entropy stays constant at every finite time but increases at $t = \infty$ due to mixing (see Section 14.7.4 above).

  It remains unclear how this view can be applied to any real-world situation, in which all times are finite.
- Some authors (e.g., Mackey, 1989) suggested considering a system $\mathscr{S}_1$ perturbed by external noise (in the same spirit Kemble, 1939). In the framework of closed Hamiltonian systems of finite size, this situation could be represented by interaction with another system $\mathscr{S}_2$. It can be argued that if $\mathscr{S}_2$ is large and if initially $\rho = \rho_1(x_1)\,\rho_2(x_2)$ with ρ_2 a thermal equilibrium distribution, then the marginal distribution

$$\rho_1^{\mathrm{marg}}(x_1, t) := \int_{\mathscr{X}_2} dx_2\, \rho(x_1, x_2, t) \tag{14.60}$$

will follow an evolution for which $S_G(\rho_1^{\mathrm{marg}})$ increases with time, and one might want to regard the latter quantity as the entropy of $\mathscr{S}_1$.

This attitude has some undesirable consequences: First, if the entropy of $\mathscr{S}_1$ increases and that of $\mathscr{S}_2$ increases then, since that of $\mathscr{S}_1 \cup \mathscr{S}_2$ is constant, entropy is not additive. Second, since it is not true of *all* ρ on $\mathscr{X}_1 \times \mathscr{X}_2$ that $S_G(\rho_1^{\mathrm{marg}})$ increases with time, the question arises for which ρ it will increase, and why real-world systems should be of that type.

• Some authors (e.g., Wallace, 2019; Tolman, 1938, Section 51) have considered a partition of (an energy shell in) phase space $\mathscr{X}$, such as the Γ_ν, and suggested coarse graining any ρ according this partition.[7] That is, the coarse-grained ρ is the function $P\rho$ whose value at $x \in \Gamma_\nu$ is the average of ρ over Γ_ν,

$$P\rho(x) = \frac{1}{\mathrm{vol}\,\Gamma_\nu} \int_{\Gamma_\nu} dy\, \rho(y). \tag{14.61}$$

P is a projection to the subspace of $L^1(\mathscr{X})$ of functions that are constant on every Γ_ν. The suggestion is that thermodynamic entropy S_t is given by

$$S_t := S_G(P\rho_t), \tag{14.62}$$

with ρ_t evolving according to the Liouville equation (14.6), and that this S_t tends to increase with time.

It is plausible that S_t indeed tends to increase, provided that the time evolution has good mixing properties (so that balls in phase space quickly become "spaghetti") and that the scale of variation of ρ_0 is not small; see also Falcioni *et al.* (2007). (The argument for the increase of (14.62) given by Tolman (1938) is without merit. After correctly showing (p. 169) that, for every distribution ρ,

$$S_G(P\rho) \geq S_G(\rho), \tag{14.63}$$

Tolman assumes that ρ_0 is uniform in each cell, i.e., $P\rho_0 = \rho_0$, and then points out that $S_G(P\rho_t) \geq S_G(P\rho_0)$ for all t. This is

---

[7]Tolman (1938) had in mind a partition into cells of equal size, but we will simply use the Γ_ν in the following.

true, but it is merely a trivial consequence of (14.63) and (14.5) owed to the special setup for which $P\rho_0 = \rho_0$, and not a general statement of entropy increase, as nothing follows about whether $S_G(P\rho_{t_2}) > S_G(P\rho_{t_1})$ for arbitrary $t_2 > t_1$.)

- Some authors have argued that the Liouville evolution (14.6) is not appropriate for evolving ρ.

  That depends on what ρ means. If we imagine a preparation procedure that produces a random phase point $X(0)$ with probability distribution ρ_0, then letting the system evolve up to time t will lead to a random phase point $X(t)$ with distribution ρ_t given by the Liouville evolution. However, if we think of ρ as representing an observer's belief or knowledge, the situation may be different. In the last two bullet points, we consider some alternative ways of evolution.

- Let L be the linear differential operator such that $L\rho$ is the right-hand side of the Liouville equation (14.6); then the Liouville evolution is $\rho_t = e^{Lt}\rho_0$. The proposal (14.62) could be re-expressed by saying that when using the Gibbs entropy, $S_t = S_G(\rho_t)$, the appropriate evolution for ρ should be

$$\rho_t = Pe^{Lt}\rho_0 \quad \text{for } t > 0. \tag{14.64}$$

- Instead of (14.64), some authors (Mackey, 1989) suggested that

$$\rho_t = \lim_{n\to\infty} (Pe^{Lt/n})^n \rho_0 \quad \text{for } t > 0. \tag{14.65}$$

This evolution corresponds to continually coarse graining ρ after each infinitesimal time step of Liouville evolution.

Then the Gibbs entropy $S_G(\rho_t)$ is indeed non-decreasing with t, in fact without exception, as a consequence of (14.63) and (14.5). If the macro evolution is deterministic, then it should agree with the Boltzmann entropy. A stochastic macro evolution would yield additional contributions to the Gibbs entropy not present in the Boltzmann entropy (contributions arising from spread over many macro sets), but these contributions are presumably tiny. Furthermore, it seems not implausible that ρ_t is often close to $Pe^{Lt}P\rho_0$, which implies that $S_G(\rho_t)$ is close to the expectation of $S_B(X_t)$ with X_0 random with distribution $P\rho_0$. However, as discussed in Section 14.4.1, one should distinguish between the entropy and the average entropy, so $S_G(\rho_t)$ is sometimes not quite the right quantity.

14.9. Further aspects of the quantum case

Much of what we have discussed in the classical setting re-appears mutatis mutandis in the quantum case, and we will focus now on the differences and specialties of the quantum case. We are considering a quantum system of a macroscopic number N (say, $>10^{23}$) of particles. Its Hilbert space can be taken to be finite-dimensional because we can take the systems we consider to be enclosed in a finite volume of 3-space, which leads to a discrete spectrum of the Hamiltonian and thus only finitely many eigenvalues in every finite energy interval. As a consequence, if we cut off high energies at an arbitrary level, or if we consider an energy shell $\mathcal{H}_{(E-\Delta E, E]}$, we are left with a finite-dimensional Hilbert space of relevant pure states.

14.9.1. Macro spaces

The orthogonal decomposition $\mathcal{H} = \oplus_\nu \mathcal{H}_\nu$ into macro spaces can be thought of as arising from macro observables as follows, following von Neumann (1929). If $\hat{A}_1, \ldots, \hat{A}_K$ are quantum observables (self-adjoint operators on $\mathcal{H}$) that should serve as macro observables, we will want to coarse grain them according to the resolution ΔM_j of macroscopic measurements by introducing $\hat{B}_j = [\hat{A}_j/\Delta M_j]\Delta M_j$ (with $[\cdot]$ again the nearest integer), thereby grouping nearby eigenvalues of $\hat{A}_j$ together into a single, highly degenerate eigenvalue of $\hat{B}_j$. Due to their macroscopic character, the commutators $[\hat{B}_j, \hat{B}_k]$ will be small; as von Neumann (1929) argued and Ogata (2013) justified, a little correction will make them commute exactly, $\hat{M}_j \approx \hat{B}_j$ and $[\hat{M}_j, \hat{M}_k] = 0$. Since a macro description ν of the system can be given by specifying the eigenvalues $(m_1, \ldots, m_K)$ of all $\hat{M}_1, \ldots, \hat{M}_K$, their joint eigenspaces are the macro spaces $\mathcal{H}_\nu$. They have dimensions of order $10^{10^{23}}$.

14.9.2. *Some historical approaches*

Pauli (1928) tried to derive entropy increase from quantum mechanics. In fact, he described several approaches. In Section 1 of his paper, he developed a modified Boltzmann equation that takes Fermi-Dirac and Bose-Einstein statistics into account and makes sense in the individualist view. In Section 2 he took an ensemblist attitude and suggested a novel dynamics in the form of a Markov chain on the macro states ν, for which he proved that the (analog of the) Gibbs entropy of the probability distribution increases with time. In Sections 3 and 4, he argued (not convincingly, to our minds) that the distribution over ν's provided by an evolving quantum state is, to a good degree of approximation, a Markov chain.

Returning to Pauli's Section 1, we note that his derivation of the modified Boltzmann equation is far less complete than Boltzmann's derivation of the Boltzmann equation from classical dynamics of hard spheres, and no valid theorem analogous to Lanford's has been proven for it, as far as we are aware. So Pauli's description has rather the status of a conjecture. According to it, the macro evolution $\nu(t)$ of a quantum gas with N particles is deterministic for large N (as with the Boltzmann equation), the expression for the entropy of ν agrees with the quantum Boltzmann entropy $S_{qB}(\nu)$ as in (14.12), and it increases with time as a mathematical consequence of the evolution equation for $\nu(t)$. This approach fits into the framework of an orthogonal decomposition $\mathscr{H} = \oplus_\nu \mathscr{H}_\nu$ that arises in a way analogous to Boltzmann's example (14.41) from an orthogonal decomposition of the 1-particle Hilbert space, $\mathscr{H}_1 = \oplus_i \mathscr{K}_i$, by taking ν to be the list of occupation numbers (within tolerances) of each $\mathscr{K}_i$ and $\mathscr{H}_\nu$ the corresponding subspaces of $\mathscr{H}$.

Von Neumann famously came up with the von Neumann entropy (14.10), the quantum analog of the Gibbs entropy. It is less known that in 1929 he defended a different formula because (von Neumann, 1929, Section 1.3)

> "[t]he expressions for entropy [of the form $-k \operatorname{tr} \hat{\rho} \log \hat{\rho}$] are not applicable here in the way they were intended, as they were computed from the perspective of an observer who can

> carry out all measurements that are possible in principle —
> i.e., regardless of whether they are macroscopic[.]"

We see here a mix of ensemblist and individualist thoughts in von Neumann's words. Be that as it may, the formula that von Neumann defended was, in our notation,

$$S(\psi) = -k \sum_\nu \|\hat{P}_\nu \psi\|^2 \log \frac{\|\hat{P}_\nu \psi\|^2}{\dim \mathscr{H}_\nu}. \tag{14.66}$$

Similar expressions were advocated recently by Safranek *et al.* (2019a, 2019b). Note that this expression has a stronger individualist character than $-k \operatorname{tr} \hat{\rho} \log \hat{\rho}$, as it ascribes a non-trivial entropy value also to a system in a pure state. The problem with the expression is that it tends to average different entropy values where no averaging is appropriate. For example, suppose that ψ has contributions from many but not too many macro spaces; since $\dim \mathscr{H}_\nu$ is very large, $\|\hat{P}_\nu \psi\|^2 \log \|\hat{P}_\nu \psi\|^2$ can be neglected in comparison, and then (14.66) is just the weighted average of the quantum Boltzmann entropies (14.12) of different macro states, weighted with $\|\hat{P}_\nu \psi\|^2$. But in many situations, such as Schrödinger's cat, we would not be inclined to regard this average value as the true entropy value valid in all cases. For comparison, we would also not be inclined to define the body temperature of Schrödinger's cat as the average over all contributions to the wave function — we would say that the quantum state is a superposition of states with significantly different body temperatures. Likewise, we should say that the quantum state is a superposition of states with significantly different entropies.

In the same paper, von Neumann (1929) proved two theorems that he called the "quantum H theorem" and the "quantum ergodic theorem." However, they are not at all quantum analogs of either the H theorem or the ergodic theorem. In fact, the two 1929 theorems are very similar to each other (almost reformulations of each other), and they assert that for a typical Hamiltonian or a typical choice of the decomposition $\mathscr{H} = \oplus_\nu \mathscr{H}_\nu$, all initial pure states ψ_0 evolve such that for most t in the long run,

$$\|\hat{P}_\nu \psi(t)\|^2 \approx \frac{\dim \mathscr{H}_\nu}{\dim \mathscr{H}}. \tag{14.67}$$

(See also Goldstein *et al.*, 2010b.) This is an interesting observation about the quantum evolution that has no classical analog; but it is far from establishing the increase of entropy in quantum mechanics.

14.9.3. *Entanglement entropy*

Let us come back to option (d) of Section 14.3.1 above: the possibility that the density matrix $\hat{\rho}$ in S_{vN} should be the system's reduced density matrix. This approach has the undesirable consequence that entropy is not extensive, i.e., the entropy of $\mathscr{S}_1 \cup \mathscr{S}_2$ is not the sum of the entropy of $\mathscr{S}_1$ and the entropy of $\mathscr{S}_2$; after all, often the entropies of $\mathscr{S}_1$ and $\mathscr{S}_2$ will each increase while that of $\mathscr{S}_1 \cup \mathscr{S}_2$ is constant if that is a closed system (because the evolution is unitary). Moreover, it is not true of *all* $\hat{\rho}$ on $\mathscr{H}_1 \otimes \mathscr{H}_2$ that $S_{vN}(\mathrm{tr}_2\,\hat{\rho})$ increases with time (think of time reversal).

A variant of option (d), and another widespread approach to defining the entropy of a macroscopic quantum system $\mathscr{S}$ with (say) a pure state Ψ, is to use the entanglement entropy as follows: Divide $\mathscr{S}$ into small subsystems $\mathscr{S}_1, \ldots, \mathscr{S}_M$, let $\hat{\rho}_j$ be the reduced density matrix of $\mathscr{S}_j$,

$$\hat{\rho}_j := \mathrm{tr}_1 \cdots \mathrm{tr}_{j-1}\,\mathrm{tr}_{j+1} \cdots \mathrm{tr}_M\,|\Psi\rangle\langle\Psi|, \qquad (14.68)$$

and add up the von Neumann entropies of $\hat{\rho}_j$,

$$S_{\mathrm{ent}} := -k \sum_{j=1}^{M} \mathrm{tr}(\hat{\rho}_j \log \hat{\rho}_j). \qquad (14.69)$$

Since the j-th term quantifies how strongly $\mathscr{S}_j$ is entangled with the rest of $\mathscr{S}$, it is known as the entanglement entropy of $\mathscr{S}_j$. We will simply call also their sum (14.69) the entanglement entropy.

In thermal equilibrium, the entanglement entropy S_{ent} actually agrees with the thermodynamic entropy. That is a consequence of microscopic thermal equilibrium as described in Section 14.3.3, and thus of canonical typicality (Gemmer *et al.*, 2004; Goldstein *et al.*, 2006; Popescu *et al.*, 2006): For most Ψ in an energy shell of $\mathscr{S}$, assuming that the $\mathscr{S}_j$ are not too large (Goldstein *et al.*, 2017b), $\hat{\rho}_j$ is approximately canonical, so its entanglement entropy agrees with its thermodynamic entropy, and extensivity implies the same for $\mathscr{S}$.

This argument also shows that for $\mathscr{S}$ in thermal equilibrium, S_{ent} does not depend much on how $\mathscr{S}$ is split into subsystems $\mathscr{S}_j$. On top of that, it also shows that S_{ent} still yields the correct value of entropy when $\mathscr{S}$ is in *local* thermal equilibrium, provided each $\mathscr{S}_j$ is so small as to be approximately in thermal equilibrium.

Many practical examples of non-equilibrium systems are in local thermal equilibrium. Still, for general non-equilibrium systems, S_{ent} may yield wrong values; an artificial example is provided by a product state $\Psi = \otimes_j \psi_j$ with each ψ_j an equilibrium state of $\mathscr{S}_j$, so $S_{\mathrm{ent}} = 0$ while the thermodynamic entropy, by extensivity, should have its equilibrium value. Artificial or not, this example at least forces us to say that S_{ent} is not correct for *all* Ψ; that S_{ent} is correct for *most* Ψ we knew already since most Ψ are in thermal equilibrium.

Another issue with S_{ent} arises when $\mathscr{S}$ is in a macroscopic superposition such as Schrödinger's cat. Then S_{ent} yields approximately the average of the entropy of a live cat and that of a dead cat, while the correct value would be *either* that of a live cat *or* that of a dead cat. In this situation, one may want to conditionalize on the case of (say) the cat being alive, and replace Ψ by the part of it corresponding to a live cat. Ultimately, this leads us to replace Ψ by its projection to one $\mathscr{H}_\nu$, and if we are so lucky that states in $\mathscr{H}_\nu$ are in local thermal equilibrium, then the conditionalized S_{ent} will agree with $S_{\mathrm{qB}}(\nu)$. So at the end of the day, S_{qB} seems to be our best choice as the fundamental definition of entropy in quantum mechanics.

14.9.4. *Increase of quantum Boltzmann entropy*

Just as the classical macro sets Γ_ν have vastly different volumes, the quantum macro spaces $\mathscr{H}_\nu$ have vastly different dimensions, and as phase space volume is conserved classically, dimensions of subspaces are conserved under unitary evolution.

It follows that if macro states ν follow an autonomous, deterministic evolution law $\nu \mapsto \nu_t$, then quantum Boltzmann entropy increases. In fact,

$$\text{if } e^{-i\hat{H}t}\mathcal{H}_\nu \subseteq \mathcal{H}_{\nu'}, \text{ then } S_{\text{qB}}(\nu') \geq S_{\text{qB}}(\nu). \tag{14.70}$$

More generally, not assuming that macro states evolve deterministically, already at this point it is perhaps not unreasonable to expect that in some sense the unitary evolution will carry a pure state ψ_t from smaller to larger subspaces $\mathcal{H}_\nu$. In fact, it was proven (Goldstein *et al.*, 2010a) that if one subspace has most of the dimensions in an energy shell, then ψ_t will sooner or later come very close to that subspace (i.e., reach thermal equilibrium) and stay close to it for an extraordinarily long time (i.e., for all practical purposes, never leave it).

Proposition 1 below is another observation in the direction of Conjecture 1, concerning only two times instead of an entire history and only two macro states ν, ν'. It is a quantum analog of the classical statement that if

$$\text{vol}\,\Gamma_{\nu'} \ll \text{vol}\,\Gamma_\nu, \tag{14.71}$$

that is, if $S_B(\nu')$ is less than $S_B(\nu)$ by an appreciable amount, then for any $t \neq 0$, most $X_0 \in \Gamma_\nu$ are such that $X_t \notin \Gamma_{\nu'}$.

Proposition 1. *Let $\hat{H}$ and $t \neq 0$ be arbitrary but fixed. If*

$$\dim \mathcal{H}_{\nu'} \ll \dim \mathcal{H}_\nu \tag{14.72}$$

(that is, if $S_{\text{qB}}(\nu')$ is less than $S_{\text{qB}}(\nu)$ by an appreciable amount), then for most $\psi_0 \in \mathbb{S}(\mathcal{H}_\nu)$,

$$\left\|\hat{P}_{\nu'}\psi_t\right\|^2 \ll 1. \tag{14.73}$$

Proof. Let $\hat{U} = e^{-i\hat{H}t}$, let $d_\nu := \dim \mathcal{H}_\nu$, and let Ψ be a random vector with uniform distribution on $\mathbb{S}(\mathcal{H}_\nu)$. Then

$$\mathbb{E}|\Psi\rangle\langle\Psi| = d_\nu^{-1}\hat{P}_\nu, \tag{14.74}$$

so

$$\mathbb{E}\|\hat{P}_{\nu'}\hat{U}\Psi\|^2 = \text{tr}(\hat{P}_{\nu'}\hat{U}d_\nu^{-1}\hat{P}_\nu\hat{U}^{-1}) \tag{14.75}$$

$$= d_\nu^{-1}\text{tr}(\hat{U}^{-1}\hat{P}_{\nu'}\hat{U}\hat{P}_\nu) \tag{14.76}$$

$$\leq d_\nu^{-1} \operatorname{tr}(\hat{U}^{-1} \hat{P}_{\nu'} \hat{U}) \qquad (14.77)$$

$$= d_\nu^{-1} \operatorname{tr}(\hat{P}_{\nu'}) \qquad (14.78)$$

$$= \frac{\dim \mathscr{H}_{\nu'}}{\dim \mathscr{H}_\nu} \ll 1. \qquad (14.79)$$

If a non-negative quantity is small on average, it must be small in most cases, so the proposition follows. □

Note that the proof only used that $\mathscr{H}_{\nu'}$ and $\mathscr{H}_\nu$ are two mutually orthogonal subspaces of suitable dimensions; as a consequence, the statement remains true if we replace $\mathscr{H}_{\nu'}$ by the sum of all macro spaces with dimension less than that of $\mathscr{H}_\nu$. That is, if $\mathscr{H}_\nu$ is much bigger than all of the smaller macro spaces together (a case that is not unrealistic), then it is atypical for entropy to be lower at time t than at time 0.

14.10. Conclusions

We have argued that entropy has nothing to do with the knowledge of observers. The Gibbs entropy (14.1) is an efficient tool for computing entropy values in thermal equilibrium when applied to the Gibbsian equilibrium ensembles ρ, but the fundamental definition of entropy is the Boltzmann entropy (14.2). We have discussed the status of the two notions of entropy and of the corresponding two notions of thermal equilibrium, the "ensemblist" and the "individualist" view. Gibbs's ensembles are very useful, in particular, as they allow the efficient computation of thermodynamic functions (Goldstein, 2019), but their role can only be understood in Boltzmann's individualist framework. We have also outlined an extension of Boltzmann's framework to quantum mechanics by formulating a definition of the quantum Boltzmann entropy and, as Conjecture 1, a statement of the second law of thermodynamics for it.

Acknowledgments

We thank Jean Bricmont and David Huse for valuable discussions. The work of JLL was supported by the AFOSR under award number

FA9500-16-1-0037. JLL thanks Stanislas Leibler for hospitality at the IAS.

References

Albert, D. (2015) *After Physics*. Cambridge, MA: Harvard University Press.

Allori, V., Goldstein, S., Tumulka, R., Zanghì, N. (2011) Many-worlds and Schrödinger's first quantum theory. *British Journal for the Philosophy of Science*, **62(1)**, 1–27. http://arxiv.org/abs/0903.2211.

Barbour, J., Koslowski, T., Mercati, F. (2013) A gravitational origin of the arrows of time. http://arxiv.org/abs/1310.5167.

Barbour, J., Koslowski, T., Mercati, F. (2014) Identification of a gravitational arrow of time. *Physical Review Letters*, **113**, 181101. http://arxiv.org/abs/1409.0917.

Barbour, J., Koslowski, T., Mercati, F. (2015) Entropy and the typicality of universes. http://arxiv.org/abs/1507.06498.

Bergmann, P. G., Lebowitz, J. L. (1955) New approach to nonequilibrium processes. *Physical Review*, **99**, 578.

Boltzmann, L. (1872) Weitere Studien über das Wärmegleichgewicht unter Gasmolekülen. *Sitzungsberichte der Akademie der Wissenschaften Wien*, **66**, 275–370.

Boltzmann, L. (1898) *Vorlesungen über Gastheorie*. Leipzig: Barth (Part I 1896, Part II 1898). English translation by S.G. Brush: *Lectures on Gas Theory*. Berkeley: University of California Press (1964).

Bricmont, J. (2019) Probabilistic explanations and the derivation of macroscopic laws. (In this volume.)

Callender, C. (1999) Reducing thermodynamics to statistical mechanics: The case of entropy. *Journal of Philosophy*, **96**, 348–373.

Carroll, S. M. (2010) *From Eternity to Here*. New York: Dutton.

Cercignani, C., Illner, R., Pulvirenti, M. (1994) *The Mathematical Theory of Dilute Gases*. Berlin: Springer-Verlag.

Cerino, L., Cecconi, F., Cencini, M., Vulpiani, A. (2016) The role of the number of degrees of freedom and chaos in macroscopic irreversibility. *Physica A: Statistical Mechanics and its Applications*, **442**, 486–497. http://arxiv.org/abs/1509.03823.

Clausius, R. (1865) Über verschiedene, für die Anwendung bequeme Formen der Hauptgleichungen der mechanischen Wärmetheorie. *Annalen der Physik und Chemie*, **125**, 353–400. English translation by J. Tyndall: On several convenient forms of the fundamental equations of the mechanical theory of heat. Pages 327–365 in R. Clausius: *The Mechanical Theory of Heat, with its Applications to the Steam Engine and to Physical Properties of Bodies*. London: van Voorst (1867).

De Roeck, W., Maes, C., Netočný, K. (2006) Quantum macrostates, equivalence of ensembles and an H-theorem. *Journal of Mathematical Physics*, **47**, 073303. http://arxiv.org/abs/math-ph/0601027.

Derrida, B. (2007) Non-equilibrium steady states: Fluctuations and large deviations of the density and of the current. *Journal of Statistical Mechanics*, **2007**, P07023. http://arxiv.org/abs/cond-mat/0703762.

Dürr, D., Goldstein, S., Tumulka, R., Zanghì, N. (2005) On the role of density matrices in Bohmian mechanics. *Foundations of Physics*, **35**, 449–467. http://arxiv.org/abs/quant-ph/0311127.

Dürr, D., Teufel, S. (2009) *Bohmian mechanics*. Berlin: Springer.

Eddington, A. S. (1931) The end of the world: From the standpoint of mathematical physics. *Nature*, **127**, 447–453. Reprinted on page 406 in D. R. Danielson (editor): *The Book of the Cosmos: Imagining the Universe from Heraclitus to Hawking*. Cambridge, MA: Perseus (2000).

Ehrenfest P., Ehrenfest, T. (1911) Begriffliche Grundlagen der statistischen Auffassung in der Mechanik. In *Enzyklopädie der Mathematischen Wissenschaften*, Vol. IV-4, Art. 32. English translation by M. J. Moravcsik in P. Ehrenfest and T. Ehrenfest: *The Conceptual Foundations of the Statistical Approach in Mechanics*. Cornell University Press (1959).

Einstein, A. (1914) Beiträge zur Quantentheorie. *Deutsche Physikalische Gesellschaft. Verhandlungen*, **16**, 820–828. English translation in *The Collected Papers of Albert Einstein*, Vol. 6, pages 20–26. Princeton University Press (1996).

Everett, H. (1957) Relative state formulation of quantum mechanics. *Reviews of Modern Physics*, **29**, 454–462.

Falcioni, M., Palatella, L., Pigolotti, S., Rondoni, L., Vulpiani, A. (2007) Initial growth of Boltzmann entropy and chaos in a large assembly of weakly interacting systems. *Physica A: Statistical Mechanics and its Applications*, **385**, 170–184. http://arxiv.org/abs/nlin/0507038.

Feynman, R. P. (1965) *The Character of Physical Law*. M.I.T. Press.

Feynman, R. P. (1995) *Lectures on Gravitation*. Edited by F. B. Morínigo, W. G. Wagner, and B. Hatfield. Westview Press.

Gallavotti, G. (2003) Nonequilibrium thermodynamics? http://arxiv.org/abs/cond-mat/0301172.

Garrido, P., Goldstein, S., Lebowitz, J. L. (2004) The Boltzmann entropy for dense fluids not in local equilibrium. *Physical Review Letters*, **92**, 050602. http://arxiv.org/abs/cond-mat/0310575.

Gemmer, J., Mahler, G., Michel, M. (2004) *Quantum Thermodynamics: Emergence of Thermodynamic Behavior within Composite Quantum Systems*. Lecture Notes in Physics, **657**. Berlin: Springer.

Gharibyan, H., Tegmark, M. (2014) Sharpening the second law of thermodynamics with the quantum Bayes' theorem. *Physical Review E*, **90**, 032125. http://arxiv.org/abs/1309.7349.

Gibbs, J. W. (1902) *Elementary Principles in Statistical Mechanics*. New York: Scribner's Sons.

Gogolin, C., Eisert, J. (2016) Equilibration, thermalisation, and the emergence of statistical mechanics in closed quantum systems. *Reports on Progress in Physics*, **79**, 056001. http://arxiv.org/abs/1503.07538.

Goldstein, S. (1998) Quantum theory without observers. *Physics Today*, Part One: March, 42–46. Part Two: April, 38–42.

Goldstein, S. (2001) Boltzmann's Approach to Statistical Mechanics. Pages 39–54 in J. Bricmont, D. Dürr, M. C. Galavotti, G. C. Ghirardi, F. Petruccione, and N. Zanghì (editors): *Chance in Physics: Foundations and Perspectives*. Lecture Notes in Physics **574**. Heidelberg: Springer-Verlag. http://arxiv.org/abs/cond-mat/0105242.

Goldstein, S. (2019) Individualist and Ensemblist Approaches to the Foundations of Statistical Mechanics. *The Monist*, **102**, 439–457.

Goldstein, S., Huse, D. A., Lebowitz, J. L., Sartori, P. (2017a) On the nonequilibrium entropy of large and small systems. Pages 581–596 in G. Giacomin, S. Olla, E. Saada, H. Spohn, and G. Stoltz (editors): *Stochastic Dynamics Out of Equilibrium*. **Springer Proceedings in Mathematics and Statistics**. Heidelberg: Springer-Verlag. http://arxiv.org/abs/1712.08961.

Goldstein, S., Huse, D. A., Lebowitz, J. L., Tumulka, R. (2017b) Macroscopic and microscopic thermal equilibrium. *Annalen der Physik*, **529**, 1600301. http://arxiv.org/abs/1610.02312.

Goldstein, S., Huse, D. A. Lebowitz, J. L., Tumulka, R. (2015) Thermal equilibrium of a macroscopic quantum system in a pure state. *Physical Review Letters*, **115**, 100402. http://arxiv.org/abs/1506.07494.

Goldstein, S., Lebowitz, J. L. (2004) On the (Boltzmann) entropy of nonequilibrium systems. *Physica D*, **193**, 53–66. http://arxiv.org/abs/cond-mat/0304251.

Goldstein, S., Lebowitz, J. L. Mastrodonato, C., Tumulka, R., Zanghì, N. (2010a) Approach to thermal equilibrium of macroscopic quantum systems. *Physical Review E*, **81**, 011109. http://arxiv.org/abs/0911.1724.

Goldstein, S., Lebowitz, J. L., Tumulka, R., Zanghì, N. (2010b) Long-time behavior of macroscopic quantum systems. *European Physical Journal H*, **35**, 173–200. http://arxiv.org/abs/1003.2129.

Goldstein, S., Lebowitz, J. L., Tumulka, R., Zanghì, N. (2006) Canonical typicality. *Physical Review Letters*, **96**, 050403. http://arxiv.org/abs/cond-mat/0511091.

Goldstein, S., Tumulka, R., Zanghì, N. (2016) Is the hypothesis about a low entropy initial state of the universe necessary for explaining the arrow of time? *Physical Review D*, **94**, 023520. http://arxiv.org/abs/1602.05601.

Griffiths, R. (1994) Statistical Irreversibility: Classical and Quantum. Pages 147–159 in J. J. Halliwell, J. Pérez-Mercader, and W. H. Zurek (editors): *Physical Origin of Time Asymmetry*. Cambridge University Press.

Jaynes, E. T. (1965) Gibbs vs Boltzmann entropies. *American Journal of Physics*, **33**, 391–398.

Kac, M. (1956) Foundations of Kinetic Theory. Pages 171–197 in J. Neyman (editor), *Proceedings of the Third Berkeley Symposium on Mathematical Statistics and Probability*, vol. III. Berkeley: University of California Press.

Kaufman, A. M. Tai, M. E., Lukin, A., Rispoli, M., Schittko, R., Preiss, P. M., Greiner, M. (2016) Quantum thermalization through entanglement in an isolated many-body system. *Science*, **353**, 794–800. http://arxiv.org/abs/ 1603.04409.

Kemble, E. C. (1939) Fluctuations, thermodynamic equilibrium and entropy. *Physical Review*, **56**, 1013–1023.

Khinchin, A. I. (1941) *Matematičeskie osnovanija statističeskoj mechaniki.* Moscow. English translation by G. Gamow: *Mathematical foundations of statistical mechanics.* New York: Dover (1949).

Krylov, N. S. (1979) *Works on the foundations of statistical physics.* Princeton: University Press.

Lanford, O. E. (1973) Entropy and Equilibrium States in Classical Statistical Mechanics. Pages 1–113 in A. Lenard (editor): *Statistical Mechanics and Mathematical Problems*, Lecture Notes in Physics vol. **2**. Berlin: Springer-Verlag.

Lanford, O. E. (1975) Time Evolution of Large Classical Systems. Pages 1–111 in J. Moser (editor): *Dynamical Systems, Theory and Applications*, Lecture Notes in Physics vol. **38**. Berlin: Springer-Verlag.

Lanford, O. E. (1976) On a derivation of the Boltzmann equation. *Astérisque*, **40**, 117–137. Reprinted in J. L. Lebowitz and E. W. Montroll: *Nonequilibrium Phenomena — The Boltzmann Equation*, North-Holland (1983).

Lazarovici, D. (2018) On Boltzmann vs. Gibbs and the Equilibrium in Statistical Mechanics. http://arxiv.org/abs/1809.04643.

Lebowitz, J. L. (2008) From Time-symmetric Microscopic Dynamics to Time-asymmetric Macroscopic Behavior: An Overview. Pages 63–88 in G. Gallavotti, W. L. Reiter, J. Yngvason (editors): *Boltzmann's Legacy*. Zürich: European Mathematical Society. http://arxiv.org/abs/0709.0724.

Lebowitz, J. L., Maes, C. (2003) Entropy — A Dialogue. Pages 269–273 in A. Greven, G. Keller, and G. Warnecke (editors): *On Entropy*. Princeton University Press.

Loewer, B. (2019) The Mentaculus Vision. (In this volume.)

Mackey, M. C. (1989) The dynamic origin of increasing entropy. *Reviews of Modern Physics*, **61**, 981–1015.

Ogata, Y. (2013) Approximating macroscopic observables in quantum spin systems with commuting matrices. *Journal of Functional Analysis*, **264**, 2005–2033. http://arxiv.org/abs/1111.5933.

Onsager, L. (1931) Reciprocal relations in irreversible processes. I. *Physical Review*, **37**, 405–426.

Pauli, W. (1928) Über das H-Theorem vom Anwachsen der Entropie vom Standpunkt der neuen Quantenmechanik. Pages 30–45 in P. Debye (editor): *Probleme der modernen Physik: Arnold Sommerfeld zum 60. Geburtstage gewidmet von seinen Schülern*. Leipzig: Hirzel.

Penrose, O. (1970) *Foundations of Statistical Mechanics*. Oxford: Pergamon.

Penrose, R. (1979) Singularities and time-asymmetry. Pages 581–638 in S. Hawking and W. Israel (editors): *General Relativity: An Einstein centenary survey*. Cambridge University Press.

Penrose, R. (1989) *The Emperor's New Mind*. Oxford University Press.

Popescu, S., Short, A. J., Winter, A. (2006) Entanglement and the foundation of statistical mechanics. *Nature Physics*, **21(11)**, 754–758.

Popper, K. R. (1976) *Unended Quest*. Glasgow: Collins.

Ruelle, D. (2007) Personal communication.

Safranek, D., Deutsch, J. M., Aguirre, A. (2019a) Quantum coarse-grained entropy and thermodynamics. *Physical Review A*, **99**, 010101. http://arxiv.org/abs/1707.09722.

Safranek, D., Deutsch, J. M., Aguirre, A. (2019b) Quantum coarse-grained entropy and thermalization in closed systems. *Physical Review A*, **99**, 012103. http://arxiv.org/abs/1803.00665.

Shannon, C. E. (1948) A mathematical theory of communication. *Bell System Technical Journal*, **27**, 379–423.

Tasaki, H. (2016) Typicality of thermal equilibrium and thermalization in isolated macroscopic quantum systems. *Journal of Statistical Physics*, **163**, 937–997. http://arxiv.org/abs/1507.06479.

Tolman, R. C. (1938) *The Principles of Statistical Mechanics*. Oxford University Press.

von Neumann, J. (1927) Thermodynamik quantenmechanischer Gesamtheiten. *Göttinger Nachrichten*, 273–291 (11 November 1927).

von Neumann, J. (1929) Beweis des Ergodensatzes und des *H*-Theorems in der neuen Mechanik. *Zeitschrift für Physik*, **57**, 30–70. English translation in *European Physical Journal H*, **35**, 201–237 (2010). http://arxiv.org/abs/1003.2133.

Wallace, D. (2012) *The Emergent Multiverse*. Oxford University Press.

Wallace, D. (2019) The Necessity of Gibbsian Statistical Mechanics. (In this volume.) http://philsci-archive.pitt.edu/15290/.

Chapter 15

The Necessity of Gibbsian Statistical Mechanics

David Wallace

Department of HPS and Department of Philosophy,
University of Pittsburgh
dmwallac@usc.edu

In discussions of the foundations of statistical mechanics, it is widely held that (a) the Gibbsian and Boltzmannian approaches are incompatible but empirically equivalent; (b) the Gibbsian approach may be calculationally preferable but only the Boltzmannian approach is conceptually satisfactory. I argue against both assumptions. Gibbsian statistical mechanics is applicable to a wide variety of problems and systems, such as the calculation of transport coefficients and the statistical mechanics and thermodynamics of mesoscopic systems, in which the Boltzmannian approach is inapplicable. And the supposed conceptual problems with the Gibbsian approach are either misconceived, or apply only to certain versions of the Gibbsian approach, or apply with equal force to both approaches. I conclude that Boltzmannian statistical mechanics is best seen as a special case of, and not an alternative to, Gibbsian statistical mechanics.

15.1. Introduction

Unlike general relativity, or non-relativistic quantum mechanics, statistical mechanics (SM) lacks a generally-accepted axiomatic foundation. Not coincidentally, also unlike those other theories, the philosophical study of statistical mechanics has little, if any, consensus as to what the main foundational *problems* are, let alone how they are to be solved.

In lieu of such a consensus, much (non-historical) study of the foundations of statistical mechanics has focussed instead on the supposed contrast between two strategies for understanding statistical mechanics: the so-called "Gibbsian" and "Boltzmannian" programs.[1] The former characterises thermodynamic systems, and thermodynamic entropy, in terms of probability distributions over microstates of a system; the latter, in terms of individual systems. (The relation of the two approaches to anything actually said by the historical Gibbs and Boltzmann is a debatable point but will not be explored here; see Uffink (2007), Myrvold (2019), and references therein, for historically sophisticated discussions.)

Furthermore, in conceptual discussions of statistical mechanics (particularly, though not exclusively, in philosophy) something close to a consensus has emerged: while Gibbsian statistical mechanics, and the Gibbsian definition of entropy (it is conceded) is a standard tool in *practical* applications of statistical mechanics, it is conceptually fatally flawed, and — unlike Boltzmannian statistical mechanics — lacks the resources to explain the key features of SM, notably the approach to equilibrium and the laws of thermodynamics. In brief summary (to be returned to later), Gibbsian statistical mechanics is criticised for trying to explain thermodynamic behaviour as a feature of our information about the world rather than as a feature of the world, for failing to identify entropy as a property of individual physical systems, and for evading rather than answering the challenges to statistical mechanics posed by time-reversibility of microdynamics and by the Poincaré recurrence theorem. (For examples of this criticism, see Albert (2000), Callender (1999), Callender (2001), Goldstein (2001), Lebowitz (2007), and Maudlin (1995).) As such, despite the acknowledged technical advantages of the Gibbsian approach, the Boltzmannian approach offers the true *explanation*

---

[1]The dichotomy is explicit in, e. g., Frigg (2007), Callender (1999, 2001), and Albert (2000).

of the successes of statistical mechanics, and in particular of thermodynamics.[2]

My thesis here is threefold. Firstly, the Gibbsian and Boltzmannian approaches are not rival approaches to the statistical mechanics and thermodynamics of the same systems: rather, the Gibbsian framework is a more general framework in which the Boltzmannian approach may be understood as a special case. Secondly, the wider applicability of the Gibbsian approach is indispensible from a naturalistic perspective, in as much as a wide range of empirically successful applications of statistical mechanics cannot be understood within the Boltzmannian approach. And thirdly, the Boltzmannian criticisms of the Gibbsian approach largely miss their mark: in the main, they apply to some of the justifications, motivations and interpretations offered for Gibbsian statistical mechanics but not to Gibbsian statistical mechanics itself.

The paper is structured as follows. In section 15.2, I briefly summarise the main commitments of the two approaches. I then consider first statistical mechanics (Sections 15.3–15.5) and then thermodynamics (Sections 15.6–15.7), arguing in each case that the two approaches coincide when it comes to thermodynamic properties of sufficiently well-behaved systems (Sections 15.3, 15.6), but that there are theoretically and experimentally relevant regimes where only the Gibbsian approach is applicable (Sections 15.4–15.5, 15.7). In Sections 15.8–15.10, I consider the main objections advanced against the Gibbsian approach. Section 15.11 is the conclusion.

I work throughout under the (of course false) assumption that the underlying microphysics is classical mechanics, so that *micro*states of a system are represented by points in a classical phase space. I do so to make contact with the contemporary literature rather than out

---

[2]This marks a rather common pattern in philosophy of physics, where philosophers predominantly study and espouse a minority position in physics which is technically less productive, but conceptually clearer, than the alternative — other examples include algebraic vs. mainstream approaches to QFT (cf Wallace (2011)), loop-space quantum gravity vs. string theory, and modificatory (e. g., Bohm or GRW) rather than pure-interpretion approaches to the quantum measurement problem (cf Wallace (2018)).

of a belief that quantum theory is irrelevant here; at the end (Section 15.12), I make some brief comments on how quantum theory affects my thesis.

15.2. Gibbs vs Boltzmann: An overview

The main features of Boltzmannian statistical mechanics[3] are:

- Phase space is divided into *macrostates*, representing (something like) "macroscopically indistinguishable states". More precisely, each energy hypersurface in phase space is divided into macrostates.
- Systems are represented by phase-space points; that is, the state of a system is a (classical) microstate. The macrostate of a system is just that macrostate in which the system's microstate lies; due to the way that the macrostate partition is constructed, to know the macroscopic properties of a system, we need know only its macrostate.
- The *Boltzmann entropy* of a system (up to a scale factor k_B) is the logarithm of the phase-space volume of the macrostate in which it lies (the "Boltzmann entropy" of the system).
- A system is in *Boltzmann equilibrium* if it lies in the largest of the macrostates (called the *equilibrium macrostate*) given the system's energy. The geometry of phase space for macroscopically large systems means that it is certain (for dilute gases) and heuristically plausible (in general) that the equilibrium macrostate for energy E is overwhelmingly larger in phase-space volume than all other macrostates of energy E. I will call a system with this property *Boltzmann-apt*.
- The approach to Boltzmann equilibrium is essentially a consequence of phase-space geometry combined with some reasonable assumptions about the dynamics: since almost all points (by phase-space volume) of a given energy lie in the equilibrium macrostate,

---

[3]My account largely follows Albert (2000) and Frigg (2007); I take it to be mostly in agreeement with, e. g., Carroll (2010), Goldstein (2001), Lebowitz (2007), North (2002), Penrose (1994).

either the dynamics or the initial state would have to be "ridiculously special" (Goldstein, 2001) for the system not to approach equilibrium. Similarly, a system in the equilibrium macrostate is exceptionally unlikely to wander out of the macrostate, given that virtually all states it could evolve into are also in the equilibrium macrostate. This conception of equilibrium makes the approach to equilibrium a statistical or probabilistic matter, and (given the Poincaré recurrence theorem[4]) systems initially away from equilibrium will eventually evolve away from equilibium again. However, for Boltzmann-apt systems the probability of equilibration is so close to unity, and the time taken for recurrence so large, that this can normally be disregarded. The *Boltzmann equilibration time* is the typical timescale after which a system reaches Boltzmann equilibrium.

- The *Boltzmann equilibrium values* of a system are just the macroscopic quantity values which specify the equilibrium macrostate. They may be measured by single measurements of the system after the equilibration timescale.

Note that at this stage, the terms 'equilibrium' and 'entropy' are being introduced as terms of art, without any prior assumption about their relationship to similarly-named thermodynamic terms. (I consider the relation between statistical-mechanical and thermodynamic concepts of equilibrium and entropy in Sections 15.6–15.7.)

At first pass, the main features of Gibbsian SM[5] are:

- Thermodynamic systems are represented by probability distributions over phase space: mathematically, that is, by positive measures on phase space assigning measure 1 to the whole space. (How are these probabilities to be interpreted? I postpone discussion to Section 15.10; for now, let me say only that it is not a core part

---

[4]See Wallace (2015b) for a review, and for emphasis of the fact that the so-called 'problem of measure zero', according to which only 'almost all' states recur, is an artifact of classical mechanics without foundational significance.

[5]At least at the foundational level, this account largely follows Sklar (1993), Callender (1999), Ridderbos (2002) and Frigg (2007).

of the Gibbsian formal framework that they must be thought of epistemically.)

- The *Gibbs entropy* of a system represented by probability measure ρ is

$$S_G(\rho) = -k_B \int \rho \ln \rho. \qquad (15.1)$$

- A system is at *Gibbs equilibrium* if ρ is time-invariant under the system's dynamics (which, again at first pass, seems fair enough: "equilibrium", after all, is understood from a macroscopic perspective as a state that is unchanging in time). It then follows from ergodic theory that, if the system is ergodic, the equilibrium distribution must be uniform on each energy hypersurface (see Malament and Zabell (1980) for more discussion of this relation between Gibbsian SM and ergodic theory). The *Gibbs equilibration time* is the typical timescale after which a system reaches Gibbs equilibrium.

  For simplicity, we will assume that this equilibrium probability function is the *microcanonical distribution*, which is restricted to a single energy hypersurface and which is uniform on that hypersurface. (Extending to the more general case does not essentially change the story, but introduces distracting technical complications.)

- The *Gibbs equilibrium values* of the system are the expected values of the various dynamical quantities (microscopic or macroscopic) evaluated with respect to the equilibrium distribution. They may be measured in the usual way, by measuring many copies of the equilibrated system and taking an average.

There is an immediate problem with this first-pass version of the Gibbsian approach: it seems to have the corollary that real systems do not increase in entropy or approach equilibrium. For if the probability measure is defined over individual microstates evolving under the system's Hamiltonian, the dynamics of the probability measure itself are determined: If the dynamics deterministically carry the points in region V at time 0 onto the points in region $V(t)$ at time t, then the conditional probability for a microstate of the system to be in region $V(t)$ at time t given that it was in region V at time 0 is unity. This suffices to uniquely determine the dynamics for ρ: they are given by Liouville's equation,

$$\frac{\mathrm{d}}{\mathrm{d}t}\rho(t) = \{H, \rho(t)\}, \tag{15.2}$$

where H is the Hamiltonian and $\{\cdot, \cdot\}$ is the Poisson bracket. This is a first-order equation, so if the first derivative of ρ is zero at some given time it must have been zero at all previous times; that is, no system that starts away from equilibrium can reach equilibrium. Nor is there any prospect that the system even gets *closer* to equilibrium, at least as measured by the Gibbs entropy, since it is an easy consequence of Liouville's equation that the Gibbs entropy is invariant over time.

One influential way to evade this problem — due to Ed Jaynes — is to drop the idea that dynamics have anything much at all to do with equilibrium or the approach to equilibrium. To Jaynes and his followers,[6] to say that a system is at equilibrium is to say simply that we know nothing about its state except its energy; to say that a system is evolving towards equilibrium is to say that we are losing information about its state. This strategy has been robustly criticised in the philosophy literature,[7] most notably by advocates of the Boltzmannian program, as simply failing to recognise the obvious fact that the approach to equilbrium is an observed, empirically confirmed physical process. (See, e. g., Albert (2000).)

A more sophisticated variant of Jaynes' approach, due to Wayne Myrvold (2014), treats the probability distribution as epistemic, takes seriously its dynamical evolution, but asserts (i) that realistic agents will lose the ability to keep track of very fine-grained features of the probability distribution, and (ii) those fine-grained features are not dynamically relevant when it comes to calculating macroscopic quantities. As such, an agent can and must keep track

---

[6]See, e. g., Jaynes (1965, 1957a,b).

[7]Though for acerbic pith in response to Jaynes, little beats physicist J. S. Rowlinson's quotation of Leslie Ellis: "Mere ignorance is no ground for any inference whatever. Ex nihilo nihil." (Rowlinson, 1970)

not of the true probability distribution but of a coarse-grained variant of it.

A quite different approach to the problem is to keep the idea that the system evolves under some dynamics, but drop the idea that those dynamics are Hamiltonian. Most radically, we could assume that explicit time asymmetry must be introduced into classical physics to understand the approach to equilibrium (the work of Ilya Prigogine is generally interpreted as exploring this option; see Bishop (2004) and references therein for philosophical discussion). Alternatively, we might appeal to the fact that no physical system short of the whole Universe is truly dynamically isolated, and so will not truly have dynamics given by a Hamiltonian for the system alone. But the first approach is in danger of looking ad hoc, and the second does not seem to have the resources to explain time asymmetry.

The main strategy to reconcile the Gibbsian approach with the Hamiltonian dynamics involves a small modification to the former, rather than the latter:

- We introduce a *coarse-graining* map J on the space of probability distributions. At least in foundational discussions, the most common form given for J works like this:

  (a) Coarse-grain phase space into cells.
  (b) Replace each probability measure ρ with a measure $J\rho$ that is uniform over each cell and which assigns the same probability to a cell as ρ did.

  This definition ensures that J is a *projection operator*; that is, $J^2 = J$. Maps like these (when linear) are sometimes called *Zwanzig projections*, and in fact can be defined by methods of coarse-graining much broader than the phase-space cell method; see, e.g., Zwanzig (2001) (or, for foundational discussion, Zeh (2007) or Wallace (2015a)) for more details. The action of J is often glossed as representing the fact that our measurements of a system have finite precision, although (as I discuss further in Section 15.9) this is not really a viable way to think about it.

- The entropy of a probability distribution ρ (with respect to a given coarse-graining J) is now defined as the Gibbs entropy of the *coarse-graining* of ρ:

$$S_{G;J}(\rho) = S_G(J\rho) = -k_B \int (J\rho) \ln(J\rho). \qquad (15.3)$$

- A system represented by probability function ρ is at equilibrium (with respect to J) if its coarse graining $J\rho$ is invariant in time.

   There is no dynamical reason to expect that the coarse-grained entropy is a constant of the motion; indeed, it is mathematically possible for the coarse-grained entropy to increase to a maximum value and then remain there indefinitely, and it is frequently possible to prove that $S_{G,J}(\rho)$ is increasing with time. Establishing equilibration for the Gibbsian then requires them to provide mathematical reasons (as usual in this field, probably heuristic reasons rather than rigorous proofs) to expect that $S_{G;J}(\rho)$ evolves over a reasonable period of time to obtain its maximum value given the constraints imposed by conservation principles, even as $S_G(\rho)$ remains a constant of the motion. (It will then follow that the equilibrium distributions are those whose coarse-grainings are the microcanonical or canonical distributions.) The details can get technical: I discuss the basic principles at length in Wallace (2010), and work through some technical examples in Wallace (2016) (see also references therein).

   Whenever it is significant, I will assume this last understanding of Gibbsian statistical mechanics — but for most of my purposes in this paper, all that matters is that there is an *effective probability distribution* for the system at any given time, which coincides with the actual probability distribution for any macroscopically relevant quantity and which evolves towards Gibbs equilibrium. Whether that distribution *is* the true probability distribution, or is simply a coarse-grained approximation to it, will be neither here nor there.

15.3. Equilibration in Boltzmann-apt systems

To begin our assessment of Boltzmannian and Gibbsian statistical mechanics, let us start by considering systems that are Boltzmann-apt; that is, each of which has a decomposition into macrostates such that, for any energy E, the equilibrium macrostate at that energy occupies the overwhelming majority of the volume of the energy hypersurface. In this context, the Gibbsian framework actually entails the Boltzmannian framework. For consider, if the Gibbs framework holds and if the system's initial energy is E, then after the Gibbs equilibration timescale, the effective probability distribution is uniform on the energy-E hypersurface. But since the system is Boltzmann-apt, that means that with overwhelmingly high probability, the system is in the Boltzmann equilibrium macrostate; that is, with overwhelmingly high probability, the system is at *Boltzmann* equilibrium.

After the equilibration timescale, the macroscopic degrees of freedom at any time will be overwhelmingly likely to possess their Boltzmann-equilibrium values. So the Gibbs and Boltzmann equilibrium values coincide to an extremely high degree of accuracy. There will be a small probability that the result of a given measurement will diverge from these values, corresponding to the small probability that the microcanonical distribution assigns to the non-equilibrum macrostates. These *fluctuations* around the Boltzmann equilibrium values can be described either are fluctuations *within* Gibbs equilibrium, or as fluctuations *into and out of* Boltzmann equilibrium, but this is simply a semantic difference, and does not correspond to any physical disagreement about the system.

So the user of Gibbsian statistical mechanics is fully entitled to use — and indeed, to accept the truth of — Boltzmannian statistical mechanics. In those systems to which the latter is applicable, the Gibbsian framework can be seen as grounding the Boltzmannian one: the assumptions required by the Boltzmannian are corollaries of Gibbsian premises.

The situation is not symmetric, for obvious conceptual reasons. The Boltzmannian framework *per se* contains no explicit notion of probability, and so does not permit us even to define the Gibbsian probability distribution. (Almost certainly the Boltzmannian framework requires some qualitative notion of probability — perhaps the notion of *typicality* advocated by Goldstein *et al.* (2006) (see Frigg (2009) and references therein for further details) — but its advocates commonly claim that the full quantitative probability distribution is not required, and indeed is an incoherent concept in classical mechanics.) Nonetheless, the Boltzmannian has a straight-forward story to tell about the practical utility of Gibbsian methods: Precisely because the equilibrium macrostate so dominates the volume of the energy hypersurface, and because macroscopic variables have almost-constant values on each macrostate, the average value of a macroscopic variable will be extremely close to its actual value at Boltzmann equilibrium. So the Boltzmannian can harmlessly use Gibbsian probabilistic methods as a calculational tool, without any commitment to their truth.

In summary: as long as we (a) confine our attention to Boltzmann-apt systems, and (b) wish to calculate only equilibrium values of macroscopic variables, the Gibbs and Boltzmann approaches are equivalent for all practical purposes. And since the Gibbsian approach introduces a problematic notion of quantitative probability, and arguably has a murkier account of how equilibration works, one can see the case for preferring the Boltzmannian approach, and for setting aside as foundationally irrelevant the incontestable fact that working physicists use the Gibbsian approach.[8]

As we shall see shortly, however, (a) and (b) by no means exhaust the range of applications of classical statistical mechanics.

15.4. Beyond Boltzmann equilibrium: Transport properties

What might we want to calculate of a (let's say Boltzmann-apt) system at equilibrium, other than the values of its macroproper-ties? One important class of properties are the *multi-time correlation functions*. For instance, consider a classical dilute gas at equilibrium, and let $X(t)$ be the position of some arbitrarily chosen particle at time t. At equilibrium, the expected value $\langle X(t) \rangle$ of that position is

---

[8]See, e. g., Callender (1999, p. 349): "[L]et me happily concede that for the practice of science, Gibbsian SM is usually to be preferred. Since the values of all the entropy functions I discuss agree at equilibrium, my arguments are necessarily philosophical in nature."

time-independent and equal to the center of mass of the gas; note that I can make this statement based on only the vaguest and most qualitative information about the gas.

But now consider the two-time correlation function

$$C(t) = \langle X(t)X(0)\rangle. \tag{15.4}$$

The form of this function *cannot* be worked out in any simple way from the equilibrium macrofeatures of the gas. Calculationally, it's clear that it involves the dynamics, since to evaluate it I need to work out how likely the system is to transition from one position to another over time t. Conceptually, we have

$$\langle (X(t) - X(0))^2 \rangle = \langle X(t)^2 \rangle + \langle X(0)^2 \rangle - 2\langle X(t)X(0)\rangle$$
$$= 2(\langle X(0)^2 \rangle - C(t)) \tag{15.5}$$

so that $C(t)$ may be recognised as a measure of how far a randomly-chosen particle diffuses through the gas after time t.

Knowing that the system is in the Boltzmannian equilibrium macrostate does not in any straightforward way provide us with enough information to calculate $C(t)$; indeed, since $C(t)$ is an explicitly probabilistic quantity, it is not even defined on the Boltzmannian approach. Of course this would not matter if $C(t)$ were simply a theoretical curiosity — but, as my discussion above may have suggested, two-time correlation functions like this are the main tool in calculating diffusion coefficients, rates of thermal conductivity, and the other quantitative properties of a system that characterise its behavior close to but not at equilibrium. Indeed, the huge subject of *transport theory* is largely concerned with calculating these two-time functions, and its methods are quite thoroughly confirmed empirically. (See, e.g., Zwanzig (2001), Altland and Simons (2010), and references therein.)

Therefore, even for Boltzmann-apt systems, there are important cases where probabilistic methods seem necessary and do not reduce to Boltzmannian methods in any simple way. (Multi-time correlation functions by no means exhaust the list of such cases — indeed, I discuss another, the modern ('BBGKY') derivation of Boltzmann-type equations through truncation of the N-particle probability distribution to 2-particle marginals, in Wallace (2016) — but they already suffice to make the point.)

15.5. Beyond Boltzmann equilibrium: Fluctuations

What about systems that are *not* Boltzmann-apt? One might *a priori* guess that statistical mechanics is inapplicable to such systems, since after all it relies, for its efficacy (doesn't it?), on being able to average over a very large number of constituents. But this is not the case. As I now illustrate, statistical mechanics is frequently and successfully applied to systems where the Boltzmannian equilibrium state does not overwhelmingly dominate the energy hypersurface. Here I give two such examples, in neither of which we can simply assume that a system — even after the equilibration timescale — is "overwhelmingly likely" to be in one particular macrostate. (I give others in Section 15.7 when I discuss the thermodynamics of small systems.)

The first example is *spontaneous symmetry breaking*, say (for definiteness) in ferrmomagnets. Here, 'the' equilibrium macrostate below a certain temperature has a non-zero expectation value of magnetic spin, corresponding to the fact that it is energetically favourable for adjacent spins to line up. But it then follows from the rotational symmetry of the underlying dynamics that there must be another macrostate, obtained by applying a rotation to each microstate in the first, of equal volume to 'the' equilibrium macrostate and equally justifiably called an equilibrium macrostate. In other words, systems with spontaneous symmetry breaking are not Boltzmann-apt, at least at energies corresponding to temperatures below the symmetry-breaking temperature.

It's possible to imagine a modification of the Boltzmannian framework to handle this case. We could generalise the definition of Boltzmann-aptness to allow for *many* equilibrium states, each related by a symmetry, and such that their *collective* volume dominates the energy hypersurface. (And notice that since Boltzmann entropy is logarithmic, replacing one equilibrium state with N reduces the entropy only by $k_B \ln N$, which will be a negligible shift if N is much smaller than the number of microscopic degrees of freedom.) We could argue that unless the dynamics are 'ridiculously special', the system is overwhelmingly likely to end up in *one* of the equilibrium macrostates, and to remain there for a very long time.

But this does not suffice to save the phenomena. We require not just that the system will end up in one such state, but that *each is equally likely*. This follows directly from the assumption of Gibbs equilibrium (each has equal volume, so each is equally likely) and is well-confirmed empirically. For instance, the pattern of symmetry breaking can be analysed in a ferromagnet, and it is clearly distributed at random. This assumption may seem obvious but it's not at all clear how Boltzmannian statistical mechanics can reproduce it, without being supplemented by an explicit probability distribution — we cannot, for instance, say 'typical states are equally likely to end up in each equilibrium macrostate', since 'being equally likely to end up in each equilibrium macrostate' is not a property that any given microstate can have in a deterministic theory.

The second example is *Brownian motion*. Here, the system consists of one large particle in a bath of smaller ones, with the latter usually taken to be at Gibbs equilibrium. The large particle has no meaningful notion of 'macrostate', or if it does, then each large-particle state corresponds to its own macrostate, so that there is no largest "equilibrium" macrostate, so the Boltzmannian concept of equilibrium does not apply to this system. The particle evolves randomly, due to fluctuations in the number of particles colliding with it from any given direction; its probability distribution will converge on the Maxwell-Boltzmann distribution, but this in no way means that the actual state of the particle is time-invariant after its 'equilibration' timescale. Applying Gibbsian methods (such as by calculating the two-time correlation functions), we can derive the stochastic equation which the particle obeys (see Zwanzig (2001) for a more detailed discussion). The resultant equation, and variants on it which apply to similar setups, has been widely applied and thoroughly confirmed empirically.

Note that in both of these examples, probability is not simply playing a foundational role (as was the case in, say, the calculation of transport coefficients from two-point functions). Rather, the predictions of the theory are themselves expressed probabilistically, and don't have any direct re-expression in terms of categorical properties. Of course, the probabilities are measured through relative frequencies, and it is always open to the Boltzmannian to insist that apparently "probabilistic" predictions should be reinterpreted as, say, claims about what is typical when an experiment is repeatedly performed on a very large number of copies of the system. But this is just a claim about the general foundations of probability in statistical mechanics (specifically, that it should be understood on frequentist lines). It in no way eliminates probability from the actual statement and use of statistical mechanics: on that reading, concrete statistical-mechanical systems would still need to be described by probability distributions, not just by macrostates and qualitative concepts of typicality.

To summarise Sections 15.3–15.5: while there is a class of statistical-mechanical systems, and a class of properties of those systems, such that Gibbsian and Boltzmannian methods are equally applicable when calculating those properties, the scope of statistical mechanics is much wider than those classes and includes many phenomena that seem treatable only by Gibbsian means. As we will now see, the same story essentially recurs when we turn to thermodynamics.

15.6. Thermodynamics in macroscopic systems

As I (Wallace, 2014) and others (see Skrzypczyk *et al.* (2014) and references therein) have argued elsewhere, thermodynamics is not a dynamical theory in the usual sense; that is, it is not a theory of how undisturbed physical systems evolve over time. (Indeed, insofar it is seen as such, thermodynamics is essentially trivial: it is concerned with equilibrium systems, and the defining feature of such systems is that they do not evolve at all over time.) Rather, thermodynamics is a *control theory* (or, alternatively, a resource theory), concerned with which transformations can or cannot be performed on a system by an external agent, given certain constraints on that agent's actions. From this perspective, the First Law of thermodynamics disallows control actions where the work done on a system (i.e. the energy cost of the control operation to the agent), plus the heat flow into the system from other systems, does not equal the change in internal energy of the system. The Second Law, meanwhile, disallows those transformations which lead to a net decrease in thermodynamic entropy.

As a more precise (though by no means completely precise) statement of the content of thermodynamics, I offer the following: There are equilbrium systems, and they can be completely characterised for thermodynamic purposes by a small number of *thermodynamic parameters*: the energy U and some externally-set parameters — in typical examples, the volume V. (It is also possible to add some conserved quantities, such as particle number; for expository simplicity, I omit this complication). The *equation of state* determines the thermodynamic entropy as a function of those parameters; schematically, we might write $S = S(U, V)$, understanding V to stand in for whatever are the actual external parameters and conserved quantities. Other thermodynamically relevant quantities can be calculated from the equation of state: thermodynamic temperature, for instance, is the rate of change of U with S at constant V; pressure is minus the rate of change of U with V at constant S. The Second Law is the requirement that any allowable control process leave the sum of all entropies non-decreasing. (For a somewhat more detailed sketch of thermodynamics on these lines, see Wallace (2018, Section 2).)

Recovering thermodynamics from an underlying mechanical theory, then, requires us to provide mechanical definitions of (inter alia) 'equilibrium', 'allowable control process', the 'thermodynamic parameters', and 'entropy' along with other reasonable mechanical posits, such that (a) these laws of thermodynamics can be derived (at least approximately) from those definitions and posits; (b) the definitions do reasonable justice to the informal, operational understanding of the thermodynamic terms (according to which, for instance, equilibrium states have macroproperties that are approximately constant in time); and (c) given that the equation of state of a system is an empirically measurable feature of that system, the definitions allow us to recover the actual, quantitative form of the equations of state of known systems.

In looking for such a recovery, we are not operating in a vacuum. After a century of statistical mechanics, it is well known how to calculate the thermodynamic entropy of a (large) system: taking the Hamiltonian of that system to be parametrised by the external parameters (so that, for instance, the Hamiltonian of a box of gas is a function of its overall volume), treat the thermodynamic energy as just directly representing the mechanical energy, and define the entropy as the log of the phase-space volume of all states with that energy (this makes the entropy indirectly a function of V as well as energy U, since the Hamiltonian depends on V). Mathematically speaking, this is the Gibbs entropy of the microcanonical distribution; empirically, this *works*, for a huge variety of systems, and so recovering it (at least to a high degree of approximation) is a *sine qua non* of any mechanical recovery of thermodynamics. (Call this the 'quantitative test' of a proposed recovery.)

We can also say something general about allowable control operations. At a minimum, such control operations ought to correspond to transformations allowable by the basic structure of classical mechanics. Any such transformations preserve phase-space volume when acting on closed systems, so we will assume that any such control operation is indeed volume-preserving. (It is possible to decrease phase-space volume by measuring the state and choosing the control operation accordingly, a strategy that leads towards Maxwell-Demon-style (apparent) counter-examples to the Second Law, but these lie beyond the scope of this paper; for discussion in the control-theory context, see Wallace (2014) and references therein.)

With all this said, let's consider what a derivation of thermodynamics for macroscopic systems might look like. In most cases (perhaps putting aside spontaneous symmetry breaking), we can reasonably assume such systems are Boltzmann-apt: that is, that for given U, V the phase space region is dominated by a single region in which the macroscopic variables are approximately constant.

Let's start with the Boltzmannian approach, in which equilibrium is defined as occupation by the system's actual microstate of the largest-volume macrostate, and in which we eschew explicit and quantitative use of probabilities. A natural choice for control operations is then

Boltzmann-equilibration operations: In which the system's thermodynamic parameters are changed, some subsystems are brought into and out of thermal contact (that is, coupled or decoupled by some Hamiltonian) and then the system is allowed to evolve such that (in an irreducibly imprecise way) it is *almost certain* to reach equilibrium. To restrict to equilibration operations is to assume an agent who has control only over a system's bulk thermodynamic parameters.

By assumption, a Boltzmannian-equilibrium operation must have the desired effect for all (or at least the vast majority in phase-space measure) of points in the equilibrium region. So it follows that the volume of the post-operation equilibrium state must exceed that of the pre-operation equilibrium state. Hence, Boltzmann entropy is non-decreasing under these operations, in accordance with the Second Law. Furthermore, since the equililbrium macrostate dominates the allowable region of the phase space, its Boltzmann entropy is numerically almost equal to the microcanonical entropy, so that the quantitative test is passed too. In short, this seems an entirely satisfactory (sketch of a) derivation of thermodynamics from Boltzmannian statistical mechancs, with the Boltzmannian notions of 'entropy' and 'equilibrium' mapping to the thermodynamic ones.

What about from the Gibbsian perspective? Here, to say that the system is at equilibrium is to say that its effective probability is uniform over the allowable region of phase space. Given that the system is Boltzmann-apt, recall that the system thus *almost certainly* has the Boltzmann-equilibrium values of the macroproperties. (Again, there is a purely semantic difference here: the Boltzmannian says that the system is almost certainly at equilibrium and, if at equilibrium, certainly has such-and-such values of the macroproperties; the Gibbsian says that the system *is* at equilibrium and, as such, *almost* certainly has those values.) The natural choice here for control operations is:

Gibbs-equilibration operations: In which the system's thermodynamic parameters are changed, some subsystems are brought into and out of thermal contact (that is, coupled or decoupled by some Hamiltonian) and then the system is allowed to evolve such that if the effective probability distribution is originally uniform over the region determined by the old parameters, after the control operation it is uniform over the region determined by the new parameters.

Given the background coarse-graining assumptions of the Gibbs approach, the Gibbs entropy of the effective probability distribution cannot go down in the control operation. And the quantitative test is trivially passed. So again, we have a (sketch of a) satisfactory derivation of thermodynamics from Gibbsian statistical mechanics, with 'equilibrium' and 'entropy' here played by the Gibbsian rather than the Boltzmannian notions. As a further difference, because the Gibbsian approach describes the system probabilistically, the Gibbsian correlates for thermodynamic energy and work are *expected* values, not categorical values — although given the assumption of Boltzmann-aptness, the actual value will be extremely close to the expected value with extremely high probability.

The situation is parallel to the statistical-mechanical case. For the Gibbsian, there is no factive difference between the two approaches: the validity of the Gibbsian approach entails that of the Boltzmannian approach, and the two strategies differ only semantically. For the Boltzmannian, the Gibbsian use of probabilities is justifiable only on pragmatic terms.

Again in parallel to the statistical-mechanical case, Gibbsian thermodynamics would be required only if there are applications of thermodynamics where the Boltzmann-apt assumption fails, where the use of actual probabilities is unavoidable, and where statistical fluctuations are non-negligible and measurable. And again, there are indeed many such cases.

15.7. Beyond macroscopic thermodynamics

A combination of theoretical and experimental advances have made
the last twenty years a golden age for the statistical mechanics of
small systems. On the theoretical side, the key advance has been *fluctuation theorems*, results derived in the Gibbsian framework which
relate the probability distributions over different transformations
between systems. For instance, in macroscopic thermodynamics, if
a system is transformed between equilibrium states while remaining
all the while in contact with a heat bath at temperature T, it is a
standard result that

$$\Delta F \geq W \qquad\qquad (15.6)$$

where ΔF is the decrease in the free energy between initial and final
states, and W is the work extracted in moving from initial to final
states. The inequality becomes an equality only in the quasi-static
limit. The *Jarzynski equality* (Jarzynski, 1997) sharpens this to:

$$\mathrm{e}^{-\Delta F/kT} = \left\langle \mathrm{e}^{-W/kT} \right\rangle \qquad\qquad (15.7)$$

where the right hand side is an expectation value over different microphysical realisations of the transition between initial and final state.
For large systems and slow changes, the fluctuations in the right-
hand-side will be negligible and we will recover the quasi-static, non-
probabilistic result, but the equality holds — according to Gibbsian
statistical mechanics — even for small systems and for rapid transformations of those systems.

The (closely related) *Crooks fluctuation theorem* (Crooks, 1998)
again concerns transitions between equilibrium states A, B of a system in thermal contact with a reservoir. For that system, the theorem
states that

$$\frac{\Pr(W|A \to B)}{\Pr(-W|B \to A)} = \exp\left(\frac{W - \Delta F}{kT}\right) \qquad\qquad (15.8)$$

where $\Pr(W|A \to B)$ is the probability that a given transition from
A to B will require work W, and $\Pr(-W|B \to A)$ is the probability
that a given transition from B to A will extract work W. In the limit

of large systems and slow processes, $W = \Delta F$ and the theorem just says that transitions in either direction are equally likely, but the result is again a mathematical consequence of Gibbsian statistical mechanics, even for fast processes on microscopic systems.

These are results that cannot be derived in full generality in the Boltzmannian framework: they are explicitly probabilistic, and the free energies are themselves defined as expectation values. This would be of only limited significance, though, if they remained purely theoretical results, experimentally untestable on systems small enough to display meaningful fluctuations.

But of course they have been tested, extensively. The most well-developed examples have involved the stretching and unstretching of RNA and polymer chains (see, for instance, Collin *et al.* (2005)). In these experiments, the work done on the chain in a given stretch-and-unstretch shows large thermal fluctuations (that is, we are way outside the Boltzmann-apt regime) but nonetheless the probability distributions over work done, and the expectation values, conform exactly to the predictions of the fluctuation theorems. Experiments have been done in a range of other small systems, and the field is moving too quickly to summarise in a foundational article like this one (for an already-dated review, see Bustamante *et al.* (2005)) but suffice it to say that the experimental evidence for the fluctuation theorems looks compelling, even (especially?) in small systems where fluctuations are large.

Tests of the fluctuation theorems do not exhaust the range of recent experiments in microscopic thermodynamics. To give one more example, recall that Richard Feynman famously argued (Feynman, 1967, pp.116–9) that a ratchet could not be used as a Maxwell demon to transfer heat between two reservoirs of equal temperature because of fluctuations. Bang *et al.* (2018) have demonstrated this result empirically, using a microscopic 'ratchet' consisting of a colloidal particle in an optical trap. They verified that although on any given run of the experiment, heat is sometimes converted into work, the expected work output (measured by averaging over many runs) is zero unless the two reservoirs are of unequal temperature.

In summary: modern physics is extensively applying, and testing, thermodynamics in the microscopic regime, where the Boltzmann-aptness assumption completely fails and predictions are explicitly probabilistic. In this regime, Boltzmannian statistical mechanics is inapplicable and the Boltzmannian conception of equilibrium is useless.

Now, it should be acknowledged that in using "thermodynamics" to apply to this regime, in which terms like 'work' and 'free energy' enter the Second Law only as expectation values, we are going well beyond their original use in nineteenth-century phenomenological thermodynamics. I could even concede that *that* subject reduces to Boltzmannian statistical mechanics as readily as to the Gibbsian version. But this is a semantic matter. Whether the subject that modern physicists call 'thermodynamics' is a precisification of that nineteenth-century subject or a genuine extension of it, it is a robust and empirically well-confirmed subject which relies for its formulation and its use on probabilistic — that is, Gibbsian — conceptions of statistical mechanics in general and equilibrium in particular.

I conclude that — even if we restrict attention to the classical case — the range of applications of modern statistical mechanics and thermodynamics vastly outstrips what can be analysed using just the methods of Boltzmannian statistical mechanics, without the explicit introduction and study of quantitative probabilities. Assuming (as I take it should be uncontroversial) that a foundation for statistical mechanics needs to be a foundation for *all* of statistical mechanics, and not just for the tiny fraction that had been developed by, say, 1900, then I don't see an alternative but to accept the Gibbsian framework as that foundation, with the Boltzmannian framework as a highly important special case of it, applicable to Boltzmann-apt systems for the calculation of certain quantities.

I could end the paper here. However, critics of Gibbsian statistical mechanics have advanced a number of objections to the effect that it is conceptually incoherent, and those objections are not magically swept away simply by a naturalistic argument that we need Gibbsian methods (even if that need might make us more confident that the objections can somehow be met). In the next three sections,

I consider what I take to be the main Boltzmannian objections to Gibbsian statistical mechanics, and argue that they either rest on misconceptions about the framework, or else apply to particular versions or developments of the Gibbsian framework but not to that framework in itself. For expository clarity, and to ensure that I am not engaging with straw men, I will concentrate on influential criticisms due to Albert (2000), Callender (1999, 2001) and Maudlin (1995).

15.8. Objections to Gibbs: Modality and probability

One of the most common objections to Gibbsian statistical mechanics is that it makes thermodynamics in general, and entropy in particular, a study of modal, in particular probabilistic, features of a system, rather than of categorical features, when the latter is what is required to do justice to the phenomena. Indeed, this is often held up as a simple and straightforward *mistake*. As representative examples, consider Callender (2001, p. 544):

> "The problem is not the use of ensembles ... The problem is instead thinking that one is *explaining* the thermal behaviour of *individual real systems* by appealing to the monotonic feature of some function, be it ensembles or not, that is not a function of the dynamical variables of individual real systems. It is impossible to calculate the intellectual cost this mistake has had on the foundations of statistical mechanics. The vast majority of projects in the field in the past century have sought to explain why my coffee (at room temperature) tends to equilibrium by proving that an ensemble has a property evincing monotonic behaviour." (Emphasis Callender's.)

Or Maudlin (1995, p.147):

> "Since phenomenological thermodynamics originally was about ... individual boxes [of gas], about their pressures and volumes and temperatures, 'saving' it by making it be about probability distributions over ensembles seems a Pyrrhic victory. It is remarkable, and not a little depressing, to see the amount of effort and ingenuity that has gone

into finding something of which the phenomenological laws
can be strictly true, while insuring that the something cannot possibly be the phenomena."

A particular concern of both authors is the Poincaré recurrence
theorem. Given recurrence, we know that a system will eventually
return arbitrarily close to its initial state. And so if entropy is indeed
a function of the microstate of a system, then entropy must eventually return to its initial value, seeming to demonstrate that any
account of entropy as monotonically increasing has lost touch with
the microfoundations of thermodynamics.

There is a great deal to say in response to these objections. Here
I identify six points that ought, jointly, to assuage such worries.

Firstly, what we want to explain in non-equilibrium statistical
mechanics is itself something modal: not that systems *invariably* go
to equilibrium but that they do so *almost certainly*. It might not be
compulsory to quantify 'almost certainly' as 'with probability very
close to 1' but, at any rate, it does not seem to involve a substantial
change of focus. A probabilistic property of a system is poorly suited
to explain why the system deterministically behaves in such-and-such
a way, but it is well suited to explain why it very probably behaves
in that way.

Secondly, in statistical mechanics (as distinct from thermodynamics), the entropy is ultimately no more than a book-keeping device
to keep track of irreversibiliy in a system's dynamics. In Gibbsian
statistical mechanics, irreversibility typically takes the form of an
increasing dispersal of a probability distribution over the constant-
energy hypersurface (which, in Boltzmann-apt systems, in turn
entails the increasing likelihood of the system being in the equilibrium macrostate). Because (an appropriately coarse-grained) Gibbs
entropy tracks this dispersal, it is a useful tool to study the approach
to equilibrium, but it has no causal or explanatory role in its own
right. (And the same is true for Boltzmannian statistical mechanics: the Boltzmann entropy, as far as non-equilibrium statistical
mechanics is concerned, is a device for tracking a system's increasing
likelihood of being found in increasingly-large macrostates.)

Thirdly, while in thermodynamics, the entropy plays a much more quantitatively significant role, that role is itself modal. As I noted in Section 15.6, the subject matter of thermodynamics is the transformations that an agent can bring about in a thermodynamic system. The word 'can' betrays the modality of this subject matter: if a system might be in one of many states, this constrains the transformations that the agent can bring about, at least without measuring that state. This modality is hidden in most applications of macroscopic, phenomenological thermodynamics — but in that context, the Gibbs entropy is just a property of the system's thermodynamic variables, and the modality is suppressed.

Fourthly, once it is recognised that in Gibbsian statistical mechanics, 'equilibrium' is a statement about the probability distribution of a system, there is no contradiction between the (classical) recurrence theorem and the claim that entropy is non-decreasing. For the former tells us that any given system has some timescale at which it has returned to its initial state, and the latter (for Boltzmann-apt systems) tells us that at any time after the equilibration timescale the system is overwhelmingly likely to be in the equilibrium macrostate, and these statements are compatible. Nor is there any contradiction between the recurrence theorem and the claim that *thermodynamic* entropy is non-decreasing, since the latter concerns the interventions we may make on a system, and it is of no use to an agent to know that any given microstate will recur, absent knowledge that the system has in fact recurred at a given time.

Fifthly, while the quantum version of the recurrence theorem has a *uniform* timescale for recurrence, and so indicates that even the coarse-grained Gibbs entropy cannot be non-decreasing for all time, this simply indicates (assuming the orthodox coarse-graining version of the Gibbsian approach) that the assumptions underpinning the validity of coarse-graining cannot apply for arbitrarily long timescales (something that can in any case be read off the master-equation or BBGKY formalisms for irreversible Gibbsian dynamics; cf the discussion in Wallace (2016)). I suppose Callender and Maudlin should be pleased at this result: properly understood, Gibbsian statistical mechanics does not after all seek an exceptionless principle of non-decreasing entropy, but only an entropy that is nondecreasing over the physically significant timescales.

Finally, *Boltzmann* entropy is itself only superficially a categorical property of a system. Yes, formally speaking, the Boltzmann entropy depends only on a system's microstate, but it relies for its definition on a partition of the energy hypersurface into macrostates, and that partition is modal in nature — most obviously because the energy hypersurface itself depends on the dynamics. To make this vivid, suppose that a system has microstate x, and then perturb its Hamiltonian such that it remains constant in some small neighborhood of x but varies sharply over the rest of x's macrostate. Then the perturbation will adjust the macrostate partition, and thus change the Boltzmann entropy of x, even though no *categorical* property of the system has been altered.

15.9. Objections to Gibbs: Coarse-graining

A second major locus of concern about the Gibbsian framework is the supposed inadequacy of any account based on coarse-graining (which, recall, is the mainstream — though not the only — approach to reconciling irreversibility with the time-invariance of fine-grained Gibbs entropy). It is variously described as ad hoc, as confusing our subjective limitations as experimenters with objective matters of fact, and as introducing some kind of spurious and empirically unsupported modification of the dynamics. Thus, Maudlin (1995, pp.146–7):

> "One can modify the underlying dynamics by adding some 'rerandomization' posit ... but these surreptitious modifications simply have no justification."

Or Callender (1999, p.360):

> "The usual response to the conservation of [the fine-grained Gibbs entropy] is to devise new notions of entropy and equilibrium, in particular, the coarse-grained entropy and a notion of equilibrium suitable for it. The motivation for these new notions is solely as a means of escaping the

> above "paradox" [time-invariance of fine-grained entropy],
> though it is usually defended with appeals to the impreci-
> sion with which we observe systems."

Here, I identify three reasons that such concerns do not undermine the Gibbsian project (properly understood).

Firstly, in statistical mechanics, the choice of coarse-graining is typically motivated neither by concerns about experimental limitations, nor through the ulterior motive of explaining irreversibility; rather, the motivation is the search for robust, autonomous higher-level dynamics. In the BBGKY approach to dilute-gas mechanics, for instance, the coarse-graining process is the discarding of three-body and higher marginals from the probability distribution; the motivation here is simply that we seem to be able to write down a well-defined and empirically successful dynamics for the so-truncated probability function. Similarly, in master-equation approaches to Brownian motion the reason for discarding information about the thermal bath in which the Brownian particle moves is not (or should not be) that we do not have that information empirically; it is that there is a robust, empirically adequate, stochastic dynamics for the Brownian particle alone. (This is perhaps a good point to observe that statistical mechanics — contrary to the way it is often discussed in the foundational literature — is not itself a foundational project: its primary goal is to find empirically-adequate values and dynamical equations for collective degrees of freedom, not conceptually-adequate foundations for thermodynamics. See Wallace (2015a) for development of this point.)

Secondly, in thermodynamics, there *is* reason to understand coarse-graining in terms of experimental limitations — but this is entirely appropriate given the control-theory understanding of thermodynamics. In the version of thermodynamics I sketched in Sections 15.6–15.7, the coarse-graining can be understood as quantifying our operational limitations: If the best we can do is alter a system's macroscopic parameters and allow it to come to equilibrium, for instance, then the right coarse-graining is the one that replaces a distribution with the equilibrium distribution at the same volume and expected energy.

D. Wallace

Thirdly, the macrostate partition at the heart of Boltzmannian statistical mechanics is just as vulnerable to these criticisms as is the Gibbsian coarse graining — indeed, it is a special case of that coarse-graining, corresponding, in Gibbsian terms to replacing a distribution with that distribution which agrees on the probability of each macrostate and is constant across macrostates. Consider some standard descriptions of the coarse-graining:

> "[W]e must partition [phase space] into compartments such that all of the microstates X in a compartment are macroscopically indistinguishable[.]"[9] (Callender, 1999, p.355)

> "Everyday macroscopic human language (that is) carves the phase space of the universe up into chunks." (Albert, 2000, p.47)

If pushed, I suspect Boltzmannians would reply that it is not the epistemic indistinguishability of macrostates that is doing the work, but rather the possibility of writing down robust higher-level dynamics in terms of macrostates, and largely abstracting over microscopic details. But of course this is exactly what the Gibbsians have in mind when they speak of coarse-graining.

15.10. Objections to Gibbs: Subjectivity and the interpretation of probability

Probably the most severe criticism made of Gibbsian statistical mechanics is that it somehow conflates the question of how much we *know* about a system, with the question of how a system will *in fact* behave. Thus, Albert (2000, p.58):

> "There's something completely insane (if you think about it) about the sort of explanation we have been imagining here ... Can anybody seriously think that our merely being

---

[9]Callender goes on to gloss this as "that is, they share the same thermodynamic features". But he does not define "thermodynamic features", and the standard definition — that thermodynamic features are restricted to the energy, volume and thermodynamic entropy, and functions thereof — does not suffice to define the macrostate partition.

> ignorant of the exact microconditions of thermodynamic systems plays some part in bringing it about, in making it the case, that (say) milk [mixes into[10]] coffee? How could that be? What can all those guys have been up to?"

Or Callender (1999, p.360):

> "Thermodynamic behaviour does not depend for its existence on the precision with which we measure systems. Even if we knew the positions and momenta of all the particles in the system, gases would still diffuse through their available volumes."

Many aspects of this concern overlap with worries about the modal or probabilistic nature of Gibbsian statistical mechanics, or the subjectivity of coarse-graining processes in particular, and so have been addressed in the previous two sections. Here, I add two further sets of observations: one aimed at statistical mechanics, one at thermodynamics.

Firstly, insofar as we accept that statistical mechanics is in the business of making probabilistic predictions (and, as we saw in Section 15.5, this is clearly implied by scientific practice), this reduces to a general concern about the *interpretation* of statistical-mechanical probability. Since if we want to explain why a deterministic system will with high probability do X, probabilistic statements about its current state are pretty much all we can expect as explananda. If probabilities are here to be interpreted epistemically, then what is to be explained is why I have a high degree of belief in the system being (say) in the equilibrium macrostate τ seconds from now, and my present beliefs about the system are quite natural explananda.

It will be objected by (e.g.) Albert and Callender that when we say 'my coffee will almost certainly cool to room temperature if I leave it', we are saying something objective about the world, not something about my beliefs. I agree, as it happens; that just tells us that the probabilities of statistical mechanics cannot be interpreted epistemically.

---

[10]Albert actually says 'dissolves in' here, but this is not strictly correct: milk, a mixture of water and water-insoluble lipids, is not itself soluble in water.

But then, how can they be interpreted? Let's admit that there is a serious puzzle here, given that the underlying dynamics is deterministic. There are a number of suggestions: perhaps they are long-run relative frequencies, perhaps they can be understood via the Lewis-Loewer best-systems approach (Lewis, 1980; Loewer, 2002), perhaps they are the decoherent limit of quantum probabilites (Wallace, 2016). Certainly there is no consensus.

But this does not advantage the Boltzmannian account; for as I have stressed throughout this paper, a large part of the *confirmed empirical predictions* of contemporary statistical mechanics are expressed in probabilistic terms (cf Sections 15.4, 15.5, 15.7), and so *any* empirically viable account of statistical mechanics must account for those probabilities. Extra-empirical virtues of one theory over another may be good tie-breakers when two theories are empirically equivalent, but they do not trump empirical success. (There is, I suppose, a sense in which classical mechanics is superior to quantum theory because it contains no probabilities, but that is scarcely a reason to adopt classical rather than quantum mechanics!)

In the thermodynamic context, by contrast, it is far less clear to me why my knowledge of a system's state cannot play an explanatory role. To be sure, that information cannot explain why the system *spontaneously* approaches equilibrium; that is the domain of statistical mechanics. But there is no paradox in supposing that the transformations I can bring about of a system's state should depend on my information about that very state. (If Albert had asked how our being ignorant of the exact microconditions of the coffee could make it the case that we are unable to un-mix the coffee and the milk, the rhetorical force is at least less clear.)

To be sure, in the phenomenological context this is largely irrelevant: information about a system is useless to me except insofar as I have access to operations which are sensitive to that information. (Even if provided with an exact readout of my coffee's microstate, I lack the manipulative precision to use that information to unmix it.) But that just brings us back to the previous section's discussion of coarse-graining.

15.11. Conclusion

The standard story in philosophy of statistical mechanics about the Gibbs and Boltzmann approaches relies on two assumptions: that the Gibbs and Boltzmann approaches are empirically equivalent and apply to the same physical systems, and that the Gibbsian framework has conceptual flaws that do not trouble the Boltzmannian.

I have argued that neither is correct. The standard objections to the Gibbsian approach either misunderstand the approach, or apply to certain ways of developing the approach but not to the approach itself, or apply with equal force to the Boltzmannian. And it is good that this is so, because physics as practiced in the 21st century requires the Gibbsian framework to handle a host of physically relevant situations in which the Boltzmannian approach is inapplicable. Correctly understood, modern statistical mechanics includes the Boltzmannian framework simply as a special case of the Gibbsian.

15.12. Epilogue: Quantum statistical mechanics

My discussion in this paper is entirely classical. But the world is not even remotely classical, and so really we should ask: what does the Gibbs/Boltzmann distinction look like in quantum theory?

The obvious thought might be: replace phase space with (projective) Hilbert space; replace phase-space points with Hilbert-space rays; replace probability distributions over phase space with probability distributions over the space of rays. But macroscopic systems need not be, and in general will not be, in pure states. And mathematically, a probability distribution over the space of mixed states cannot be distinguished from a single element of that space.

So a hypothetical "Gibbsian" quantum statistical mechanics works with density operators *understood as probability distributions over mixed states*; a "Boltzmannian" statistical mechanics instead works with density operators *understood as individual mixed states*. But nothing at the level of the mathematics will distinguish the two approaches (see Wallace (2016) for further discussion). And I have been arguing that the *machinery* of the Gibbsian approach, not a hypothetical *interpretation* of that machinery, is compatible with the Boltzmannian conception of statistical mechanics as objective. In the case of quantum theory, this seems to follow automatically: at the level of machinery, there *is* no difference between the two conceptions. In reality, modern quantum statistical mechanics simply studies the evolution of mixed states and coarse-grainings of those states, with no need within the formalism for any additionally statistical-mechanical conception of probability. Mathematically speaking, the methods used are continuous with Gibbsian, rather than Boltzmannian statistical mechanics, but it is mere semantics whether those methods should be called 'Gibbsian' or 'Boltzmannian'.

Acknowledgements

I thank David Albert, Harvey Brown, Sean Carroll, Sheldon Goldstein, Bixin Guo, Chris Jarzynski, Owen Maroney, Wayne Myrvold, Simon Saunders, and Chris Timpson for various conversations on this material; thanks also to several generations of Balliol undergraduates.

References

Albert, D. Z. (2000) *Time and Chance.* Harvard University Press, Cambridge, MA.

Altland, A., Simons, B. D. (2010) *Condensed Matter Field Theory* (2nd ed.). Cambridge University Press, Cambridge.

Bang, J., Pan, R., Hoang, T., Ahn, J., Jaryznski, C., Quan, H. T., Li, T.(2018) Experimental realization of Feynman's ratchet, *New Journal of Physics*, **20**, 103032.

Bishop, R. C. (2004) Nonequilibrium statistical mechanics Brussels-Austin style, *Studies in the History and Philosophy of Modern Physics*, **35**, 1–30.

Bustamante, C., Liphardt, J., Ritort, F. (2005) The nonequilibrium thermodynamics of small systems. https://arxiv.org/abs/cond-mat/0511629.

Callender, C. (1999) Reducing thermodynamics to statistical mechanics: The case of entropy, *Journal of Philosophy*, **96**, 348–373.

Callender, C. (2001) Taking thermodynamics too seriously, *Studies in the History and Philosophy of Modern Physics*, **32**, 539–553.

Carroll, S. (2010) *From Eternity to Here: The Quest for the Ultimate Theory of Time.* Dutton.

Collin, D., Ritort, F., Jarzynski, C., Smith, S. B., Tinoco, I., Bustamante, C. (2005) Verification of the Crooks fluctuation theorem and recovery of RNA folding free energies, *Nature*, **437**, 231–234.

Crooks, G. E. (1998) Nonequilibrium measurements of free energy differences for microscopically reversible Markovian systems markovian systems, *Journal of Statistical Physics*, **90**, 1481.

Feynman, R. P. (1967) *The Character of Physical Law*. MIT Press, Cambridge, Mass.

Frigg, R. (2007) A field guide to recent work on the foundations of thermodynamics and statistical mechanics. In D. Rickles (Ed.), *The Ashgate Companion to the New Philosophy of Physics*, pp. 99–196. London: Ashgate.

Frigg, R. (2009) Typicality and the approach to equilibrium in Boltzmannian statistical mechanics, *Philosophy of Science*, **76**, 997–1008. Available online at http://philsci-archive.pitt.edu.

Goldstein, S. (2001) Boltzmann's approach to statistical mechanics. In J. Bricmont, D. Dürr, M. Galavotti, F. Petruccione, and N. Zanghi (Eds.), *In: Chance in Physics: Foundations and Perspectives*, Berlin, pp. 39. Springer. Available online at http://arxiv.org/abs/cond-mat/0105242.

Goldstein, S., Lebowitz, J. L., Tumulka, R., Zanghi, N. (2006) Canonical typicality, *Physical Review Letters*, **96**, 050403.

Jarzynski, C. (1997) Nonequilibrium equality for free energy differences, *Physical Review Letters*, **78**(2690)

Jaynes, E. (1957a) Information theory and statistical mechanics, *Physical Review*, **106**, 620.

Jaynes, E. (1957b) Information theory and statistical mechanics ii, *Physical Review*, **108**, 171.

Jaynes, E. (1965) Gibbs vs boltzmann entropies, *American Journal of Physics*, **5**, 391–398.

Lebowitz, J. (2007) From time-symmetric microscopic dynamics to time-asymmetric macroscopic behavior: An overview. Available online at http://arxiv.org/abs/0709.0724.

Lewis, D. (1980) A subjectivist's guide to objective chance. In R. C. Jeffrey (Ed.), *Studies in Inductive Logic and Probability, volume II*. Berkeley: University of California Press. Reprinted, with postscripts, in David Lewis, *Philosophical Papers*, Volume II (Oxford University Press, Oxford, 1986); page numbers refer to this version.

Loewer, B. (2002) Determinism and chance, *Studies in the History and Philosophy of Modern Physics*, **32**, 609–620.

Malament, D., Zabell, S. L. (1980) Why Gibbs phase space averages work: The role of ergodic theory, *Philosophy of Science*, **47**, 339–349.

Maudlin, T. (1995) Review [of L. Sklar, *Time and Chance*, and L. Sklar, *Philosophy of Physics*], *British Journal for the Philosophy of Science*, **46**, 145–149.

Myrvold, W. (2014) Probabilities in statistical mechanics. In C. Hitchcock and A. Hajek (Eds.), *Oxford Handbook of Probability and Philosophy*. Oxford University Press.

Myrvold, W. (2019) Beyond chance and credence. Forthcoming.

North, J. (2002) What is the problem about the time-asymmetry of thermodynamics? - a reply to Price, *British Journal for the Philosophy of Science*, **53**,
121–136.

Penrose, R. (1994) On the second law of thermodynamics, *Journal of Statistical
Physics*, **77**, 217–221.

Ridderbos, K. (2002) The coarse-graining approach to statistical mechanics: How
blissful is our ignorance? *Studies in the History and Philosophy of Modern
Physics*, **33**, 65–77.

Rowlinson, J. S. (1970) Probability, information, and entropy, *Nature*, **225**, 1196.

Sklar, L. (1993) *Physics and Chance: Philosophical Issues in the Foundations of
Statistical Mechanics*, Cambridge University Press, Cambridge.

Skrzypczyk, P., Short, A. J., Popescu, S. (2014) Work extraction and thermodynamics for individual quantum systems, *Nature Communications*, **5**,
4185.

Uffink, J. (2007) Compendium of the foundations of classical statistical physics.
In J. Butterfield and J. Earman (Eds.), *Handbook for Philosophy of Physics*.
Elsevier. Available online at philsci-archive.pitt.edu.

Wallace, D. (2010)   The logic of the past hypothesis.   http://philsci-
archive.pitt.edu/8894/.

Wallace, D. (2011) Taking particle physics seriously: A critique of the algebraic
approach to quantum field theory, *Studies in the History and Philosophy
of Modern Physics*, **42**, 116–125.

Wallace, D. (2014)   Thermodynamics as control theory,   *Entropy*, **16**,
699–725.

Wallace, D. (2015a) The quantitative content of statistical mechanics, *Studies
in the History and Philosophy of Modern Physics*, **52**, 285–293. Originally
published online under the title "What statistical mechanics actually does".

Wallace, D. (2015b) Recurrence theorems: A unified account, *Journal of Mathematical Physics*, **56**, 022105.

Wallace, D. (2016) Probability and irreversibility in modern statistical mechanics:
Classical and quantum. To appear in D. Bedingham, O. Maroney and
C. Timpson (eds.), *Quantum Foundations of Statistical Mechanics* (Oxford
University Press, forthcoming).

Wallace, D. (2018) The case for black hole thermodynamics, part I: phenomenological thermodynamics. *Studies in the History and Philosophy of Modern
Physics*, **64**, pp. 52–67.

Wallace, D. (2018) On the plurality of quantum theories: Quantum theory as
a framework, and its implications for the quantum measurement problem.
Forthcoming in S. French and J. Saatsi (eds.), *Scientific Metaphysics and
the Quantum* (Oxford: OUP, forthcoming).

Zeh, H. D. (2007) *The Physical Basis of the Direction of Time* (5th ed.). Springer,
Berlin.

Zwanzig, R. (2001)   *Nonequilibrium Statistical Mechanics*.   Oxford University
Press, Oxford.

Chapter 16

Taming Abundance:
On the Relation between Boltzmannian
and Gibbsian Statistical Mechanics

Charlotte Werndl and Roman Frigg

Theoreticians working in statistical mechanics seem to be spoilt for choice. The theory offers two different theoretical approaches, one associated with Boltzmann and the other with Gibbs. These approaches are neither theoretically equivalent nor in any obvious way inter-translatable. This raises the question about the relation between them. We argue that Boltzmannian statistical mechanics (BSM) is a fundamental theory while the Gibbisan approach is an effective theory, meaning that the former provides a true description of the systems within its scope while the latter offers an algorithm to calculate the values of physical quantities defined by the fundamental theory. This algorithm is often easier to handle than the fundamental theory and provides results where the fundamental theory is intractable. Being an effective theory, the Gibbsian approach works only within a certain domain of application. We provide a characterisation of the limits of the approach and argue that BSM provides correct results in cases in which the two theories disagree.

16.1. Introduction

Theoreticians working in statistical mechanics (SM) seem to be spoilt for choice. The theory offers two different theoretical approaches, one associated with Ludwig Boltzmann and the other with J. Willard Gibbs. There are significant differences between the two approaches, which offer distinct descriptions of the same physical system. We refer to them as Boltzmannian SM (BSM) and Gibbsian SM (GSM) respectively. While one can cherish theoretical pluralism as a virtue, it does raise the question about the relation between the two approaches because they are neither theoretically equivalent nor in any obvious way inter-translatable. The question of the relation between BSM and GSM becomes even more acute when we realise what functions they perform in the practice of SM. GSM is the drudge of SM. It provides the tools and methods to carry out a wide range of equilibrium calculations, and it is the formalism that yields the results in applications. However, as Lavis (2005) notes, when the question arises 'what is actually going on' in a physical system, physicists are often quick to desert GSM and offer an account of 'why SM works' in terms of BSM. And discrepancies are not restricted to foundational issues. In non-equilibrium situations, BSM is usually the theory of choice because despite many attempts to extend GSM to non-equilibrium, no workable Gibbsian non-equilibrium theory has emerged (see Frigg (2008), Sklar (1993), and Uffink (2007) for reviews). But the practice of using one approach for everyday equilibrium calculations while explaining the non-equilibrium behaviour of physical systems and giving a foundational account of SM using the other approach is of questionable legitimacy as long as the relation between the two approaches remains unclear. What we need is an account of how the two approaches relate, and the account must be such that it justifies the customary division of labour. Abundance must be tamed.

Unfortunately, attempts to give explicit accounts of the relation between BSM and GSM are few and far between. In part, this is due to the fact that there is a strand of arguments that downplays the problem. What drives such views is the claim that the two approaches are empirically equivalent, at least as far as equilibrium calculations are concerned.[1] Theoretical differences can then be brushed aside because discrepancies concerning foundational issues are something that one can live with. This argument is problematic for two reasons. First, while it is true that GSM and BSM produce the same

---

[1]See, for instance, Davey (2009, pp. 566–567) and Wallace (2015, p. 289). Arguments for special cases are given in Lavis (2005).

predictions in many cases, agreement is not universal. In fact, there are important cases where GSM and BSM make conflicting predictions, which implies that GSM and BSM are not empirically equivalent (Werndl and Frigg, 2017b, 2019a). This forecloses the escape route of non-committal theoretical pluralism.

Where the status of one theory vis-à-vis the other is explicitly discussed, either the view is that GSM and BSM have to be reconciled (Lavis, 2005), or it is suggested that GSM is the preferred formulation of SM (Wallace, 2015). We are taking a different route and claim that BSM is a fundamental theory while GSM is an effective theory. This means that BSM provides a true description of the systems within the scope of SM; GSM offers an algorithm to calculate values defined by the fundamental theory. The algorithm is often easier to handle than the fundamental theory and provides result where the fundamental theory is intractable. Like every effective theory, GSM works only within a certain domain of application. We provide a characterisation of the limits of GSM, and argue that BSM provides the correct results in cases in which the two theories disagree.

We discuss both approaches in the setting of classical systems. For want of space, we state definitions and theorems only for the deterministic case. The generalisation to stochastic classical systems is straightforward,[2] and in Section 16.3, we briefly discuss an example that has a stochastic time evolution. There is an interesting question, whether our approach generalises to quantum statistical mechanics. While we are optimistic that it does, we acknowledge that, given the current state of the discussion, this claim is largely speculative. The reason for this is that no generally accepted quantum formulation of BSM is currently available,[3] and so we lack the theoretical basis to compare BSM with GSM.

The paper is structured as follows. In Section 16.2 we introduce BSM and GSM; and in Section 16.3 we note that they are not empirically equivalent. In Section 16.4 we draw a contrast between

---

[2]Statements of the relevant definitions and results can be found in Werndl and Frigg (2017a, 2019a).
[3]See Dizadji-Bahmani (2011) for a discussion.

fundamental and effective theories and argue that GSM is an effective theory while BSM is a fundamental theory. Effective theories are not universally applicable, and the most useful effective theories are ones for which we know the domain of applicability. In Section 16.5 we offer sufficient conditions for GSM to provide correct results. In Section 16.6 we briefly summarise our results and point out that regarding GSM as an effective theory has important repercussions for a number of projects, in particular, attempts to turn GSM into a non-equilibrium theory.

16.2. A primer on BSM and GSM

SM studies physical systems like a gas in a container, a magnet on a laboratory table and a liquid in jar. Described mathematically, these systems have the structure of a *measure-preserving dynamical system*, i.e. a quadruple (X, Σ_X, T_t, μ). X is the state space of the system, i.e. a set containing all possible micro-states the system can be in. For a gas with n molecules, X has $6n$ dimensions: three dimensions for the position of each particle and three dimensions for the corresponding momenta. Σ_X is a σ-algebra on X and μ is a measure on (X, Σ_X) (it is required to be invariant under the dynamics, meaning that $\mu_X(T_t(A)) = \mu_X(A)$ for all $A \in \Sigma_X$ and all t). The dynamics of the model is given by an *evolution function* $T_t : X \to X$, where $t \in \mathbb{R}$ if time is continuous and $t \in \mathbb{Z}$ if time is discrete. T_t is assumed to be measurable in (t, x) and to satisfy the requirement $T_{t_1+t_2}(x) = T_{t_2}(T_{t_1}(x))$ for all $x \in X$ and all $t_1, t_2 \in \mathbb{R}$ or $\mathbb{Z}$. If at a certain point of time t_0 the system is in micro-state x_0, then it will be in state $T_t(x_0)$ at a later time t. For systems that are governed by an equation of motion such as Newton's equation, T_t corresponds to the solutions of this equation. The *solution* (or *trajectory*) through a point x in X is the function $s_x : \mathbb{R} \to X$, $s_x(t) = T_t(x)$ (and *mutatatis mutandis* for discrete time).

At the macro-level, the system is characterised by a set of *macro-variables*, which are measurable functions $v_i : X \to \mathbb{V}_i$, associating a value with each point in state space. Examples of macro-variables include volume, internal energy, and magnetisation. Mathematically speaking, macro-variables are real-valued functions on the state space, i.e. $f : X \to \mathbb{R}$. For example, if f is the magnetisation of the system and the system is in micro-state x, then $f(x)$ is the magnetisation of the system.

Both BSM and GSM share this characterisation of systems. What they disagree about is how statistical assumptions are introduced into SM and about what the observables of the theory are. We now introduce each theory in turn and describe how they differ.

In BSM, a system is in a particular *macro-state* at any given time. A macro-state is defined by the values of a set of *macro-variables* $\{v_1, \ldots, v_l\}$ ($l \in \mathbb{N}$). We use capital letters V_i to denote the values of v_i. A macro-state is then defined by a particular set of values $\{V_1, \ldots, V_l\}$. That is, the model is in macro-state $M_{V_1, \ldots, V_l}$ iff $v_1 = V_1, \ldots, v_l = V_l$.[4] A central posit of BSM is that macro-states *supervene* on micro-states, implying that at a system's micro-state uniquely determines its macro-state. This determination relation is normally many-to-one. Therefore, every macro-state M is associated with a *macro-region* X_M consisting of all micro-states for which the system is in M. For a complete set of macro-states the macro-regions form a partition of X (i.e. the different X_M do not overlap and jointly cover X).

One of these macro-states is the equilibrium macro-state of the system. Intuitively speaking, a system is in equilibrium when its properties do not change. This intuition is built into thermodynamics, where a system is said to be in equilibrium when all change has come to a halt and the thermodynamic properties of the system remain constant over time (Fermi, 2000, 4). However, such a definition of equilibrium cannot be implemented in SM because measure-preserving dynamical systems are time reversal invariant and, if they are bounded, also exhibit Poincaré recurrence. As a consequence, when the time evolution of a system unfolds without any outside influence, the system will eventually return arbitrarily close to the micro-state in which it started. Hence a system starting outside equilibrium (for instance, when the gas was confined to one half of

---

[4]Sometimes it is also useful to define macro-states by interval ranges, i.e. by the macro-variables taking values in a certain range or interval. One can then say that the model is in macro-state $M_{[A_1,B_1],\ldots,[A_l,B_l]}$ iff $V_1 \in [A_1, B_1], \ldots, V_l \in [A_l, B_l]$ for suitably chosen intervals.

the container) will eventually return to that macro-state. So in SM no system will remain in any state forever. Obviously, this precludes a definition of equilibrium as the state which the system never leaves once it has reached it. Different formulations of BSM offer different prescriptions of how exactly to define equilibrium in SM. We base our discussion on the *long-run residence time definition of equilibrium*, which aims to come as close to the thermodynamic definition of equilibrium as the mathematical constraints imposed by measure-preserving dynamical systems permit (Werndl and Frigg, 2015); we briefly comment on the typicality version of BSM below and point out that our main conclusion can equally be reached from this alternative point of view (and hence does not depend on which reading of BSM one adopts).

To give a formal statement of this definition, we first need the concept of the long-run fraction of time $LF_A(x)$ that a system spends in a subset A of X^5:

$$LF_A(x) = \lim_{t \to \infty} \frac{1}{t} \int_0^t 1_A(T_\tau(x))d\tau, \qquad (16.1)$$

where $1_A(x)$ is the characteristic function of A: $1_A(x) = 1$ for $x \in A$ and 0 otherwise. Note that long-run fractions depend on the initial condition.

The notion of 'most of the time' can be read in two different ways, giving rise to two different notions of equilibrium. The first introduces a lower bound of $1/2$ for the fraction of time; it then stipulates that whenever a model spends more than half of the time in a particular macro-state, this is the equilibrium state of the model. Mathematically, let α be a real number in $(\frac{1}{2}, 1]$, and let ε be a very small positive real number. If there is a macro-state $M_{V_1^*,...,V_l^*}$ satisfying the following condition, then that state is the system's α-ε-*equilibrium state*:

> There exists a set $Y \subseteq X$ such that $\mu_X(Y) \geq 1 - \varepsilon$, and all initial states $x \in Y$ satisfy $LF_{X_{M_{V_1^*,...,V_l^*}}}(x) \geq \alpha$. A

---

[5]We state the definitions for continuous time. The corresponding definitions for discrete time are obtained simply by replacing the integrals by sums.

system is in equilibrium at time t iff its micro-state at t, x_t, is in $X_{M_{V_1^*,...,V_l^*}}$.

According to the second reading, 'most of the time' refers to the fact that the model spends more time in the equilibrium state than in any other state (and this can be less than 50% of its time). Mathematically, let γ be a real number in $(0, 1]$ and let ε be a small positive real number. If there is a macro-state $M_{V_1^*,...,V_l^*}$ satisfying the following condition, then that state is the system's γ-ε-*equilibrium state*:

> There exists a set $Y \subseteq X$ such that $\mu_X(Y) \geq 1 - \varepsilon$ and for all initial conditions $x \in Y$: $LF_{X_{M_{V_1^*,...,V_l^*}}}(x) \geq LF_{Z_M}(x) + \gamma$ for all macro-states $M \neq X_{V_1^*,...,V_l^*}$. Again, a system is in equilibrium at time t iff its micro-state at t, x_t, is in $X_{M_{V_1^*,...,V_l^*}}$.

It should come as no surprise that these two notions are not equivalent. More specifically, an α-ε-equilibrium is strictly stronger than a γ-ε-equilibrium in the sense that the existence of the former implies the existence of the latter but not vice versa.

These definitions are about the *time* a model spends in the equilibrium state. Hence it is not immediately clear what they imply about the *size* of the equilibrium macro-regions. It turns out that equilibrium regions, thus defined, are the largest macro-regions. More specifically, a macro-region is called β-*dominant* if its measure is greater or equal to β for a particular $\beta \in (\frac{1}{2}, 1]$. A macro-region is called δ-*prevalent* if its measure is larger than the measure of any other macro-region by a margin of at least $\delta > 0$. The following theorems can then be proved (Werndl and Frigg, 2015b, 2017b):

> *Dominance Theorem*: If $M_{\alpha\text{-}\varepsilon\text{-}eq}$ is an α-ε-equilibrium, then the following holds for $\beta = \alpha(1 - \varepsilon)$: $\mu_X(X_{M_{\alpha\text{-}\varepsilon\text{-}eq}}) \geq \beta$.[6]

> *Prevalence Theorem*: If $M_{\gamma\text{-}\varepsilon\text{-}eq}$ is a γ-ε-equilibrium, then the following holds for $\delta = \gamma - \varepsilon$: $\mu_X(X_{M_{\gamma\text{-}\varepsilon\text{-}eq}}) \geq \mu_X(X_M) + \delta$.[7]

---

[6]We assume that ε is small enough so that $\alpha(1 - \varepsilon) > \frac{1}{2}$.
[7]We assume that $\varepsilon < \gamma$.

It is a consequence of these definitions that a system is not always in equilibrium and that it does fluctuate away from equilibrium. This is a radical departure from thermodynamics. It is therefore worth pointing out that this is not merely a concession to the demands of measure-preserving dynamical systems. Having no fluctuations at all is also physically undesirable. There are experimental results that show that equilibrium is not the immutable state that classical thermodynamics presents us with because systems exhibit fluctuations away from equilibrium (MacDonald, 1962; Wang *et al.*, 2002). Hence adopting a notion of equilibrium that allows for fluctuations increases the empirical adequacy of the theory.

Before turning to GSM, we would like to comment briefly on an alternative version of BSM, namely the typicality approach. In his seminal 1877 paper, Boltzmann introduced what is today known as the *combinatorial argument*. It is a consequence of this argument that the equilibrium macro-region of an ideal gas is by far the largest macro-region. This prevalence of the equilibrium macro-region can be described in terms of *typicality*: equilibrium micro-states are typical in X because they occupy a region that is much larger than the region occupied by non-equilibrium states. Typicality then affords an explanation of the thermodynamic behaviour of a gas. As Goldstein puts it, because the phase space "consists almost entirely of phase points in the equilibrium macrostate", "[f]or a non-equilibrium phase point $[x]$ of energy E, the Hamiltonian dynamics governing the motion $[x(t)]$ would have to be ridiculously special to avoid reasonably quickly carrying $[x(t)]$ into [the equilibrium macro-region] and keeping it there for an extremely long time" (2001, pp. 43–44).[8]

So the typicality approach reaches the same conclusion as the long-run residence time approach, namely that the system spends most of its time in equilibrium. The two approaches differ in how

---

[8]The combinatorial argument is introduced in Boltzmann's (1877); for discussions of this argument, see Albert's (2000), Frigg's (2008) and Uffink's (2007). The typicality approach originates in Goldstein's (2001) and Lebowitz's (1993a, 1993b). Discussions of typicality and its use in GSM can be found in Frigg's (2009, 2010), Frigg and Werndl's (2011), Uffink's (2007), Volchan's (2007), and Wilhelm's (forthcoming).

they reach the conclusion — in fact, they reach them in reverse order. The typicality approach takes as its point of departure the fact that the equilibrium macro-region is the largest macro-region and argues that this has the consequence that the system spends most of its time in equilibrium; the long-run residence time approach defines equilibrium as the state in which the system spends most of its time and then establishes that this implies that the equilibrium macro-region is also the largest region (either in the sense of prevalence or in the sense of dominance). At this point, we are not concerned with the pros and cons of these approaches; nor are we concerned with their relative advantages vis-à-vis each other. What matters at this point is that both see equilibrium as the state in which the system spends most of its time, and hence the main point that we are making in this paper — that BSM is the fundamental theory while GSM is an effective theory — can equally be made from both perspectives.

The core object that is studied in GSM is a probability density (or distribution) $\rho(x, t)$ over X.[9] The density $\rho(x, t)$ describes the probability of finding the state of a system in a region $R \subseteq X$ at time t:

$$p_t(R) = \int_R \rho(x, t)dx. \qquad (16.2)$$

On physical grounds, the probability density must be conserved, meaning that for every region $R(t)$ of X that is moving forward under the time evolution T_t, the probability must be constant. If the time evolution is generated by Hamiltonian equations of motion, this is the case if and only if the Liouville's equation holds (Tolman, 1938).

Gibbs (1902, p. 8) introduces what he refers to as the *condition of statistical equilibrium*: a probability density is in statistical equilibrium iff it is stationary. That is, if it does not change under the

---

[9]In Gibbs' (1902) original presentation, $\rho(x)$ is described as representing an ensemble, an infinite collection of independent systems that are governed by the same laws of motion but are in different initial states. There are alternative presentations that endeavour to avoid reference to ensembles; they regard GSM simply as probabilistic algorithm. What follows does not depend on these interpretational issues and hence we set this question aside. Various different interpretations of GSM are discussed in Frigg and Werndl (2019).

dynamics of the system: $\rho(x,t) = \rho(x)$ for all t. There are usually a large number of stationary density functions for a given T_t. Hence, the question arises, which of these is the best to characterise a given physical situation. According to Gibbs, the so-called *microcanonical distribution* describes the equilibrium of a physical system which is completely isolated from its environment. It is the constant distribution on the system's energy hypersurface $H(x) = E$. The *canonical distribution* should be used when the system is in contact with a heat bath. It is given by $e^{-H(x)/kT}/\zeta_T$, where H is the system's Hamiltonian, T is the temperature, k is the Boltzmann constant, and ζ_T is the so-called *partition function*. For a discussion how to justify the choice of these distributions, see Frigg and Werndl's (2019) and Myrvold's (2016).

How do Gibbsian probability densities connect to observations on physical systems? That is, what does an experimentalist observe when measuring, say, the magnetisation of a sample of iron? To reply to this question, we first introduce the *phase average* $\langle f \rangle$ of a macro-variable f:

$$\langle f \rangle = \int_X f(x)\rho(x,t)dx. \tag{16.3}$$

When the system is in statistical equilibrium, it follows that $\langle f \rangle$ is time-independent. Standardly, a connection between the Gibbsian probability density and observable results is established by appealing to the *averaging principle* (AP). According to this principle, when observing the physical quantity associated with f on a system in equilibrium, then the observed equilibrium value of f is the phase average $\langle f \rangle$. When reviewing textbooks of statistical mechanics, it becomes clear that many textbooks explicitly state and endorse this principle (for a more detailed discussion of the principle, see Werndl and Frigg's (2019a), and references therein). For example, Chandler calls AP "[t]he primary assumption of statistical mechanics" (1987, p. 58), and Pathria and Beale state that they regard AP as the "the most important result" in SM (2011, p. 31). For this reason our discussion in Section 16.4 is based on a version of GSM that incorporates AP. However, there are alternative interpretations of

GSM that do not accept AP. We will comment on how our arguments carry over to these alternative versions at the end of Section 16.5.

These brief summaries of BSM and GSM should make it clear how distinct the two theories are; in particular, their characterisations of equilibrium are entirely different. BSM first introduces macro-states and then defines the equilibrium macro-state as the macro-state in which the system spends most of its time. In doing so, it explicitly allows systems to fluctuate away from the equilibrium state every now and then. In GSM, by contrast, equilibrium is a property of a probability distribution. More specifically, it is defined as a stationary probability distribution. Observable equilibrium properties are equated with the phase averages of macro-variables, and these phase averages stay constant over time if the distribution is in equilibrium.

So we seem to find ourselves in the disconcerting situation that when we talk about 'statistical mechanics', it is unclear whether we mean BSM or GSM (or both), and that the two theories are different in important respects.

As a first reaction, one might try to downplay the problem by arguing that despite their theoretical differences, the formalisms are empirically equivalent, at least as far as equilibrium properties are concerned. This immediately raises a prior question: what does it mean for BSM and GSM to be empirically equivalent? The Boltzmannian notion of equilibrium is formulated to mirror the thermodynamic notion of equilibrium; the Gibbsian notion of statistical equilibrium connects to thermodynamic equilibrium through the averaging principle. Hence it is natural to think that Gibbsian phase averages, Boltzmannian equilibrium values, and thermodynamic equilibrium values should all coincide. This suggests that the following is a necessary condition for the empirical equivalence of BSM and GSM:

$$F \approx \langle f \rangle \tag{16.4}$$

holds for *all* macro-variables f in all systems that fall within the scope of both theories, where F is the Boltzmannian equilibrium value of f and '$\approx$' means that the two values are approximately equal. We refer to Equation (16.4) as the *mechanical averaging equation* ('mechanical' because the principle connects two mechanical quantities, namely equilibrium values in BSM and equilibrium values in GSM). The question now is whether this principle holds true.

16.3. When the mechanical averaging equation fails

BSM and GSM turn out not to be empirically equivalent. Boltzmannian equilibrium values and Gibbsian phase averages do agree for paradigmatic examples, such as the dilute gas with macro-variables that assign the same value to all states that are in the Maxwell-Boltzmann distribution (or in a distribution that is very close to the Maxwell-Boltzmann distribution). Yet there are important cases where F and $\langle f \rangle$ are substantially different. This shows that the mechanical averaging equation does not hold generally. We will now present the six vertex model with the internal energy macro-variable as an example in which Gibbsian and Boltzmannian calculations come apart for any finite number of particles N. Further examples where the Boltzmannian equilibrium values and the Gibbsian phase averages disagree are discussed in Werndl and Frigg's (2019a).

Consider a two-dimensional quadratic lattice with N grid points that lies on a two-dimensional torus (which allows us to neglect border effect because on a torus every grid point has exactly four nearest neighbours). The 'vertices' are the grid points, and each vertex is connected to its four nearest neighbours by an edge. On each edge, there is an arrow pointing either toward the vertex or away from it. A rule known as the 'ice-rule' is now imposed: the arrows have to be distributed such that at each vertex in the lattice there are exactly two outward and two inward pointing arrows. As shown in Figure 16.1, at every vertex there are exactly six configurations of the arrows that satisfy the ice-rule. The ice rule is satisfied by water ice and several crystals including potassium dihydrogen phosphate (Baxter, 1982; Lavis and Bell, 1999).

The micro-state of the six-vertex model $\kappa = (\kappa_1, \ldots, \kappa_N)$ is specified by assigning to each vertex in the model one of the six types of configurations of the arrows permitted by the ice rule. Each of the six configurations has a certain energy ϵ_i, $1 \leq i \leq 6$. Denote by $\epsilon(\kappa_j)$,

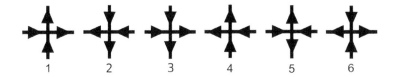

Fig. 16.1.   The configurations of the six-vertex model.

the energy or the jth vertex (thus the $\epsilon(\kappa_j)$ range over the ϵ_i). Then, the energy of the state κ is:

$$E(\kappa) = \sum_{j=1}^{N} \epsilon(\kappa_j). \qquad (16.5)$$

We make the common assumption that the energy of the different configurations is $\epsilon_1 = \epsilon_2 = 0$ and $\epsilon_3 = \epsilon_4 = \epsilon_5 = \epsilon_6 = 1$ (cf. Lavis and Bell, 1999, p. 299). The canonical distribution

$$\rho(\kappa) = e^{-E(\kappa)/kT}/\zeta_T \qquad (16.6)$$

to be used in calculations with

$$\zeta_T = \sum_{\kappa} e^{-E(\kappa)/kT} \qquad (16.7)$$

is usually taken to be the probability distribution. There are many versions of the six-vertex model, but most versions work with a stochastic dynamics that is assumed to be an irreducible Markov model. Note that the canonical distribution is historically associated with Gibbs. Yet this should not mislead us to believe that we treat the model in a Gibbsian way right from the start. The canonical distribution *per se* is neither Gibbsian nor Boltzmannian and is simply used here as a probability distribution that can figure in either BSM or GSM.[10]

---

[10] As noted in the Introduction, BSM can be formulated with stochastic rather than a deterministic time evolution. In most basic terms, this means that one replaces T_t with a stochastic algorithm, and such algorithms can be formulated using the canonical distribution. For details, see Werndl and Frigg (2017a).

Let us now consider the *internal energy* defined in Equation (16.5) as the relevant macro-variable. From the *Boltzmannian* perspective, the state space consist of all possible states κ which satisfy the ice rule. $E = 0$ is the lowest energy value and it defines a macro-state M_0 with the corresponding macro-region $\bar{X}_{M_0} = \{\kappa^*, \kappa^+\}$, where κ^* is the state for which all vertices are in the first configuration, and κ^+ is the state for which all vertices are in the second configuration. Suppose now that the number of vertices N is a sufficiently large but finite number. Due to the fact that for sufficiently low values of the temperature T, the probability mass is concentrated on the two lowest energy states, it is the case that $\bar{X}_{M_0}$ is the largest macro-region for sufficiently low temperatures. For an irreducible Markov process, the model spends more time in the largest macro-region than in any other of the macro-regions. For this reason, M_0 is a Boltzmannian γ-0-equilibrium, and the Boltzmannian equilibrium value is $E = 0$.

In the Gibbsian treatment, $\rho(\kappa)$ is the stationary equilibrium distribution and the observable f is of course the internal energy. We know that the internal energy assumes its lowest value $E = 0$ only for the two specific micro-states κ^* and κ^+, and that it will assume higher values for all other micro-states (and we know that for any $T > 0$ there will be a non-zero probability assigned to these higher energy states). Consequently, the phase average $\langle E \rangle$ is greater than 0; thus it is higher than the Boltzmannian equilibrium value, implying that the mechanical averaging equation fails.

It is not difficult to see that this difference can be significant: choose a T such that $\{\kappa^*, \kappa^+\}$ is the largest macro-region while its probability is less than 0.5.[11] In such a case, the Boltzmannian equilibrium value is still $E = 0$. For the six vertex model, the second lowest macro-value is $E = \sqrt{N}$, which corresponds to the energy of micro-states where all columns except one are taken up by states with the first or the second configuration, and the states in the exceptional

---

[11]This is possible because the higher the temperature, the more uniform the probability distribution. Consequently, for sufficiently high values of T, the largest macro-region is different from $\{\kappa^*, \kappa^+\}$. Because of the continuity of the canonical distribution, there has to be a T such that $\{\kappa^*, \kappa^+\}$ is the largest equilibrium macro-region while its probability is less than 0.5.

row are all states of the third or fourth configuration.[12] Consequently, $\langle E \rangle$ is higher than $\sqrt{N}/2$, implying that the Gibbsian phase average and the Boltzmannian equilibrium value will differ by at least $\sqrt{N}/2$. This is clearly not a negligible difference, in particular for large N, and hence the mechanical averaging equation fails. This is underscored by the fact that the Boltzmannian macro-value that is closest to the value obtained from Gibbsian phase averaging is higher or equal to $\sqrt{N}$. And a Boltzmannian macro-value of higher or equal to $\sqrt{N}$ is different from 0, the Boltzmannian macro-value in equilibrium.[13]

16.4. GSM as an effective theory

We argued that GSM and BMS are not empirically equivalent, which forces upon us the question of how the two approaches relate to one another and of which prediction is correct if they disagree. The answer to these question, we submit, lies in the realisation that BSM and GSM are not alternative theories that are on par with each other. Rather, GSM is an *effective theory* while BSM is the fundamental theory. This implies that in situations where Boltzmannian and Gibbsian equilibrium values come apart, the Boltzmannian values are the correct values.

---

[12]This is the smallest possible departure from states with zero energy: it can be shown that the number of downward pointing arrows is the same for all rows. From this it follows that there has to be a perturbation in each row. Therefore $\sqrt{N}$ is the second lowest value of the internal energy (Lavis and Bell, 1999, Chapter 10).
[13]We note that the internal energy macro-variable considered here is an *extensive* macrovariable, i.e. it depends on the number of constituents of the system. If one instead considers the energy density, an *intensive* macro-variable, then one finds that the difference between the Gibbsian and Boltzmannian equilibrium calculations tends toward zero as $N \to \infty$ because $\sqrt{N}/2N \to 0$. This illustrates the point that whether or not Gibbsian and Boltzmannian calculations agree crucially depends on the macro-variable. Yet we note that the problems we are discussing cannot be dissolved simply by restricting attention to intensive variables. Extensive variables are important (cf. Lieb and Yngvason, 1999), and, as discussed in Werndl and Frigg (2019a), in some cases there differences between Gibbsian and Boltzmannian values both for intensive and extensive variables.

To add specificity to the claim that GSM is an effective theory we have to characterise effective theories in more detail. Physicist James Wells says that effective theories "are theories because they are able to organise phenomena under an efficient set of principles, and they are effective because it is not impossibly complex to compute outcomes" (2012, p. 1). The ability of effective theories to provide results comes at the cost of incompleteness. As Wells puts it, "[t]he only way a theory can be effective is if it is manifestly incomplete [...] Any good Effective Theory systematises what is irrelevant for the purposes at hand. In short, an Effective Theory enables a useful prediction with a finite number of input parameters." (*ibid.*) His examples of effective theories are Galileo's law of falling bodies, the harmonic oscillator, classical gravity, and effective theories of particle masses.

Our suggestion is that GSM should be added to this list because it fits Wells' criteria. First, by offering a characterisation of equilibrium in terms of stationary distributions, GSM offers an *organisation* of phenomena under the umbrella of small set of principles. Second, GSM offers actionable principles and tractable methods to calculate equilibrium values of large array of materials, which makes it an efficient tool for *computations*. Third, GSM is *incomplete* in a number of ways. As noted in Section 16.2, GSM is unconcerned with the dynamics of the model. The role of the system's dynamics in GSM is reduced to ensuring that the proposed equilibrium distribution is stationary, and no other properties of the dynamics play any role in the theory. GSM considers neither equations of motion nor dynamical laws; it completely disregards trajectories; no time averages along trajectories are studied; and the initial conditions are left unspecified.[14] The system's dynamics is considered immaterial to understanding equilibrium as long as it — somehow — produces the stationary distribution that enters into the calculations. The system's Hamiltonian is used in formulating the most common Gibbsian distributions — the microcanonical and the canonical distributions — but a Hamiltonian does

---

[14]Notions of this kind are sometimes considered in attempts to justify the Gibbsian formalism, but they are not part of the formalism itself. For a discussion of justificatory endeavours see, for instance, Frigg's (2008).

not, by itself, provide dynamical information. The Hamiltonian of the system becomes relevant to the dynamics only when combined with equations of motions or when used in the formulation of a stochastic process, which is not part of the Gibbsian formalism. The Gibbsian phase averages are the same for all time evolutions that are such that ρ is invariant over time, no matter how different they may otherwise be. Finally, GSM is explicit about what it omits, and it thereby *systematises* what it regards as irrelevant.

These features of GSM stand in stark contrast with BSM, where dynamical considerations occupy centre stage. As we saw in Section 16.2, GSM introduces macro-states with corresponding macro-regions, and then defines equilibrium in explicitly dynamical terms (namely, as the macro-state whose macro-region is such that, in the long run, the system's state spends most of its time in that macro-region). Under the assumption (adopted in this paper) that the world is classical and that systems are governed by classical laws of motion, the dynamics considered in BSM is the true dynamics at the fundamental level: the unabridged and unidealised dynamics with all interactions between all micro-constituents of the system. Equilibrium results from macro-states that are defined in terms of macro-variables that supervene on the true micro-dynamics of the system, and where a system fluctuates away from equilibrium it does so as a result of the true underlying dynamics. The theory gives a full account of what happens in a classical world — nothing is left out and nothing is averaged over. BSM therefore is the fundamental theory.

Since a true and complete fundamental theory cannot be wrong, the results of BSM are the correct results in cases where BSM and GSM disagree. Consider the example of magnet with the macro-variable of total magentization m as discussed in Werndl and Frigg's (2019a). Such a system can be represented by the Ising model. Experiments show a phase transition as the temperature is varied. Calculations in the Boltzmannian framework show this phase transition as one would expect. Yet the Gibbsian account fails to describe the phase transition successfully because it yields that $\langle m \rangle = 0$ for all temperature values. Hence, the results of BSM are in agreement with the experimental results but the Gibbsian results obtained by phase averaging are not.

The flip side of fundamentality often is intractability, and BSM is no exception. If one wants to find out whether a Boltzmannian equilibrium exists, and if so, determine the equilibrium state, then one has to explicitly specify the macro-state structure of the system and determine the macro-states' macro-regions; one has to know enough about the underlying dynamics to be able to calculate the long-run fractions of time that a model spends in each macro-region; and one has to be able to estimate the measure of the set of initial conditions that lie on trajectories that do not have well-behaved long-run fractions of time. This is feasible only in very special cases (for instance, if the dynamics is ergodic); in general, it is a dead end because it requires more information than we have. The omissions in effective theories are designed to eliminate these intractabilities and deliver workable recipes. So the omissions and simplifications are what make effective theories effective!

Relegating a host of things to the realm of irrelevance comes at a cost. Wells is careful in pointing out that whenever we recognise a theory as an effective theory we will also have to "confront a theory's flaws, its incompletenesses, and its domain of applicability as an integral part of the theory enterprise" (*ibid.*). In other words, effective theories have limited domains of applicability, and using the theory correctly requires scientists to know where the limits are. For this reason, Wells notes that "[t]he most useful Effective Theories are ones where we know well their domains of applicability" (*ibid.*).

So if we view GSM as an effective theory, we have to be able to delimit its range of application. That is, we have to be able to say when it yields trustworthy results and when its procedures fail to deliver. This is the task to which we turn now.

16.5. The boundaries of effectiveness

Recall that for BSM and GSM to agree on a system's equilibrium properties it must be the case that $F \approx \langle f \rangle$ (where F is the Boltzmannian equilibrium value), and agreement is a necessary condition for BSM to be effective. This can be the case under several different conditions. We will now discuss four conditions that are individually sufficient for $F \approx \langle f \rangle$ to hold: the Khinchin condition, the requirement that fluctuations be small, the averaging equivalence theorem and the cancelling out theorem. However, note that these conditions are not necessary and there could be other conditions that guarantee that the phase average equals the Boltzmannian equilibrium value.

It is trivial that phase averages and Boltzmannian equilibrium values are identical if the macro-variables under considerations take the same value everywhere, i.e. $f(x) = c$ for all x in X and a constant c. Then, $F = \langle f \rangle = c$. Clearly, such macro-variables are not interesting; yet, they raise a useful question: how far does one have to move away from this uninteresting case to obtain a useful condition while still retain the basic idea? The Khinchin condition provides an answer to this question.[15]

First of all, we need the notion of a *fluctuation*. For an arbitrary micro-state x consider the difference between the value $f(x)$ (the true value of the observable if the model is in state x) and the phase average:

$$\Delta f(x) = f(x) - \langle f \rangle. \tag{16.8}$$

$\Delta f(x)$ is the fluctuation when the system is in micro-state x, and $|\Delta f(x)|$ is the *magnitude* of the fluctuation. The *Khinchin condition* then requires that there is a subset $\bar{X}$ of X with $\mu(\bar{X}) = 1 - \delta$ for a very small $\delta \geq 0$ such that $|\Delta f(x)| = 0$ for all x in $\bar{X}$.

If the condition is satisfied, we have $F = \langle f \rangle$, which can be seen as follows. Assume that a Boltzmannian equilibrium exists and that F is

---

[15]The condition owes its name to Khinchin (1949), who engaged in a systematic study of functions that satisfy strong symmetry requirements and therefore have small fluctuations for systems with a large number of constituents. There are at least two versions of the condition. We here focus on the first version, which is the one appealed to, amongst others, by Wallace (2015, p. 289), Lavis (2005, pp. 267–268); Malament and Zabell (1980, pp. 344–345), and Vranas (1998, p. 693). The second version is appealed to in Ehrenfest and Ehrenfest-Afanassjewa's (1959, pp. 46–52) and discussed in Werndl and Frigg's (2019c).

the Boltzmannian equilibrium value of f. According to the Khinchin condition, then there are only a few states (of at most measure δ) whose macro-values differ from $\langle f \rangle$. It is clear that these 'exceptional' states cannot form the Boltzmannian equilibrium macro-state because the macro-region corresponding to the Boltzmannian macro-state is the largest macro-region. For this reason the set of micro-states for which $f(x) = F$ has to be the macro-region of the Boltzmannian macro-state, and for the states in that region it follows that $F = \langle f \rangle$. We conclude that if the Khinchin condition is satisfied, then BSM equilibrium value and the Gibbsian phase average agree. The paradigmatic example for the Khinchin condition is the dilute gas with macro-variables that assign the same value to all states that are in the Maxwell-Boltzmann distribution, or in a distribution that is very close to the Maxwell-Boltzmann distribution (Ehrenfest and Ehrenrest-Afanassjewa, 1959).

An alternative approach focuses on the statistics of fluctuation patterns, and then shows that GSM reproduces the fluctuation patterns of BSM under certain conditions. More specifically, let us start by looking at fluctuations from a Gibbsian perspective. The core idea here is to use the probabilities of GSM as given in Equation (16.2) to calculate the probability that a fluctuation of a certain magnitude occurs. In more detail: given an interval $\delta := [\delta_1, \delta_2]$, where δ_1 and δ_2 are real numbers such that $0 \leq \delta_1 \leq \delta_2$, Equation (16.2) can then be used to arrive at the probability for a fluctuation of a magnitude between δ_1 and δ_2 to occur:

$$p(\delta) = \int_D \rho(x)dx, \qquad (16.9)$$

where $D = \{x \in X : \delta_1 \leq |f(x)| \leq \delta_2\}$.

It is important to interpret the scope of this equation correctly. Sometimes the probabilities in Equation (16.2) are interpreted as holding universally; that is, ρ is seen as providing the correct probabilities for the state of a system to be in region R at time t for *all* R in X and for *any* time t. Under such an interpretation, the fluctuation probabilities in Equation (16.9) are then seen as universal in the sense that for any magnitude and for any time t, $p(\delta)$ gives the probability for a fluctuation of a certain magnitude to occur at t. Yet universality of this kind is a very strong demand and fails in general. A careful study of GSM reveals that at least one of two conditions have to be met in order for this to be the case (for more on those two conditions, see Frigg and Werndl, 2019). First, the *masking condition* requires either that the system has access to all parts of phase space, or, if that is not the case, that f must be such that the proportion of states for which f assumes a particular value is the same in each invariant subset of X. Second, f-*independence* (roughly) states that the dynamics of the system must be such that the probability of finding a specific value of f in two consecutive yet sufficiently temporally distant measurements have to be (approximately) independent of each other. The Gibbsian ρ can be used to calculate correct fluctuation probabilities only if at least one of these conditions is satisfied. These conditions limit the scope of GSM in determining fluctuations. Since both conditions are strong conditions on the dynamics and the macro-variables, their satisfaction cannot be taken for granted and they limit the scope of GSM in determining fluctuations.

Let us now turn to BSM and first focus on the masking condition and explain how, from the perspective of BSM, the fluctuation probabilities of Equation (16.9) turn out to be correct. The starting point for the masking condition is to observe the behaviour of the same system over time and to consider the fluctuations that arise in this way. What this amounts to is to track the system over an infinite period of time when the system starts in a particular initial condition and its state evolves under the dynamics of the system. According to the masking condition, either the system can access all parts of X or the proportion of states for which f assumes a particular value is the same in each invariant subset of X. From this it follows immediately that from a Boltzmannian perspective, the fluctuations that arise in the same system over an infinite period of time are equal to the probabilities assigned to the fluctuations by the measure ρ, implying that Equation (16.9) holds. For example, suppose that a system spends, say, β of its time in a certain macro-state for which the function f assumes the value F'. Then, for this macro-state, the magnitude of the fluctuation away from the phase average is $|F' - \langle f \rangle|$. Suppose that δ_0 is the interval that consists only of $|F' - \langle f \rangle|$. Then the probability $p(\delta_0)$ has to be β.

Let us now turn to the second condition, i.e. f-independence, and again explain how in the Boltzmannian framework the fluctuation probabilities of Equation (16.9) turn out to be the correct ones. For an observable f with a finite number of macro-states, suppose that the dynamics of the system is such that for two points of time t_1 and t_2 that are sufficiently far apart f-independence holds, meaning that the probability of finding a specific value of f in the two measurements are approximately independent of each other. From a Boltzmannian perspective, this immediately implies that, given a specific macro-value at t_1, the probability of finding the system in a macro-value at t_2 is given by the probability measure ρ, and hence again Equation (16.9) holds. For example, suppose that the measure assigns β to a certain macro-state for which the function f assumes the value F'. For this macro-state, the magnitude of the fluctuations away from the phase average is $|F' - \langle f \rangle|$. Suppose δ_0 is the interval consisting only of $|F' - \langle f \rangle|$. It then follows that, assuming that the system was in a certain macro-state at t_1, the probability of obtaining the fluctuation δ_0 at t_2 (where t_1 and t_2 are sufficiently far apart) is given by the probability $p(\delta_0) = \beta$.

The third condition that is sufficient to guarantee the equality of the Boltzmannian equilibrium value and phase averages is given by the Average Equivalence Theorem (Werndl and Frigg 2017).[16] The conditions of this theorem are referred to as the 'Average Equivalence Conditions'.

> **Average Equivalence Theorem (AET).** Assume that a model is composed of $N \geq 1$ components. That is, the state $x \in X$ is given by the N coordinates $x = (x_1, \ldots, x_N)$; $X = X_1 \times X_2 \ldots \times X_N$, where $X_i = X_{oc}$ for all i, $1 \leq i \leq N$ (X_{oc} is the one-component space). Let

---

[16]In Werndl and Frigg's (2017) paper, the theorem was referred to as 'equilibrium equivalence theorem'. This name turned out to be potentially misleading because the theorem concerns the largest macro-region and *not* the a Boltzmannian equilibrium *per se*. For this reason, we now use the label 'average equivalence theorem'.

μ_X be the product measure $\mu_{X_1} \times \mu_{X_2} \cdots \times \mu_{X_N}$, where $\mu_{X_i} = \mu_{X_{oc}}$ is the measure on X_{oc}.[17] Suppose that the macro-variable K is the sum of the one-component variable, i.e. $K(x) = \sum_{i=1}^{N} \kappa(x_i)$ (the assumptions made here is that all sums of possible values of the one-component variable are possible values of the macro-variable). Then it follows that the value corresponding to the largest macro-region as well as the value obtained by Gibbsian phase averaging is $\frac{N}{k}(\kappa_1 + \kappa_2 + \cdots \kappa_N)$.

For a proof of the theorem, see Werndl and Frigg (2017). Note that if the theorem is used to make claims about Boltzmannian equilibrium (as we do here), dynamical assumptions have to come in. Assuming now that a Boltzmannian equilibrium exists, it follows from the dominance/prevalence theorems that the Boltzmannian equilibrium value is equal to the value of the largest macro-region. From the Average Equivalence Theorem, it then follows that this Boltzmannian equilibrium value is equal to the Gibbsian phase average.

The most important assumptions of the theorem are (i) that the macro-variable is a sum of variables on the one-component space (where all sums of possible values of the one-component variable are possible values of the macro-variable); (ii) that the macro-variable on the one-component space corresponds to a partition with cells of equal probability; and (iii) that the measure on the state space is the product measure of the measure on the one-component space. To some extent, these assumptions are restrictive; still, a number of standard applications of SM fall within the scope of the theorem. One instance of the AET is the baker's gas; another instance of the AET is the Kac ring with the standard macro-state structure; and yet another example is the ideal gas with N particles (see Werndl and Frigg (2019) for details).[18]

The fourth sufficient condition to guarantee the equality of the Boltzmannian equilibrium value and the Gibbisan phase average is

---

[17]It is assumed here that N is a multiple of k, i.e. $N = k * s$ for some $s \in \mathbb{N}$.

[18]One might think that there is a similarity between AET and the weak law of large numbers (LLN), which states that given independent and identically distributed random variables (which we consider in the AET theorem) for any $\varepsilon > 0$

given by the Cancelling Out Theorem (the conditions of this theorem will be refereed to as the 'Cancelling Out Conditions'). Let us first state the theorem (for the proof, see Werndl and Frigg, 2019):

> **Cancelling Out Theorem (COT).** Consider a deterministic model with Boltzmannian equilibrium macro-state M_{equ} with equilibrium value V_{equ} and other macro-states $M_1, \ldots, M_q$, $q \in \mathbb{N}$, with corresponding macro-values $V_{M_1}, \ldots, V_{M_q}$. Further, suppose that for any macro-state $M_i \neq M_{equ}$ there is a macro-state M_j such that (i) $\mu_X(X_{M_i}) = \mu_X(X_{M_j})$ and (ii) $V_{M_i} + V_{M_j} = 2V_{M_{equ}}$. Then, the Boltzmannian equilibrium value as well as the value obtained by phase averaging is V_{equ}.

Intuitively, the theorem states the following: if the state space is divided up in such a way that next to the largest macro-region (corresponding to the Boltzmannian equilibrium), there are always two macro-states of equal measure such that their average equals the Boltzmannian equilibrium value, then it follows that the Boltzmannian equilibrium value equals the Gibbsian phase average, and GSM can serve as an effective theory. Needless to say, the assumptions of the theorem are to some extent restrictive because it requires the macro-state structure to be of a special kind. Still, a number of standard cases in SM such as the six-vertex model with the polarisation macro-variable for sufficiently high temperatures or the

---

(cf. Meester, 2003, Section 4.1):

$$\mu\left(\left\{x : \left|\frac{\sum_{i=1}^N \kappa(x_i)}{N} - \frac{(\kappa_1 + \kappa_2 + \cdots \kappa_N)}{k}\right| < \varepsilon\right\}\right) \geq 1 - \frac{\sigma^2}{\varepsilon^2 N}. \qquad (16.10)$$

This similarity is superficial, and the theorems are different. First, the LLN does *not* say whether the values of the *extensive* macro-variables we consider in the AET, $\sum_{i=1}^N \kappa(x_i)$, are close to $N(\kappa_1 + \kappa_2 + \cdots \kappa_N)/k$. All one obtains from the LLN is that their values are within $N\varepsilon$, but $N\varepsilon$ can be very large. Second, AET and LLN are results about *different* macro-variables. AET is a result about the macro-variable $\sum_{i=1}^N \kappa(x_i)$ or, if it is divided by N, about $\sum_{i=1}^N \kappa(x_i)/N$. By contrast, LLN is a statement about the probability of states that are close or equal to $(\kappa_1 + \cdots + k_k)/k$. Hence, it can tell us something about the *different* macro-variables that are defined by assigning the same macro-value to all states that are close or equal to $(\kappa_1 + \cdots + k_k)/k$.

Ising model with the magnetisation macro-variable for sufficiently high temperatures fall under the scope of the theorem (cf. Cipra, 1987; Lavis and Bell, 1999, Chapter 3).

As noted in Section 16.2, there are alternative interpretations of GSM that do not include AP. One such interpretation is that the theoretical core of GSM contains only ρ, while Equation (16.4), the mechanical averaging equation, has the status of a pragmatic rule that is adopted *only when it provides correct results*. When this equation fails, GSM is simply silent about the correct equilibrium values. This move immunises GSM against arriving at calculations that disagree with the calculations of BSM, but it does so at the cost of further restricting the scope of GSM. This is not *per se* objectionable, but it changes nothing fundamentally in our argument. On this alternative interpretation, GSM is still an effective theory with a limited range of applicability (and the limits are identical to the limits of the standard interpretation). The only difference is that in cases in which there would be disagreement, GSM is now seen not as giving wrong results but rather as providing no results at all.[19]

16.6. Conclusion

We argued that GSM is an effective theory while BSM is a fundamental theory. This clarifies the relation between the two approaches. We presented an account of effective theories and showed that GSM fits the relevant criteria. The range of application of effective theories is limited, which raises the question under what circumstances the calculations of GSM are correct. We presented four conditions under which this happens; these are individually sufficient but not necessary. This means that other, yet unknown, conditions could

---

[19]Furthermore, as argued by Frigg and Werndl (2019), there is no single reasonable interpretation of Gibbs that can make sense of all the successful applications of GSM. Reasonable interpretations of GSM such as the fluctuation account can always only explain some of the applications of GSM, that there is no single reasonable interpretation of Gibbs that can account for all successful applications of GSM further strenghtens the view that GSM is an effective theory.

exist. It is, however, an open question whether such conditions do exist and whether there is a complete list of such conditions.

Classifying GSM as an effective theory has implications for foundational debates. It implies that GSM does not address foundational questions and that such questions should not be discussed in that theory. Asking whether GSM provides a correct fundamental description of the world, or, if the answer to this question is negative, trying to revise GSM so that it does provide such a description, is a mistaken endeavour. Effective theories do not offer fundamental descriptions; they are calculatory devices of instrumental value; no more and no less.

There have been (and still are) sustained attempts to turn GSM into a fundamental theory. Probably the most important programme of that kind aims to extend GSM to non-equilibrium processes. It is common to characterise the approach to equilibrium as a process of increasing entropy. This means that if a system is prepared in a non-equilibrium macro-state of low entropy, then, as soon as the constraints are lifted and the system evolves freely, the entropy should increase until it reaches a maximum. This does not happen in GSM because the Gibbs entropy, defined as $\int_X \rho \ln(\rho) dx$, is a constant of motion. This undermines attempts to describe the approach to equilibrium as a process of increasing entropy. This problem is the starting point of a research programme that aims to revise GSM in such a way that the Gibbs entropy increases over time. Coarse-graining combined with a mixing dynamics, interventionism, and attempts to redefine Gibbsian equilibrium in way that avoids reference to stationary distributions are but the most prominent proposals in that programme.[20]

For those who regard GSM as an effective theory, this research programme started on the wrong foot. If the Gibbs entropy does not change over time, we should conclude that GSM does not offer an effective description of non-equilibrium processes and limit its range of applicability to equilibrium situations, rather than trying

---

[20]For a review and discussion of these proposals see Frigg (2008), Sklar (1993) and Uffink (2007).

to turn GSM into a correct description of non-equilibrium processes. Such a programme would be justified only if it turned GSM into an effective theory of non-equilibrium processes. But, at least so far, this does not seem to have happened. Non-equilibrium versions of GSM are not effective non-equilibrium theories. Not only do they not offer manageable algorithms to compute outcomes (thereby violating Wells' first criterion); they often also are not empirically adequate (spin echo experiments are a case in point).

Foundational questions concerning GSM remain important when they concern the empirical adequacy of the theory or its connection to the fundamental theory, BSM. One such question is the one we addressed in the previous section, namely under what circumstances Gibbsian and Boltzmannian equilibrium values coincide. Another is the problem of the justification of maximum entropy methods. The choice of the outcome distribution is often guided by maximum entropy considerations, and there is a legitimate question why these considerations work, and why they deliver distributions that provide correct equilibrium values; see Uffink (1995, 1996) for a discussion. But these are questions concerning the instrumental efficiency of the theory and are not aimed at turning GSM into a fundamental theory.

Acknowledgments

We would like to thank Valia Allori for inviting us to participate in this project. Thanks also to an anonymous referee for helpful comments. In the process of researching this paper, we benefited from discussions with Jeremy Butterfield, David Lavis, Stephan Hartmann, Wayne Myrvold, Patricia Palacios, Jos Uffink and Giovanni Valente.

References

Albert, David. (2000) *Time and Chance*, Cambridge, MA, Harvard University Press, London.

Baxter, Rodney J. (1982) *Exactly Solved Models in Statistical Mechanics*, Academic Press, London.

Boltzmann, Ludwig. (1877) Über die Beziehung zwischen dem zweiten Hauptsatze der mechanischen Wärmetheorie und der Warhscheinlichkeitsrechnung resp. den Sätzen über des Wärmegleichgewicht, *Wiener Berichte*, **76**, 373–435.

Chandler, David. (1987) *Introduction to Modern Statistical Mechanics*, Oxford University Press, New York.

Davey, Kevin. (2009) What is Gibbs's canonical distribution? *Philosophy of Science* **76**(5), 970–783.

Dizadji-Bahmani, Foad. (2011) The Aharonov approach to equilibrium, *Philosophy of Science* **78**(5), 976–988.

Ehrenfest Paul, Tatjana Ehrenfest-Afanassjewa. (1959) *The Conceptual Foundations of the Statistical Approach in Mechanics*, Cornell University Press, Ithaka.

Fermi, Enrico. (2000) *Thermodynamics*, Dover, Mineola.

Frigg, Roman. (2008) *A Field Guide to Recent Work on the Foundations of Statistical Mechanics*, in Rickles, ed. The Ashgate Companion to Contemporary Philosophy of Physics, London: Ashgate, 99–196.

———. (2009) Typicality and the approach to equilibrium in Boltzmannian statistical mechanics, *Philosophy of Science*, **76**, 2009, 997–1008.

———. (2010) *Why Typicality Does Not Explain the Approach to Equilibrium*, in Mauricio Surez (ed.) Probabilities, Causes and Propensities in Physics. Synthese Library. Dordrecht: Springer, 2010, 77–93.

———. (2010) *Probability in Boltzmannian Statistical Mechanics*, in: Ernst and Hüttemann, eds. Time, Chance and Reduction. Philosophical Aspects of Statistical Mechanics. Cambridge: Cambridge University Press, 92–118.

Frigg, Roman, Charlotte Werndl. (2011) Demystifying typicality, *Philosophy of Science* **79**(5), 2012, 917–929.

———. (2019) *Can Somebody Please Say What Gibbsian Statistical Mechanics Says?* forthcoming in The British Journal for Philosophy of Science.

Gibbs, Josiah W. (1902) *Elementary Principles in Statistical Mechanics*, Scribner, New York.

Goldstein, Sheldon. (2001) Boltzmann's Approach to Statistical Mechanics, in: In Jean Bricmont, Detlef Drr, Maria Carla Galavotti, Gian Carlo Ghirardi, Francesco Petruccione and Nino Zangh (eds.), Chance in Physics: Foundations and Perspectives (pp. 39–54). Berlin and New York: Springer, 39-54.

Isihara, A. (1971) *Statistical Physics*, Academic Press, Cambridge MA.

Khinchin, Alexandr. I. (1949) Mathematical foundations of statistical mechanics. Dover Publications, Mineola/New York.

Landau, Lev D., Evgeny M. Lifshitz. (1980) Statistical Physics, Pergamon Press, Oxford.

Lavis, David. (2005) Boltzmann and Gibbs: An attempted reconciliation, in *Studies in History and Philosophy of Modern Physics*, **36**, 245–73.

Lavis, David, Bell, G. M. (1999) *Statistical Mechanics of Lattice Systems, Volume 1: Closed Form and Exact Solutions*. Ellis Horwood.

Lebowitz, Joel L. (1993a) Boltzmann's entropy and time's arrow, *Physics Today*, September Issue, 32–38.

Lebowitz, Joel L. (1993b) Macroscopic laws, microscopic dynamics, time's arrow and Boltzmann's entropy, *Physica A*, **194**, 1–27.

Lieb, Elliott H., Yngvason, Jakob. (1999) The physics and mathematics of the second law of thermodynamics, *Physics Reports*, **310**, 1–96.

MacDonald, Donald K. C. (1962) Noise and Fluctuations. An Introduction, Wiley, New York.

Malament, David B., Zabell, Sandy, L. (1980) Why Gibbs phase averages work. The role of ergodic theory, in *Philosophy of Science*, **56**, 339–49.

Meester, Ronald. (2003) *A Natural Introduction to Probability Theory*. Springer.

Myrvold, Wayne C. (2016) Probabilities in Statistical Mechanics, in Hájek and Hitchcock, eds. The Oxford Handbook of Probability and Philosophy, Oxford University Press, 573–600.

Pathria, Raj K., Paul D. Beale (2011) Statistical Mechanics, Elsevier, Oxford.

Sklar, Lawrence. (1993) Physics and Chance. Philosophical Issues in the Foundations of Statistical Mechanics, Cambridge University Press, Cambridge.

Tolman, Richard C. (1938) The Principles of Statistical Mechanics, Dover Publications, Mineola, NY.

Uffink, Jos. (1995) Can the maximum entropy principle be explained as a consistency requirement? *Studies in History and Philosophy of Modern Physics*, **26**(3), 223–261.

———. (1996) The constraint rule of the maximum entropy principle, *Studies in History and Philosophy of Modern Physics* **27**(1), 47–79.

———. (2007) Compendium of the Foundations of Classical Statistical Physics, in Butterfield and Earman, eds. Philosophy of Physics, Amsterdam: North Holland, 923–1047.

Volchan, Sérgio B. (2007) Probability as Typicality, *Studies in History and Philosophy of Modern Physics*, **38**, 801–814

Vranas, Peter B. (1998) Epsilon-ergodicity and the success of equilibrium statistical mechanics. *Philosophy of Science*, **65**, 688–708.

Wallace, David. (2015) The quantitative content of statistical mechanics. *Studies in History and Philosophy of Modern Physics*, **52**, 285–293.

Wang, Genmiao, Edie Sevick, Emil Mittag, Debra, J., Searles, Denis J. Evans. (2002) Experimental demonstration of violations of the second law of thermodynamics for small systems and short time scales, in *Physical Review Letters*, **89**, 050601.

Wells, James D. (2012) Effective Theories in Physics. From Planetary Orbits to Elementary Particle Masses. Springer, Heidelberg and New York.

Werndl, Charlotte, Roman Frigg. (2015) Reconceptionalising equilibrium in Boltzmannian statistical mechanics, *Studies in History and Philosophy of Modern Physics*, **49**, 19–31.

———. (2017a) Boltzmannian Equilibrium in Stochastic Systems, in Massimi and Romeijn, eds. Proceedings of the EPSA15 Conference. Berlin and New York: Springer, 243–254.

———. (2017b) Mind the gap: Boltzmannian versus Gibbsian equilibrium, *Philosophy of Science* **84**(5), 2017, 1289–1302.

———. (2019a) When Does Gibbsian Phase Averaging Work? Manuscript.

————. (2019b) When Does a Boltzmannian Equilibrium Exist?, forthcoming in Bendingham, Maroney, and Timpson, eds. Quantum Foundations of Statistical Mechanics, Oxford University Press.

————. (2019c) Ehrenfest and Ehrenfest-Afanassjewa on Why Boltzmannian and Gibbsian Calculations Agree, forthcoming in Uffink, Valente, Werndl and Zuchowski, eds. Tatjana Afanjassewa and Her Legacy: Philosophical Insights from the Work of an Original Physicist and Mathematician, Springer, Berlin.

Wilhelm, Isaac. (fortcoming) Typical, forthcoming in The British Journal for Philosophy of Science.

Index

CPSIA information can be obtained
at www.ICGtesting.com
Printed in the USA
LVHW011231230420
653775LV00004B/10

9 789811 211713